Advances in Industrial Control

Other titles published in this series:

Digital Controller Implementation and Fragility
Robert S.H Istepanian and James F. Whidborne (Eds.)

Optimisation of Industrial Processes at Supervisory Level
Doris Sáez, Aldo Cipriano and Andrzej W. Ordys

Robust Control of Diesel Ship Propulsion
Nikolaos Xiros

Hydraulic Servo-systems
Mohieddine Jelali and Andreas Kroll

Model-based Fault Diagnosis in Dynamic Systems Using Identification Techniques
Silvio Simani, Cesare Fantuzzi and Ron J. Patton

Strategies for Feedback Linearisation
Freddy Garces, Victor M. Becerra, Chandrasekhar Kambhampati and Kevin Warwick

Robust Autonomous Guidance
Alberto Isidori, Lorenzo Marconi and Andrea Serrani

Dynamic Modelling of Gas Turbines
Gennady G. Kulikov and Haydn A. Thompson (Eds.)

Control of Fuel Cell Power Systems
Jay T. Pukrushpan, Anna G. Stefanopoulou and Huei Peng

Fuzzy Logic, Identification and Predictive Control
Jairo Espinosa, Joos Vandewalle and Vincent Wertz

Optimal Real-time Control of Sewer Networks
Magdalene Marinaki and Markos Papageorgiou

Process Modelling for Control
Benoît Codrons

Computational Intelligence in Time Series Forecasting
Ajoy K. Palit and Dobrivoje Popovic

Modelling and Control of Mini-Flying Machines
Pedro Castillo, Rogelio Lozano and Alejandro Dzul

Ship Motion Control
Tristan Perez

Hard Disk Drive Servo Systems (2nd Ed.)
Ben M. Chen, Tong H. Lee, Kemao Peng and Venkatakrishnan Venkataramanan

Measurement, Control, and Communication Using IEEE 1588
John C. Eidson

Piezoelectric Transducers for Vibration Control and Damping
S.O. Reza Moheimani and Andrew J. Fleming

Manufacturing Systems Control Design
Stjepan Bogdan, Frank L. Lewis, Zdenko Kovačić and José Mireles Jr.

Windup in Control
Peter Hippe

Nonlinear H_2/H_∞ Constrained Feedback Control
Murad Abu-Khalaf, Jie Huang and Frank L. Lewis

Practical Grey-box Process Identification
Torsten Bohlin

Control of Traffic Systems in Buildings
Sandor Markon, Hajime Kita, Hiroshi Kise and Thomas Bartz-Beielstein

Wind Turbine Control Systems
Fernando D. Bianchi, Hernán De Battista and Ricardo J. Mantz

Advanced Fuzzy Logic Technologies in Industrial Applications
Ying Bai, Hanqi Zhuang and Dali Wang (Eds.)

Practical PID Control
Antonio Visioli

(continued after Index)

Khac Duc Do • Jie Pan

Control of Ships and Underwater Vehicles

Design for Underactuated and Nonlinear Marine Systems

Springer

Khac Duc Do, PhD
Jie Pan, PhD

School of Mechanical Engineering
The University of Western Australia
35 Stirling Highway
Crawley, WA 6009
Australia

duc@mech.uwa.edu.au
pan@mech.uwa.edu.au

ISSN 1430-9491
ISBN 978-1-4471-2672-0 e-ISBN 978-1-84882-730-1
DOI 10.1007/978-1-84882-730-1
Springer Dordrecht Heidelberg London New York

British Library Cataloguing in Publication Data
A catalogue record for this book is available from the British Library

Softcover reprint of the hardcover 1st edition 2009

Cover design: eStudioCalamar, Figueres/Berlin

Printed on acid-free paper

Springer is part of Springer Science+Business Media (www.springer.com)

Advances in Industrial Control

Professor (Emeritus) O.P. Malik
Department of Electrical and Computer Engineering
University of Calgary
2500, University Drive, NW
Calgary, Alberta
T2N 1N4
Canada

Professor K.-F. Man
Electronic Engineering Department
City University of Hong Kong
Tat Chee Avenue
Kowloon
Hong Kong

Professor G. Olsson
Department of Industrial Electrical Engineering and Automation
Lund Institute of Technology
Box 118
S-221 00 Lund
Sweden

Professor A. Ray
Department of Mechanical Engineering
Pennsylvania State University
0329 Reber Building
University Park
PA 16802
USA

Professor D.E. Seborg
Chemical Engineering
3335 Engineering II
University of California Santa Barbara
Santa Barbara
CA 93106
USA

Doctor K.K. Tan
Department of Electrical and Computer Engineering
National University of Singapore
4 Engineering Drive 3
Singapore 117576

Professor I. Yamamoto
Department of Mechanical Systems and Environmental Engineering
The University of Kitakyushu
Faculty of Environmental Engineering
1-1, Hibikino,Wakamatsu-ku, Kitakyushu, Fukuoka, 808-0135
Japan

The first author dedicates this book to his parents, Do Thi Duyen and Do Khac Yen, and his little daughter, Do Thu Trang.

The second author dedicates this book to his parents, Liyou Pan and Hungxiu Diao.

Series Editors' Foreword

The series *Advances in Industrial Control* aims to report and encourage technology transfer in control engineering. The rapid development of control technology has an impact on all areas of the control discipline. New technology, new controllers, actuators, sensors, new industrial processes, computer methods, new applications, new philosophies..., new challenges. Much of this development work resides in industrial reports, feasibility study papers, and the reports of advanced collaborative projects. The series offers an opportunity for researchers to present an extended exposition of such new work in all aspects of industrial control for wider and rapid dissemination.

When advances are made in industrial control technology, for example, sensors, actuators, controllers, communications, and computing power, there are at least three possible consequences for system control. The control engineering may decide to use the new hardware to make an existing control system perform better. Another possibility is that the new hardware advances make a proposed but previously impractical control method feasible. Alternatively, it may be necessary to devise a completely new control technique in order to exploit the new hardware. However, in many cases, it is often economically impossible to advance an aspect of the system's control hardware until there is a major upgrade of the system and the industrial control engineer has to grapple with the limitations of the system as it exists. (This is where control engineering science becomes an art!) Underactuated marine vessels are a case in point. In most configurations, a vessel's main actuators are propellers and rudders, yet a marine vessel has six degrees of freedom in its motion, and the marine control engineer simply has to work with the control surfaces and sensors available. One area that may advance control performance is the use of better control designs, and recently control engineers have become more interested in what nonlinear control might have to offer.

Researchers Khac Duc Do and Jie Pan have published a sequence of journal and conference papers on new control algorithms for underactuated systems. Most of this work has used models of marine systems (surface ships and underwater vehicles), but some of the work has been with other mechanical system models (wheeled mobile robots, vertical take-off and landing (VTOL) aircraft). Now they have taken the opportunity to capture this research and development work in a monograph entitled *Control of Ships and Underwater Vehicles: Design for Underactuated and Nonlinear Marine Systems* for the *Advances in Industrial Control* series. This will enable industrial control engineers and control researchers to read and study a systematic presentation of their ideas and work. The book opens with chapters that introduce the appropriate nonlinear control theory and marine vessel models used in the research, proceeds to controller derivation and development, and finally details simulation results for a series of nonlinear control schemes devised for marine vessel control problems like point-to-point navigation and path-following. Also presented are some field results of a laboratory-scale vessel on a local river. In one chapter other applications fields are considered and the nonlinear control results for a simple wheeled robots and a simple VTOL aircraft model are given. Each of the applications chapters contains illustrative simulation studies and control results.

The monograph will interest control researchers, graduate students, and industrial control engineers alike, particularly those involved in marine and wheeled robotic motion control problems. The Series Editors, being based in Glasgow, Scotland, have always had an interest in marine control problems and have endeavoured to ensure that the *Advances in Industrial Control* series has useful volumes from this field of control application. Past volumes have included: *Ship Motion Control* by Tristan Perez (ISBN 978-1-85233-959-3, 2005), *Compressor Surge and Rotating Stall* by Jan Tommy Gravdahl and Olav Egeland (ISBN 978-1-85233-067-5, 1998), and *Robust Control of Diesel Ship Propulsion* by Nikolaos Xiros (ISBN 978-1-85233-543-4, 2002), and we are pleased to welcome this new volume, *Control of Ships and Underwater Vehicles* into the series.

Industrial Control Center
Glasgow
Scotland, UK
2009

M.J. Grimble
M.A. Johnson

Preface

Control of ocean vessels including ships and underwater vehicles is an active field due to its theoretical challenges and important applications such as passenger and goods transportation, environmental surveying, undersea cable inspection, offshore oil installations, and many others. Most ocean vessels are underactuated meaning that they have more degrees of freedom to be controlled than the number of independent control inputs. Ships do not usually have an independent sway actuator while for underwater vehicles there are often no independent sway and heave actuators. As a result, motion control of underactuated ocean vessels opened a new territory in applied nonlinear control, and attracted special attention from both marine technology and control engineering communities.

If classical motion control systems designed for fully or overactuated vessels are directly used on underactuated vessels, the resulting performance of controlled systems is very poor or control objectives cannot be achieved. For example, the traditional approach, in which a combination of a conventional autopilot and a line-of-sight algorithm is used to steer an underactuated ship from one point to another on a straight line, does not impose on minimizing the lateral distance. Consequently, the shortest traveling distance is not achieved. Another example is that underactuated ocean vessels cannot be stabilized by any time-invariant continuous state feedback controllers although they are open loop controllable. This fact resulted from a direct application of the Brockett necessary condition to feedback stabilization of underactuated ocean vessels.

Inspired by progress in the field, we present this monograph on control of underactuated ocean vessels including ships and underwater vehicles to senior and postgraduate students, researchers and practitioners of marine technology, control engineering, mechanical engineering, electrical engineering, and mechatronics. This is the first book in the literature that offers various solutions to advanced feedback control topics of practical importance including stabilization, trajectory-tracking, path-tracking, and path-following for underactuated ocean vessels. In the control development and stability analysis of the controlled systems, practical motivations as well as nontrivial techniques are carefully detailed. The techniques presented in the book can be readily applied to other underactuated mechanical systems such as aircraft, spacecraft, mobile robots, and robot arms.

Acknowledgements

The first author would like to thank the Rectoral Board and his colleagues at Thai Nguyen University of Technology and the University of Western Australia for providing him with a friendly and efficient working environment during his time of writing this book. He is grateful to H.L. Nguyen for her support and encouragement in different ways. He also thanks Z.P. Jiang for inviting him to Polytechnic University for a period of three months from October 2001 to January 2002. Both authors thank the anonymous reviewers and editorial staff of various journals and conferences for their helpful comments on the authors' research papers, which are the main source of research and results for this book. The authors' special thanks go to M.A. Johnson, Joint Editor of the *Advances in Industrial Control* series, O. Jackson, Editor (Engineering), and A. Bunning, Editorial Assistant (Engineering) of Springer, UK, for their support and constructive comments in the process of publishing the book.

The writing of this book was supported in part by the Australian Research Council under grants DP0453294, LP0219249, DP0774645, and DP0988424.

Perth, Australia — *Khac Duc Do*
Thai Nguyen, Vietnam
Perth, Australia — *Jie Pan*
May, 2009

Contents

Chapter 1
Introduction

The subject of this book is the application of nonlinear control theory to control underactuated ocean vessels and to analyze stability of their controlled systems.

A vessel is said to be underactuated if it has more degrees of freedom to be controlled than the number of independent control inputs. Many ocean ships are equipped with propellers and rudders for surge and yaw motions only, but without any actuators for direct control of their sway motion. The control of underactuated ocean vessels has become an active area of research in recent years not only because it poses many challenging questions in applied nonlinear control theory, but also because of its practical importance.

This chapter starts with a brief review of the development in nonlinear control theory and its applications. Next, difficulties in control of underactuated ocean vessels are discussed. The chapter ends with a description of the organization of the book.

1.1 Overview of Nonlinear Control Developments

Over the last two decades nonlinear control theory has received considerable research effort and has undergone a period of significant progress. This is motivated from the fact that real world problems are often inherently nonlinear. Conventionally, control engineers studied a nonlinear system by using a linearized model around selected operating points, and suitable techniques for linear control systems founded on the basis of the superposition principle. A fundamental limitation of this linearization approach is that stability and performance can only be guaranteed in the neighborhood of the selected operating points. Consequently, studying nonlinear systems based on linear models is severely limited since real world systems are usually required to operate over a large number of operating points. In addition, many phenomena such as the existence of multiple equilibria or operating points, periodic variation of state variables or limit cycles, finite escape time, and bifurcations cannot be described or predicted by using linear models. The above-mentioned limitations led control researchers to directly study nonlinear systems. Moving from linear to nonlinear systems, we are faced with a much more difficult situation. The superposition principle no longer holds and analysis tools involve more advanced mathematics. In addition, the separation principle does not hold for nonlinear systems in general.

Currently, the techniques that are available for nonlinear control system design include linearization and gain-scheduling techniques, feedback linearization, circle Popov criteria, small-gain theorems, passivity, averaging, singular perturbation, sliding mode control, Lyapunov design and redesign, backstepping and forwarding, adaptive control, input-to-state stability, integral input-to-state stability, and differential geometric approaches. For details of the above-mentioned approaches, the reader is referred to [1–10]. Nonlinear control methods are increasingly implemented in practice with great success such as electrical motors, diesel engines, ocean vessels, jet engine compressors and electromechanical systems. For some of these applications, the reader is referred to [3, 11–15].

1.2 Difficulties in Control of Underactuated Ocean Vessels

The main difficulty in the control of an underactuated system is that the system has more degrees of freedom (outputs) to be controlled than the number of independent actuators (inputs). A direct application of the Brockett necessary condition indicates that an underactuated system cannot be stabilized by any time-invariant continuous state feedback controllers although it is open loop controllable. This topic is detailed in Section 2.7.4. Moreover, if classical motion control systems designed for fully or overactuated systems are directly used on the underactuated ones, the resulting performance of the controlled systems is very poor or the control objectives cannot be achieved. The above observations are also true for underactuated ocean vessels. Below we illustrate the aforementioned difficulties through a simple example.

In Chapter 3, the equations of motion of an underactuated ocean vessel without environmetal disturbances are given by (see (3.31)):

$$\begin{aligned}\dot{\eta} &= \boldsymbol{J}(\eta)\boldsymbol{v},\\ \boldsymbol{M}\dot{\boldsymbol{v}} &= -\boldsymbol{C}(\boldsymbol{v})\boldsymbol{v} - \boldsymbol{D}(\boldsymbol{v})\boldsymbol{v} - \boldsymbol{g}(\eta) + \boldsymbol{\tau},\end{aligned} \tag{1.1}$$

where

$$\begin{aligned}\eta &= [x, y, z, \phi, \theta, \psi]^T,\\ \boldsymbol{v} &= [u, v, w, p, q, r]^T\end{aligned} \tag{1.2}$$

are the vectors of position/Euler angles and velocities, respectively. The matrices $\boldsymbol{M}$, $\boldsymbol{C}(\boldsymbol{v})$, and $\boldsymbol{D}(\boldsymbol{v})$ denote inertia, Coriolis, and damping, respectively, $\boldsymbol{g}(\eta)$ is the vector of gravitational/buoyancy forces, $\boldsymbol{\tau} \in \mathbb{R}^6$ is the vector of the forces and moments provided by the actuators. Depending on the vessel's configuration, several elements of $\boldsymbol{\tau}$ are zero. For example, for the case where an underwater vehicle does not have independent actuators in the heave and sway axes, the second and third elements of $\boldsymbol{\tau}$ are zero. This gives an underactuated situation. We will show that an underactuation causes serious problems for classical control design techniques.

We now assume that we want to design the control input vector $\boldsymbol{\tau}$ to force the vector of position/Euler angles $\boldsymbol{\eta}$ to track a reference vector of position/Euler angles $\boldsymbol{\eta}_d$ with bounded $\dot{\boldsymbol{\eta}}_d$ and $\ddot{\boldsymbol{\eta}}_d$. As a standard application of the control design techniques in [3, 6] for such a strict feedback system like (1.1), we define the tracking error vector $\boldsymbol{\eta}_e$ as

$$\boldsymbol{\eta}_e = \boldsymbol{\eta} - \boldsymbol{\eta}_d \tag{1.3}$$

and differentiate $\boldsymbol{\eta}_e$ twice along the solutions of the system (1.1) to obtain

$$\ddot{\boldsymbol{\eta}}_e = \dot{\boldsymbol{J}}(\boldsymbol{\eta})\boldsymbol{v} + \boldsymbol{J}(\boldsymbol{\eta})\boldsymbol{M}^{-1}\Big(-\boldsymbol{C}(\boldsymbol{v})\boldsymbol{v} - \boldsymbol{D}(\boldsymbol{v})\boldsymbol{v} - \boldsymbol{g}(\boldsymbol{\eta}) + \boldsymbol{\tau}\Big) - \ddot{\boldsymbol{\eta}}_d. \tag{1.4}$$

Letting

$$\dot{\boldsymbol{J}}(\boldsymbol{\eta})\boldsymbol{v} + \boldsymbol{J}(\boldsymbol{\eta})\boldsymbol{M}^{-1}\Big(-\boldsymbol{C}(\boldsymbol{v})\boldsymbol{v} - \boldsymbol{D}(\boldsymbol{v})\boldsymbol{v} - \boldsymbol{g}(\boldsymbol{\eta}) + \boldsymbol{\tau}\Big) - \ddot{\boldsymbol{\eta}}_d = -\boldsymbol{K}_1\boldsymbol{\eta}_e - \boldsymbol{K}_2\dot{\boldsymbol{\eta}}_e, \tag{1.5}$$

where $\boldsymbol{K}_1 \in \mathbb{R}^{6\times6}$ and $\boldsymbol{K}_2 \in \mathbb{R}^{6\times6}$ are positive definite symmetric control gain matrices, results in

$$\ddot{\boldsymbol{\eta}}_e = -\boldsymbol{K}_1\boldsymbol{\eta}_e - \boldsymbol{K}_2\dot{\boldsymbol{\eta}}_e, \tag{1.6}$$

where $\dot{\boldsymbol{\eta}}_e = \boldsymbol{J}(\boldsymbol{\eta})\boldsymbol{v} - \dot{\boldsymbol{\eta}}_d$, which is globally exponentially stable (GES) at the origin. We now need to solve (1.5) for $\boldsymbol{\tau}$ for implementation. From (1.5), we have

$$\begin{aligned}\boldsymbol{\tau} &= \boldsymbol{M}\boldsymbol{J}^{-1}(\boldsymbol{\eta})\big(-\boldsymbol{K}_1\boldsymbol{\eta}_e - \boldsymbol{K}_2\dot{\boldsymbol{\eta}}_e - \dot{\boldsymbol{J}}(\boldsymbol{\eta})\boldsymbol{v} + \ddot{\boldsymbol{\eta}}_d\big) + \boldsymbol{C}(\boldsymbol{v})\boldsymbol{v} + \boldsymbol{D}(\boldsymbol{v})\boldsymbol{v} + \boldsymbol{g}(\boldsymbol{\eta}) \\ &:= \boldsymbol{\Psi}(\boldsymbol{\eta}_d, \dot{\boldsymbol{\eta}}_d, \ddot{\boldsymbol{\eta}}_d, \boldsymbol{\eta}, \boldsymbol{v}).\end{aligned} \tag{1.7}$$

Let us assume that the actuators available are independent and assume that they number m_c, where $m_c < 6$. Consequently we can propose a decomposition of $\boldsymbol{\tau}$ as:

H1. $\boldsymbol{\tau}_a \in \mathbb{R}^6$, where $\boldsymbol{\tau}_a$ has zeros in the element positions identifying that there are no actuator inputs and where $\boldsymbol{\tau}_a$ contains the elements of $\boldsymbol{\tau}$ associated with actuator locations.

H2. $\boldsymbol{\tau}_{na} \in \mathbb{R}^6$. This vector complements $\boldsymbol{\tau}_a$ by having zeros in the locations associated with actuator inputs and contains the elements of $\boldsymbol{\tau}$ in the locations associated with no actuator inputs.

H3. Similarly, the vector $\boldsymbol{\Psi}(\cdot)$ can also be decomposed into vectors $\boldsymbol{\Psi}_a(\cdot)$ and $\boldsymbol{\Psi}_{na}(\cdot)$ according to the rules of schemes (H1) and (H2).

Thus, the computed control vector $\boldsymbol{\tau}$ can be decomposed as

$$\boldsymbol{\tau} = \boldsymbol{\tau}_a + \boldsymbol{\tau}_{na} = \boldsymbol{\Psi}_a(\cdot) + \boldsymbol{\Psi}_{na}(\cdot) \in \mathbb{R}^6 \tag{1.8}$$

and

$$\boldsymbol{\tau_a} = \boldsymbol{\Psi_a}(\cdot) \in \mathbb{R}^6,$$
$$\boldsymbol{\tau_{na}} = \boldsymbol{\Psi_{na}}(\cdot) \in \mathbb{R}^6. \tag{1.9}$$

In practice the actuator configuration only allows $\boldsymbol{\tau_a}$ to be implemented, and the response of the vessel (system) will result from this input alone.

If the vessel response is given by $\boldsymbol{\eta}$, where

$$\boldsymbol{\eta} = \boldsymbol{H}(\boldsymbol{\tau}), \tag{1.10}$$

the full response would have been $\boldsymbol{\eta}_{\text{full}}$, where

$$\boldsymbol{\eta}_{\text{full}} = \boldsymbol{H}(\boldsymbol{\tau}) = \boldsymbol{H}(\boldsymbol{\tau_a} + \boldsymbol{\tau_{na}}), \tag{1.11}$$

but the actual response is $\boldsymbol{\eta}_{\text{partial}}$, where

$$\boldsymbol{\eta}_{\text{partial}} = \boldsymbol{H}(\boldsymbol{\tau_a}) = \boldsymbol{H_s}(\boldsymbol{\tau_{ra}}) \tag{1.12}$$

and $\boldsymbol{\tau_a} \in \mathbb{R}^6$, $\boldsymbol{\eta}_{\text{partial}} \in \mathbb{R}^6$, $\boldsymbol{H}(\cdot) : \mathbb{R}^6 \to \mathbb{R}^6$ but $\boldsymbol{\tau_{ra}}$ is the m_c vector of nonzero actuator inputs, $\boldsymbol{\tau_{ra}} \in \mathbb{R}^{m_c}$, and $\boldsymbol{H_s}(\cdot)$ is the reduced functional representation with $\boldsymbol{H_s} : \mathbb{R}^{m_c} \to \mathbb{R}^6$.

Therefore the fully actuated vessel is represented by $\boldsymbol{H}(\cdot)$ and the underactuated vessel is represented by $\boldsymbol{H_s}(\cdot)$. These two representations may have very different properties. The implications of not being able to implement the control action $\boldsymbol{\tau_{na}}$, and having a vessel whose representation is $\boldsymbol{H_s}(\cdot)$ are potentially:

1. Degraded performance achieved by $\boldsymbol{\eta}_{\text{partial}}$ when compared to $\boldsymbol{\eta}_{\text{full}}$.
2. Total loss of performance and inability to meet the control objectives in any useful way.
3. Loss of guaranteed properties like closed loop stability with possible catastrophic system failure.

Ways to overcome these problems include:

1. Reduction of performance specification as a control objective that can be achieved by $\boldsymbol{\eta}_{\text{partial}}$.
2. Modification of the control design procedure to accommodate the reduced degrees of freedom available for control.
3. Procedures to use the available actuators to minimize the deleterious effects caused by not being able to implement the $\boldsymbol{\tau_{na}}$ component.
4. The use of advanced control procedures like those described in this book.
5. The installation of additional actuators able to create access to effect and control the missing degrees of freedom.

1.3 Organization of the Book

The rest of the book is organized in 14 chapters, see Table 1.1. A brief summary of each chapter is given below.

Chapter 2 presents mathematical tools like Lyapunov stability theory, Barbalat-like lemmas, and the backstepping technique to be used in control design and stability analysis in the book.

Chapter 3 sets out the basic material that will be used in the subsequent chapters. Basic motion control tasks for ocean vessels and modeling of vessels are given.

Chapter 4 presents control properties of ocean vessels. Then, the existing literature on the control of underactuated ocean vessels is reviewed. Through the review of the previous work in the area of stabilization, trajectory-tracking, path-following, and output feedback control of underactuated ocean vessels, challenging questions are raised. These questions motivate contributions of the book to new solutions for the motion control of underactuated ships and underwater vehicles.

Chapter 5 addresses the problem of trajectory-tracking control of underactuated ships. These ships do not have independent actuators in the sway axis. The reference trajectory is generated by a suitable virtual ship. The control development is based on an elegant coordinate transformation, Lyapunov's direct method, and the backstepping technique, and utilizes passivity properties of the ship dynamics and their interconnected structure.

Chapter 6 examines the problem of designing a single controller that achieves stabilization and trajectory-tracking simultaneously for underactuated ships. In comparison with the preceding chapter, a path approaching the origin and a set-point can also be included in the reference trajectory, i.e., stabilization/regulation is also considered. The control development is based on several special coordinate transformations plus the techniques in the preceding chapter.

Chapter 7 presents global partial-state feedback and output feedback control schemes for trajectory-tracking control of underactuated ships. For the case of partial-state feedback, measurements of the ship sway and surge velocities are not needed, while for the case of output feedback, no ship velocities are required for feedback. Global nonlinear coordinate changes are introduced to transform the ship dynamics to a system affine in ship velocities to design observers to globally exponentially estimate unmeasured velocities. These observers plus the techniques in Chapters 5 and 6 facilitate the controllers' development.

Chapter 8 deals with the problem of path-tracking control of underactuated ships. In comparison with Chapters 5, 6, and 7, the requirement of the reference trajectory generated by a suitable virtual ship is relaxed. Both full state feedback and output feedback cases are considered. The control development is based on a series of nontrivial coordinate transformations plus the techniques in the previous chapters.

Chapter 9 addresses the problem of way-point tracking control of underactuated ships. Both full state feedback and output feedback controllers are designed. The controllers in this chapter can be considered as an advanced version of the conventional course-keeping controllers in the sense that in addition to maintaining the desired heading, both the sway displacement (lateral distance) and sway velocity are controlled.

Chapter 10 develops full state feedback and output feedback controllers that force underactuated ships to follow a predefined path. The control development is motivated by an observation that it is practical to steer a vessel such that the vessel is on the reference path and its total velocity is tangent to the reference path, and that the vessel's forward speed is controlled separately by the main thruster control system. The techniques in the previous chapters plus the use of the Serret–Frenet frame facilitate the results.

Chapter 11 presents a different approach from Chapter 10 to solve a path-following problem for underactuated ships. Unlike in Chapter 10, here the control development is based on the method of generating reference paths by the helmsman. The path-following errors are first interpreted in polar coordinates, the techniques used in the previous chapters are then used to design path-following controllers. Interestingly, some situations of practical importance such as parking and point-to-point navigation are covered in this chapter as a by-product of the developed path-following system.

Chapter 12 addresses the problem of trajectory-tracking control of underactuated underwater vehicles. These vehicles do not have independent actuators in the sway and heave axes. The control development is built on the techniques developed for underactuated ships in Chapters 5, 6, and 7. Due to complex dynamics of underwater vehicles in comparison with that of the surface ships, the control design and stability analysis require more complicated coordinate transformations and techniques than those developed for underactuated ships in Chapters 5, 6, and 7.

Chapter 13 extends the results of Chapter 11 to the design of a path-following control system for underactuated underwater vehicles. A series of path-following strategies for the vehicles is first discussed. A practical approach is then chosen to facilitate the control development. We also address the parking and point-to-point navigation problems of the vehicles in this chapter.

Chapter 14 illustrates several applications of the observer and control design techniques developed in the previous chapters to control of other underactuated mechanical systems, which are common in practice. These systems include mobile robots and VTOL aircraft. For mobile robots, a global exponential observer is first designed based on the observer design for underactuated ships in Chapter 7. Output feedback simultaneous stabilization and trajectory-tracking and path-following controllers are then developed using the control design techniques proposed for underactuated ships in Chapters 5 and 10. For VTOL aircraft, the observer and control design strategies used for underactuated ships in Chapters 5 and 6 are utilized to design a global output feedback trajectory-tracking controller.

Finally, Chapter 15 concludes the book by briefly summarizing the main results presented in the previous chapters and presenting related open problems for further investigation.

Table 1.1 Organization of the book

1. Introduction	
Part I: Mathematical Tools	2. Mathematical Preliminaries
Part II: Modeling and Control Properties of Ocean Vessels	3. Modeling of Ocean Vessels 4. Control Properties and Previous Work on Control of Ocean Vessels
Part III: Control of Underactuated Ships	5. Trajectory-tracking Control of Underactuated Ships 6. Simultaneous Stabilization and Trajectory-tracking Control of Underactuated Ships 7. Partial-state and Output Feedback Trajectory-tracking Control of Underactuated Ships 8. Path-tracking Control of Underactuated Ships 9. Way-point Tracking Control of Underactuated Ships 10. Path-following of Underactuated Ships Using Serret–Frenet Coordinates 11. Path-following of Underactuated Ships Using Polar Coordinates
Part IV: Control of Underactuated Underwater Vehicles	12. Trajectory-tracking Control of Underactuated Underwater Vehicles 13. Path-following of Underactuated Underwater Vehicles
Part V: Control of Other Underactuated Mechanical Systems	14. Control of Other Underactuated Mechanical Systems
15. Conclusions and Perspectives	

Part I
Mathematical Tools

Chapter 2
Mathematical Preliminaries

This chapter presents mathematical tools, which will be used in control design and stability analysis in the subsequent chapters. Some standard theorems, lemmas and corollaries, which are available in references, are sometimes given without a proof.

2.1 Lyapunov Stability

Stability theory is important in system theory and engineering. For a controlled system to be usable in practice, the least requirement is that the system is stable under unknown disturbances or noise. There are various types of stability problems that arise in the study of dynamical systems such as input–output stability, Lyapunov stability, absolute stability and stability of periodic solutions. These stability concepts have been studied extensively over the last 100 years. For control design and stability analysis in this book, Lyapunov stability plays an important role for the following reasons. First, Lyapunov's direct method uses an energy-like function called the Lyapunov function to study the behaviors of dynamical systems analytically. This function reflects physical properties of the systems under study. Second, Lyapunov's second method is applicable to both linear and nonlinear systems. Third, many results in input–output stability can be obtained directly using Lyapunov stability theory. Lyapunov stability theory generally includes Lyapunov's first and second methods. The first method and the center manifold theory developed later are techniques based on the lowest-order approximation around a given point or a nominal trajectory. The stability results obtained using these two approximation methods are inherently local, and the stability regions are hard to estimate and often quite small. Since the objective of this book is to study nonlinear dynamics of ocean vessels, we omit the aforementioned two approximation techniques. Our primary interest is in stability based on Lyapunov's second method for systems described by ordinary differential equations.

This section is concerned with stability of equilibrium points in the sense of Lyapunov. An equilibrium point is stable if all solutions starting at nearby points stay nearby; otherwise it is unstable. It is asymptotically stable if in addition, all solutions tend to the equilibrium point as time approaches infinity. These notions will be made mathematically rigorous in this section.

Consider the following nonautonomous system:

$$\dot{x} = f(t,x), \tag{2.1}$$

where $f : [0,\infty) \times D \to \mathbb{R}^n$ is piecewise continuous in t and locally Lipschitz in x on $[0,\infty) \times D$ and $D \subseteq \mathbb{R}^n$ is a domain that contains the origin $x = 0$.

2.1.1 Definitions

Definition 2.1. The origin $x = 0$ is the equilibrium point of (2.1) if

$$f(t,0) = 0, \ \forall t \geq 0. \tag{2.2}$$

Definition 2.2. A continuous function $\alpha : [0,a) \to \mathbb{R}^+$ is said to belong to class K if it is strictly increasing and $\alpha(0) = 0$. It is said to belong to class K_∞ if $a = \infty$ and $\alpha(r) \to \infty$ as $r \to \infty$.

Definition 2.3. A continuous function $\beta : [0,a) \times \mathbb{R}^+ \to \mathbb{R}^+$ is said to belong to class KL if, for each fixed s, the mapping $\beta(r,s)$ belongs to class K with respect to r and, for each fixed r, the mapping $\beta(r,s)$ is decreasing with respect to s and $\beta(r,s) \to 0$ as $s \to \infty$. It is said to belong to class KL_∞ if, in addition, for each fixed s the mapping $\beta(r,s)$ belongs to class K_∞ with respect to r.

Definition 2.4. The equilibrium point $x = 0$ of (2.1) is

1. stable if, for each $\varepsilon > 0$, there is $\delta = \delta(\varepsilon, t_0) > 0$ such that

$$\|x(t_0)\| < \delta \Rightarrow \|x(t)\| < \varepsilon, \ \forall t \geq t_0 \geq 0, \tag{2.3}$$

2. uniformly stable if, for each $\varepsilon > 0$, there is $\delta = \delta(\varepsilon) > 0$ independent of t_0 such that (2.3) is satisfied,
3. unstable if it is not stable,
4. asymptotically stable if it is stable and there is a positive constant $c = c(t_0)$ such that $x(t) \to 0$ as $t \to \infty$, for all $\|x(t_0)\| < c$,
5. uniformly asymptotically stable if it is uniformly stable and there is a positive constant c, independent of t_0, such that for all $\|x(t_0)\| < c$, $x(t) \to 0$ as $t \to \infty$, uniformly in t_0; that is, for each $\eta > 0$, there is $T = T(\eta) > 0$ such that

$$\|x(t)\| < \eta, \ \forall t \geq t_0 + T(\eta), \ \forall \, \|x(t_0)\| < c, \tag{2.4}$$

6. globally uniformly asymptotically stable (GUAS) if it is uniformly stable, $\delta(\varepsilon)$ can be chosen to satisfy $\lim_{\varepsilon\to\infty} \delta(\varepsilon) = \infty$, and, for each pair of positive numbers η and c, there is $T = T(\eta, c) > 0$ such that

$$\|x(t)\| < \eta, \; \forall t \geq t_0 + T(\eta, c), \; \forall \|x(t_0)\| < c. \tag{2.5}$$

Definition 2.5. The equilibrium point $x = 0$ of (2.1) is exponentially stable if there exist positive constants c, k, and λ such that

$$\|x(t)\| \leq k \|x(t_0)\| e^{-\lambda(t-t_0)}, \; \forall t \geq t_0 \geq 0, \; \forall \; \|x(t_0)\| < c \tag{2.6}$$

and GES if (2.6) is satisfied for any initial state $x(t_0)$.

Definition 2.6. The equilibrium point $x = 0$ of (2.1) is K-exponentially stable if there exist positive constants c and λ and a class K function α such that

$$\|x(t)\| \leq \alpha (\|x(t_0)\|) e^{-\lambda(t-t_0)}, \; \forall t \geq t_0 \geq 0, \; \forall \; \|x(t_0)\| < c \tag{2.7}$$

and globally K-exponentially stable if (2.7) is satisfied for any initial state $x(t_0)$.

Definition 2.7. The solutions of (2.1) are as follows:

1. Uniformly bounded if there exists a positive constant c, independent of $t_0 \geq 0$, and for every $a \in (0, c)$, there is $\beta = \beta(a) > 0$, independent of t_0, such that

$$\|x(t_0)\| \leq a \Rightarrow \|x(t)\| \leq \beta, \; \forall t \geq t_0. \tag{2.8}$$

2. Globally uniformly bounded if (2.8) holds for an arbitrarily large a.
3. Uniformly ultimately bounded with ultimate bound b if there exist positive constants b and c, independent of $t_0 \geq 0$, and for every $a \in (0, c)$ there is $T = T(a, b) \geq 0$, independent of t_0, such that

$$\|x(t_0)\| \leq a \Rightarrow \|x(t)\| \leq b, \; \forall t \geq t_0 + T. \tag{2.9}$$

4. Globally uniformly ultimately bounded if (2.9) holds for an arbitrarily large a.

2.1.2 Lemmas and Theorems

The following lemma provides equivalent definitions of uniform stability and uniform asymptotic stability by using class K and class KL functions.

Lemma 2.1. *The equilibrium point $x = 0$ of (2.1) is*

1. *uniformly stable if and only if there exist a class K function α and a positive constant c, independent of t_0, such that*

$$\|x(t)\| \leq \alpha (\|x(t_0)\|), \; \forall t \geq t_0 \geq 0, \; \forall x(t_0) \big| \, \|x(t_0)\| < c, \tag{2.10}$$

2. *uniformly asymptotically stable if and only if there exist a class KL function β and a positive constant c, independent of t_0, such that*

$$\|x(t)\| \leq \beta (\|x(t_0)\|, t - t_0), \; \forall t \geq t_0 \geq 0, \; \forall x(t_0) \big| \, \|x(t_0)\| < c, \tag{2.11}$$

3. *GUAS if and only if inequality (2.11) is satisfied with $\beta \in KL_\infty$ for any initial state $x(t_0)$.*

Proof. See [6]. □

Lemma 2.2. *Assume that $d : \mathbb{R}^n \to \mathbb{R}^n$ satisfies*

$$P\left[\frac{\partial d}{\partial x}\right]+\left[\frac{\partial d}{\partial x}\right]^T P \geq 0,\ \forall\, x \in \mathbb{R}^n, \tag{2.12}$$

where $P = P^T > 0$. Then

$$(x-y)^T P(d(x)-d(y)) \geq 0,\ \ \forall\, x,\, y \in \mathbb{R}^n. \tag{2.13}$$

Proof. See [16]. □

Lemma 2.3. *The following nonlinear interconnected system:*

$$\begin{aligned} \dot{x}_1 &= f_1(t,x_1,x_2)+g_1(t,x_1,x_2)u, \\ \dot{x}_2 &= f_2(t,x_1,x_2)+g_2(t,x_1,x_2)u, \end{aligned} \tag{2.14}$$

where $x_i \in \mathbb{R}$, $i = 1,2$, $f_i(t,x_1,x_2)$ are locally Lipschitz in x_i and piecewise continuous in t; $u \in \mathbb{R}$ is the control input, and $g_2(t,x_1,x_2) \neq 0$, $\forall\, t \geq 0$, $x_i \in \mathbb{R}$, can be transformed to the following system

$$\begin{aligned} \dot{z}_1 &= \gamma_1(t,z_1,x_2), \\ \dot{x}_2 &= \gamma_2(t,z_1,x_2)+\varphi_2(t,z_1,x_2)u. \end{aligned} \tag{2.15}$$

Proof. Define

$$z_1 = x_1+\pi(t,x_1,x_2), \tag{2.16}$$

where $\pi(t,x_1,x_2)$ is to be determined. Differentiating both sides of (2.16) along the solutions of (2.14) yields

$$\begin{aligned} \dot{z}_1 = &\left(\frac{\partial \pi(t,x_1,x_2)}{\partial x_1}+1\right) f_1(t,x_1,x_2)+ \\ &\frac{\partial \pi(t,x_1,x_2)}{\partial x_2} f_2(t,x_1,x_2)+\frac{\partial \pi(t,x_1,x_2)}{\partial t}+ \\ &\left[\left(\frac{\partial \pi(t,x_1,x_2)}{\partial x_1}+1\right) g_1(t,x_1,x_2)+\frac{\partial \pi(t,x_1,x_2)}{\partial x_2} g_2(t,x_1,x_2)\right]u. \end{aligned} \tag{2.17}$$

Now choosing the function $\pi(t,x_1,x_2)$ such that

$$\left(\frac{\partial \pi(t,x_1,x_2)}{\partial x_1}+1\right) g_1(t,x_1,x_2)+\frac{\partial \pi(t,x_1,x_2)}{\partial x_2} g_2(t,x_1,x_2)=0 \tag{2.18}$$

results in (2.15), where the functions $\gamma_1(t,z_1,x_2)$, $\gamma_2(t,z_1,x_2)$, and $\varphi_2(t,z_1,x_2)$ are defined as

$$\gamma_1(t,z_1,x_2) := \left(\frac{\partial \pi(t,x_1,x_2)}{\partial x_1}+1\right) f_1(t,x_1,x_2)+ \frac{\partial \pi(t,x_1,x_2)}{\partial x_2} f_2(t,x_1,x_2)+\frac{\partial \pi(t,x_1,x_2)}{\partial t}, \quad (2.19)$$
$$\gamma_2(t,z_1,x_2) := f_2(t,x_1,x_2),\ \varphi_2(t,z_1,x_2) := g_2(t,x_1,x_2),$$

with x_1 being solved from (2.16) and substituted in. □

Remark 2.1.

1. The success of the above lemma depends on whether a solution to the partial differential equation (2.18) can be found. Solving this partial differential equation is difficult in general but might be possible in some specific cases such as the vehicle systems in this book.
2. In some cases, designing a control input u for the transformed system (2.15) is simpler than that for the original system (2.14).

The main Lyapunov stability theorem, which has a number of applications in studying stability of (2.1), is given below.

Theorem 2.1. *Let $D=\{x\in\mathbb{R}^n \,|\, \|x\|<r\}$ and $x=0$ be an equilibrium point of (2.1). Let $V: D\times\mathbb{R}^n\to\mathbb{R}^+$ be a continuously differentiable function such that $\forall t\ge 0,\ \forall x\in D$,*

$$\gamma_1(\|x\|)\le V(x,t)\le \gamma_2(\|x\|),$$
$$\frac{\partial V}{\partial t}+\frac{\partial V}{\partial x}f(t,x)\le -\gamma_3(\|x\|). \quad (2.20)$$

Then the equilibrium point $x=0$ is

1. *uniformly stable, if γ_1 and γ_2 are class K functions on $[0,r)$ and $\gamma_3\ge 0$ on $[0,r)$,*
2. *uniformly asymptotically stable, if γ_1, γ_2 and γ_3 are class K functions on $[0,r)$,*
3. *exponentially stable if $\gamma_i(\rho)=k_i\rho^\alpha$ on $[0,r)$, $k_i>0,\ \alpha>0,\ i=1,2,3$,*
4. *globally uniformly stable if $D=\mathbb{R}^n$, γ_1 and γ_2 are class K_∞ functions, and $\gamma_3\ge 0$ on $\mathbb{R}^+$,*
5. *GUAS if $D=\mathbb{R}^n$, γ_1 and γ_2 are class K_∞ functions, and γ_3 is a class K function on $\mathbb{R}^+$,*
6. *GES, if $D=\mathbb{R}^n$, $\gamma_i(\rho)=k_i\rho^\alpha$ on $\mathbb{R}^+$, $k_i>0,\alpha>0,\ i=1,2,3$.*

Proof. See [3]. □

Theorem 2.2. *Let $x=0$ be an equilibrium point of (2.1) and suppose that f is locally Lipschitz in x and uniformly continuous in t. Let $V:\mathbb{R}^n\times\mathbb{R}^+\to\mathbb{R}^+$ be a continuously differentiable function such that*

$$\gamma_1(\|x\|)\le V(x,t)\le \gamma_2(\|x\|),$$
$$\dot V=\frac{\partial V}{\partial t}+\frac{\partial V}{\partial x}f(t,x)\le -W(x)\le 0, \quad (2.21)$$

for all $t \geq 0$ and $x \in \mathbb{R}^n$, where γ_1 and γ_2 are class K_∞ functions, and W is a continuous function. Then all solutions of (2.1) are globally uniformly bounded and satisfy

$$\lim_{t\to\infty} W(x(t)) = 0. \tag{2.22}$$

In addition, if $W(x)$ is positive definite, then the equilibrium point $x = 0$ is GUAS.

Proof. See [3]. □

The following Lyapunov-like theorem is useful for showing uniform boundedness and ultimate boundedness.

Theorem 2.3. *Let $V : [0,\infty)\times D \to \mathbb{R}$ be a continuously differentiable function and $D \subset \mathbb{R}^n$ be a domain that contains the origin such that*

$$\begin{aligned} &\alpha_1(\|x\|) \leq V(x,t) \leq \alpha_2(\|x\|), \\ &\frac{\partial V}{\partial t} + \frac{\partial V}{\partial x} f(t,x) \leq -W(x), \ \forall \|x\| \geq \mu > 0 \end{aligned} \tag{2.23}$$

for all $t \geq 0$ and $x \in D$ where α_1 and α_2 are class K functions, and W is a continuous positive definite function. Take $r > 0$ such that $B_r \subset D$ and suppose that

$$\mu < \alpha_2^{-1}(\alpha_1(r)). \tag{2.24}$$

Then, there exists a class KL function β and for every initial state $x(t_0)$, satisfying $\|x(t_0)\| < \alpha_2^{-1}(\alpha_1(r))$, there is $T \geq 0$ (dependent on $x(t_0)$ and μ) such that the solutions of (2.1) satisfies

$$\begin{aligned} &\|x(t)\| \leq \beta(\|x(t_0)\|, t-t_0), \ \forall\, t_0 \leq t \leq t_0 + T, \\ &\|x(t)\| < \alpha_1^{-1}(\alpha_2(\mu)), \ \forall\, t \geq t_0 + T. \end{aligned} \tag{2.25}$$

Moreover, if $D = \mathbb{R}^n$ and α_1 belongs to class K_∞, then (2.25) holds for any initial state $x(t_0)$ with no restriction on how large μ is.

Proof. See [6]. □

2.1.3 Stability of Cascade Systems

Consider the following cascade system:

$$\begin{aligned} \dot{z}_1 &= f_1(t,z_1) + g(t,z_1,z_2)z_2, \\ \dot{z}_2 &= f_2(t,z_2), \end{aligned} \tag{2.26}$$

where $z_1 \in \mathbb{R}^n$, $z_2 \in \mathbb{R}^m$, $f_1(t,z_1)$ is continuously differentiable in (t,z_1), and $f_2(t,z_2)$ and $g(t,z_1,z_2)$ are continuous and locally Lipschitz in z_2 and (z_1,z_2), respectively.

If we set $z_2 = 0$, then the first equation of (2.26) becomes $\dot{z}_1 = f_1(t, z_1)$. Therefore we can view the first equation of (2.26) as the system

$$\Omega_1 : \dot{z}_1 = f_1(t, z_1), \tag{2.27}$$

which is perturbed by the output of the system

$$\Omega_2 : \dot{z}_2 = f_2(t, z_2). \tag{2.28}$$

Now assume that the systems Ω_1 and Ω_2 are asymptotically stable at the origin, i.e., (2.27) and (2.28) yield $\lim_{t\to\infty} z_1(t) = 0$ and $\lim_{t\to\infty} z_2(t) = 0$, respectively. Based on these assumptions, it is plausible to conclude that the system (2.26) is asymptotically stable at the origin in general. In many cases, the solution $z_1(t)$ of the system (2.26) goes to infinity in finite time as can be seen from the simple cascade system:

$$\begin{aligned} \dot{z}_1 &= -k_1 z_1 + z_1^2 z_2, \\ \dot{z}_2 &= -k_2 z_2, \end{aligned} \tag{2.29}$$

where k_1 and k_2 are strictly positive constants. It is obvious that the subsystems $\dot{z}_1 = -k_1 z_1$ and $\dot{z}_2 = -k_2 z_2$ are GES at the origin. It is straightforward to show that the solution of (2.29) is

$$\begin{aligned} z_1(t) &= \frac{z_1(t_0)(k_1 + k_2)}{z_1(t_0)z_2(t_0)e^{-k_2(t-t_0)} + (k_1 + k_2 - z_1(t_0)z_2(t_0))e^{k_1(t-t_0)}}, \\ z_2(t) &= z_2(t_0)e^{-k_2(t-t_0)}. \end{aligned} \tag{2.30}$$

It can be seen from (2.30) that if $z_1(t_0)z_2(t_0) < k_1 + k_2$, then both $z_1(t)$ and $z_2(t)$ are bounded and converge exponentially to zero, respectively. If $z_1(t_0)z_2(t_0) = k_1 + k_2$, then $z_2(t)$ is still bounded and converges exponentially to zero but $z_1(t)$ tends to infinity exponentially fast. If $z_1(t_0)z_2(t_0) > k_1 + k_2$, the situation is catastrophic, that is $z_1(t)$ tends to infinity when $t \to t_0 + t_f$ with

$$t_f = \frac{1}{k_1 + k_2} \ln\left(\frac{z_1(t_0)z_2(t_0)}{z_1(t_0)z_2(t_0) - (k_1 + k_2)}\right). \tag{2.31}$$

The following theorem gives sufficient conditions of stability of the cascade system (2.26) based on stability of (2.27) and (2.28), and the connected term $g(t, z_1, z_2)$.

Theorem 2.4. *Consider the following assumptions:*

1. *The systems (2.27) and (2.28) are both GUAS and we know explicitly a C^1 Lyapunov function $V(t, z_1)$, two class-K_∞ functions α_1 and α_2, a class-K function α_4, and a positive semidefinite function $W(z_1)$ such that*

$$\begin{aligned} &\alpha_1(\|z_1\|) \leq V(t, z_1) \leq \alpha_2(\|z_1\|), \\ &\frac{\partial V}{\partial t} + \frac{\partial V}{\partial z_1} f_1(t, z_1) \leq -W(z_1), \ \left\|\frac{\partial V}{\partial z_1}\right\| \leq \alpha_4(\|z_1\|). \end{aligned} \tag{2.32}$$

2. *For each fixed z_2, there exists a continuous function $\lambda : \mathbb{R}^+ \to \mathbb{R}^+$ such that*

$$\lim_{s\to\infty} \lambda(s) = 0,$$
$$\left\| \frac{\partial V}{\partial z_1} g(t, z_1, z_2) \right\| \le \lambda(\|z_1\|) W(z_1). \tag{2.33}$$

3. *There exist continuous functions $\theta : \mathbb{R}^+ \to \mathbb{R}^+$ and $\alpha_5 : \mathbb{R}^+ \to \mathbb{R}^+$ such that*

$$\|g(t, z_1, z_2)\| \le \theta(\|z_2\|) \alpha_5(\|z_1\|) \tag{2.34}$$

and a continuous nondecreasing function $\alpha_6 : \mathbb{R}^+ \to \mathbb{R}^+$, and a nonnegative constant a such that

$$\alpha_6(s) \ge \alpha_4\left(\alpha_1^{-1}(s)\right) \alpha_5\left(\alpha_1^{-1}(s)\right),$$
$$\int_a^\infty \frac{\mathrm{d}s}{\alpha_6(s)} = \infty. \tag{2.35}$$

4. *For each $r > 0$, there exist constants $\chi > 0$ and $\eta > 0$ such that for all $t \ge 0$ and all $\|z_2\| < r$*

$$\left\| \frac{\partial V}{\partial z_1} g(t, z_1, z_2) \right\| \le \chi W(z_1), \forall \|z_1\| \ge \eta. \tag{2.36}$$

5. *There exists a class K function ϕ such that the solution $z_2(t)$ of (2.28) satisfies*

$$\int_{t_0}^\infty \|z_2(t)\| \mathrm{d}t \le \phi(\|z_2(t_0)\|). \tag{2.37}$$

Then we can conclude that if

- *Assumptions 1 and 2, or*
- *Assumptions 1, 3 and 4, or*
- *Assumptions 1, 3 and 5*

hold then the cascade system (2.26) is GUAS.

Proof. See [17]. □

2.2 Input-to-state Stability

Definition 2.8. The system

$$\dot{x} = f(t, x, u), \tag{2.38}$$

where f is piecewise continuous in t and locally Lipschitz in x and u, is said to be input-to-state stable (ISS) if there exist a class KL function β and a class K function

γ, such that, for any $x(t_0)$ and for any input $u(\cdot)$ continuous and bounded on $[0,\infty)$, the solution exists for all $t \geq t_0 \geq 0$ and satisfies

$$\|x(t)\| \leq \beta(\|x(t_0)\|, t-t_0) + \gamma\left(\sup_{t_0 \leq \tau \leq t} \|u(\tau)\|\right). \tag{2.39}$$

The following theorem establishes the equivalence between the existence of a Lyapunov-like function and the input-to-state stability.

Theorem 2.5. *Suppose that for the system (2.38) there exists a C^1 function $V : \mathbb{R}^+ \times \mathbb{R}^n \to \mathbb{R}^+$ such that for all $x \in \mathbb{R}^n$ and $u \in \mathbb{R}^m$,*

$$\begin{aligned} &\gamma_1(\|x\|) \leq V(t,x) \leq \gamma_2(\|x\|), \\ &\|x\| \geq \rho(\|u\|) \Rightarrow \frac{\partial V}{\partial t} + \frac{\partial V}{\partial x} f(t,x,u) \leq -\gamma_3(\|x\|), \end{aligned} \tag{2.40}$$

where γ_1, γ_2, and ρ are class K_∞ functions and γ_3 is a class-K function. Then the system (2.38) is ISS with $\gamma = \gamma_1^{-1} \circ \gamma_2 \circ \rho$.

Proof. If $x(t_0)$ is in the set

$$R_{t_0} = \left\{x \in \mathbb{R}^n \,\middle|\, \|x\| \leq \rho\left(\sup \|u(\tau)\|_{\tau \geq t_0}\right)\right\}, \tag{2.41}$$

then $x(t)$ remains within the set

$$S_{t_0} = \left\{x \in \mathbb{R}^n \,\middle|\, \|x\| \leq \gamma_1^{-1} \circ \gamma_2 \circ \rho\left(\sup \|u(\tau)\|_{\tau \geq t_0}\right)\right\}, \tag{2.42}$$

for all $t \geq t_0$. Define $B = [t_0, T)$ as the time interval before $x(t)$ enters R_{t_0} for the first time. In view of the definition of R_{t_0} we have

$$\dot{V} \leq -\gamma_3 \circ \gamma_2^{-1}(V), \ \forall t \in B. \tag{2.43}$$

Then, there exists a class-KL function β_v such that $V(t) \leq \beta_v(V(t_0), t-t_0)$, $\forall t \in B$, which implies

$$\|x(t)\| \leq \gamma_1^{-1}(\beta_v(\gamma_2(\|x(t_0)\|, t-t_0)) := \beta(\|x(t_0)\|, t-t_0), \ \forall t \in B. \tag{2.44}$$

On the other hand, by (2.42), we conclude that

$$\|x(t)\| \ \leq \gamma_1^{-1} \circ \gamma_2 \circ \rho(\sup \|u(\tau)\|_{\tau \geq t_0}) := \gamma(\sup \|u(\tau)\|_{\tau \geq t_0}), \tag{2.45}$$

for all $t \in [t_0, \infty] \backslash B$. Then by (2.44) and (2.45),

$$\|x(t)\| \leq \beta(\|x(t_0)\|, t-t_0) + \gamma(\sup \|u(\tau)\|_{\tau \geq t_0}), \quad \forall t \geq t_0 \geq 0. \tag{2.46}$$

By causality, we have

$$\|x(t)\| \leq \beta(\|x(t_0)\|, t-t_0) + \gamma(\sup \|u(\tau)\|_{t_0 \leq \tau \leq t}), \quad \forall t \geq t_0 \geq 0. \tag{2.47}$$

A function V satisfying conditions (2.40) is called an ISS Lyapunov function. □

2.3 Control Lyapunov Functions

This section presents an extension of the Lyapunov function concept, which is a useful tool to design an adaptive controller for nonlinear systems. Assuming that the problem is to design a feedback control law $\alpha(x)$ for the time-invariant system:

$$\dot{x} = f(x,u), \ x \in \mathbb{R}^n, \ u \in \mathbb{R}, \ f(0,0) = 0, \tag{2.48}$$

such that the equilibrium $x = 0$ of the closed loop system:

$$\dot{x} = f(x,\alpha(x)) \tag{2.49}$$

is globally asymptotically stable (GAS). We can take a function $V(x)$ as a Lyapunov candidate function, and require that its derivative along the solutions of (2.49) satisfy $\dot{V}(x) \le -W(x)$, where $W(x)$ is a positive definite function. We therefore need to find $\alpha(x)$ to guarantee that for all $x \in \mathbb{R}^n$ such that

$$\frac{\partial V(x)}{\partial x} f(x,\alpha(x)) \le -W(x). \tag{2.50}$$

This is a difficult problem. A stabilizing control law for (2.48) may exist but we may fail to satisfy (2.50) because of a poor choice of $V(x)$ and $W(x)$. A system for which a good choice of $V(x)$ and $W(x)$ exists is said to possess a control Lyapunov function (CLF). For systems affine in the control:

$$\dot{x} = f(x) + g(x)u, \ f(0) = 0, \tag{2.51}$$

the CLF inequality (2.50) becomes

$$\frac{\partial V}{\partial x} f(x) + \frac{\partial V}{\partial x} g(x)\alpha(x) \le -W(x). \tag{2.52}$$

If $V(x)$ is a CLF for (2.51), then a particular stabilizing control law $\alpha(x)$, smooth for all $x \neq 0$, is given by

$$u = \alpha(x) = \begin{cases} -\dfrac{\dfrac{\partial V}{\partial x} f(x) + \sqrt{\left(\dfrac{\partial V}{\partial x} f(x)\right)^2 + \left(\dfrac{\partial V}{\partial x} g(x)\right)^4}}{\dfrac{\partial V}{\partial x} g(x)}, & \dfrac{\partial V}{\partial x} g(x) \neq 0, \\ 0, & \dfrac{\partial V}{\partial x} g(x) = 0. \end{cases} \tag{2.53}$$

It should be noted that (2.52) can be satisfied only if

$$\frac{\partial V}{\partial x} g(x) = 0 \Rightarrow \frac{\partial V}{\partial x} f(x) < 0, \qquad \forall x \neq 0 \tag{2.54}$$

and that in this case (2.53) gives

$$W(x) = \sqrt{\left(\frac{\partial V}{\partial x} f\right)^2 + \left(\frac{\partial V}{\partial x} g\right)^4} > 0, \quad \forall x \neq 0. \tag{2.55}$$

The main drawback of the CLF concept as a design tool is that for most nonlinear systems a CLF is not known. The task of finding an appropriate CLF may be as complex as that of designing a stabilizing feedback law. The backstepping technique [3] presented in the next section gives a constructive way to construct an appropriate CLF for various systems.

2.4 Backstepping

Assumption 2.1. *Consider the system*

$$\dot{x} = f(x) + g(x)u, \ \ f(0) = 0, \tag{2.56}$$

where $x \in \mathbb{R}^n$ is the state and $u \in \mathbb{R}$ is the control input. There exist a continuously differentiable feedback control law

$$u = \alpha(x), \ \alpha(0) = 0 \tag{2.57}$$

and a smooth, positive definite, radially unbounded function $V : \mathbb{R}^n \to \mathbb{R}$ such that

$$\frac{\partial V}{\partial x}[f(x) + g(x)\alpha(x)] \leq -W(x) \leq 0, \forall x \in \mathbb{R}^n, \tag{2.58}$$

where $W : \mathbb{R}^n \to \mathbb{R}$ is positive semidefinite.

It should be noted that under this assumption, the control law (2.57) applied to the system (2.56) guarantees global boundedness of $x(t)$, and the regulation of $W(x)$: $\lim_{t\to\infty} W(x) = 0$. If $W(x)$ is positive definite, the control law (2.57) renders the global asymptotic stable equilibrium of (2.56).

Theorem 2.6. *Let the system (2.56) be augmented by an integrator:*

$$\begin{aligned} \dot{x} &= f(x) + g(x)\xi, \\ \dot{\xi} &= u, \end{aligned} \tag{2.59}$$

and suppose that the first equation of (2.59) satisfies Assumption 2.1 with ξ as its control.

1. *If* $W(x)$ *is positive definite then*

$$V_a = V(x) + \frac{1}{2}[\xi - \alpha(x)]^2 \tag{2.60}$$

is a CLF for the system (2.59), that is, there exists a feedback control law $u = \alpha_a(x,\xi)$*, which renders* $x = 0$, $\xi = 0$ *the GAS equilibrium of (2.59). One such control law choice is*

$$u = -c(\xi - \alpha(x)) + \frac{\partial \alpha}{\partial x}[f(x) + g(x)\xi] - \frac{\partial V}{\partial x} g(x), \; c > 0. \tag{2.61}$$

2. *If* $W(x)$ *is only positive semidefinite, then there exists a feedback control law that renders* $\dot{V}_a \leq -W_a(x,\xi) \leq 0$*, such that* $W_a(x,\xi) > 0$ *whenever* $W(x) > 0$ *or* $\xi \neq \alpha(x)$*. This guarantees global boundedness and convergence of* $[x(t)\ \xi(t)]^T$ *to the largest invariant set* M_a *contained in the set*

$$E_a = \left\{ \begin{bmatrix} x \\ \xi \end{bmatrix} \in \mathbb{R}^{n+1} \,|\, W(x) = 0, \; \xi = \alpha(x) \right\}.$$

Proof. We only prove the first part of the theorem. Proof of the second part is trivial. Introducing an error variable

$$z = \xi - \alpha(x) \tag{2.62}$$

and differentiating with respect to time, (2.59) can be written as

$$\begin{aligned} \dot{x} &= f(x) + g(x)[\alpha(x) + z], \\ \dot{z} &= u - \frac{\partial \alpha}{\partial x}[f(x) + g(x)(\alpha(x) + z)]. \end{aligned} \tag{2.63}$$

Using (2.58), the first time derivative of (2.60) along the solutions of (2.63) satisfies

$$\dot{V}_a \leq -W(x) + z\left[u - \frac{\partial \alpha}{\partial x}(f(x) + g(x)(\alpha(x) + z)) + \frac{\partial V}{\partial x} g(x)\right]. \tag{2.64}$$

Any control law, such as (2.61), which renders $\dot{V}_a \leq -W_a(x,\xi) \leq -W(x)$, with W_a positive definite in z, guarantees global boundedness of x and z, and regulation of $W(x)$ and $z(t)$. □

Corollary 2.1. *Let the system (2.56) satisfy Assumption 2.1 with* $\alpha(x) = \alpha_0(x)$ *being augmented by a chain of k integrators so that u is replaced by* ξ_1*, the state of the last integrator in the chain is*

$$\begin{aligned} \dot{x} &= f(x) + g(x)\xi_1, \\ \dot{\xi}_1 &= \xi_2, \\ &\vdots \\ \dot{\xi}_{k-1} &= \xi_k, \\ \dot{\xi}_k &= u. \end{aligned} \tag{2.65}$$

For this system, repeated application of Theorem 2.6 with $\xi_1,\ldots,\xi_k$ as virtual controls, results in the Lyapunov function

$$V_a(x,\xi_1,\ldots,\xi_k) = V(x) + \frac{1}{2}\sum_{i=1}^{k}[\xi_i - \alpha_{i-1}(x,\xi_1,\ldots,\xi_{i-1})]^2. \tag{2.66}$$

Any choice of feedback control which renders $\dot{V}_a \le -W_a(x,\xi_1,\ldots,\xi_k) \le 0$, with $W_a(x,\xi_1,\ldots,\xi_k) = 0$ only if $W(x) = 0$ and $\xi_i \neq \alpha_{i-1}(x,\xi_1,\ldots,\xi_{i-1})$, $i = 1,\ldots,k$, guarantees that $\left[x^T(t), \xi_1(t),\ldots,\xi_k(t)\right]^T$ is globally bounded and converges to the largest invariant set M_a contained in the set

$$E_a = \left\{\left[x^T,\xi_1,\ldots,\xi_k\right]^T \in \mathbb{R}^{n+k} \middle| W(x) = 0,\, \xi_i = \alpha_{i-1}(x,\xi_1,\ldots,\xi_{i-1})\right\}$$

for all $i = 1,\ldots,k$. Furthermore, if $W(x)$ is positive definite, that is, if $x = 0$ can be rendered GAS through ξ_1, then (2.66) is a CLF for (2.65) and the equilibrium $x = 0, \xi_1 = 0,\ldots,\xi_k = 0$ can be rendered GAS through u.

Proof. See [3]. □

2.5 Stabilization Under Uncertainties

The power of adaptive control is exhibited in the presence of uncertain nonlinearities and unknown parameters. Such uncertainties in linear systems make the control design procedure difficult and become more serious in control of nonlinear systems. For nonlinear systems, the states can easily escape to infinity in a finite time. The following theorem introduces the use of a term in the control law called nonlinear damping to stabilize the system (2.56) in the presence of disturbance.

Theorem 2.7. *Consider the system (2.56) satisfying Assumption 2.1 which is perturbed as*

$$\dot{x} = f(x) + g(x)\left[u + \varphi(x)^T \Delta(x,u,t)\right], \tag{2.67}$$

where $\varphi(x)$ is a $(p\times 1)$ vector of known smooth nonlinear functions, and $\Delta(x,u,t)$ is a $(p\times 1)$ vector of uncertain nonlinearities, which are uniformly bounded for all values of x, u, and t. If Assumption 2.1 satisfies with $W(x)$ being positive definite and radially unbounded, then the control

$$u = \alpha(x) - k\frac{\partial V}{\partial x}(x)g(x)\left\|\varphi(x)\right\|^2, \;\; k > 0, \tag{2.68}$$

when applied to the system (2.67) renders the closed loop system ISS with respect to the disturbance input $\Delta(x,u,t)$ and hence guarantees global uniform boundedness of $x(t)$ and convergence to the residual set

$$R = \left\{ x : \|x\| \le \gamma_1^{-1} \circ \gamma_2 \circ \gamma_3^{-1} \left(\frac{\|\Delta\|_\infty^2}{4k} \right) \right\}, \tag{2.69}$$

where γ_1, γ_2, γ_3 are class K_∞ functions such that

$$\begin{aligned} \gamma_1(\|x\|) &\le V(x) \le \gamma_2(\|x\|), \\ \gamma_3(\|x\|) &\le W(x). \end{aligned} \tag{2.70}$$

Proof. By using (2.68) and (2.70), the first time derivative of $V(x)$ is

$$\begin{aligned} \dot{V} &= \frac{\partial V}{\partial x}[f + gu] + \frac{\partial V}{\partial x} g \varphi^T \Delta \\ &= \frac{\partial V}{\partial x}[f + g\alpha] - k \left(\frac{\partial V}{\partial x} g \right)^2 \|\varphi\|^2 + \frac{\partial V}{\partial x} g \varphi^T \Delta \\ &\le -W(x) - k \left(\frac{\partial V}{\partial x} g \right)^2 \|\varphi\|^2 + \frac{\partial V}{\partial x} g \varphi^T \Delta \\ &\le -W(x) - k \left(\frac{\partial V}{\partial x} g \right)^2 \|\varphi\|^2 + \left\| \frac{\partial V}{\partial x} g \right\| \|\varphi\| \, \|\Delta\|_\infty \le -W(x) + \frac{\|\Delta\|_\infty^2}{4k}. \end{aligned} \tag{2.71}$$

From (2 71), it follows that $\dot{V}$ is negative whenever $W(x) \ge \frac{\|\Delta\|_\infty^2}{4k}$. Combining this with the second equation of (2.70), we conclude that

$$\|x(t)\| > \gamma_3^{-1} \left(\frac{\|\Delta\|_\infty^2}{4k} \right) \Rightarrow \dot{V} < 0. \tag{2.72}$$

This means that if $\|x(0)\| \le \gamma_3^{-1} \left(\frac{\|\Delta\|_\infty^2}{4k} \right)$, then

$$V(x(t)) \le \gamma_2 \circ \gamma_3^{-1} \left(\frac{\|\Delta\|_\infty^2}{4k} \right), \tag{2.73}$$

which implies that

$$\|x(t)\| \le \gamma_1^{-1} \circ \gamma_2 \circ \gamma_3^{-1} \left(\frac{\|\Delta\|_\infty^2}{4k} \right). \tag{2.74}$$

If, on the other hand, $\|x(0)\| \ge \gamma_3^{-1} \left(\frac{\|\Delta\|_\infty^2}{4k} \right)$, then $V(x) \le V(x(0))$, which implies

$$\|x(t)\| \le \gamma_1^{-1} \circ \gamma_2 (\|x(0)\|). \tag{2.75}$$

Combining (2.74) and (2.75) leads to global uniform boundedness of $x(t)$:

$$\|x\|_\infty \le \max \left\{ \gamma_1^{-1} \circ \gamma_2 \circ \gamma_3^{-1} \left(\frac{\|\Delta\|_\infty^2}{4k} \right), \gamma_1^{-1} \circ \gamma_2(\|x(0)\|) \right\}, \tag{2.76}$$

while (2.72) and the first equation of (2.70) prove the convergence of $x(t)$ to the residual set defined in (2.69). □

Combining the above theorem with Theorem 2.6 leads to the following corollary.

Corollary 2.2. *Consider the following system*

$$\dot{x} = f(x) + g(x)u + F(x)\Delta_1(x,u,t), \tag{2.77}$$

where $x \in \mathbb{R}^n$, $u \in \mathbb{R}$, $F(x)$ is an $(n \times q)$ matrix of known smooth nonlinear functions, and $\Delta_1(x,u,t)$ is a $(q \times 1)$ vector of uncertain nonlinearities, which is uniformly bounded for all values of x, u and t. Suppose that there exists a feedback control law $u = \alpha(x)$ that renders $x(t)$ globally uniformly bounded, and that this is established via positive definite and radially unbounded functions $V(x)$, $W(x)$, and a constant b such that

$$\frac{\partial V}{\partial x}[f(x) + g(x)\alpha(x) + F(x)\Delta_1(x,u,t)] \leq -W(x) + b. \tag{2.78}$$

Now consider the augmented system

$$\begin{aligned} \dot{x} &= f(x) + g(x)\xi + F(x)\Delta_1(x,u,t), \\ \dot{\xi} &= u + \varphi^T(x,\xi)\Delta_2(x,\xi,u,t), \end{aligned} \tag{2.79}$$

where $\varphi(x,\xi)$ is a $(p \times 1)$ vector of known smooth nonlinear functions, $\Delta_2(x,u,\xi,t)$ is a $(p \times 1)$ vector of uncertain nonlinearities, which are uniformly bounded for all values of x, u, ξ and t. For this system, the feedback control law

$$\begin{aligned} u = \; & -c(\xi - \alpha(x)) + \frac{\partial \alpha}{\partial x}[f(x) + g(x)\xi] - \\ & \frac{\partial V}{\partial x}g(x) - k(\xi - \alpha(x))\left\{ \|\varphi(x,\xi)\|^2 + \left\| \frac{\partial \alpha}{\partial x}F(x) \right\|^2 \right\} \end{aligned} \tag{2.80}$$

guarantees global uniform boundedness of $x(t)$ and $\xi(t)$ with any $c > 0$ and $k > 0$.

Proof. See [3]. □

2.6 Barbalat-like Lemmas

This section presents lemmas that are useful in investigating the convergence of time-varying systems.

Lemma 2.4. *(Barbalat's lemma) Consider the function $\phi : \mathbb{R}^+ \to \mathbb{R}$. If ϕ is uniformly continuous and* $\lim_{t\to\infty} \int_0^t \phi(\tau)\mathrm{d}\tau$ *exists and is finite, then*

$$\lim_{t\to\infty} \phi(t) = 0. \tag{2.81}$$

Proof. See [6]. □

Lemma 2.5. *Assume that a nonnegative scalar differentiable function $f(t)$ enjoys the following conditions*

$$\begin{aligned} &1. \left|\frac{\mathrm{d}}{\mathrm{d}t} f(t)\right| \le k_1 f(t), \\ &2. \int_0^\infty f(t)\mathrm{d}t \le k_2 \end{aligned} \tag{2.82}$$

for all $t \ge 0$, where k_1 and k_2 are positive constants, then $\lim_{t\to\infty} f(t) = 0$.

Proof. Integrating both sides of 1) in (2.82) gives

$$\begin{aligned} f(t) &\le f(0) + k_1 \int_0^t f(s)\mathrm{d}s \le f(0) + k_1 k_2, \\ f(t) &\ge f(0) - k_1 \int_0^t f(s)\mathrm{d}s \ge f(0) - k_1 k_2. \end{aligned} \tag{2.83}$$

These inequalities imply that $f(t)$ is a uniform bounded function. From (2.83) and the second condition in (2.82), we have that $f(t)$ is also bounded on the half axis $[0,\infty]$, i.e., $f(t) \le k_3$ with k_3 a positive constant. Hence $\left|\frac{\mathrm{d}}{\mathrm{d}t} f(t)\right| \le k_1 k_3$. Now assume that $\lim_{t\to\infty} f(t) \ne 0$. Then there exists a sequence of points t_i and a positive constant ϵ such that $f(t_i) \ge \epsilon$, $t_i \to \infty$, $i \to \infty$, $|t_i - t_{i-1}| > 2\epsilon/(k_1k_3)$ and moreover $f(s) \ge \epsilon/2$, $s \in L_i = [t_i - \epsilon/(2k_1k_3), t_i + \epsilon/(2k_1k_3)]$. Since the segments L_i and L_j do not intersect for any i and j with $i \ne j$, we have

$$\int_0^\infty f(t)\mathrm{d}t \ge \int_0^T f(t)\mathrm{d}t \ge \sum_{t_i \le T} \int_{L_i} f(t)\mathrm{d}t \ge \frac{\epsilon}{2}\frac{\epsilon}{k_1k_3} M(T) \tag{2.84}$$

where $M(T)$ is the number of points t_i not exceeding T. Since $\lim_{T\to\infty} M(T) = \infty$, the integral $\int_0^\infty f(t)\mathrm{d}t$ is divergent. This contradicts Condition 2 in (2.82). This contradiction proves the lemma. □

Remark 2.2. Lemma 2.5 is different from Barbalat's lemma 2.4. While Barbalat's lemma assumes that $f(t)$ is uniformly continuous, Lemma 2.5 assumes that $|\frac{\mathrm{d}}{\mathrm{d}t} f(t)|$ is bounded by $k_1 f(t)$.

Lemma 2.6. *Consider a scalar system*

$$\dot{x} = -cx + p(t), \tag{2.85}$$

where $c > 0$ and $p(t)$ is a bounded and uniformly continuous function. If, for any initial time $t_0 \ge 0$ and any initial condition $x(t_0)$, the solution $x(t)$ is bounded and converges to 0 as $t \to \infty$, then

$$\lim_{t\to\infty} p(t) = 0. \tag{2.86}$$

Proof. See [18]. □

Lemma 2.7. *Consider a first-order differential equation of the form*

$$\dot{x} = -(a(t) + f_1(\xi(t)))x + f_2(\xi(t)), \tag{2.87}$$

where f_1 and f_2 are continuous functions, and $\xi : [0,\infty) \to \mathbb{R}^m$ is a time-varying vector-valued signal that exponentially converges to zero and, for all $t \geq t_0 \geq 0$, satisfies

$$|f_i(\xi(t))| \leq \gamma_i(\|\xi(t_0)\|)e^{-\sigma_i(t-t_0)}, \tag{2.88}$$

where $\sigma_i > 0, i = 1,2$ and γ_i are class-K functions. If $a(t)$ enjoys the property that there is a constant σ_3 such that

$$\int_{t_1}^{t_2} a(\tau)\mathrm{d}\tau \geq \sigma_3(t_2 - t_1),\ \forall t_2 \geq t_1 \geq 0, \tag{2.89}$$

then there exist a class-K function γ and a constant $\sigma > 0$ such that

$$|x(t)| \leq \gamma(\|(x(t_0), \xi(t_0))\|)e^{-\sigma(t-t_0)}. \tag{2.90}$$

Proof. See [19]. □

2.7 Controllability and Observability

2.7.1 Controllability and Observability of Linear Time-invariant Systems

This section deals with the controllability and observability properties of systems described by linear time-invariant state-space representations. In particular, consider a linear and time-invariant system defined by the state–space representation:

$$\begin{aligned}\dot{x}(t) &= \boldsymbol{A}x(t) + \boldsymbol{B}u(t),\\ y(t) &= \boldsymbol{C}x(t) + \boldsymbol{D}u(t),\end{aligned} \tag{2.91}$$

where $\boldsymbol{A}$ has n distinct eigenvalues $\lambda_1, \lambda_2, \ldots, \lambda_n$, which define the poles and the corresponding modes $e^{\lambda_i t}$.

2.7.1.1 Controllability

Conditions for controllability of the system (2.91) are given in the following lemmas.

Lemma 2.8. *The system (2.91) is controllable if and only if any one of the following (equivalent) conditions holds:*

1. *The controllability matrix*

$$\boldsymbol{C_c} := [\boldsymbol{B}, \boldsymbol{AB}, \ldots, \boldsymbol{A}^{n-1}\boldsymbol{B}] \tag{2.92}$$

has full rank n.

2. *The n rows of $e^{\boldsymbol{A}t}\boldsymbol{B}$, where $e^{\boldsymbol{A}t}$ represents the unique state transition matrix of the system, are linearly independent over the real field $\mathbb{R}$ for all t.*
3. *The controllability Grammian*

$$\boldsymbol{G_c}(t_0, t_f) := \int_{t_0}^{t_f} e^{-\boldsymbol{A}\tau}\boldsymbol{B}\boldsymbol{B}^{\boldsymbol{T}} e^{-\boldsymbol{A}^T\tau} \mathrm{d}\tau \tag{2.93}$$

is nonsingular for all $t_f > t_0$.

4. *The $n \times (n+m)$ matrix $[\lambda \boldsymbol{I} - \boldsymbol{A}, \boldsymbol{B}]$ has rank n at all eigenvalues λ_i of $\boldsymbol{A}$ or equivalently, $\lambda \boldsymbol{I} - \boldsymbol{A}$ and $\boldsymbol{B}$ are left comprime polynomial matrices.*

Proof. See [20], Chapter 8. □

Since the solution to the system (2.91) is given by

$$x(t) = e^{\boldsymbol{A}(t-t_0)}x(t_0) + \int_{t_0}^{t} e^{\boldsymbol{A}(t-\tau)}\boldsymbol{B}u(\tau)\mathrm{d}\tau, \tag{2.94}$$

it follows that the controllability Grammian-based control input

$$u(t) = \boldsymbol{B}^{\boldsymbol{T}} e^{-\boldsymbol{A}^T t}\boldsymbol{G_c}^{-1}(t_0, t_f)\left[e^{-\boldsymbol{A}t_f}x(t_f) - e^{-\boldsymbol{A}t_0}x(t_0)\right] \tag{2.95}$$

transfers any initial state $x(t_0)$ to any arbitrarily chosen final state $x(t_f)$ at any arbitrary $t_f > 0$. This observation is consistent with the more traditional definition of controllability.

2.7.1.2 Observability

Conditions for observability of the system (2.91) are given in the following lemmas.

Lemma 2.9. *The system (2.91) is observable if and only if any one of the following (equivalent) conditions holds:*

1. *The observability matrix*

$$C_o := \begin{bmatrix} C \\ CA \\ \vdots \\ CA^{n-1} \end{bmatrix} \tag{2.96}$$

has full rank n.

2. *The n columns of* Ce^{At}, *where* e^{At} *represents the unique state transition matrix of the system, are linearly independent over the real field* $\mathbb{R}$ *for all* t.
3. *The observability Grammian*

$$G_o(t_0, t_f) := \int_{t_0}^{t_f} e^{A^T\tau} C^T C e^{A\tau} d\tau \tag{2.97}$$

is nonsingular for all $t_f > t_0$.

4. *The* $(n+p) \times n$ *matrix* $\begin{bmatrix} \lambda I - A \\ C \end{bmatrix}$ *has rank n at all eigenvalues* λ_i *of* A *or equivalently* $\lambda I - A$ *and* C *are right comprime polynomial matrices.*

Proof. See [20], Chapter 8. □

If the system (2.91) is observable, it then follows that the system initial state can be determined by

$$x(t_0) = e^{At_0} G_o{}^{-1}(t_0, t_f) e^{A^T t_0} \int_{t_0}^{t_f} e^{A^T(t-t_0)} C^T f(t) dt, \tag{2.98}$$

where

$$f(t) := y(t) - C \int_{t_0}^{t} e^{A(t-\tau)} Bu(\tau) d\tau - Du(t). \tag{2.99}$$

This observation is consistent with the more traditional definition of observability.

2.7.2 Controllability and Observability of Linear Time-varying Systems

This section deals with controllability and observability properties of the following n-dimensional linear time-varying system

$$\begin{aligned} \dot{x}(t) &= A(t)x(t) + B(t)u(t), \\ y(t) &= C(t)x(t) + D(t)u(t). \end{aligned} \tag{2.100}$$

We assume that the matrices $A(t)$, $B(t)$, $C(t)$, and $D(t)$ are at least continuous functions of t.

2.7.2.1 Controllability

The system (2.100) is said to be controllable on the interval $[t_0, t_f]$, where $t_f > t_0 \geq 0$, if for any states x_0 and x_f, a continuous input $u(t)$ exists that drives the system to the state $x(t_f) = x_f$ at time $t = t_f$ starting from the state $x(t_0) = x_0$ at time $t = t_0$. Conditions for controllability of the system (2.100) are given in the following lemma:

Lemma 2.10. *Assuming that the matrix $\boldsymbol{A}(t)$ is $n-2$ times differentiable and $\boldsymbol{B}(t)$ is $n-1$ times differentiable, a sufficient condition for the system (2.100) to be controllable on the interval $[t_0, t_f]$ is that the following matrix has rank n for at least one $t \in [t_0, t_f]$:*

$$\boldsymbol{K}(t) = [\boldsymbol{K}_0(t)\ \boldsymbol{K}_1(t)\ \ldots\ \boldsymbol{K}_{n-1}(t)], \tag{2.101}$$

where $\boldsymbol{K}_i$, $i = 1, \ldots, n-1$ are defined by

$$\begin{aligned} \boldsymbol{K}_0(t) &= \boldsymbol{B}(t), \\ \boldsymbol{K}_i(t) &= -\boldsymbol{A}(t)\boldsymbol{K}_{i-1}(t) + \dot{\boldsymbol{K}}_{i-1}(t). \end{aligned} \tag{2.102}$$

Moreover, if the matrix $\boldsymbol{K}(t)$ has rank n for all $t \in [t_0, t_f]$, then the system (2.100) is uniformly controllable on the interval $[t_0, t_f]$.

Proof. See [20], Chapter 25. □

2.7.2.2 Observability

Now suppose that the control input is zero, i.e., the system (2.100) is given by

$$\begin{aligned} \dot{x}(t) &= \boldsymbol{A}(t)x(t), \\ y(t) &= \boldsymbol{C}(t)x(t). \end{aligned} \tag{2.103}$$

The system (2.103) is said to be observable on the interval $[t_0, t_f]$ if any initial state $x(t_0) = x_0$ can be determined from the output $y(t)$ for $t \in [t_0, t_f]$. Conditions for observability of the system (2.103) are given in the following lemma.

Lemma 2.11. *Assuming that the matrix $\boldsymbol{A}(t)$ is $n-2$ times differentiable and $\boldsymbol{B}(t)$ is $n-1$ times differentiable, a sufficient condition for the system (2.100) to be observable on the interval $[t_0, t_f]$ is that the following matrix has rank n for at least one $t \in [t_0, t_f]$*

$$\boldsymbol{L}(t) = \begin{bmatrix} \boldsymbol{L}_0(t) \\ \boldsymbol{L}_1(t) \\ \vdots \\ \boldsymbol{L}_{n-1}(t) \end{bmatrix}, \tag{2.104}$$

where $\boldsymbol{L}_i$, $i = 1, \ldots, n-1$ are defined by

$$\begin{aligned} \boldsymbol{L}_0(t) &= \boldsymbol{C}(t), \\ \boldsymbol{L}_i(t) &= \boldsymbol{L}_{i-1}(t)\boldsymbol{A}(t) + \dot{\boldsymbol{L}}_{i-1}(t). \end{aligned} \tag{2.105}$$

Moreover, if the matrix $\boldsymbol{L}(t)$ *has rank n for all* $t \in [t_0, t_f]$, *then the system (2.100) is uniformly observable on the interval* $[t_0, t_f]$.

Proof. See [20], Chapter 25. □

2.7.3 Controllability and Observability of Nonlinear Systems

2.7.3.1 Controllability

This subsection deals with controllability of the following nonlinear system

$$\dot{x} = f(x) + \sum_{i=1}^{m} g_i(x)u_i, \qquad x \in \Omega_x \subset \mathbb{R}^n, \tag{2.106}$$

where $u = [u_1, u_2, ..., u_m]^T \in \Omega_u \subset \mathbb{R}^m$ is the input vector. The system (2.106) is controllable if there exists an admissible input vector $u(t)$ such that the state $x(t)$ can travel from an initial point $x(t_0) = x_0 \in \Omega_x$ to $x(t_f) \in \Omega_x$ within a finite time interval $T = t_f - t_0$. The controllability reveals whether the control system has a set of "healthy" input channels through which the input can excite the states effectively to reach the destination x_f. Based on this interpretation, the controllability of (2.106) should clearly depend on the function forms of all $f(x)$ and $g_i(x)$. The controllability of the nonlinear system (2.106) is based on a useful mathematical concept called Lie algebra, which is defined as follows.

Definition 2.9. A Lie algebra over the real field $\mathbb{R}$ or the complex field $\mathbb{C}$ is a vector space $\mathbb{G}$ for which a bilinear map $(X,Y) \to [X,Y]$ is defined from $\mathbb{G} \times \mathbb{G} \to \mathbb{G}$ such that

1. $[X,Y] = -[Y,X],\ X,\ Y \in \mathbb{G}$,
2. $[X,[Y,Z]] + [Y,[Z,X]] + [Z,[X,Y]] = 0$ for $X,Y,Z \in \mathbb{G}$.

According to this definition, a Lie algebra is a vector space where an operator $[.,.]$ is installed. Such a general operator, conventionally called a Lie bracket, can be defined arbitrarily as long as it satisfies the preceding two specified conditions simultaneously. The first condition is often called a skew symmetric relation and obviously implies that $[X,X] = 0$. The second condition is called the Jacobi identity, which reveals a closed loop cyclic relation among any three elements in a Lie algebra.

Let us now define a special Lie algebra E that collects all n-dimensional differentiable vector fields in $\mathbb{R}^n$ along with a commutative derivative relation: For any two vector fields f and $g \in \mathbb{R}^n$, which are functions of $x \in \mathbb{R}^n$,

$$[f,g] = \frac{\partial g}{\partial x} f - \frac{\partial f}{\partial x} g. \tag{2.107}$$

It can be immediately shown that this Lie bracket satisfies the two conditions of a Lie algebra.

To extend the above Lie bracket between two vector fields to higher order derivatives, a more compact notation may be defined based on an *adjoint operator*, that is, $[f,g] = \mathrm{ad}_f g$. This new notation treats the Lie bracket $[f,g]$ as vector field g operated on by an adjoint operator $\mathrm{ad}_f = [f,.]$. Therefore, for an n-order Lie bracket $(n > 1)$, one can simply write

$$[f,\underbrace{\ldots}_{n}[f,g]\underbrace{\ldots}_{n}] = \mathrm{ad}_f^n g.$$

For a general control system given by (2.106), we now specifically define a control Lie algebra E_c, which is spanned by all up to order $(n-1)$ Lie brackets among f and g_1 through g_m as

$$E_c = \mathrm{span}\{g_1, \ldots, g_m, \mathrm{ad}_f g_1, \ldots, \mathrm{ad}_f g_m, \ldots, \mathrm{ad}_f^{n-1} g_1, \ldots, \mathrm{ad}_f^{n-1} g_m\}. \tag{2.108}$$

With the control Lie algebra concept, we can show that the following theorem is true and is also a general effective testing criterion for system controllability.

Theorem 2.8. *The control system (2.106) is controllable if and only if* $\dim(E_c) = \dim(\Omega_x) = n.$

Proof. See [2]. □

Note that because each element in E_c is a function of x, the dimension of E_c may be different from one point to another. Thus, if the preceding condition of dimension is valid only in a neighborhood of a point in $\Omega_x \subset \mathbb{R}^n$, we say that the system (2.106) is locally controllable. On the other hand, if the condition of dimension can cover all of region Ω_x, then it is globally controllable. Moreover, it is not hard to show that Theorem 2.8 covers the controllability condition for a linear time-invariant system (2.91).

2.7.3.2 Observability

We address the observability for the following nonlinear system

$$\begin{aligned} \dot{x} &= f(x), \\ y &= h(x), \end{aligned} \tag{2.109}$$

where $y \in \Omega_y \subset \mathbb{R}^p$ is the output vector. This system is said to be observable if for each pair of distinct states x_1 and x_2, the corresponding outputs y_1 and y_2 are also distinguishable. Clearly, the observability can be interpreted as a testing criterion to check whether the entire system has sufficient output channels to measure (or observe) each internal state change. Intuitively, the observability should depend on the function forms of both $f(x)$ and $h(x)$.

We now introduce a Lie derivative, which is virtually a *directional derivative* for a scalar field $\lambda(x)$, with $x \in \mathbb{R}^n$ along the direction of an n-dimensional vector field $f(x)$. The mathematical expression is given as

$$L_f \lambda(x) = \frac{\partial \lambda(x)}{\partial x} f(x). \tag{2.110}$$

Since $\frac{\partial \lambda(x)}{\partial x}$ is a $1 \times n$ gradient vector of the scalar $\lambda(x)$ and the norm of a gradient vector represents the maximum rate of function value changes, the product of the gradient and the vector field $f(x)$ in (2.109) becomes the directional derivative of $\lambda(x)$ along $f(x)$. Therefore, the Lie derivative of a scalar field defined by (2.110) is also a scalar field.

If each component of a vector field $h(x) \in \mathbb{R}^p$ is considered to take a Lie derivative along $f(x) \in \mathbb{R}^n$, then all components can be acted on concurrently and the result is a vector field that has the same dimension as $h(x)$; its ith element is the Lie derivative of the ith component of $h(x)$. Namely, if $h(x) = [h_1(x), ..., h_p(x)]^T$ and each component $h_i(x)$, $i = 1, ..., p$ is a scalar field, then the Lie derivative of the vector field $h(x)$ is defined as

$$L_f h(x) = \begin{bmatrix} L_f h_1(x) \\ \vdots \\ L_f h_p(x) \end{bmatrix}. \tag{2.111}$$

With the Lie derivative concept, we now define an observation space Ω_o over $\mathbb{R}^n$ as

$$\Omega_o = \text{span}\{h(x), L_f h(x), ..., L_f^{n-1} h(x)\}. \tag{2.112}$$

In other words, this space is spanned by all up to order $(n-1)$ Lie derivatives of the output function $h(x)$. Then, we further define an observability distribution, denoted by $d\Omega_o$, which collects the "gradient" vector of every component in Ω_o. Namely,

$$d\Omega_o = \text{span}\left\{ \frac{\partial \phi}{\partial x} \middle| \phi \in \Omega_o \right\}. \tag{2.113}$$

With these definitions, we can present the following theorem for testing the observability.

Theorem 2.9. *The system (2.109) is observable if and only if* $\dim(d\Omega_o) = n$.

Proof. See [2]. □

Similarly to the controllability case, this testing criterion also has locally observable and globally observable cases, depending on whether the condition of dimension in the theorem is valid only in a neighborhood of a point or over the entire state region.

2.7.4 Brockett's Theorem on Feedback Stabilization

The following theorem, which is due to Brockett [21], gives a necessary condition for the existence of a stabilizing control law for the system

$$\dot{x} = f(x,u) \tag{2.114}$$

at an equilibrium point x_0 with x being the state and u being the control input.

Theorem 2.10. *Let the system (2.114) be given with $f(x_0,0) = 0$ and $f(\cdot,\cdot)$ continuously differentiable in a neighborhood of $(x_0,0)$. A necessary condition for the existence of a continuously differentiable control law that makes $(x_0,0)$ asymptotically stable is that*

1. *the linearized system should have no uncontrollable modes associated with eigenvalues whose real part is positive,*
2. *there exists a neighborhood Ω of $(x_0,0)$ such that for each $\xi \in \Omega$ there exists a control $u_\xi(\cdot)$ defined on $[0,\infty)$ such that this control steers the solution of $\dot{x} = f(x,u_\xi)$ from $x = \xi$ at $t = 0$ to $x = x_0$ at $t = \infty$,*
3. *the mapping*
$$\gamma : A \times \mathbb{R}^m \to \mathbb{R}^n$$
defined by $\gamma : (x,u) \to f(x,u)$ should be onto an open set containing 0.

Proof. See [21]. □

Remark 2.3. If the system (2.114) is of the form

$$\dot{x} = f(x) + \sum_{i=1}^{m} u_i g_i(x); \quad x(t) \in \Omega \subset \mathbb{R}^n, \tag{2.115}$$

then condition 3 of Theorem 2.10 implies that the stabilization problem cannot have a solution if there is a smooth distribution D containing $f(\cdot)$ and $g_1(\cdot),\ldots,g_m(\cdot)$ with dim $D < n$. One further special case: If the system (2.114) is of the form

$$\dot{x} = \sum_{i=1}^{m} u_i g_i(x); \quad x(t) \in \Omega \subset \mathbb{R}^n \tag{2.116}$$

with the vectors $g_i(x)$ being linearly independent at x_0, then there exists a solution to the stabilization problem if and only if $m = n$. In this case, we must have as many control parameters as we have dimensions of x.

2.8 Conclusions

This chapter briefly provides the fundamental tools that will be used for control design and stability analysis of underactuated vehicles in the coming chapters. The interested reader is referred to [3–6, 22–24] for a detailed and comprehensive coverage of the topics discussed in this chapter.

Part II
Modeling and Control Properties of Ocean Vessels

Chapter 3
Modeling of Ocean Vessels

In this chapter, we classify the basic motion tasks for ocean vessels and their mathematical models, which will be used for the design of various control systems in the subsequent chapters.

3.1 Introduction

In automatic control, feedback improves system performance by allowing the successful completion of a task even in the presence of external disturbances and initial errors, and inaccuracy of the system parameters. To this end, real-time sensor measurements are used to reconstruct the vehicle state. Throughout this study, the latter is assumed to be available at every instant, as provided by local/global position and orientation measurement sensors. In some cases, we also assume that the vehicle velocities are measurable or constructible from position measurements.

We will concentrate on the case of a vessel workspace free of obstacles. In fact, we implicitly consider the vessel controller to be embedded in a hierarchical architecture in which a higher-level planner solves the obstacle avoidance problem and provides a series of motion goals to the lower control layer. In this perspective, the controller deals with the basic issue of converting ideal plans into actual motion execution. The nonholonomic nature of the ocean vessels is related to the fact that the vessel does not usually have independent actuators in the sway and heave axes. This implies the presence of a nonintegrable set of second-order differential constraints on the configuration variables. While these nonholonomic constraints reduce the instantaneous motions that the vessel can perform, they still allow almost global controllability in the configuration space. This feature leads to some challenging problems in the synthesis of feedback controllers, which parallel the new research issues arising in nonholonomic motion planning. Indeed, the ocean vessel application has triggered the search for innovative types of feedback controllers that can be used also for more general nonlinear systems that describe the motion of more complicated vessel systems such as ocean vessels and air vehicles working in a group.

3.2 Basic Motion Tasks

In order to derive the most suitable feedback controllers for each case, it is convenient to classify the possible motion tasks as follows:

- Point-to-point motion: The vessel must reach a desired goal configuration starting from a given initial configuration, see Figure 3.1a.
- Path-following: The vessel must reach and follow a geometric reference path in the Cartesian space starting from a given initial configuration (on or off the path), see Figure 3.1b.
- Trajectory-tracking and path-tracking: The vessel must reach and follow a reference trajectory/path in the Cartesian space (i.e., a geometric path with an associated timing law) starting from a given initial configuration (on or off the trajectory/path), see Figure 3.1c. Trajectory-tracking is referred to as the case where the reference trajectory is generated by a suitable virtual vessel whereas the reference path is not required to be generated by a virtual vessel for the path-tracking.

The above tasks for an ocean vessel are sketched in Figure 3.1. Execution of these tasks can be achieved using either feedforward commands, or feedback control, or a combination of the two. Indeed, feedback solutions exhibit an intrinsic degree of robustness.

Using a more control-oriented terminology, the point-to-point motion task is a stabilization problem for a (equilibrium) point in the vessel state space. When using a feedback strategy, the point-to-point motion task leads to a state regulation control problem for a point in the vessel state space. Posture stabilization is another frequently used term. Without loss of generality, the goal can be taken as the origin of the n-dimensional vessel configuration space. Contrary to the usual situation, trajectory-tracking, path-tracking, and path-following are easier than regulation for a nonholonomic vessel. An intuitive explanation of this can be given in terms of a comparison between the number of controlled variables (outputs) and the number of control inputs. For the ship or underwater vehicle moving in a horizontal plane, two input commands are available while three variables (position and orientation) are needed to determine its configuration. Thus, regulation of the surface ship or the underwater vehicle in a horizontal position to a desired configuration implies zeroing three independent configuration errors.

In the path-following task, the controller is given a geometric description of the assigned Cartesian path. This information is usually available in a parameterized form expressing the desired motion in terms of a path parameter, which may be in particular the arc length along the path. For this task, time dependence is not relevant because one is concerned only with the geometric displacement between the vessel

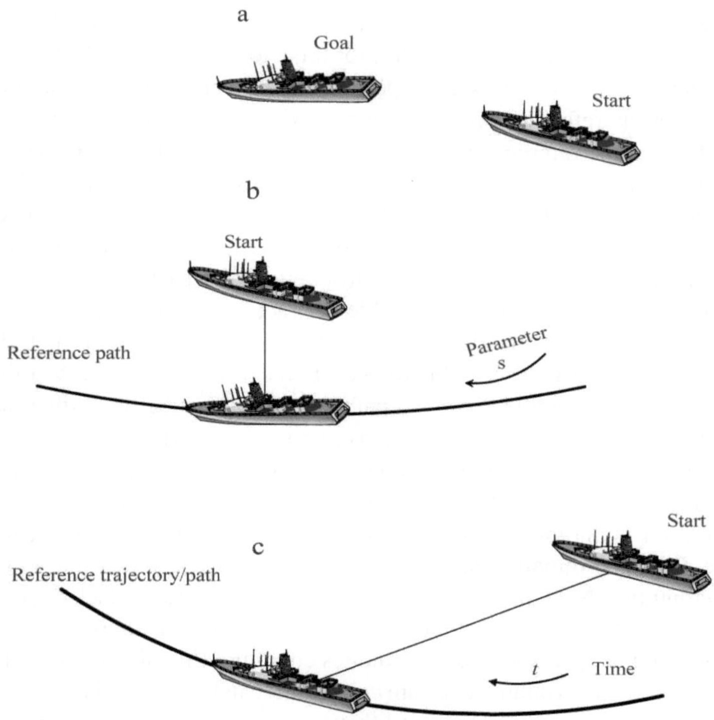

Figure 3.1 Basic motion tasks for an ocean vessel

and the path. In this context, the time evolution of the path parameter is usually free and, accordingly, the command inputs can be arbitrarily scaled with respect to time without changing the resulting vessel path. It is then customary to set the vessel forward velocity (one of the inputs) to an arbitrary constant or time-varying value, leaving the other input variables for control. The path-following problem is thus rephrased as the stabilization to zero of a suitable scalar path error function using only the rest of the control inputs.

In the trajectory-tracking and path-tracking tasks, the vessel must follow the desired Cartesian path with a specified timing law. Although the reference trajectory/path can be split into a parameterized geometric path and a timing law for the parameter, such separation is not strictly necessary. Often, it is simpler to specify the workspace trajectory as the desired time evolution for the position of some representative point of the vessel. The trajectory-tracking and path-tracking problems consist then in the stabilization to zero of the Cartesian errors using all the available control inputs.

The point stabilization problem can be formulated in a local or in a global sense, the latter meaning that we allow for initial configurations that are arbitrarily far from the destination. The same is true also for path-following, trajectory-tracking, and path-tracking, although locality has two different meanings in these tasks. For path-following, a local solution means that the controller works properly, provided that we start sufficiently close to the path; for trajectory-tracking and path-tracking, closeness should be evaluated with respect to the current position of the reference vessel and a current reference point, which moves on the reference path with a specified time law, on the reference path, respectively. The amount of information that should be provided by a high-level motion planner varies for each control task. In point-to-point motion, information is reduced to a minimum (i.e., the goal configuration only) when a globally stabilizing feedback control solution is available. However, if the initial error is large, such a control may produce erratic behavior and/or large control effort, which are unacceptable in practice. On the other hand, a local feedback solution requires the definition of intermediate subgoals at the task planning level in order to get closer to the final desired configuration. For the other motion tasks, the planner should provide a path that is kinematically feasible (namely, that complies with the nonholonomic constraints of the specific vessel), so as to allow its perfect execution in nominal conditions. While for a fully or overactuated vessel in which any path is feasible, some degree of geometric smoothness is in general required for nonholonomic vessels. Nevertheless, the intrinsic feedback structure of the driving commands enables it to recover transient errors due to isolated path discontinuities. Note also that the infeasibility arising from a lack of continuity in some higher-order derivative of the path may be overcome by appropriate motion timing. For example, paths with discontinuous curvature (like the Reeds and Shepp optimal paths under maximum curvature constraint) can be executed by choosing an appropriate point on the vessel provided that the vessel is allowed to stop, whereas paths with discontinuous tangent are not feasible. In this analysis, the selection of the vessel representative point for path/trajectory planning is critical. The timing profile is the additional item needed in trajectory-tracking and path-tracking control tasks. This information is seldom provided by current motion planners, also because the actual dynamics of the specific vessel are typically neglected at this level. The above example suggests that it may be reasonable at the planning stage to enforce requirements such as "move slower where the path curvature is higher".

3.3 Modeling of Ocean Vessels

Modeling of the ocean vessels is usually based on mechanics, principles of statics and dynamics. Statics is concerned with the equilibrium of bodies at rest or moving with a constant velocity. Dynamics deals with bodies having accelerated motion resulting from disturbances or/and control forces. Since we are interested in a mathematical model of the ocean vessels for the purpose of designing the control systems, this section focuses on dynamics of the vessels rather than statics. The following briefly presents the ocean vessel equations of motion based on the results in [11]. The resulting nonlinear model presented in this section is mainly intended for designing control systems in the next chapters. For a detailed and comprehensive derivation of the model, the reader is referred to [11,12,25]. The physical and control properties of the model are also presented for control design and stability analysis. In this section, we use the notation, see Table 3.1 and Figure 3.2, that complies with the Society of Naval Architects and Marine Engineers (SNAME) [26].

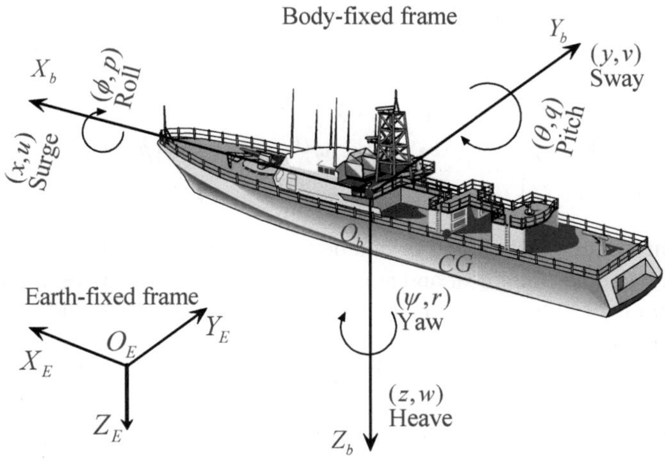

Figure 3.2 Motion variables for an ocean vessel

For an ocean vessel moving in six degrees of freedom, six independent coordinates are required to determine its position and orientation. The first three coordinates (x, y, z) and their first time derivatives correspond to the position and translational motion along the x-, y- and z-axes, while the last three coordinates (ϕ, θ, ψ) and their first time derivatives describe orientation and rotational motion.

Table 3.1 SNAME Notation for ocean vessels

Degree of freedom		Force and moment	Linear and angular velocity	Position and Euler angles
1	Surge	X	u	x
2	Sway	Y	v	y
3	Heave	Z	w	z
4	Roll	K	p	ϕ
5	Pitch	M	q	θ
6	Yaw	N	r	ψ

According to SNAME, the six different motion components are defined as *surge, sway, heave, roll, pitch*, and *yaw*. To determine the equations of motion, two reference frames are considered: the inertial or fixed to earth frame $O_E X_E Y_E Z_E$ that may be taken to coincide with the vessel fixed coordinates in some initial condition and the body-fixed frame $O_b X_b Y_b Z_b$ see Figure 3.2. Since the motion of the Earth hardly affects ocean vessels (different from air vehicles), the earth-fixed frame $O_E X_E Y_E Z_E$ can be considered to be inertial. For ocean vessels in general, the most commonly adopted position for the body-fixed frame is such that it gives hull symmetry about the $O_bX_bZ_b$-plane and approximate symmetry about the $O_bY_bZ_b$-plane. In this sense, the body axes O_bX_b, O_bY_b, and O_bZ_b coincide with the principal axes of inertia and are usually defined as follows: O_bX_b is the longitudinal axis (directed from aft to fore); O_bY_b is the transverse axis (directed to starboard); and O_bZ_b is normal axis (directed from top to bottom. Based on the notion in Table 3.1, the general motion of an ocean vessel can be described by the following vectors:

$$\begin{aligned} \boldsymbol{\eta} &= [\boldsymbol{\eta}_1\ \boldsymbol{\eta}_2]^T, & \boldsymbol{\eta}_1 &= [x\ y\ z]^T, & \boldsymbol{\eta}_2 &= [\phi\ \theta\ \psi]^T,\\ \boldsymbol{v} &= [\boldsymbol{v}_1\ \boldsymbol{v}_2]^T, & \boldsymbol{v}_1 &= [u\ v\ w]^T, & \boldsymbol{v}_2 &= [p\ q\ r]^T,\\ \boldsymbol{\tau} &= [\boldsymbol{\tau}_1\ \boldsymbol{\tau}_2]^T, & \boldsymbol{\tau}_1 &= [X\ Y\ Z]^T, & \boldsymbol{\tau}_2 &= [K\ M\ N]^T, \end{aligned}$$

where $\boldsymbol{\eta}$ denotes the position and orientation vector with coordinates in the earth-fixed frame, $\boldsymbol{v}$ denotes the linear and angular velocity vector with coordinates in the body-fixed frame, and $\boldsymbol{\tau}$ denotes the forces and moments acting on the vessel in the body-fixed frame.

In deriving equations of motion of the ocean vessels, we divide the study of vessel dynamics into two parts *kinematics*, which treats only geometrical aspects of motion, and *kinetics*, which is the analysis of the forces resulting in the motion.

3.3.1 Kinematics

The first time derivative of the position vector $\boldsymbol{\eta}_1$ is related to the linear velocity vector $\boldsymbol{v}_1$ via the following transformation:

$$\dot{\boldsymbol{\eta}}_1 = \boldsymbol{J}_1(\boldsymbol{\eta}_2)\boldsymbol{v}_1, \tag{3.1}$$

where $\boldsymbol{J}_1(\boldsymbol{\eta}_2)$ is a transformation matrix, which is related through the functions of the Euler angles: roll (ϕ), pitch (θ), and yaw (ψ). This matrix is given by

$$\boldsymbol{J}_1(\boldsymbol{\eta}_2) = \begin{bmatrix} \cos(\psi)\cos(\theta) & -\sin(\psi)\cos(\phi)+\sin(\phi)\sin(\theta)\cos(\psi) & \sin(\psi)\sin(\phi)+\sin(\theta)\cos(\psi)\cos(\phi) \\ \sin(\psi)\cos(\theta) & \cos(\psi)\cos(\phi)+\sin(\phi)\sin(\theta)\sin(\psi) & -\cos(\psi)\sin(\phi)+\sin(\theta)\sin(\psi)\cos(\phi) \\ -\sin(\theta) & \sin(\phi)\cos(\theta) & \cos(\phi)\cos(\theta) \end{bmatrix}. \tag{3.2}$$

It is noted that the matrix $\boldsymbol{J}_1(\boldsymbol{\eta}_2)$ is globally invertible since $\boldsymbol{J}_1^{-1}(\boldsymbol{\eta}_2) = \boldsymbol{J}_1^T(\boldsymbol{\eta}_2)$.

On the other hand, the first time derivative of the Euler angle vector $\boldsymbol{\eta}_2$ is related to the body-fixed velocity vector $\boldsymbol{v}_2$ through the following transformation:

$$\dot{\boldsymbol{\eta}}_2 = \boldsymbol{J}_2(\boldsymbol{\eta}_2)\boldsymbol{v}_2, \tag{3.3}$$

where the transformation matrix $\boldsymbol{J_2}(\boldsymbol{\eta_2})$ is given by

$$\boldsymbol{J_2}(\boldsymbol{\eta_2}) = \begin{bmatrix} 1 & \sin(\phi)\tan(\theta) & \cos(\phi)\tan(\theta) \\ 0 & \cos(\phi) & -\sin(\phi) \\ 0 & \sin(\phi)/\cos(\theta) & \cos(\phi)/\cos(\theta) \end{bmatrix}. \tag{3.4}$$

Note that the transformation matrix $\boldsymbol{J_2}(\boldsymbol{\eta_2})$ is singular at $\theta = \pm\frac{\pi}{2}$. However, during practical operations ocean vessels are not likely to enter the neighborhood of $\theta = \pm\frac{\pi}{2}$ because of the metacentric restoring forces. For the case where it is essential to consider a region containing $\theta = \pm\frac{\pi}{2}$, a four-parameter description based on Euler parameters can be used instead. The interested reader is referred to [12] for more details. Combining (3.1) and (3.3) results in the kinematics of the ocean vessels:

$$\begin{bmatrix} \dot{\boldsymbol{\eta}}_1 \\ \dot{\boldsymbol{\eta}}_2 \end{bmatrix} = \begin{bmatrix} \boldsymbol{J_1}(\boldsymbol{\eta_2}) & 0_{3\times 3} \\ 0_{3\times 3} & \boldsymbol{J_2}(\boldsymbol{\eta_2}) \end{bmatrix} \begin{bmatrix} \boldsymbol{v_1} \\ \boldsymbol{v_2} \end{bmatrix} \quad \Leftrightarrow \quad \dot{\boldsymbol{\eta}} = \boldsymbol{J}(\boldsymbol{\eta})\boldsymbol{v}. \tag{3.5}$$

3.3.2 Kinetics

3.3.2.1 Rigid Body Equations of Motion

Let us define the following vectors:

- $\boldsymbol{f_{Ob}} = [X\ Y\ Z]^T$: force decomposed in the body-fixed frame.
- $\boldsymbol{m_{Ob}} = [K\ M\ N]^T$: moment decomposed in the body-fixed frame.
- $\boldsymbol{v_{Ob}} = [u\ v\ w]^T$: linear velocity decomposed in the body-fixed frame.
- $\boldsymbol{\omega}^E_{Ob} = [p\ q\ r]^T$: angular velocity of the body-fixed frame relative to the earth-fixed frame.
- $\boldsymbol{r_{Ob}} = [x_g\ y_g\ z_g]^T$: vector from O_b to CG (center of gravity of the vessel) decomposed in the body-fixed frame.

By the Newton–Euler formulation for a rigid body with a mass of m, we have the following balancing forces and moments:

$$\begin{aligned} &m[\dot{\boldsymbol{v}}_{Ob} + \dot{\boldsymbol{\omega}}^E_{Ob} \times \boldsymbol{r}_{Ob} + \boldsymbol{\omega}^E_{Ob} \times \boldsymbol{v}_{Ob} + \boldsymbol{\omega}^E_{Ob} \times (\boldsymbol{\omega}^E_{Ob} \times \boldsymbol{r}_{Ob})] = \boldsymbol{f}_{Ob}, \\ &\boldsymbol{I}_o \dot{\boldsymbol{\omega}}^E_{Ob} + \boldsymbol{\omega}^E_{Ob} \times \boldsymbol{I}_o \boldsymbol{\omega}^E_{Ob} + m\boldsymbol{r}_{Ob} \times (\dot{\boldsymbol{v}}_{Ob} + \boldsymbol{\omega}^E_{Ob} \times \boldsymbol{v}_{Ob}) = \boldsymbol{m}_{Ob}, \end{aligned} \tag{3.6}$$

where $\boldsymbol{I_o}$ is the inertia matrix about O_b defined by

$$\boldsymbol{I_o} = \begin{bmatrix} I_x & -I_{xy} & -I_{xz} \\ -I_{yx} & I_y & -I_{yz} \\ -I_{zx} & -I_{zy} & I_z \end{bmatrix}. \tag{3.7}$$

Here I_x, I_y, and I_z are the moments of inertia about the O_bX_b, O_bY_b, and O_bZ_b axes, and $I_{xy} = I_{yx}$, $I_{xz} = I_{zx}$, and $I_{yz} = I_{zy}$ are the products of inertia. These quantities are defined as

$$
\begin{aligned}
I_x &= \int_V (y^2+z^2)\rho_m \mathrm{d}V, \; I_{xy} = \int_V xy\rho_m \mathrm{d}V, \\
I_y &= \int_V (x^2+z^2)\rho_m \mathrm{d}V, \; I_{xz} = \int_V xz\rho_m \mathrm{d}V, \\
I_z &= \int_V (x^2+y^2)\rho_m \mathrm{d}V, \; I_{zy} = \int_V zy\rho_m \mathrm{d}V,
\end{aligned}
\tag{3.8}
$$

where ρ_m and V are, respectively, the mass density and the volume of the rigid body. Substituting the definitions of $\boldsymbol{f}_{Ob}$, $\boldsymbol{m}_{Ob}$, $\boldsymbol{v}_{Ob}$, $\boldsymbol{\omega}_{Ob}^{E}$, and $\boldsymbol{r}_{Ob}$ into (3.6) results in the following equations of motion of a rigid body:

$$
\boldsymbol{M_{RB}}\dot{\boldsymbol{v}} + \boldsymbol{C_{RB}}(\boldsymbol{v})\boldsymbol{v} = \boldsymbol{\tau_{RB}}, \tag{3.9}
$$

where $\boldsymbol{v} = [u \; v \; w \; p \; q \; r]^T$ is the generalized velocity vector decomposed in the body-fixed frame, $\boldsymbol{\tau_{RB}} = [X \; Y \; Z \; K \; M \; N]^T$ is the generalized vector of external forces and moments, the rigid body system inertia matrix $\boldsymbol{M_{RB}}$ is given by

$$
\boldsymbol{M_{RB}} = \begin{bmatrix}
m & 0 & 0 & 0 & mz_g & -my_g \\
0 & m & 0 & -mz_g & 0 & mx_g \\
0 & 0 & m & my_g & -mx_g & 0 \\
0 & -mz_g & my_g & I_x & -I_{xy} & -I_{xz} \\
mz_g & 0 & -mx_g & -I_{yx} & I_y & -I_{yz} \\
-my_g & mx_g & 0 & -I_{zx} & -I_{zy} & I_z
\end{bmatrix}, \tag{3.10}
$$

and the rigid body Coriolis and centripetal matrix $\boldsymbol{C_{RB}}(\boldsymbol{v})$ is given by

$$
\boldsymbol{C_{RB}}(\boldsymbol{v}) = \left[\begin{matrix}
0 & 0 & 0 \\
0 & 0 & 0 \\
0 & 0 & 0 \\
-m(y_g q + z_g r) & m(y_g p + w) & m(z_g p - v) \\
m(x_g q - w) & -m(z_g r + x_g p) & m(z_g q + u) \\
m(x_g r + v) & m(y_g r - u) & -m(x_g p + y_g q)
\end{matrix}\right.
$$

$$
\left.\begin{matrix}
m(y_g q + z_g r) & -m(x_g q - w) & -m(x_g r + v) \\
-m(y_g p + w) & m(z_g r + x_g p) & -m(y_g r - u) \\
-m(z_g p - v) & -m(z_g q + u) & m(x_g p + y_g q) \\
0 & -I_{yz} q - I_{xz} p + I_z r & I_{yz} r + I_{xy} p - I_y q \\
I_{yz} q + I_{xz} p - I_z r & 0 & -I_{xz} r - I_{xy} q + I_x p \\
-I_{yz} r - I_{xy} p + I_y q & I_{xz} r + I_{xy} q - I_x p & 0
\end{matrix}\right]. \tag{3.11}
$$

The generalized external force and moment vector, $\boldsymbol{\tau_{RB}}$, is a sum of hydrodynamic force and moment vector $\boldsymbol{\tau_H}$, external disturbance force and moment vector $\boldsymbol{\tau_E}$, and propulsion force and moment vector $\boldsymbol{\tau}$. Each of these vectors is detailed in the following sections.

3.3.2.2 Hydrodynamic Forces and Moments

In hydrodynamics, it is usually assumed that the hydrodynamic forces and moments on a rigid body can be linearly superimposed, see [27]. The hydrodynamic forces and moments are forces and moments on the body when the body is forced to oscillate with the wave excitation frequency and there are no incident waves. These forces and moments can be identified as the sum of three components: (1) added mass due to the inertia of the surrounding fluid, (2) radiation-induced potential damping due to the energy carried away by generated surface waves, and (3) restoring forces due to Archimedean forces (weight and buoyancy). The hydrodynamic force and moment vector $\boldsymbol{\tau_H}$ is given by

$$\boldsymbol{\tau_H} = -\boldsymbol{M_A}\dot{\boldsymbol{v}} - \boldsymbol{C_A}(\boldsymbol{v})\boldsymbol{v} - \boldsymbol{D}(\boldsymbol{v})\boldsymbol{v} - \boldsymbol{g}(\boldsymbol{\eta}), \tag{3.12}$$

where $\boldsymbol{M_A}$ is the added mass matrix, $\boldsymbol{C_A}(\boldsymbol{v})$ is the hydrodynamic Coriolis and centripetal matrix, $\boldsymbol{D}(\boldsymbol{v})$ is the damping matrix, and $\boldsymbol{g}(\boldsymbol{\eta})$ is the position and orientation depending vector of restoring forces and moments.

The added mass matrix $\boldsymbol{M_A}$ is given by

$$\boldsymbol{M_A} = -\begin{bmatrix} X_{\dot{u}} & X_{\dot{v}} & X_{\dot{w}} & X_{\dot{p}} & X_{\dot{q}} & X_{\dot{r}} \\ Y_{\dot{u}} & Y_{\dot{v}} & Y_{\dot{w}} & Y_{\dot{p}} & Y_{\dot{q}} & Y_{\dot{r}} \\ Z_{\dot{u}} & Z_{\dot{v}} & Z_{\dot{w}} & Z_{\dot{p}} & Z_{\dot{q}} & Z_{\dot{r}} \\ K_{\dot{u}} & K_{\dot{v}} & K_{\dot{w}} & K_{\dot{p}} & K_{\dot{q}} & K_{\dot{r}} \\ M_{\dot{u}} & M_{\dot{v}} & M_{\dot{w}} & M_{\dot{p}} & M_{\dot{q}} & M_{\dot{r}} \\ N_{\dot{u}} & N_{\dot{v}} & N_{\dot{w}} & N_{\dot{p}} & N_{\dot{q}} & N_{\dot{r}} \end{bmatrix}, \tag{3.13}$$

where the SNAME notation has been used. For example, the hydrodynamic added mass force Y along the y-axis due to an acceleration $\dot{u}$ in the x-direction is written as

$$Y = -Y_{\dot{u}}\dot{u}, \; Y_{\dot{u}} := \frac{\partial Y}{\partial \dot{u}}. \tag{3.14}$$

The hydrodynamic Coriolis and centripetal matrix is given by

$$\boldsymbol{C_A}(\boldsymbol{v}) = \begin{bmatrix} 0 & 0 & 0 & 0 & -a_3 & a_2 \\ 0 & 0 & 0 & a_3 & 0 & -a_1 \\ 0 & 0 & 0 & -a_2 & a_1 & 0 \\ 0 & -a_3 & a_2 & 0 & -b_3 & b_2 \\ a_3 & 0 & -a_1 & b_3 & 0 & -b_1 \\ -a_2 & a_1 & 0 & -b_2 & b_1 & 0 \end{bmatrix}, \tag{3.15}$$

where

$$\begin{aligned} a_1 &= X_{\dot{u}}u + X_{\dot{v}}v + X_{\dot{w}}w + X_{\dot{p}}p + X_{\dot{q}}q + X_{\dot{r}}r, \\ a_2 &= Y_{\dot{u}}u + Y_{\dot{v}}v + Y_{\dot{w}}w + Y_{\dot{p}}p + Y_{\dot{q}}q + Y_{\dot{r}}r, \\ a_3 &= Z_{\dot{u}}u + Z_{\dot{v}}v + Z_{\dot{w}}w + Z_{\dot{p}}p + Z_{\dot{q}}q + Z_{\dot{r}}r, \\ b_1 &= K_{\dot{u}}u + K_{\dot{v}}v + K_{\dot{w}}w + K_{\dot{p}}p + K_{\dot{q}}q + K_{\dot{r}}r, \end{aligned}$$

$$
\begin{aligned}
b_2 &= M_{\dot{u}}u + M_{\dot{v}}v + M_{\dot{w}}w + M_{\dot{p}}p + M_{\dot{q}}q + M_{\dot{r}}r, \\
b_3 &= N_{\dot{u}}u + N_{\dot{v}}v + N_{\dot{w}}w + N_{\dot{p}}p + N_{\dot{q}}q + N_{\dot{r}}r. \qquad (3.16)
\end{aligned}
$$

In general, hydrodynamic damping for ocean vessels is mainly caused by potential damping, skin friction, wave drift damping, and damping due to vortex shedding. It is difficult to give a general expression of the hydrodynamic damping matrix $\boldsymbol{D}(\boldsymbol{v})$. However, it is common to write the hydrodynamic damping matrix $\boldsymbol{D}(\boldsymbol{v})$ as

$$\boldsymbol{D}(\boldsymbol{v}) = \boldsymbol{D} + \boldsymbol{D_n}(\boldsymbol{v}). \qquad (3.17)$$

Here the linear damping matrix $\boldsymbol{D}$ is given by

$$\boldsymbol{D} = -\begin{bmatrix} X_u & X_v & X_w & X_p & X_q & X_r \\ Y_u & Y_v & Y_w & Y_p & Y_q & Y_r \\ Z_u & Z_v & Z_w & Z_p & Z_q & Z_r \\ K_u & K_v & K_w & K_p & K_q & K_r \\ M_u & M_v & M_w & M_p & M_q & M_r \\ N_u & N_v & N_w & N_p & N_q & N_r \end{bmatrix}. \qquad (3.18)$$

The nonlinear damping matrix $\boldsymbol{D_n}(\boldsymbol{v})$ is usually modeled by using a third-order Taylor series expansion or modulus functions (quadratic drag). If the xz-plane is a plane of symmetry (starboard/port symmetry) an odd Taylor series expansion containing first-order and third-order terms in velocity can be sufficient to describe most manoeuvres. An approximate expression of each of this matrices will be given in the next section when specific vessels are considered.

3.3.2.3 Restoring Forces and Moments

In this section, a model for $\boldsymbol{g}(\eta)$ is described. Let ∇ be the volume of fluid displaced by the vessel, g the acceleration of gravity (positive downwards), and ρ the water density. The submerged weight of the body and buoyancy force are defined as

$$
\begin{aligned}
W &= mg, \\
B &= \rho g \nabla. \qquad (3.19)
\end{aligned}
$$

With the above definition, the restoring force and moment vector $\boldsymbol{g}(\eta)$ is due to gravity and buoyancy forces, and is given by

$$\boldsymbol{g}(\eta) = \begin{bmatrix} (W-B)\sin(\theta) \\ -(W-B)\cos(\theta)\sin(\phi) \\ -(W-B)\cos(\theta)\cos(\phi) \\ -(y_g W - y_b B)\cos(\theta)\cos(\phi) + (z_g W - z_b B)\cos(\theta)\sin(\phi) \\ (z_g W - z_b B)\sin(\theta) + (x_g W - x_b B)\cos(\theta)\cos(\phi) \\ -(x_g W - x_b B)\cos(\theta)\sin(\phi) - (y_g W - y_b B)\sin(\theta) \end{bmatrix}, \qquad (3.20)$$

where (x_b, y_b, z_b) denote coordinates of the center of buoyancy.

3.3.2.4 Environmental Disturbances

In this section, we detail the vector, $\boldsymbol{\tau}_E$, of forces and moments induced by environmental disturbances including ocean currents, waves (wind generated) and wind, i.e., we can write

$$\boldsymbol{\tau}_E = \boldsymbol{\tau}_E^{cu} + \boldsymbol{\tau}_E^{wa} + \boldsymbol{\tau}_E^{wi}, \tag{3.21}$$

where $\boldsymbol{\tau}_E^{cu}$, $\boldsymbol{\tau}_E^{wa}$, and $\boldsymbol{\tau}_E^{wi}$ are vectors of forces and moments induced by ocean currents, waves and wind, respectively.

Current-induced Forces and Moments

The vector $\boldsymbol{\tau}_E^{cu}$ of the current-induced forces and moments is given by

$$\boldsymbol{\tau}_E^{cu} = (\boldsymbol{M}_{RB} + \boldsymbol{M}_A)\dot{\boldsymbol{v}}_c + \boldsymbol{C}(\boldsymbol{v}_r)\boldsymbol{v}_r - \boldsymbol{C}(\boldsymbol{v})\boldsymbol{v} + \boldsymbol{D}(\boldsymbol{v}_r)\boldsymbol{v}_r - \boldsymbol{D}(\boldsymbol{v})\boldsymbol{v}, \tag{3.22}$$

where $\boldsymbol{v}_r = \boldsymbol{v} - \boldsymbol{v}_c$ and $\boldsymbol{v}_c = [u_c, v_c, w_c, 0, 0, 0]^T$ is a vector of irrotational body-fixed current velocities. Let the earth-fixed current velocity vector be denoted by $[u_c^E, v_c^E, w_c^E]^T$. Then, the body-fixed components $[u_c, v_c, w_c]^T$ can be computed as

$$\begin{bmatrix} u_c \\ v_c \\ w_c \end{bmatrix} = \boldsymbol{J}_1^T(\eta_2) \begin{bmatrix} u_c^E \\ v_c^E \\ w_c^E \end{bmatrix}. \tag{3.23}$$

Wave-induced Forces and Moments

The vector $\boldsymbol{\tau}_E^{wa}$ of the wave-induced forces and moments is given by

$$\boldsymbol{\tau}_E^{wi} = \begin{bmatrix} \sum_{i=1}^N \rho g B L T \cos(\beta) s_i(t) \\ \sum_{i=1}^N -\rho g B L T \sin(\beta) s_i(t) \\ 0 \\ 0 \\ 0 \\ \sum_{i=1}^N \frac{1}{24} \rho g B L (L^2 - B^2) \sin(2\beta) s_i^2(t) \end{bmatrix}, \tag{3.24}$$

where β is the vessel's heading (encounter) angle, see Figure 3.3, ρ is the water density, L is the length of the vessel, B is the breadth of the vessel, and T is the draft of the vessel. Ignoring the higher-order terms of the wave amplitude, the wave slope $s_i(t)$ for the wave component i is defined as:

$$s_i(t) = A_i \frac{2\pi}{\lambda_i} \sin(\omega_{ei} t + \phi_i), \tag{3.25}$$

where A_i is the wave amplitude, λ_i is the wave length, ω_{ei} is the encounter frequency, and ϕ_i is a random phase uniformly distributed and constant with time in $[0\ 2\pi)$ corresponding to the wave component i.

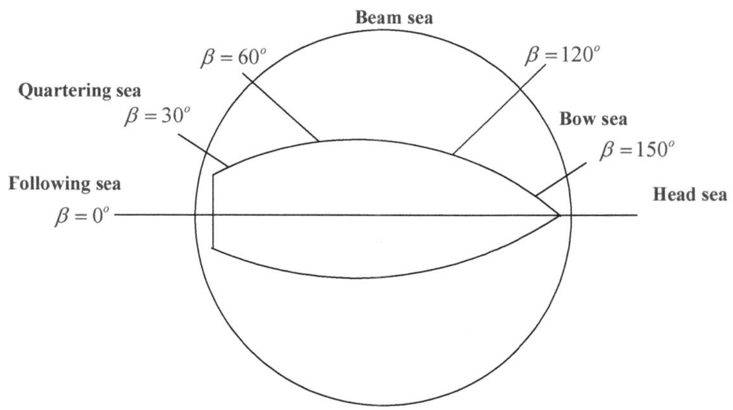

Figure 3.3 Definition of a vessel's heading (encounter) angle

Wind-induced Forces and Moments

For the case where the vessel is at rest (zero speed), the vector $\boldsymbol{\tau}_E^{wi}$ of the wind-induced forces and moments is given by

$$\boldsymbol{\tau}_E^{wi} = \frac{1}{2}\rho_a V_w^2 \begin{bmatrix} C_X(\gamma_w) A_{Fw} \\ C_Y(\gamma_w) A_{Lw} \\ C_Z(\gamma_w) A_{Fw} \\ C_K(\gamma_w) A_{Lw} H_{Lw} \\ C_M(\gamma_w) A_{Fw} H_{Fw} \\ C_N(\gamma_w) A_{Lw} L_{oa} \end{bmatrix}, \tag{3.26}$$

where V_w is the wind speed, ρ_a is the air density, A_{Fw} is the frontal projected area, A_{Lw} is the lateral projected area, H_{Fw} is the centroid of A_{Fw} above the water line, H_{Lw} is the centroid of A_{Lw} above the water line, L_{oa} is the over all length of the vessel, γ_w is the angle of relative wind of the vessel bow, see Figure 3.4, and is given by

$$\gamma_w = \psi - \beta_w - \pi, \tag{3.27}$$

with β_w being the wind direction. All the wind coefficients (look-up tables) $C_X(\gamma_w)$, $C_Y(\gamma_w)$, $C_Z(\gamma_w)$, $C_K(\gamma_w)$, $C_M(\gamma_w)$, and $C_N(\gamma_w)$ are computed numerically or by experiments in a wind tunnel, see [28].

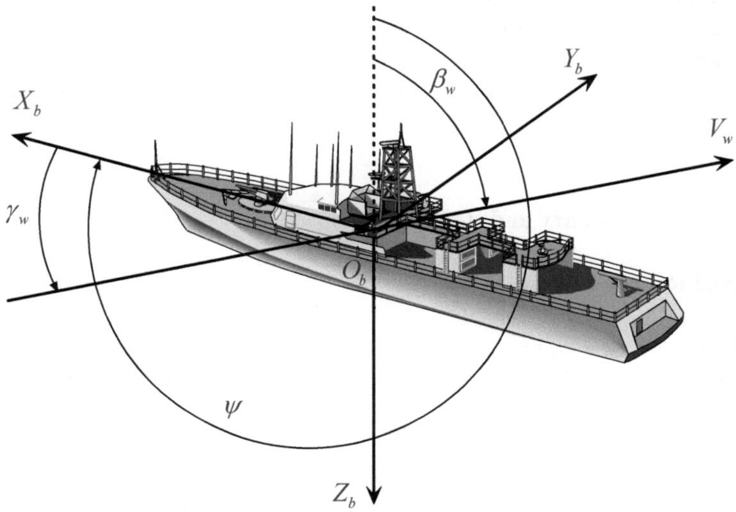

Figure 3.4 Definition of wind speed and direction

For the case where the vessel is moving, the vector $\boldsymbol{\tau}_E^{wi}$ is given by

$$\boldsymbol{\tau}_E^{wi} = \frac{1}{2}\rho_a V_{rw}^2 \begin{bmatrix} C_X(\gamma_{rw})A_{Fw} \\ C_Y(\gamma_{rw})A_{Lw} \\ C_Z(\gamma_{rw})A_{Fw} \\ C_K(\gamma_{rw})A_{Lw}H_{Lw} \\ C_M(\gamma_{rw})A_{Fw}H_{Fw} \\ C_N(\gamma_{rw})A_{Lw}L_{oa} \end{bmatrix} \tag{3.28}$$

where

$$\begin{aligned} V_{rw} &= \sqrt{u_{rw}^2 + v_{rw}^2}, \\ \gamma_{rw} &= -\text{arctan2}(v_{rw}, u_{rw}), \end{aligned} \tag{3.29}$$

with

$$\begin{aligned} u_{rw} &= u - V_w \cos(\beta_w - \psi), \\ v_{rw} &= v - V_w \sin(\beta_w - \psi). \end{aligned} \tag{3.30}$$

3.3.2.5 Propulsion Forces and Moments

The vector, $\boldsymbol{\tau}$, of propulsion forces and moments depends on a specific configuration of actuators such as propellers, rudders, and water jets on a particular vessel. In the next section where $\boldsymbol{\tau}$ is specified, we consider some classes of the ocean vessels that are common in practice. In this book, we neglect the dynamics of the actuators that provide the propulsion forces and moments since the response of the actuators such as hydraulic systems and electrical motors is much faster than the response of the vessel.

3.3.2.6 Model Summary and its Properties

Body-fixed Representation

Now substituting $\boldsymbol{\tau_{RB}} = \boldsymbol{\tau_H} + \boldsymbol{\tau_E} + \boldsymbol{\tau}$ into (3.9) and combining it with (3.5) results in the equations of motion of an ocean vessel in six degrees of freedom as follows:

$$\begin{aligned} \dot{\boldsymbol{\eta}} &= \boldsymbol{J}(\boldsymbol{\eta})\boldsymbol{v}, \\ \boldsymbol{M}\dot{\boldsymbol{v}} &= -\boldsymbol{C}(\boldsymbol{v})\boldsymbol{v} - \boldsymbol{D}(\boldsymbol{v})\boldsymbol{v} - \boldsymbol{g}(\boldsymbol{\eta}) + \boldsymbol{\tau} + \boldsymbol{\tau_E}, \end{aligned} \tag{3.31}$$

where

$$\begin{aligned} \boldsymbol{M} &= \boldsymbol{M_{RB}} + \boldsymbol{M_A}, \\ \boldsymbol{C}(\boldsymbol{v}) &= \boldsymbol{C_{RB}}(\boldsymbol{v}) + \boldsymbol{C_A}(\boldsymbol{v}). \end{aligned} \tag{3.32}$$

Under the assumption that the body is at rest (or at most is moving at low speed) in ideal fluid, the matrix $\boldsymbol{M}$ is always symmetric positive definite, i.e.,

$$\boldsymbol{M} = \boldsymbol{M}^T > \boldsymbol{0}. \tag{3.33}$$

For a rigid body moving in fluid, the Coriolis and centripetal matrix $\boldsymbol{C}(\boldsymbol{v})$ can always be parameterized such that it is skew-symmetric, i.e.,

$$\boldsymbol{C}(\boldsymbol{v}) = -\boldsymbol{C}^T(\boldsymbol{v}), \ \forall \boldsymbol{v} \in \mathbb{R}^6. \tag{3.34}$$

For a rigid body moving in an ideal fluid, the hydrodynamic damping matrix $\boldsymbol{D}(\boldsymbol{v})$ is real, non-symmetric and strictly positive, i.e.,

$$\boldsymbol{D}(\boldsymbol{v}) > \boldsymbol{0}, \ \forall \boldsymbol{v} \in \mathbb{R}^6. \tag{3.35}$$

Earth-fixed Representation

The mathematical model (3.31) can also be written using a representation of the earth-fixed coordinates by applying the following kinematic transformations (with the assumption that $J^{-1}(\eta)$ exists, i.e., $\theta \neq \pm\frac{\pi}{2}$):

$$\begin{aligned}\boldsymbol{v} &= \boldsymbol{J}^{-1}(\eta)\dot{\eta},\\ \dot{\boldsymbol{v}} &= \boldsymbol{J}^{-1}(\eta)\Big[\ddot{\eta} - \dot{\boldsymbol{J}}(\eta)\boldsymbol{J}^{-1}(\eta)\dot{\eta}\Big].\end{aligned} \tag{3.36}$$

Now substituting (3.36) into the second equation of (3.31) results in

$$\boldsymbol{M}^{\star}(\eta)\ddot{\eta} = -\boldsymbol{C}^{\star}(\boldsymbol{v},\eta)\dot{\eta} - \boldsymbol{D}^{\star}(\boldsymbol{v},\eta)\dot{\eta} - \boldsymbol{g}^{\star}(\eta) + \boldsymbol{J}^{T}(\eta)(\boldsymbol{\tau} + \boldsymbol{\tau}_{E}), \tag{3.37}$$

where

$$\begin{aligned}\boldsymbol{M}^{\star}(\eta) &= \boldsymbol{J}^{-T}(\eta)\boldsymbol{M}\boldsymbol{J}^{-1}(\eta),\\ \boldsymbol{C}^{\star}(\boldsymbol{v},\eta) &= \boldsymbol{J}^{-T}(\eta)\big[\boldsymbol{C}(\boldsymbol{v}) - \boldsymbol{M}\boldsymbol{J}^{-1}(\eta)\dot{\boldsymbol{J}}(\eta)\big]\boldsymbol{J}^{-1}(\eta),\\ \boldsymbol{D}^{\star}(\boldsymbol{v},\eta) &= \boldsymbol{J}^{-T}(\eta)\boldsymbol{D}(\boldsymbol{v})\boldsymbol{J}^{-1}(\eta),\\ \boldsymbol{g}^{\star}(\eta) &= \boldsymbol{J}^{-T}(\eta)\boldsymbol{g}(\eta).\end{aligned} \tag{3.38}$$

Under the same assumptions used in the body-fixed representation, the model (3.37) using the earth-fixed representation has the following properties:

$$\begin{aligned}&\boldsymbol{M}^{\star}(\eta) = \boldsymbol{M}^{\star}(\eta)^{T},\ \forall\,\eta \in \mathbb{R}^{6},\\ &\boldsymbol{s}^{T}\big[\dot{\boldsymbol{M}}^{\star}(\eta) - 2\boldsymbol{C}^{\star}(\boldsymbol{v},\eta)\big]\boldsymbol{s} = \boldsymbol{0},\ \forall\,\eta \in \mathbb{R}^{6},\ \boldsymbol{v} \in \mathbb{R}^{6},\ \boldsymbol{s} \in \mathbb{R}^{6},\\ &\boldsymbol{D}^{\star}(\boldsymbol{v},\eta) > \boldsymbol{0},\quad \forall\,\eta \in \mathbb{R}^{6},\ \boldsymbol{v} \in \mathbb{R}^{6}.\end{aligned} \tag{3.39}$$

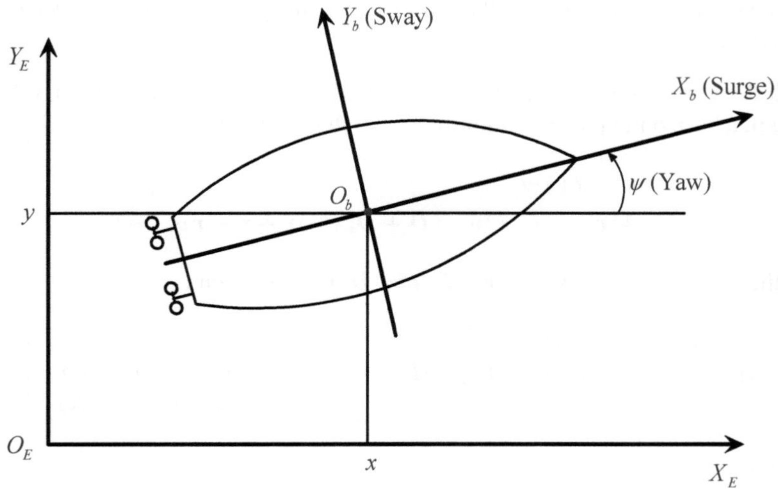

Figure 3.5 Motion variables for an ocean vessel moving in a horizontal plane

3.4 Standard Models for Ocean Vessels

In this section, the main results of the previous section are simplified to give a set of standard models for surface ships and underwater vehicles. These standard models will be extensively used for the control design in the coming chapters.

3.4.1 Three Degrees of Freedom Horizontal Model

3.4.1.1 Standard Three Degrees of Freedom Horizontal Model

The horizontal motion of a surface ship or an underwater vehicle moving in a horizontal plane is often described by the motion components in surge, sway, and yaw. Figure 3.5 illustrates the motion variables in this case. Therefore, we choose $\boldsymbol{\eta} = [x\ y\ \psi]^T$ and $\boldsymbol{v} = [u\ v\ r]^T$. This model is obtained from the general model (3.31) under the following assumption.

Assumption 3.1.

1. *The motion in roll, pitch, and heave is ignored. This means that we ignore the dynamics associated with the motion in heave, roll, and pitch, i.e.,* $z = 0$, $w = 0$, $\phi = 0$, $p = 0$, $\theta = 0$, *and* $q = 0$.
2. *The vessel has homogeneous mass distribution and* xz*-plane of symmetry so that*

$$I_{xy} = I_{yz} = 0. \tag{3.40}$$

3. *The center of gravity* CG *and the center of buoyancy,* CB*, are located vertically on the* z*-axis.*

With Assumption 3.1, the dynamics of a surface ship or an underwater vehicle moving in a horizontal plane is simplified from the general model (3.31) as follows:

$$\begin{aligned} \dot{\boldsymbol{\eta}} &= \boldsymbol{J}(\boldsymbol{\eta})\boldsymbol{v}, \\ \boldsymbol{M}\dot{\boldsymbol{v}} &= -\boldsymbol{C}(\boldsymbol{v})\boldsymbol{v} - (\boldsymbol{D} + \boldsymbol{D_n}(\boldsymbol{v}))\boldsymbol{v} + \boldsymbol{\tau} + \boldsymbol{\tau_E}, \end{aligned} \tag{3.41}$$

where the matrices $\boldsymbol{J}(\boldsymbol{\eta})$, $\boldsymbol{M}$, $\boldsymbol{C}(\boldsymbol{v})$, $\boldsymbol{D}$, and $\boldsymbol{D_n}(\boldsymbol{v})$ are given by

$$\boldsymbol{J}(\boldsymbol{\eta}) = \begin{bmatrix} \cos(\psi) & -\sin(\psi) & 0 \\ \sin(\psi) & \cos(\psi) & 0 \\ 0 & 0 & 1 \end{bmatrix}, \quad \boldsymbol{M} = \begin{bmatrix} m - X_{\dot{u}} & 0 & 0 \\ 0 & m - Y_{\dot{v}} & mx_g - Y_{\dot{r}} \\ 0 & mx_g - Y_{\dot{r}} & I_z - N_{\dot{r}} \end{bmatrix},$$

$$C(v) = \begin{bmatrix} 0 & 0 & -m(x_g r + v) + Y_{\dot{v}} v + Y_{\dot{r}} r \\ 0 & 0 & mu - X_{\dot{u}} u \\ m(x_g r + v) - Y_{\dot{v}} v - Y_{\dot{r}} r & -mu + X_{\dot{u}} u & 0 \end{bmatrix},$$

$$D = -\begin{bmatrix} X_u & 0 & 0 \\ 0 & Y_v & Y_r \\ 0 & N_v & N_r \end{bmatrix},$$

$$D_n(v) = -\begin{bmatrix} X_{|u|u}|u| & 0 & 0 \\ 0 & Y_{|v|v}|v| + Y_{|r|v}|r| & Y_{|v|r}|v| \\ 0 & N_{|v|v}|v| + N_{|r|v}|r| & N_{|v|r}|v| + N_{|r|r}|r| \end{bmatrix}. \quad (3.42)$$

The propulsion force and moment vector $\boldsymbol{\tau}$ is given by

$$\boldsymbol{\tau} = \begin{bmatrix} \tau_u \\ 0 \\ \tau_r \end{bmatrix}. \quad (3.43)$$

The above propulsion force and moment vector $\boldsymbol{\tau}$ implies that we are considering a surface vessel, which does not have an independent actuator in the sway, i.e., an underactuated vessel is under consideration. Such a vessel can be one equipped with a pair of water jets or a pair of propellers.

The environmental disturbance vector $\boldsymbol{\tau}_E$ is given by

$$\boldsymbol{\tau}_E = \begin{bmatrix} \tau_{uE} \\ \tau_{vE} \\ \tau_{rE} \end{bmatrix}, \quad (3.44)$$

where τ_{uE} and τ_{vE} are disturbance forces acting in surge and sway respectively, and τ_{rE} is the disturbance moment acting in yaw.

3.4.1.2 Simplified Three Degrees of Freedom Horizontal Model

In some cases, in addition to Assumption 3.1 we ignore the off-diagonal terms of the matrices $\boldsymbol{M}$ and $\boldsymbol{D}$, all elements of the nonlinear damping matrix $\boldsymbol{D_n}(\boldsymbol{v})$. These assumptions hold when the vessel has three planes of symmetry, for which the axes of the body-fixed reference frame are chosen to be parallel to the principal axis of the displaced fluid, which are equal to the principal axis of the vessel. Most ships have port/starboard symmetry, and moreover, bottom/top symmetry is not required for horizontal motion. Ship fore/aft nonsymmetry implies that the off-diagonal terms of the inertia and damping matrices are nonzero. However, these terms are small compared to the main diagonal terms. Furthermore, disturbances induced by waves, wind, and ocean currents are ignored. Under the just-mentioned assumptions, the dynamics of a surface ship or an underwater vehicle moving in a horizontal plane is simplified from the three degrees of freedom model (3.41) as follows:

$$\begin{aligned}\dot{\boldsymbol{\eta}} &= \boldsymbol{J}(\boldsymbol{\eta})\boldsymbol{v},\\ \boldsymbol{M}\dot{\boldsymbol{v}} &= -\boldsymbol{C}(\boldsymbol{v})\boldsymbol{v} - \boldsymbol{D}\boldsymbol{v} + \boldsymbol{\tau},\end{aligned} \tag{3.45}$$

where the matrices $\boldsymbol{J}(\boldsymbol{\eta})$, $\boldsymbol{M}$, $\boldsymbol{C}(\boldsymbol{v})$ and $\boldsymbol{D}$ are given by

$$\begin{aligned}\boldsymbol{J}(\boldsymbol{\eta}) &= \begin{bmatrix}\cos(\psi) & -\sin(\psi) & 0\\ \sin(\psi) & \cos(\psi) & 0\\ 0 & 0 & 1\end{bmatrix}, \quad \boldsymbol{M} = \begin{bmatrix}m_{11} & 0 & 0\\ 0 & m_{22} & 0\\ 0 & 0 & m_{33}\end{bmatrix},\\ \boldsymbol{C}(\boldsymbol{v}) &= \begin{bmatrix}0 & 0 & -m_{22}v\\ 0 & 0 & m_{11}u\\ m_{22}v & -m_{11}u & 0\end{bmatrix}, \quad \boldsymbol{D} = \begin{bmatrix}d_{11} & 0 & 0\\ 0 & d_{22} & 0\\ 0 & 0 & d_{33}\end{bmatrix},\end{aligned} \tag{3.46}$$

with

$$\begin{aligned}&m_{11} = m - X_{\dot{u}},\ m_{22} = m - Y_{\dot{v}},\ m_{33} = I_z - N_{\dot{r}},\\ &d_{11} = -X_u,\ d_{22} = -Y_v,\ d_{33} = -N_r.\end{aligned} \tag{3.47}$$

The propulsion force and moment vector $\boldsymbol{\tau}$ is still given by (3.43), i.e., $\boldsymbol{\tau} = [\tau_u\ 0\ \tau_r]^T$.

3.4.1.3 Spherical Three Degrees of Freedom Horizontal Model

In addition to the assumptions made in Subsection 3.4.1.2, we assume that the vessel has bottom/top symmetry. An example of this type of vessels is an ODIN moving in a horizontal plane, see Figure 3.6. In this case, the model is further simplified to

$$\begin{aligned}\dot{\boldsymbol{\eta}} &= \boldsymbol{J}(\boldsymbol{\eta})\boldsymbol{v},\\ \boldsymbol{M}\dot{\boldsymbol{v}} &= -\boldsymbol{C}(\boldsymbol{v})\boldsymbol{v} - \boldsymbol{D}\boldsymbol{v} + \boldsymbol{\tau}\end{aligned} \tag{3.48}$$

where the matrices $\boldsymbol{J}(\boldsymbol{\eta})$, $\boldsymbol{M}$, $\boldsymbol{C}(\boldsymbol{v})$ and $\boldsymbol{D}$ are given by

$$\begin{aligned}\boldsymbol{J}(\boldsymbol{\eta}) &= \begin{bmatrix}\cos(\psi) & -\sin(\psi) & 0\\ \sin(\psi) & \cos(\psi) & 0\\ 0 & 0 & 1\end{bmatrix}, \quad \boldsymbol{M} = \begin{bmatrix}m_{xy} & 0 & 0\\ 0 & m_{xy} & 0\\ 0 & 0 & m_{33}\end{bmatrix},\\ \boldsymbol{C}(\boldsymbol{v}) &= \begin{bmatrix}0 & 0 & -m_{xy}v\\ 0 & 0 & m_{xy}u\\ m_{xy}v & -m_{xy}u & 0\end{bmatrix}, \quad \boldsymbol{D} = \begin{bmatrix}d_{xy} & 0 & 0\\ 0 & d_{xy} & 0\\ 0 & 0 & d_{33}\end{bmatrix},\end{aligned} \tag{3.49}$$

with

$$m_{xy} = m - X_{\dot{u}} = m - Y_{\dot{v}},\ m_{33} = I_z - N_{\dot{r}},$$
$$d_{xy} = -X_u = -Y_v,\ d_{33} = -N_r. \tag{3.50}$$

The propulsion force and moment vector $\boldsymbol{\tau}$ is still given by (3.43), i.e. $\boldsymbol{\tau} = [\tau_u\ 0\ \tau_r]^T$.

Figure 3.6 An omnidirectional intelligent navigator (ODIN).
Courtesy http://www.math.hawaii.edu/~ryan/STOMP/Photos 20ODIN.html

3.4.2 Six Degrees of Freedom Model

3.4.2.1 Standard Model

In addition to the assumptions made in Section 3.3, we assume that the center of gravity and the center of buoyancy are located vertically on the $O_b Z_b$-axis, and that there are no couplings (off-diagonal terms) in the matrices $\boldsymbol{M}$, $\boldsymbol{D}$, and $\boldsymbol{D_n}(\boldsymbol{v})$. In this case, the model presented in Section 3.3 is simplified to

$$\begin{aligned}
\dot{\boldsymbol{\eta}}_1 &= \boldsymbol{J}_1(\boldsymbol{\eta}_2)\boldsymbol{v}_1,\\
\boldsymbol{M}_1\dot{\boldsymbol{v}}_1 &= -\boldsymbol{C}_1(\boldsymbol{v}_1)\boldsymbol{v}_2 - \boldsymbol{D}_1\boldsymbol{v}_1 - \boldsymbol{D}_{n1}(\boldsymbol{v}_1)\boldsymbol{v}_1 + \boldsymbol{\tau}_1 + \boldsymbol{\tau}_{1E},\\
\dot{\boldsymbol{\eta}}_2 &= \boldsymbol{J}_2(\boldsymbol{\eta}_2)\boldsymbol{v}_2,\\
\boldsymbol{M}_2\dot{\boldsymbol{v}}_2 &= -\boldsymbol{C}_1(\boldsymbol{v}_1)\boldsymbol{v}_1 - \boldsymbol{C}_2(\boldsymbol{v}_2)\boldsymbol{v}_2 - \boldsymbol{D}_2\boldsymbol{v}_2 - \boldsymbol{D}_{n2}(\boldsymbol{v}_2)\boldsymbol{v}_2 -\\
&\qquad \boldsymbol{g}_2(\boldsymbol{\eta}_2) + \boldsymbol{\tau}_2 + \boldsymbol{\tau}_{2E},
\end{aligned} \tag{3.51}$$

where $J_1(\eta_2)$ and $J_2(\eta_2)$ are given in (3.2) and (3.4). The matrices M_1 and M_2 are

$$M_1 = \begin{bmatrix} m_{11} & 0 & 0 \\ 0 & m_{22} & 0 \\ 0 & 0 & m_{33} \end{bmatrix},$$

$$M_2 = \begin{bmatrix} m_{44} & 0 & 0 \\ 0 & m_{55} & 0 \\ 0 & 0 & m_{66} \end{bmatrix}, \tag{3.52}$$

where

$$\begin{aligned} m_{11} &= m - X_{\dot{u}},\ m_{22} = m - Y_{\dot{v}}, \\ m_{33} &= m - Z_{\dot{w}},\ m_{44} = I_x - K_{\dot{p}}, \\ m_{55} &= I_y - M_{\dot{q}},\ m_{66} = I_z - N_{\dot{r}}. \end{aligned}$$

The matrices $C_1(v_1)$ and $C_2(v_2)$ are

$$C_1(v_1) = \begin{bmatrix} 0 & m_{33}w & -m_{22}v \\ -m_{33}w & 0 & m_{11}u \\ m_{22}v & -m_{11}u & 0 \end{bmatrix},$$

$$C_2(v_2) = \begin{bmatrix} 0 & m_{66}r & -m_{55}q \\ -m_{66}r & 0 & m_{44}p \\ m_{55}q & -m_{44}p & 0 \end{bmatrix}. \tag{3.53}$$

The linear damping matrices D_1 and D_2 are

$$D_1 = \begin{bmatrix} d_{11} & 0 & 0 \\ 0 & d_{22} & 0 \\ 0 & 0 & d_{33} \end{bmatrix},$$

$$D_2 = \begin{bmatrix} d_{44} & 0 & 0 \\ 0 & d_{55} & 0 \\ 0 & 0 & d_{66} \end{bmatrix}, \tag{3.54}$$

where

$$\begin{aligned} d_{11} &= -X_u, \\ d_{22} &= -Y_v, \\ d_{33} &= -Z_w, \\ d_{44} &= -K_p, \\ d_{55} &= -M_q, \\ d_{66} &= -N_r. \end{aligned}$$

The nonlinear damping matrices $\boldsymbol{D}_{n1}(\boldsymbol{v}_1)$ and $\boldsymbol{D}_{n2}(\boldsymbol{v}_2)$ are

$$\boldsymbol{D}_{n1}(\boldsymbol{v}_1) = \begin{bmatrix} \sum_{i=2}^{3} d_{ui}|u|^{i-1} & 0 & 0 \\ 0 & \sum_{i=2}^{3} d_{vi}|v|^{i-1} & 0 \\ 0 & 0 & \sum_{i=2}^{3} d_{wi}|w|^{i-1} \end{bmatrix},$$

$$\boldsymbol{D}_{n2}(\boldsymbol{v}_2) = \begin{bmatrix} \sum_{i=2}^{3} d_{pi}|p|^{i-1} & 0 & 0 \\ 0 & \sum_{i=2}^{3} d_{qi}|q|^{i-1} & 0 \\ 0 & 0 & \sum_{i=2}^{3} d_{ri}|r|^{i-1} \end{bmatrix}, \tag{3.55}$$

where d_{ui}, d_{vi}, d_{wi}, d_{pi}, d_{qi}, and d_{ri} with $i = 2, 3$ are the nonlinear hydrodynamic damping coefficients.

The restoring force and moment vector $\boldsymbol{g}_2(\boldsymbol{\eta}_2)$ is given by

$$\boldsymbol{g}_2(\boldsymbol{\eta}_2) = \begin{bmatrix} \rho g \nabla \overline{GM_T} \sin(\phi)\cos(\theta) \\ \rho g \nabla \overline{GM_L} \sin(\theta) \\ 0 \end{bmatrix}, \tag{3.56}$$

where ρ, g, ∇, $\overline{GM_T}$ and $\overline{GM_L}$ are the water density, gravity acceleration, displaced volume of water, transverse metacentric height and longitudinal metacentric height, respectively.

The propulsion force and moment vectors $\boldsymbol{\tau}_1$ and $\boldsymbol{\tau}_2$ are

$$\boldsymbol{\tau}_1 = \begin{bmatrix} \tau_u \\ 0 \\ 0 \end{bmatrix}, \quad \boldsymbol{\tau}_2 = \begin{bmatrix} \tau_p \\ \tau_q \\ \tau_r \end{bmatrix}, \tag{3.57}$$

which imply that the vessel under consideration does not have independent actuators in the sway and heave.

The environmental disturbance vectors $\boldsymbol{\tau}_{1E}$ and $\boldsymbol{\tau}_{2E}$ are given by

$$\boldsymbol{\tau}_{1E} = \begin{bmatrix} \tau_{Eu} \\ \tau_{Ev} \\ \tau_{Ew} \end{bmatrix}, \quad \boldsymbol{\tau}_{2E} = \begin{bmatrix} \tau_{Ep} \\ \tau_{Eq} \\ \tau_{Er} \end{bmatrix}, \tag{3.58}$$

where τ_{Eu}, τ_{Ev}, τ_{Ew}, τ_{Ep}, τ_{Eq}, and τ_{Er} are the environmental disturbance forces or moments acting on the surge, sway, heave, roll, pitch, and yaw axes, respectively.

3.4.2.2 Ignoring Nonlinear Damping Terms and Roll Model

In addition to the assumptions made in Section 3.4.2.1, it is sometimes reasonable to ignore nonlinear hydrodynamic damping terms and roll, and environmental disturbances. This holds when the vessel is operating at low speed and is equipped with independent internal/external roll actuators. As such, the model (3.51) is simplified to:

1. Kinematics

$$\begin{aligned}
\dot{x} &= \cos(\psi)\cos(\theta)u - \sin(\psi)v + \sin(\theta)\cos(\psi)w,\\
\dot{y} &= \sin(\psi)\cos(\theta)u + \cos(\psi)v + \sin(\theta)\sin(\psi)w,\\
\dot{z} &= -\sin(\theta)u + \cos(\theta)w,\\
\dot{\theta} &= q,\\
\dot{\psi} &= \frac{r}{\cos(\theta)}.
\end{aligned} \tag{3.59}$$

2. Kinetics

$$\begin{aligned}
\dot{u} &= \frac{m_{22}}{m_{11}}vr - \frac{m_{33}}{m_{11}}wq - \frac{d_{11}}{m_{11}}u + \frac{1}{m_{11}}\tau_u,\\
\dot{v} &= -\frac{m_{11}}{m_{22}}ur - \frac{d_{22}}{m_{22}}v,\\
\dot{w} &= \frac{m_{11}}{m_{33}}uq - \frac{d_{33}}{m_{33}}w,\\
\dot{q} &= \frac{m_{33}-m_{11}}{m_{55}}uw - \frac{d_{55}}{m_{55}}q - \frac{\rho g \nabla \overline{GM_L}\sin(\theta)}{m_{55}} + \frac{1}{m_{55}}\tau_q,\\
\dot{r} &= \frac{m_{11}-m_{22}}{m_{66}}uv - \frac{d_{66}}{m_{66}}r + \frac{1}{m_{66}}\tau_r.
\end{aligned} \tag{3.60}$$

3.5 Conclusions

This chapter sets out material about the basic motion tasks and mathematical models of the ocean vessels that will be used in the subsequent chapters. More details on deriving the mathematical models of the ocean vessels are given in [11, 12, 25]. It has also been pointed out that regulation/stabilization is much more difficult than trajectory-tracking, path-tracking, and path-following for underactuated ocean vessels.

Chapter 4
Control Properties and Previous Work on Control of Ocean Vessels

In this chapter, first control properties of ocean vessels are presented. Then, the existing literature on the control of underactuated ocean vessels is reviewed. Through the review of the previous work in the areas of stabilization, trajectory-tracking, path-following, and output feedback control of underactuated ocean vessels, challenging questions are raised. Illustration of the background and process of solutions of those questions, as well as an explanation of the solutions in terms of their physical insights and practical applications are then presented in subsequent chapters.

4.1 Controllability Properties

4.1.1 Acceleration Constraints

The number (m_c) of independent control inputs (the number of nonzero elements of the propulsion force and moment vector $\boldsymbol{\tau}$) is smaller than the number (n_c) of degrees of freedom to be controlled for a standard model of the ocean vessels. As such, we remove all zero elements of $\boldsymbol{\tau}$ and denote the resulting vector by $\boldsymbol{\tau_a}$. Thus, if $\boldsymbol{\tau_a} \in \mathbb{R}^{m_c}$ and $\boldsymbol{\eta} \in \mathbb{R}^{n_c}$, then $m_c < n_c$. For example, for the case of the vessels with six degrees of freedom to be controlled we have $m_c < 6$, for the case of the vessels with five degrees of freedom to be controlled $m_c < 5$, and $m_c < 3$ for the case of the vessels with three degrees of freedom to be controlled. For clarity, we ignore the environmental disturbance forces and moments to investigate acceleration constraints on the aforementioned ocean vessels. Let $\boldsymbol{M_u}$, $\boldsymbol{C_u}(\boldsymbol{v})$, $\boldsymbol{D_u}(\boldsymbol{v})$, and $\boldsymbol{g_u}(\boldsymbol{\eta})$ denote the rows of $\boldsymbol{M}$, $\boldsymbol{C}(\boldsymbol{v})$, $\boldsymbol{D}(\boldsymbol{v})$, and $\boldsymbol{g}(\boldsymbol{\eta})$ that correspond to those rows without propulsion forces or moments, i.e.,

$$\boldsymbol{M_u}\dot{\boldsymbol{v}} + \boldsymbol{C_u}(\boldsymbol{v})\boldsymbol{v} + \boldsymbol{D_u}(\boldsymbol{v})\boldsymbol{v} + \boldsymbol{g_u}(\boldsymbol{\eta}) = \mathbf{0}. \tag{4.1}$$

The above equation describes the acceleration constraints, i.e., second-order constraints. The following results give the conditions whether the constraints given in (4.1) are partially integrable or totally integrable.

Lemma 4.1. *The constraints (4.1) are partially integrable if and only if the following conditions hold:*

1. $\boldsymbol{g_u}(\eta)$ *is a constant vector.*
2. $(\boldsymbol{C_u}(\boldsymbol{v}) + \boldsymbol{D_u}(\boldsymbol{v}))$ *is a constant matrix.*
3. *The distribution* $\Omega^{\perp}(\eta) = \ker\left((\boldsymbol{C_u}(\boldsymbol{v}) + \boldsymbol{D_u}(\boldsymbol{v}))\boldsymbol{J}^{-1}(\eta)\right)$ *is completely integrable.*

Proof. See [29]. □

Lemma 4.2. *The constraints (4.1) are totally integrable if and only if the following conditions hold:*

1. *The constraints are partially integrable.*
2. $(\boldsymbol{C_u}(\boldsymbol{v}) + \boldsymbol{D_u}(\boldsymbol{v}))=0$.
3. *The distribution* $\Delta(\eta) = \ker\left(\boldsymbol{M_u}\boldsymbol{J}^{-1}(\eta)\right)$ *is completely integrable.*

Proof. See [29]. □

The following lemma gives a result on the stabilizability of an underactuated ocean vessel.

Lemma 4.3. *Consider the system (3.31) with* $\boldsymbol{\tau_E} = \boldsymbol{0}^{n_c \times 1}$. *Assume that the elements of the restoring force and moment vector* $\boldsymbol{g}(\eta)$ *corresponding to the unactuated dynamics are zero, i.e., the vector* $\boldsymbol{g}(\eta)$ *can be written in the form of*

$$\boldsymbol{g}(\eta) = \begin{bmatrix} \boldsymbol{g_a}(\eta) \\ \boldsymbol{0}^{(n_c - m_c)\times 1} \end{bmatrix}, \tag{4.2}$$

where $\boldsymbol{g_a}(\eta) \in \mathbb{R}^{m_c}$ *is the restoring force and moment vector corresponding to the actuated dynamics. Let* $(\eta, \boldsymbol{v}) = (\eta^e, \boldsymbol{0}^{n_c - m_c})$ *be an equilibrium. There is no* C^1 *state feedback law* $\boldsymbol{\alpha}(\eta, \boldsymbol{v}) : \mathbb{R}^{n_c} \times \mathbb{R}^{n_c} \to \mathbb{R}^{m_c}$ *that makes the equilibrium* $(\eta^e, \boldsymbol{0}^{n_c - m_c})$ *asymptotically stable.*

Proof. See [29]. □

4.1.2 Kinematic Constraints

In this section, we address controllability properties of ocean vessels. Since the vessel under consideration has a number of degrees of freedom to be controlled greater than control inputs (e.g., underwater vehicles do not have independent actuators in the heave and sway axes, see Section 3.4.2 and surface ships do not have an independent actuator in the sway axis, see Section 3.4.1), we can address the controllability issue of the vessel kinematic by considering the kinematics with the linear velocity vector

$$\boldsymbol{v}_1 = \begin{bmatrix} u \\ 0 \\ 0 \end{bmatrix}. \tag{4.3}$$

We analyze controllability properties of six degrees of freedom vessels. The case of three degrees of freedom vessels can be obtained directly from the results for the six degrees of freedom vessels. With (4.3), we now write the kinematics of the vessel as follows:

$$\begin{gathered} \dot{\eta} = \gamma_1(\eta)u + \gamma_2(\eta)p + \gamma_3(\eta)q + \gamma_4(\eta)r \\ \Updownarrow \\ \dot{\eta} = \boldsymbol{\Upsilon}(\eta)\boldsymbol{u}, \end{gathered} \tag{4.4}$$

where

$$\gamma_1(\eta) = \begin{bmatrix} \cos(\theta)\cos(\psi) \\ \cos(\theta)\sin(\psi) \\ -\sin(\theta) \\ 0 \\ 0 \\ 0 \end{bmatrix}, \quad \gamma_2(\eta) = \begin{bmatrix} 0 \\ 0 \\ 0 \\ 1 \\ 0 \\ 0 \end{bmatrix},$$

$$\gamma_3(\eta) = \begin{bmatrix} 0 \\ 0 \\ 0 \\ \sin(\phi)\tan(\theta) \\ \cos(\phi) \\ \sin(\phi)\sec(\theta) \end{bmatrix}, \quad \gamma_4(\eta) = \begin{bmatrix} 0 \\ 0 \\ 0 \\ \cos(\phi)\tan(\theta) \\ -\sin(\phi) \\ \cos(\phi)\sec(\theta) \end{bmatrix}, \tag{4.5}$$

and

$$\begin{aligned} \boldsymbol{\Upsilon}(\eta) &= \left[\, \gamma_1(\eta)\ \gamma_2(\eta)\ \gamma_3(\eta)\ \gamma_4(\eta) \,\right], \\ \boldsymbol{u} &= [u\ p\ q\ r]^T. \end{aligned} \tag{4.6}$$

From (4.4) and (4.5), a calculation shows that the following nonhonolomic (non-integrable) constraints are satisfied:

$$\begin{aligned} &\big(\cos(\psi)\sin(\theta)\sin(\phi) - \sin(\psi)\cos(\phi)\big)\dot{x} + \\ &\qquad \big(\sin(\psi)\sin(\theta)\sin(\phi) + \cos(\psi)\cos(\phi)\big)\dot{y} + \cos(\theta)\sin(\phi)\dot{z} = 0, \\ &\big(\sin(\psi)\sin(\theta)\cos(\phi) - \sin(\psi)\sin(\phi)\big)\dot{x} + \\ &\qquad \big(\sin(\psi)\sin(\theta)\cos(\phi) - \cos(\psi)\sin(\phi)\big)\dot{y} + \cos(\theta)\cos(\phi)\dot{z} = 0. \end{aligned} \tag{4.7}$$

Based on (4.4), we will address the following controllability issues: Controllability about a point (i.e., stabilization) and controllability about a trajectory (i.e., trajectory-tracking).

4.1.3 Controllability at a Point

We will first consider a linear approximation of the system (4.4) at an equilibrium point η_e. Let the error associated with the equilibrium point η_e be as follows:

$$\tilde{\eta} = \eta - \eta_e . \tag{4.8}$$

With (4.8), we can write the tangent linearization of (4.4) at the equilibrium point η_e as

$$\dot{\tilde{\eta}} = \boldsymbol{\Upsilon}(\eta_e)\boldsymbol{u}, \tag{4.9}$$

which is not controllable because the rank of the matrix $\boldsymbol{\Upsilon}(\eta_e)$ is 4. This implies that a linear controller will never achieve posture stabilization, not even in a local sense. In order to study the controllability of the vessel in question, we need to use some tools (the Lie algebra rank condition and nilpotent concepts) from nonlinear control theory [4].

Given a set of generators or basis vector fields $\gamma_1, \gamma_2, ..., \gamma_{m_c}$, the length of a Lie product recursively defined as

$$\begin{aligned} &\ell\{\gamma_i\} = 1, \quad i = 1, 2, ..., m_c \\ &\ell([A, B]) = \ell[A] + \ell[B], \end{aligned} \tag{4.10}$$

where A and B are themselves Lie products. Alternatively, $\ell[A]$ is the number of generators in the expansion for A. A Lie algebra or basis is nilpotent if there exists an integer k such that all Lie products of length greater than k are zero. The integer k is called the order of nilpotency. The use of the nilpotent basis eliminates the need for cumbersome computations as we see that all higher order Lie brackets above some particular order are zero.

The above concepts and conditions imply that Lie algebra $L\{\gamma_1, \gamma_2, \gamma_3, \gamma_4\}$ is nilpotent algebra of order $k = 2$, i.e., the vector fields $\gamma_1, \gamma_2, \gamma_3$, and γ_4 are the nilpotent basis. Thus all Lie brackets of order greater than two are zero. The only independent Lie brackets computed from the four basis vector fields are $[\gamma_1, \gamma_3]$ and $[\gamma_1, \gamma_4]$. Therefore, for our system the Lie algebra rank condition becomes

$$\text{rank}[C_c] = 6 \Leftrightarrow \text{rank}[\gamma_1, \gamma_2, \gamma_3, \gamma_4, [\gamma_1, \gamma_3], [\gamma_1, \gamma_4]] = 6, \tag{4.11}$$

where $[\gamma_1, \gamma_3]$ and $[\gamma_2, \gamma_4]$ are the two independent Lie brackets computed from the four vector fields $(\gamma_1, \gamma_2, \gamma_3, \gamma_4)$ and C_c is called the controllability matrix. For two vector fields $g(x)$ and $h(x)$, a Lie bracket is computed based on the following formula:

$$[g,h](x) = \frac{\partial h}{\partial x}g - \frac{\partial g}{\partial x}h. \tag{4.12}$$

Using the definition (4.12), Lie brackets $[\gamma_1, \gamma_3]$ and $[\gamma_1, \gamma_4]$ are given by

$$[\gamma_1, \gamma_3] = \frac{\partial \gamma_3}{\partial \eta}\gamma_1 - \frac{\partial \gamma_1}{\partial \eta}\gamma_3 = \begin{bmatrix} \cos(\psi)\sin(\theta)\cos(\phi) + \sin(\psi)\sin(\phi) \\ \sin(\psi)\sin(\theta)\cos(\phi) - \cos(\psi)\sin(\phi) \\ \cos(\theta)\cos(\phi) \\ 0 \\ 0 \\ 0 \end{bmatrix},$$

$$[\gamma_1, \gamma_4] = \frac{\partial \gamma_4}{\partial \eta}\gamma_1 - \frac{\partial \gamma_1}{\partial \eta}\gamma_4 = \begin{bmatrix} -\cos(\psi)\sin(\theta)\sin(\phi) + \sin(\psi)\cos(\phi) \\ -\sin(\psi)\sin(\theta)\sin(\phi) - \cos(\psi)\cos(\phi) \\ -\cos(\theta)\sin(\phi) \\ 0 \\ 0 \\ 0 \end{bmatrix}. \tag{4.13}$$

Therefore, the controllability matrix C_c is given by

$$C_c = \begin{bmatrix} \cos(\psi)\cos(\theta) & 0 & 0 & 0 & \cos(\psi)\sin(\theta)\cos(\phi) + \sin(\psi)\sin(\phi) & -\cos(\psi)\sin(\theta)\sin(\phi) + \sin(\psi)\cos(\phi) \\ \sin(\psi)\cos(\theta) & 0 & 0 & 0 & \sin(\psi)\sin(\theta)\cos(\phi) - \cos(\psi)\sin(\phi) & -\sin(\psi)\sin(\theta)\sin(\phi) - \cos(\psi)\cos(\phi) \\ -\sin(\theta) & 0 & 0 & 0 & \cos(\theta)\cos(\phi) & -\cos(\theta)\sin(\phi) \\ 0 & 1 & \sin(\phi)\tan(\theta) & \cos(\phi)\tan(\theta) & 0 & 0 \\ 0 & 0 & \cos(\phi) & -\sin(\phi) & 0 & 0 \\ 0 & 0 & \sin(\phi)\sec(\theta) & \cos(\phi)\sec(\theta) & 0 & 0 \end{bmatrix}. \tag{4.14}$$

It can be seen that the above matrix C_c has one nonzero minor of order 6. Therefore, this matrix is full rank provided that $\theta \neq \frac{\pi}{2}$. This implies that the vessel is locally controllable and also globally controllable as long as the singular condition $\theta \neq \frac{\pi}{2}$ is avoided. As for the stabilizability of system (4.4) to a point, the failure of the previous linear analysis indicates that exponential stability cannot be achieved by smooth feedback [21]. Things turn out to be even worse: If smooth (in fact, even continuous) time-invariant feedback laws are used, Lyapunov stability cannot be used directly. This negative result is established on the basis of a necessary condition due to Brockett [21], see Section 2.7.4: Smooth stabilizability of a driftless regular system (i.e., such that the input vector fields are well defined and linearly independent at $\boldsymbol{\eta}_e$) requires that the number of inputs be equal to the number of states. The above difficulty has a deep impact on the control design. In fact, to obtain a posture stabilizing controller it is either necessary to give up the continuity requirement and/or to resort to time-varying control laws.

4.1.4 Controllability About a Trajectory

For the system (4.4), let the reference trajectory $\boldsymbol{\eta}_d$ and the reference trajectory input $\boldsymbol{u}_d$ be

$$\boldsymbol{\eta}_d = \begin{bmatrix} x_d(t) \\ y_d(t) \\ z_d(t) \\ \phi_d(t) \\ \theta_d(t) \\ \psi_d(t) \end{bmatrix}, \quad \boldsymbol{u}_d = \begin{bmatrix} u_d(t) \\ p_d(t) \\ q_d(t) \\ r_d(t) \end{bmatrix}. \tag{4.15}$$

Indeed, the reference trajectory $\boldsymbol{\eta}_d$ and the reference trajectory input $\boldsymbol{u}_d$ should satisfy the nonhonolomic constraints (4.7), i.e.,

$$\begin{aligned} &\big(\cos(\psi_d)\sin(\theta_d)\sin(\phi_d)-\sin(\psi_d)\cos(\phi_d)\big)\dot{x}_d+ \\ &\qquad \big(\sin(\psi_d)\sin(\theta_d)\sin(\phi_d)+\cos(\psi_d)\cos(\phi_d)\big)\dot{y}_d+\cos(\theta_d)\sin(\phi_d)\dot{z}_d=0, \\ &\big(\sin(\psi_d)\sin(\theta_d)\cos(\phi_d)-\sin(\psi_d)\sin(\phi_d)\big)\dot{x}_d+ \\ &\qquad \big(\sin(\psi_d)\sin(\theta_d)\cos(\phi_d)-\cos(\psi_d)\sin(\phi_d)\big)\dot{y}_d+\cos(\theta_d)\cos(\phi_d)\dot{z}_d=0. \end{aligned} \tag{4.16}$$

Let the errors associated with the reference trajectory and the reference input trajectory be

$$\begin{aligned} \boldsymbol{\eta}_e &= \boldsymbol{\eta}-\boldsymbol{\eta}_d, \\ \boldsymbol{u}_e &= \boldsymbol{u}-\boldsymbol{u}_d. \end{aligned} \tag{4.17}$$

Using (4.17), we can write (4.4) as

$$\dot{\boldsymbol{\eta}} = \boldsymbol{\Upsilon}(\boldsymbol{\eta}_d+\boldsymbol{\eta}_e)\big(\boldsymbol{u}_d+\boldsymbol{u}_e\big). \tag{4.18}$$

The Taylor series expansion of $\boldsymbol{\Upsilon}(\boldsymbol{\eta}_d+\boldsymbol{\eta}_e)$ about the nominal solution $\boldsymbol{\eta}_d$ is given by

$$\dot{\boldsymbol{\eta}} = \left(\boldsymbol{\Upsilon}(\boldsymbol{\eta}_d, t) + \left. \frac{\partial \boldsymbol{\Upsilon}(\boldsymbol{\eta})}{\partial \boldsymbol{\eta}} \right|_{\boldsymbol{\eta}=\boldsymbol{\eta}_d} \boldsymbol{\eta}_e(t) + \mathrm{HOT} \right) \left(\boldsymbol{u}_d(t) + \boldsymbol{u}_e(t) \right). \tag{4.19}$$

Since the reference trajectory $\boldsymbol{\eta}$ and the reference input trajectory $\boldsymbol{u}_d$ satisfy the nonholonomic constraints (4.16), we have

$$\dot{\boldsymbol{\eta}}_d = \boldsymbol{\Upsilon}(\boldsymbol{\eta}_d, t) \boldsymbol{u}_d(t). \tag{4.20}$$

Subtracting (4.19) by (4.20) and ignoring the high-order terms (HOT) gives

$$\begin{aligned} \dot{\boldsymbol{\eta}}_e &= \left(\left. \frac{\partial \boldsymbol{\Upsilon}(\boldsymbol{\eta})}{\partial \boldsymbol{\eta}} \right|_{\boldsymbol{\eta}=\boldsymbol{\eta}_d} \boldsymbol{\eta}_e(t) \right) \boldsymbol{u}_d(t) + \boldsymbol{\Upsilon}(\boldsymbol{\eta}_d, t) \boldsymbol{u}_e(t) \\ &:= \boldsymbol{A}(t) \boldsymbol{\eta}_e(t) + \boldsymbol{B}(t) \boldsymbol{u}_e(t), \end{aligned} \tag{4.21}$$

where

$$\boldsymbol{A}(t) = \begin{bmatrix} 0_{3\times3} & \boldsymbol{A}_1(t) \\ 0_{3\times3} & \boldsymbol{A}_2(t) \end{bmatrix}, \quad \boldsymbol{B}(t) = \begin{bmatrix} \boldsymbol{J}_{d1}(t) & 0_{3\times3} \\ 0_{3\times1} & \boldsymbol{J}_{d2}(t) \end{bmatrix}, \tag{4.22}$$

with $\boldsymbol{A}_1$ and $\boldsymbol{A}_2$ given by

$$\boldsymbol{A}_1(t) = \begin{bmatrix} 0 & -\cos(\psi_d)\sin(\theta_d)u_d & -\sin(\psi_d)\cos(\theta_d)u_d \\ 0 & -\sin(\psi_d)\sin(\theta_d)u_d & \cos(\psi_d)\cos(\theta_d)u_d \\ 0 & -\cos(\theta_d)u_d & 0 \end{bmatrix},$$

$$\boldsymbol{A}_2(t) = \left[\begin{matrix} \cos(\phi_d)\tan(\phi_d)q_d - \sin(\phi_d)\tan(\phi_d)r_d \\ -\sin(\phi_d)q_d - \cos(\phi_d)r_d \\ \cos(\phi_d)\sec(\theta_d)q_d - \sin(\phi_d)\sec(\theta_d)r_d \end{matrix}\right. \tag{4.23}$$
$$\left.\begin{matrix} \sin(\phi_d)\sec^2(\theta_d)q_d + \cos(\phi_d)\sec^2(\theta_d)r_d & 0 \\ 0 & 0 \\ \sin(\phi_d)\sec(\theta_d)\tan(\theta_d)q_d + \cos(\phi_d)\sec(\theta_d)\tan(\theta_d)r_d & 0 \end{matrix}\right],$$

and $\boldsymbol{J}_{d1}(t)$ and $\boldsymbol{J}_{d2}(t)$ given by

$$\begin{aligned} \boldsymbol{J}_{d1}(t) &= \begin{bmatrix} \cos(\theta_d)\cos(\psi_d) \\ \cos(\theta_d)\sin(\psi_d) \\ -\sin(\theta_d) \end{bmatrix}, \\ \boldsymbol{J}_{d2}(t) &= \begin{bmatrix} 1 & \sin(\phi_d)\tan(\theta_d) & \cos(\phi_d)\tan(\theta_d) \\ 0 & \cos(\phi_d) & -\sin(\phi_d) \\ 0 & \sin(\phi_d)\sec(\theta_d) & \cos(\phi_d)\sec(\theta_d) \end{bmatrix}. \end{aligned} \tag{4.24}$$

In (4.23) and (4.24), the argument t of ϕ_d, θ_d, ψ_d, u_d, p_d, and r_d is omitted for simplicity.

The system (4.21) is linear time-varying. The controllability condition becomes

$$\text{rank}\{\boldsymbol{B}, \boldsymbol{AB}, \boldsymbol{A}^2\boldsymbol{B}, \boldsymbol{A}^3\boldsymbol{B}, \boldsymbol{A}^4\boldsymbol{B}, \boldsymbol{A}^5\boldsymbol{B}\} = 6. \tag{4.25}$$

A calculation shows that the above matrix has a nonzero minor of order 6 provided that ($u_d \neq 0$, $p_d \neq 0$, $q_d \neq 0$, $r_d \neq 0$) and ($\theta_d \neq \frac{\pi}{2}$). Therefore, we conclude that the kinematic system (4.21) can be locally stabilized by linear feedback about trajectories consisting of linear or circular or helix paths, which do not collapse to a point.

4.2 Previous Work on Control of Underactuated Ocean Vessels

This section starts with a brief review on the control of nonholonomic systems, due to their relevance to the control of underactuated ocean vessels. Next, the existing methods on control of underactuated ocean vessels are reviewed. Limitations of the existing methods are then pointed out and hence motivate the contributions of the book.

4.2.1 Control of Nonholonomic Systems

The term "nonholonomic system" originates from classical mechanics and has its widely accepted meaning as a "Lagrange system with linear constraints being nonintegrable". A mechanical system is said to be nonholonomic if its generalized velocity satisfies an equality condition that cannot be written as an equivalent condition on the generalized position, see [30]. Control of nonholonomic dynamic systems has formed an active area in the control community – see surveys by Kolmanovsky and McClamroch in [31], Canudas de Wit et al. in [15], Murray and Sastry in [32], and references therein for an overview and interesting introductory examples in this expanding area.

Nonholonomic systems have inherent difficulties in feedback stabilization at the origin or at a given equilibrium point since the tangent linearization of these systems is uncontrollable. In fact, a direct application of Brockett's necessary condition, see Section 2.7.4 for more details, for feedback stabilization implies that nonholonomic systems cannot be stabilized by any stationary continuous state feedback although they are open loop controllable. As a consequence, the classical smooth control theory cannot be applied. This motivates researchers to seek novel approaches. These approaches can be roughly classified into discontinuous feedback, see for example [33–45] and time-varying feedback, see for example, [15, 32, 46–48]. The discontinuous feedback approach often uses the state scaling originated from the σ-process [49] and a switching control strategy to overcome the difficulty due to the loss of controllability. This approach results in a fast transient response and usually an exponential convergence can be achieved. The drawback is discontinuity in the control input. On the other hand, the time-varying feedback approach provides a smooth/continuous controller, i.e., no switching is required, however the price is slow convergence. The stability analysis is often based on linear time-varying system theory and Barbalat's lemma. The backstepping technique [3] is usually used for high-order chained form systems in both discontinuous and time-varying approaches. Those aforementioned systems are either driftless or have weak nonlinear drifts. When nonholonomic systems are perturbed by drifts with uncertainties, robust and adaptive control approaches are often applied. The robust control design schemes are based on the size domination concept [50]. The control is conservative when a priori knowledge of uncertainties is poor. A class of nonholonomic systems with strong nonlinear uncertainties was recently considered in [51]. Discontinuous state feedback and output feedback controllers were designed to achieve global exponential stability. However the x_0-subsystem is required to be Lipschitz since a constant control input u_0 is used to get around the difficulty due to the loss of controllability. The adaptive approach [38, 40, 46] provides less conservative control input but increases the dynamics of the closed loop system. The systems studied in these papers do not allow drifts in the x_0-subsystem. A difficulty in designing adaptive stabilization controllers for chained systems with drifts is that the state of the x_0-subsystem can have several zero crossings due to transient behavior of the unknown parameter estimate. This phenomenon causes difficulties in applying the state scaling. For a solution of the stabilization of nonholonomic systems in a chained form with strong nonlinear drifts and unknown parameters, the reader is referred to [52].

4.2.2 Control of Underactuated Ships and Underwater Vehicles

Control of underactuated ships and autonomous underwater vehicles (AUVs) is an active field due to its important applications such as passenger and goods transportation, environmental surveying, undersea cable inspection, and offshore oil installations.

Based on its practical requirement, motion control of underactuated ocean vessels has been divided into three areas: Stabilization, trajectory-tracking, and path-following. These control problems are challenging due to the fact that the motion of underactuated surface ships and AUVs possesses more degrees of freedom to be controlled than the number of the independent controls under some nonintegrable second-order nonholonomic constraints [29, 53, 54]. In particular, underactuated ships do not usually have an actuator in the sway axis while in the case of AUVs there are no actuators in the sway and heave directions. This configuration is by far the most common among marine vessels. Therefore, Brockett's condition indicates that any continuous time-invariant feedback control law does not make a null solution of the underactuated surface ship and AUV dynamics asymptotically stable in the sense of Lyapunov. Furthermore as observed in [22, 54], the underactuated ship and AUV system is not transformable into a standard chain system. Consequently, existing control schemes [15, 32–48] developed for chained systems cannot be applied directly. Nevertheless, in the past decade, stabilization, trajectory-tracking control, and path-following of underactuated ocean vessels have been studied separately from different viewpoints.

4.2.2.1 Stabilization

An underactuated ocean vessel belongs to a class of underactuated mechanical systems subject to some nonintegrable second-order nonholonomic constraints, see [29,53,54]. Therefore, design of a feedback stabilizer using linear and classical nonlinear control theories is not possible. There are two main approaches to deal with stabilization of an underactuated ocean vessel. They are (time-invariant and time-varying) discontinuous feedback and time-varying continuous/smooth feedback. We here mention some typical results of both approaches.

Time-invariant and Time-varying Discontinuous Approach

A discontinuous state feedback control law was proposed in [55] using the σ-process to exponentially stabilize an underactuated ship at the origin where the ship model is discontinuously transformed to an extended chained form system. The dynamics of an underactuated ship is considered in [55], see also Section 3.4.1:

$$\begin{aligned}
&\dot{\boldsymbol{\eta}} = \boldsymbol{J}(\boldsymbol{\eta})\boldsymbol{v}, \\
&\boldsymbol{M}\dot{\boldsymbol{v}} = -\boldsymbol{C}(\boldsymbol{v})\boldsymbol{v} - \boldsymbol{D}\boldsymbol{v} + \boldsymbol{\tau}, \\
&\boldsymbol{\eta} = [x,\ y,\ \psi]^T,\ \boldsymbol{v} = [u,\ v,\ r]^T,\ \boldsymbol{\tau} = [\tau_u,\ 0,\ \tau_r]^T,
\end{aligned} \tag{4.26}$$

where (x, y) denotes the earth-fixed position of the center of mass of the ship, ψ denotes the orientation angle, (u, v) and r are the linear and angular velocities in the body-fixed frame, and (τ_u, τ_r) are the surge force and yaw moment. The matrices $\boldsymbol{J}(\boldsymbol{\eta})$, $\boldsymbol{M}$, $\boldsymbol{C}(\boldsymbol{v})$, and $\boldsymbol{D}$ are given by

$$\boldsymbol{J}(\boldsymbol{\eta}) = \begin{bmatrix} \cos(\psi) & -\sin(\psi) & 0 \\ \sin(\psi) & \cos(\psi) & 0 \\ 0 & 0 & 1 \end{bmatrix},\ \boldsymbol{M} = \begin{bmatrix} m_{11} & 0 & 0 \\ 0 & m_{22} & 0 \\ 0 & 0 & m_{33} \end{bmatrix},$$

$$\boldsymbol{C}(\boldsymbol{v}) = \begin{bmatrix} 0 & 0 & -m_{22}v \\ 0 & 0 & m_{11}u \\ m_{22}v & -m_{11}u & 0 \end{bmatrix},\ \boldsymbol{D} = \begin{bmatrix} d_{11} & 0 & 0 \\ 0 & d_{22} & 0 \\ 0 & 0 & d_{33} \end{bmatrix}, \tag{4.27}$$

where m_{11}, m_{22}, and m_{33} denote the ship inertia including added mass, and d_{11}, d_{22}, and d_{33} are hydrodynamic damping constants, see Chapter 3 for more details. The control objective is to design the control inputs τ_u and τ_r to stabilize (4.26) asymptotically at the origin. In [55], the coordinate transformation

$$
\begin{bmatrix} x_1 \\ x_2 \\ x_3 \\ x_4 \\ x_5 \\ x_6 \end{bmatrix} = \begin{bmatrix} \psi \\ x\cos(\psi) + y\sin(\psi) \\ -x\sin(\psi) + y\cos(\psi) \\ v \\ r \\ u \end{bmatrix} \tag{4.28}
$$

is used to transform the ship model (4.26) to the following system

$$
\begin{aligned}
\dot{x}_1 &= x_5, \\
\dot{x}_2 &= x_6 + x_3 x_5, \\
\dot{x}_3 &= x_4 - x_2 x_5, \\
\dot{x}_4 &= -\alpha x_4 - \beta x_5 x_6, \\
\dot{x}_5 &= \Omega_1, \\
\dot{x}_6 &= \Omega_2,
\end{aligned} \tag{4.29}
$$

where $\alpha = d_{22}/m_{22}$, $\beta = m_{11}/m_{22}$, and

$$
\Omega_1 = \frac{\tau_r - d_{33} r + (m_{11} - m_{22}) u v}{m_{33}}, \; \Omega_2 = \frac{\tau_u + m_{22} v r - d_{11} u}{m_{11}}. \tag{4.30}
$$

It can be seen that the system (4.29) consists of two subsystems, namely (x_1, x_2, x_3, x_4) and (x_5, x_6), connected to each other in a strict feedback form [3]. The control design can be simply carried out in two steps as follows.

Step 1

In this step, the author of [55] considers the first four equations of (4.29), and (x_5, x_6) as controls (v_1, v_2). With the assumption of $x_1 \neq 0$, the coordinate transformation (σ-process)

$$
y = x_1, \; z_1 = x_2, \; z_2 = \frac{x_3}{x_1}, \; z_3 = \frac{x_4}{x_1} \tag{4.31}
$$

results in

$$
\begin{aligned}
\dot{y} &= v_1, \\
\dot{z}_1 &= v_2 + y z_2 v_1, \\
\dot{z}_2 &= z_3 - \frac{z_1 + z_2}{y} v_1, \\
\dot{z}_3 &= -\alpha z_3 - \frac{z_3 + \beta v_2}{y} v_1.
\end{aligned} \tag{4.32}
$$

The feedback control law is designed as

$$\begin{aligned} v_1 &= -k_1 y, \\ v_2 &= -k_{21} z_1 - k_{22} z_2 - k_{23} z_3, \end{aligned} \tag{4.33}$$

where k_1, k_{21}, k_{22}, and k_{23} are the control gains chosen such that the matrix

$$A_1 = \begin{bmatrix} -k_{21} & -k_{22} & -k_{23} \\ k_1 & k_1 & 1 \\ -k_1\beta k_{21} & -k_1\beta k_{22} & k_1 - \alpha - k_1\beta k_{23} \end{bmatrix} \tag{4.34}$$

is Hurwitz.

Step 2

At this step, the last two equations of (4.29) are considered. Using the standard backstepping technique results in the following control law

$$\begin{aligned} \Omega_1 &= -k_3(x_5 - v_1) - k_1 x_5, \\ \Omega_2 &= -k_4(x_6 - v_2) - k_{21}(x_6 + x_3 x_5) \\ &\quad -k_{22}\frac{x_4 - x_2 x_5}{x_1} + k_{23}\frac{\alpha x_4 + \beta x_5 x_6}{x_1} + k_{22}\frac{x_3 x_5}{x_1^2} + k_{23}\frac{x_4 x_5}{x_1^2}, \end{aligned} \tag{4.35}$$

where $k_3 > k_1$ and k_4 are positive constants. The actual controls τ_u and τ_r can be found from (4.35) and (4.30). In [55] it is shown that if the initial conditions $x_1(t_0) \neq 0$ and $x_1(t_0)(x_5(t_0) + k_1 x_1(t_0)) \geq 0$ then $(x_1(t), x_2(t), x_3(t), x_4(t), x_5(t), x_6(t))$ is bounded for all $t \geq t_0 \geq 0$, and exponentially converges to zero. If the above conditions do not hold, the controls

$$\begin{aligned} \Omega_1 &= -|x_1 - \epsilon|^a \mathrm{sign}(x_1 - \epsilon) - |x_5|^b \mathrm{sign}(x_5), \\ \Omega_2 &= 0, \end{aligned} \tag{4.36}$$

with $\epsilon \neq 0$, $b \in (0,1)$, and $a > b/(2-b)$ being constants, can be used to make the above conditions hold in finite time. For more details, the reader is referred to [55].

Remark 4.1. The aforementioned discontinuous stabilizer provides a fast convergence of the stabilizing errors to zero. However, the control inputs τ_u and τ_r are discontinuous. Moreover, under arbitrarily small nonvanishing environmental disturbances induced by waves, wind, and ocean currents, the closed loop system consisting of (4.35) and (4.29) can be unstable in the sense that the states $(x_1(t), x_2(t), x_3(t), x_4(t), x_5(t), x_6(t))$ can go to infinity exponentially fast.

The work mentioned in [22,53,54,56–58] can also be grouped in the discontinuous approach. The authors of [22] developed a discontinuous time-varying feedback stabilizer for a nonholonomic system and applied it to underactuated ships. Some local exponential stabilization results were reported in [53,54] based on the time-varying homogeneous control approach. An application of averaging and backstepping techniques was proposed in [59] to design a global practical controller for stabilization and tracking control of surface ships. Experimental results on dynamic positioning of underactuated ships were reported in [56]. By transforming the underactuated ship kinematics and dynamics into the so-called skew form, some dynamic feedback results on stabilization were given in [57]. In [58], the authors proposed a discontinuous solution to the problem of steering an underactuated AUV to a point with desired orientation using the polar coordinate transformation motivated from the work in [60].

Time-varying Continuous/Smooth Approach

A typical result on stabilization of the underactuated ship (4.26) in the time-varying continuous/smooth approach is given in [61]. In [61], the coordinate transformations (similar to the ones given in (4.30) and (4.28))

$$\begin{aligned}
z_1 &= \cos(\psi)x + \sin(\psi)y, \\
z_2 &= -\sin(\psi)x + \cos(\psi)y, \\
z_3 &= \psi, \\
\Omega_1 &= \frac{\tau_r - d_{33}r + (m_{11} - m_{22})uv}{m_{33}}, \\
\Omega_2 &= \frac{\tau_u + m_{22}vr - d_{11}u}{m_{11}}
\end{aligned} \tag{4.37}$$

are first used to transform the ship system (4.26) to

$$\begin{aligned}
\dot{z}_1 &= u + z_2 r, \\
\dot{z}_2 &= v - z_1 r, \\
\dot{z}_3 &= r, \\
\dot{u} &= \Omega_2, \\
\dot{v} &= -cur - dv, \\
\dot{r} &= \Omega_1,
\end{aligned} \tag{4.38}$$

where $c = m_{11}/m_{22}$ and $d = d_{22}/m_{22}$. Then the following nontrivial coordinate transformations

$$\begin{aligned}
Z_2 &= z_2 + \frac{v}{d}, \\
u &= -\frac{d}{c}z_1 - \frac{d}{c}\mu, \\
\Omega_{2\mu} &= \frac{d}{c}z_1 + \frac{d}{c}\mu - Z_2 r + \frac{v}{d}r - \frac{c}{d}\Omega_2
\end{aligned} \tag{4.39}$$

are applied to (4.38) to obtain the system

$$\begin{aligned}
\dot{z}_1 &= -\frac{d}{c}z_1 - \frac{d}{c}\mu + Z_2 r - \frac{v}{d}r,\\
\dot{Z}_2 &= \mu r,\\
\dot{z}_3 &= r,\\
\dot{v} &= -dv + d(z_1+\mu)r,\\
\dot{\mu} &= \Omega_{2\mu},\\
\dot{r} &= \Omega_1.
\end{aligned} \tag{4.40}$$

Let k_2, k_3, k_μ, and k_r be strictly positive constants such that $1 \geq k_2 \geq k_3$. The controls Ω_1 and $\Omega_{2\mu}$ are designed in [61] as

$$\begin{aligned}
\Omega_1 &= -k_r(r-r_f) + \dot{r}_f - \lambda\big(Z_2\mu_f + 2Z_3 + 2Z_2k_2\cos(t)\mu_f\big),\\
\Omega_{2\mu} &= -k_\mu(\mu-\mu_f) + \dot{\mu}_f - \lambda\big(Z_2 + 2Z_3k_2\cos(t)\big)r,
\end{aligned} \tag{4.41}$$

where

$$\begin{aligned}
\lambda &= 2 + \frac{k_3}{3} - \frac{k_3\sin(2t)}{6}\frac{2V_1+V_1^2}{(1+V_1)^2},\\
Z_3 &= z_3 + k_2\cos(t)Z_2,\\
V_1 &= Z_2^2 + 2Z_3^2,\\
\mu_f &= -\frac{\sin(t)Z_2^2}{2(0.001+Z_2^2)},\\
r_f &= \frac{-k_3Z_3 + k_2\sin(t)Z_2}{1+k_2\cos(t)\mu_f}.
\end{aligned} \tag{4.42}$$

In [61], it is proven that the closed loop system consisting of (4.41), (4.39), (4.37), and (4.26) is GAS at the origin.

Remark 4.2. The design of the feedback given in (4.41) is nontrivial. Overall, convergence of the stabilizing errors to zero is slow. This is a well-known phenomenon of the continuous/smooth time-varying approach applying not to only underactuated ships but also to mobile robots. Moreover, since the stabilizer design mentioned above is nontrivial, it is difficult to extend the control design scheme to solve a trajectory-tracking problem, see the next section. In addition, the physical meaning of the feedback is not clear.

In addition to the aforementioned results on stabilization of underactuated vessels, the following results are also related to the topic under discussion. In [62], several control configurations were considered, and a technique for synthesizing open loop controls was given. The first control scheme with the dynamic AUV model taken into account was proposed in [63]. A kinematic drift free model of the underwater vehicles with four control inputs was used to design a regulation controller in [64]. The authors of [65] proposed a controller that is able to stabilize an AUV to some equilibria based on the interconnection and damping assignment passivity-based control approach, which has been successfully applied to many other mechanical systems [66]. See also [67] for stabilization results of underactuated mechanical systems on Riemannian manifolds.

4.2.2.2 Trajectory-tracking

Trajectory-tracking is here defined as a control problem of forcing an underactuated surface ship or AUV to track a reference trajectory generated by a suitable vessel model, i.e., the vessel model that has the same parameters as the real one. There are two main approaches to solve the trajectory-tracking control problems. The first approach is based on linear time-varying control system theory while the second approach relies on the Lyapunov direct method. We here briefly describe typical results of the two approaches.

Linear Time-varying Approach

A typical work belonging to this approach is given in [68] on a global K-exponential tracking result for the underactuated ship (4.26). In [68], the authors consider a problem of designing the control τ_u and τ_r to force the position (x, y) and orientation ψ of the ship (4.26) to track the reference position (x_d, y_d) and orientation ψ_d generated by the reference ship model

$$\begin{aligned}
\dot{\boldsymbol{\eta}}_d &= \boldsymbol{J}(\boldsymbol{\eta}_d)\boldsymbol{v}_d,\\
\boldsymbol{M}\dot{\boldsymbol{v}}_d &= -\boldsymbol{C}(\boldsymbol{v}_d)\boldsymbol{v}_d - \boldsymbol{D}\boldsymbol{v}_d + \boldsymbol{\tau}_d,\\
\boldsymbol{\eta}_d &= \begin{bmatrix} x_d \\ y_d \\ \psi_d \end{bmatrix}, \ \boldsymbol{v}_d = \begin{bmatrix} u_d \\ v_d \\ r_d \end{bmatrix}, \ \boldsymbol{\tau}_d = \begin{bmatrix} \tau_{ud} \\ 0 \\ \tau_{rd} \end{bmatrix}.
\end{aligned} \tag{4.43}$$

In [68], the coordinate transformations

$$\begin{cases} z_1 = \cos(\psi)x + \sin(\psi)y, \\ z_2 = -\sin(\psi)x + \cos(\psi)y, \\ z_3 = \psi, \end{cases} \qquad \begin{cases} z_{1d} = \cos(\psi_d)x_d + \sin(\psi_d)y_d, \\ z_{2d} = -\sin(\psi_d)x_d + \cos(\psi_d)y_d, \\ z_{3d} = \psi_d, \end{cases} \tag{4.44}$$

and the tracking errors

$$\begin{aligned}
&u_e = u - u_d, \ v_e = v - v_d, \ r_e = r - r_d,\\
&z_{1e} = z_1 - z_{1d}, \ z_{2e} = z_2 - z_{2d}, \ z_{3e} = z_3 - z_{3d}
\end{aligned} \tag{4.45}$$

are used to obtain the tracking error dynamics of a chained form

$$\dot{u}_e = \frac{m_{22}}{m_{11}}\left(v_e r_e + v_e r_d + v_d r_e\right) - \frac{d_{11}}{m_{11}} u_e + \frac{1}{m_{11}}(\tau_u - \tau_{ud}),$$

$$
\begin{aligned}
\dot{v}_e &= -\frac{m_{11}}{m_{22}}\left(u_e r_e + u_e r_d + u_d r_e\right) - \frac{d_{22}}{m_{22}} v_e, \\
\dot{r}_e &= \frac{m_{11}-m_{22}}{m_{33}}\left(u_e v_e + u_e v_d + u_d v_e\right) - \frac{d_{33}}{m_{33}} r_e + \frac{1}{m_{33}}(\tau_r - \tau_{rd}), \\
\dot{z}_{1e} &= u_e + z_{2e} r_e + z_{2e} r_d + z_{2d} r_e, \\
\dot{z}_{2e} &= v_e - z_{1e} r_e - z_{1e} r_d - z_{1d} r_e, \\
\dot{z}_{3e} &= r_e.
\end{aligned}
\tag{4.46}
$$

Assuming that u_d, v_d, z_{1d}, and z_{2d} are bounded, and that $r_d(t)$ is persistently exciting, the controls

$$
\begin{aligned}
\tau_u &= \tau_{ud} - k_1 u_e + k_2 r_d v_e - k_3 z_{1e} + k_4 r_d z_{2e}, \\
\tau_r &= \tau_{rd} - (m_{11} - m_{22})(u_e v_e + v_d u_e + u_d v_e) - \\
&\quad k_5 r_e - k_6 z_{3e},
\end{aligned}
\tag{4.47}
$$

where the control gains k_i, $i = 1,\ldots,6$ satisfy

$$
\begin{aligned}
& k_1 > d_{22} - d_{11}, \\
& k_2 = \frac{m_{22} k_4 (k_4 + k_1 + d_{11} - d_{22})}{d_{22} k_4 + m_{11} k_3}, \\
& 0 < k_3 < (k_1 + d_{11} - d_{22}) \frac{d_{22}}{m_{11}}, \\
& k_4 > 0, \\
& k_5 > -d_{33}, \\
& k_6 > 0,
\end{aligned}
\tag{4.48}
$$

make the closed loop system consisting of (4.47), (4.43), and (4.26), that is,

$$
\begin{bmatrix} \dot{u}_e \\ \dot{v}_e \\ \dot{z}_{1e} \\ \dot{z}_{2e} \end{bmatrix}
=
\begin{bmatrix}
-\frac{k_1 + d_{11}}{m_{11}} & \frac{k_2 + m_{22}}{m_{11}} r_d(t) & -\frac{k_3}{m_{11}} & \frac{k_4}{m_{11}} r_d(t) \\
-\frac{m_{11}}{m_{22}} r_d(t) & -\frac{d_{22}}{m_{22}} & 0 & 0 \\
1 & 0 & 0 & r_d(t) \\
0 & 1 & -r_d(t) & 0
\end{bmatrix}
\begin{bmatrix} u_e \\ v_e \\ z_{1e} \\ z_{2e} \end{bmatrix}
+
\begin{bmatrix}
\frac{m_{22}}{m_{11}}(v_e + v_d) & 0 \\
-\frac{m_{11}}{m_{22}}(u_e + u_d) & 0 \\
z_{2e} + z_{2d} & 0 \\
-(z_{1e} + z_{1d}) & 0
\end{bmatrix}
\begin{bmatrix} r_e \\ z_{3e} \end{bmatrix},
$$

$$\begin{bmatrix} \dot{r}_e \\ \dot{z}_{3e} \end{bmatrix} = \begin{bmatrix} -\dfrac{d_{33}+k_5}{m_{33}} & -\dfrac{k_6}{m_{33}} \\ 1 & 0 \end{bmatrix} \begin{bmatrix} r_e \\ z_{3e} \end{bmatrix}, \tag{4.49}$$

globally K-exponentially stable at the origin. Proof of K-exponential stability of the above closed loop system is straightforward using the results on stability of cascade systems in [17] and [69], and those of linear time-varying system theory in [6], see [68] for details. It should be noted that the persistently exciting condition on the yaw reference velocity, r_d, is required to prove K-exponential stability of the closed loop system (4.49).

Remark 4.3. The persistent exciting condition on the yaw reference velocity r_d excludes a straight-line reference trajectory. In comparison with the direct Lyapunov approach summarized below, it is difficult to deal with any external disturbances and/or actuator dynamics using the linear time-varying approach.

Direct Lyapunov Approach

The direct Lyapunov method has been widely used in designing controllers for underactuated ocean vessels. However, the use of the Lyapunov direct method for designing control systems for underactuated ocean vessels is not straightforward due to the underactuated nature of ocean vessels. We here describe typical results of trajectory-tracking control based on the Lyapunov direct method.

Local Trajectory-tracking Results

An application of the recursive technique proposed in [70] for the standard chain-form systems yields a high-gain based local tracking result in [59] for surface ships. The experimental results of this proposed controller were reported in [71]. The control design in [59,71] starts from (4.46) as follows. First of all, the following restrictive assumption is made on the yaw reference velocity:

$$0 < r_{d\,\min} < |r_d(t)| < r_{d\,\max}, \tag{4.50}$$

where $r_{d\,\min}$ and $r_{d\,\max}$ are strictly positive constants. Motivated by the work in [70], the authors define new error variables as

$$\begin{aligned} \omega_1 &= z_{1e} - z_2 z_{3e}, \\ \omega_2 &= z_{2e} + z_1 z_{3e}, \\ y_1 &= v_e + c u z_{3e} + k_2 \omega_2, \\ y_2 &= u_e + k_1 \omega_1 - \frac{k_2(d-k_2)}{c r_d}\omega_2, \\ y_3 &= z_{3e}, \end{aligned} \tag{4.51}$$

where $c = m_{11}/m_{22}$, $d = d_{22}/m_{22}$, and k_2 is a parameter to be determined later. It should be stressed that the condition (4.50) on the yaw reference velocity, r_d is required so that the error transformations (4.51) are valid. With (4.51), the tracking error system (4.46) is written in a triangular-like structure as follows:

$$\begin{aligned}
\dot{\omega}_1 &= y_2 - k_1\omega_1 + \frac{k_2(d-k_2)}{cr_d}\omega_2 + \omega_2 r_d - (v - z_1 r_e)y_3,\\
\dot{\omega}_2 &= y_1 - k_2\omega_2 - \omega_1 r_d + ((1-c)u + z_2 r_e)y_3,\\
\dot{y}_1 &= -cy_2 r_d + (ck_1 - k_2)r_d\omega_1 - (d-k_2)y_1 + (c\Omega_2 + cdu +\\
&\quad k_2((1-c)u + z_2 r_e))y_3,\\
\dot{y}_2 &= \Omega_2 - \Omega_{2d} + k_1 y_2 - k_1^2\omega_1 + \frac{1}{cr_d}k_1k_2(d-k_2)\omega_2 + k_1 r_d\omega_2 +\\
&\quad \frac{\dot{r}_d}{cr_d^2}k_2(d-k_2)\omega_2 - \frac{1}{cr_d}k_2(d-k_2)(y_1 - k_2\omega_2 - r_d\omega_1) -\\
&\quad (k_1(v - z_1 r_e) + \frac{1}{cr_d}k_2(d-k_2)((1-c)u + z_2 r_e))y_3,\\
\dot{y}_3 &= r_e,\\
\dot{r}_e &= \Omega_1 - \Omega_{1d},
\end{aligned} \tag{4.52}$$

where Ω_1 and Ω_2 are given in (4.30), and

$$\begin{aligned}
\Omega_{1d} &= \frac{\tau_{rd} - d_{33}r_d + (m_{11} - m_{22})u_d v_d}{m_{33}},\\
\Omega_{2d} &= \frac{\tau_{ud} + m_{22}v_d r_d - d_{11}u_d}{m_{11}}.
\end{aligned} \tag{4.53}$$

The triangular structure (4.52) allows us to use the backstepping technique [3] to design the controls Ω_1 and Ω_2. In [71], the the controls Ω_1 and Ω_2 are designed as

$$\begin{aligned}
\Omega_1 &= -a_3(r - \alpha_r) + \dot{\alpha}_r - \kappa y_3,\\
\Omega_2 &= -a_1 y_2 - \gamma\omega_1 + car_d y_1 - \Big(-\Omega_{2d} + k_1 y_2 - k_1^2\omega_1 +\\
&\quad \frac{1}{cr_d}k_1k_2(d-k_2)\omega_2 + k_1 r_d\omega_2 + \frac{\dot{r}_d}{cr_d^2}k_2(d-k_2)\omega_2 - \frac{1}{cr_d}\times\\
&\quad k_2(d-k_2)(y_1 - k_2\omega_2 - r_d\omega_1)\Big),
\end{aligned} \tag{4.54}$$

where

$$\begin{aligned}
\alpha_r = &\Big(\lambda + \gamma(\omega_1 z_1 + \omega_2 z_2) + ak_2 y_1 z_2 + k_1 y_2 z_1 - \frac{1}{cr_d}k_2(d-k_2)y_2 z_2\Big)^{-1} \times\\
&\Big(-a_2 y_3 + \gamma\omega_1 v - \gamma\omega_2(1-c)u - ay_1(c(\Omega_2 + du) + k_2(1-c)u) +
\end{aligned}$$

$$k_1 y_2 v + \frac{1}{cr_d} k_2(d-k_2)(1-c)uy_2 \Big) + r_d,$$

$$\kappa = \lambda + \gamma\omega_1 z_1 + \gamma\omega_2 z_2 + ak_2 y_1 z_2 + k_1 y_2 z_1 - \frac{1}{cr_d} k_2(d-k_2) y_2 z_2. \quad (4.55)$$

In (4.54) and (4.55), the control parameters $k_1, k_2, a, a_1, a_2, a_3$, and λ are positive constants, and are chosen such that

$$k_2 < d,\ k_1 > \frac{k_2(d-k_2)^2}{c^2 r_{d\min}^2},$$
$$\frac{1}{k_2(d-k_2)} < \frac{a}{\gamma} < \frac{k_1(d-k_2)}{(ck_1-k_2)^2 r_{d\max}^2}. \quad (4.56)$$

It is noted that the virtual control α_r given in (4.55) is solvable if and only if

$$\lambda > -\Big(\gamma\omega_1 z_1 + \gamma\omega_2 z_2 + ak_2 y_1 z_2 + k_1 y_2 z_1 - \frac{1}{cr_d} k_2(d-k_2) y_2 z_2\Big). \quad (4.57)$$

Proof of local exponential stability of the closed loop system consisting of (4.54), (4.46), and (4.30) can be carried out by using the Lyapunov function

$$V = \frac{1}{2}\gamma\omega_1^2 + \frac{1}{2}\gamma\omega_2^2 + \frac{1}{2}ay_1^2 + \frac{1}{2}y_2^2 + \frac{\lambda}{2}y_3^2 + \frac{1}{2}(r-\alpha_r)^2. \quad (4.58)$$

Remark 4.4. There are two limitations of the aforementioned tracking controllers. These limitations are described in conditions (4.50) and (4.57). The condition (4.50) implies that the reference yaw velocity r_d cannot be zero at any time. This restrictive condition excludes a straight-line reference trajectory. The condition (4.57) implies that the aforementioned trajectory-tracking result is inherently local. One can argue that by the control parameters k_1, k_2, a, a_1, a_2, a_3, and λ may increase the size of the attraction region. However, it is very hard to ensure this property since the control parameters must satisfy various conditions specified in (4.56). In fact, this is true, as said in [71].

Global Trajectory-tracking Results

Based on Lyapunov's direct method and the passivity approach [72], two restricted tracking solutions of an underactuated surface ship were proposed in [19]. We here briefly summarize the result based on the passivity approach in [19]. The result based on the standard backstepping technique [3] is discussed later. In [19], the starting point is the tracking error system (4.46). The passivity based method consists of two steps as follows.

Step 1

Design of the surge force τ_u: This force is designed based on the Lyapunov function

$$V_1 = \frac{1}{2}\left(z_{1e} - \lambda_1 z_{2e} r_d\right)^2 + \frac{1}{2} z_{2e}^2 + \frac{\lambda_0}{2} v_e^2 + \frac{1}{2} \bar{u}_e^2, \tag{4.59}$$

where $\bar{u}_e = u_e - \alpha_0$ with

$$\alpha_0 = -\lambda_2 (z_{1e} - \lambda_1 z_{2e} r_d). \tag{4.60}$$

In (4.59) and (4.60), the control parameters λ_0, λ_1 and λ_2 are chosen such that

$$c(t) = \min\left(2\epsilon(\lambda_2 - \lambda_1 r_d^2(t)), 2\left(\lambda_1 r_d^2(t) - \frac{m_{22}}{(1-\epsilon)\lambda_0 d_{22}}\right), 2\epsilon \frac{d_{22}}{m_{22}}, 2c_1\right) \geq c^*, \tag{4.61}$$

where c_1 is a positive constant, $0 < \epsilon < 1$, and c^* is strictly positive. In (4.59) and (4.60), α_0 is understood as a virtual control of u_e. From the first time derivative of the Lyapunov function V_1 given in (4.59) along the solutions of (4.46), a choice of the surge force τ_u

$$\begin{aligned}
\tau_u = {} & \tau_{ud} + m_{11}\Bigg[-\frac{m_{22}}{m_{11}}(vr - v_d r_d) + \frac{d_{11}}{m_{11}} u_e - c_1(u_e + \lambda_2(z_{1e} - \lambda_1 z_{2e} r_d)) - \\
& \left((z_{1e} - \lambda_1 z_{2e} r_d) - \frac{\lambda_0 m_{11}}{m_{22}} r_d v_e\right) - \lambda_2(u_e + z_{2e} r_d + z_2 r_e) + \\
& \lambda_1 \lambda_2 \dot{r}_d z_{2e} + \lambda_1 \lambda_2 r_d (v_e - z_{1e} r_d - z_1 r_e)\Bigg]
\end{aligned} \tag{4.62}$$

gives

$$\dot{V}_1 \leq -c(t) V_1 + \left[(z_{1e} - \lambda_1 z_{2e} r_d)(z_2 + \lambda_1 r_d z_1) - z_{2e} z_1 - \frac{\lambda_0 m_{11}}{m_{22}} v_e u\right] r_e, \tag{4.63}$$

where $c(t)$ is given in (4.61).

Step 2

Design of the yaw moment τ_r: This moment is designed based on the Lyapunov function

$$V_2 = V_1 + \frac{1}{2} z_{3e}^2 + \frac{1}{2} \bar{r}_e^2, \tag{4.64}$$

where $\bar{r}_e = r_e - \alpha_1$ and

$$\alpha_1 = -c_2\left[(z_{1e} - \lambda_1 z_{2e} r_d)(z_2 + \lambda_1 r_d z_1) - z_{2e} z_1 - \frac{\lambda_0 m_{11}}{m_{22}} v_e u + z_{3e}\right], \tag{4.65}$$

with $c_2 > 0$. From the first time derivative of the Lyapunov function V_2 given in (4.64) along the solutions of (4.46), a choice of the yaw moment

$$\tau_r = \tau_{rd} + m_{33}\Bigg[-\frac{m_{11}-m_{22}}{m_{33}}(uv - u_d v_d) + \frac{d_{33}}{m_{33}} r_e - c_3 \bar{r}_e + \dot{\alpha}_1 - \left((z_{1e} - \lambda_1 z_{2e} r_d)(z_2 + \lambda_1 r_d z_1) - z_{2e} z_1 - \frac{\lambda_0 m_{11}}{m_{22}} v_e u + z_{3e} \right) \Bigg] \tag{4.66}$$

where $c_3 > 0$ results in

$$\dot{V}_2 \leq -c(t)V_1 - c_2\Big((z_{1e} - \lambda_1 z_{2e} r_d)(z_2 + \lambda_1 r_d z_1) - z_{2e} z_1 - \frac{\lambda_0 m_{11}}{m_{22}} v_e u + z_{3e} \Big)^2 - c_3 \bar{r}_e^2. \tag{4.67}$$

This implies global asymptotic stability of the closed loop system at the origin as long as the control parameters λ_0, λ_1, and λ_2 are chosen such that (4.61) holds. Note that this condition is feasible only when the reference yaw velocity r_d satisfies the following restrictive condition

$$0 < r_\star \leq |r_d(t)| \leq r^\star, \tag{4.68}$$

where $r_\star$ and $r^\star$ are positive constants. In [19], the result based on the standard backstepping technique also consists of two steps. The first step is to design the surge force τ_u. This step is the same as Step 1 mentioned above. The second step is to design the yaw moment τ_r. This step is slightly different from Step 2. In this step, a simple controller to stabilize the (z_{3e}, r_e)-subsystem, that is the third and last equations of (4.46), is designed as

$$\tau_r = \tau_{rd} + m_{33}\left[-\frac{m_{11}-m_{22}}{m_{33}}(uv - u_d v_d) + \frac{d_{33}}{m_{33}} r_e - k_1 z_{3e} - k_2 r_e \right]. \tag{4.69}$$

With the surge force τ_u and the yaw moment τ_r designed as in (4.62) and (4.70), it is proven in [19] that the tracking errors $(z_{1e}, z_{2e}, z_{3e}, u_e, v_e, r_e)$ exponentially converge (*not exponential stability of the closed loop system*) to zero as long as the following restrictive condition on the reference yaw velocity r_d holds

$$\int_{t_0}^{t} r_d^2(\tau)\mathrm{d}\tau \geq \sigma_r(t - t_0), \quad \forall 0 \leq t_0 \leq t < \infty, \tag{4.70}$$

where σ_r is a strictly positive constant.

Remark 4.5. In comparison with the trajectory-tracking results in [71], we see that the control design in [19] is much simpler and gives global solutions. However, the restrictive conditions on the yaw reference velocity cannot be relaxed, see (4.68) and (4.70). Moreover, it is only possible to find the control parameters such that $c(t)$ given in (4.61) is strictly positive for vessels with a small ratio $\frac{m_{22}}{d_{22}}$, i.e., the vessels with large damping in the sway axis.

Remark 4.6. A common restriction of the above results on trajectory-tracking control of underactuated ships is that the reference yaw velocity has to satisfy various kinds of persistently exciting conditions. This implies that the reference trajectory must be curved, and indeed excludes a straight-line reference trajectory, hence, it substantially limits the practical use of the aforementioned control systems. A curious question is why all the above controllers suffer from the *must-be-curved* reference trajectory restriction. An answer is that the design of the above controllers starts from the chained form (4.46). The reader will find that this book provides various solutions for trajectory-tracking control of underactuated ships without imposing a persistent exciting condition on the yaw reference velocity. As such, we will not use the chained form (4.46) but will project the tracking errors, $x - x_d$, $y - y_d$, and $\psi - \psi_d$, on the body-fixed frame.

Apart from the aforementioned results on trajectory-tracking control of underactuated ships there are a few more results that are worth reviewing. Using sliding mode control, output redefinition and results on tracking of nonlinear nonminimum phase system [73], a path controller for surface ships was proposed in [74]. However, the convergence of the combined output does not guarantee convergence of its components. A continuous time-invariant state feedback controller was developed in [75] to achieve global exponential position tracking under the assumption that the reference surge velocity is always positive. Unfortunately, the orientation of the ship was not controlled. In [76, 77], (see also [78]), the authors developed a high-gain dynamic feedback control law to achieve global ultimate regulation and tracking of underactuated ships. The dynamics of the closed loop system is increased due to the controller designed to make the state of the transformed system track the auxiliary signals generated by some oscillator. The same approach was extended to the case of adaptive tracking control in [77]. It is worth mentioning that in [47], a time-varying velocity feedback controller was proposed to achieve both stabilization and tracking of unicycle mobile robots at the kinematics level motivated by the work in [18]. However this controller cannot be extended directly to the case of underactuated ships or AUVs due to the nonintegrable second-order constraint. Some related independent work includes [79, 80] on local H_∞ tracking control and output redefinition, and the trajectory planning approach, see [81, 82].

4.2.2.3 Path-following

Path-following is here defined as a control problem of forcing an underactuated ship or AUV to follow a specified path at a desired forward speed. Due to the high dependence on the reference model and complicated control laws of the trajectory-tracking approach, several researchers have studied the path-following problem, which is more suitable for practical implementation. The problem of path-following for air and underwater vehicles was introduced in [83] where some local results were obtained using linearization techniques. In [84], a feedforward cancelation of simplified vessel dynamics scheme followed by a linear quadratic regulator design was proposed to obtain local results on "track-keeping". A fourth-order ship model in Serret–Frenet frame was used in [85] to develop a control strategy to track both a straight line and a circumference under constant ocean current disturbance. The ocean-current direction was assumed to be known. A path-following controller was proposed in [86] by using a kinematic model written in polar coordinates, which is inspired by the solution for mobile robots in [60]. However, the controller was designed at the kinematic level with an assumption of constant ocean current and its direction known to be to achieve an adjustable boundedness of the path-following error. Since ocean vessels do not have direct control over velocities, a static mapping implementation might result in an unstable closed loop system due to nonvanishing environmental disturbances. Recently, a path-following controller based on a transformation of the ship kinematics to the Serret–Frenet frame, which was used for mobile robot control [44], on the path was proposed in [87], where an acceleration feedback and linearization of ship dynamics were used. It is worth mentioning that in [88, 89], a simple control scheme was proposed to make mobile robots follow a specified path using a polar coordinate transformation. Since underactuated surface ships have fewer numbers of actuators than the to-be-controlled degrees of freedom and are subject to nonintegrable acceleration constraints, their dynamic models are not transformable into a system without drifts. Therefore, the above control scheme is not directly applicable. In [54], a continuous, periodic time-varying feedback control law was proposed to locally exponentially stabilize an underactuated underwater vehicle at the origin. When the hydrodynamic restoring force in roll is large enough, this controller can be used without a roll control torque. However the closed loop system exhibits undesired oscillatory motions.

In [90], a linearization technique with an assumption of reference trajectories of underwater vehicles, which are helices parameterized by the vehicles' linear speed, yaw rate, and path angle, was introduced to develop the so-called time-invariant generalized vehicle error dynamics and kinematics. Various controllers were then designed based on the gain-scheduling technique to yield some local stability result about the trimming trajectories.

4.2.2.4 Output Feedback

Output feedback control of an underactuated ocean vessel is here defined as a control problem of forcing the vessel to achieve the aforementioned tasks (stabilization, trajectory-tracking, and path-following) without using measurements of the vessel's velocities for feedback. For ocean vessels, output feedback control usually consists of two stages. The first stage is to design an observer to reconstruct unmeasured states. Using the reconstructed states, a controller is designed to achieve control objectives in the second stage. In the literature, there are two main approaches to designing an observer for ocean vessels.

The first approach is based on the output-injection method applied directly to the vessel's equations of motion. This approach is simple and usually results in a semiglobal observer due to the quadratic terms of the vessel's velocities. Belonging to this approach are the results presented in [11, 14, 91–94] on output feedback control of fully actuated ocean vessels or Lagrange systems. In addition, the reader is referred to [93] for an exponential observer and output feedback controller for a special class of multi-degree of freedom Lagrange systems without cross terms of quadratic velocities, and to [14, 95–98] for output feedback control of robot manipulators and rigid body without measurements of angular velocities. Another method to design an observer is the use of contraction theory, see for example, [99, 100]. This method has been applied to Lagrange systems with monotonic velocity terms but without any quadratic velocity terms. Below, we summarize the aforementioned results on an observer design. We will show that a standard observer design cannot be used to obtain a global exponential/asymptotical observer for the ocean vessel system (1.1). Assume that the vessel velocity vector $\boldsymbol{v}$ is not measurable for feedback. We would then design an output injection observer to estimate $\boldsymbol{v}$ as follows:

$$\begin{aligned}\dot{\hat{\boldsymbol{\eta}}} &= \boldsymbol{J}(\boldsymbol{\eta})\hat{\boldsymbol{v}} + \boldsymbol{K}_{\mathbf{01}}(\boldsymbol{\eta} - \hat{\boldsymbol{\eta}}),\\ \boldsymbol{M}\dot{\hat{\boldsymbol{v}}} &= -\boldsymbol{C}(\hat{\boldsymbol{v}})\hat{\boldsymbol{v}} - \boldsymbol{D}(\hat{\boldsymbol{v}})\hat{\boldsymbol{v}} - \boldsymbol{g}(\boldsymbol{\eta}) + \boldsymbol{\tau} + \boldsymbol{K}_{\mathbf{02}}(\boldsymbol{\eta} - \hat{\boldsymbol{\eta}}),\end{aligned} \tag{4.71}$$

where $\hat{\boldsymbol{\eta}}$ and $\hat{\boldsymbol{v}}$ are estimates of $\boldsymbol{\eta}$ and $\boldsymbol{v}$, respectively, and the positive definite symmetric matrices $\boldsymbol{K}_{\mathbf{01}} \in \mathbb{R}^{6\times 6}$ and $\boldsymbol{K}_{\mathbf{02}} \in \mathbb{R}^{6\times 6}$ are the observer gain matrices. It is noted that in some of the aforementioned work, the observer gain matrices $\boldsymbol{K}_{\mathbf{01}}$ and $\boldsymbol{K}_{\mathbf{02}}$ depend on the measurable state $\boldsymbol{\eta}$. Letting the observer errors be

$$\begin{aligned}\tilde{\boldsymbol{\eta}} &= \boldsymbol{\eta} - \hat{\boldsymbol{\eta}},\\ \tilde{\boldsymbol{v}} &= \boldsymbol{v} - \hat{\boldsymbol{v}}\end{aligned} \tag{4.72}$$

and differentiating (4.72) along the solutions of (1.1) and (4.71) results in

$$\begin{aligned}\dot{\tilde{\boldsymbol{\eta}}} &= -\boldsymbol{K}_{\mathbf{01}}\tilde{\boldsymbol{\eta}} + \boldsymbol{J}(\boldsymbol{\eta})\tilde{\boldsymbol{v}},\\ \boldsymbol{M}\dot{\tilde{\boldsymbol{v}}} &= -\boldsymbol{K}_{\mathbf{02}}\tilde{\boldsymbol{\eta}} - \big(\boldsymbol{C}(\boldsymbol{v})\boldsymbol{v} - \boldsymbol{C}(\hat{\boldsymbol{v}})\hat{\boldsymbol{v}}\big) - \big(\boldsymbol{D}(\boldsymbol{v})\boldsymbol{v} - \boldsymbol{D}(\hat{\boldsymbol{v}})\hat{\boldsymbol{v}}\big).\end{aligned} \tag{4.73}$$

The term $\big(\boldsymbol{D}(\boldsymbol{v})\boldsymbol{v} - \boldsymbol{D}(\hat{\boldsymbol{v}})\hat{\boldsymbol{v}}\big)$ does not cause a problem if the damping matrix $\boldsymbol{D}(\boldsymbol{v})$ is monotonic, i.e., $\big((\boldsymbol{v} - \hat{\boldsymbol{v}})^T\big(\boldsymbol{D}(\boldsymbol{v})\boldsymbol{v} - \boldsymbol{D}(\hat{\boldsymbol{v}})\hat{\boldsymbol{v}}\big)$ is nonnegative for all $\boldsymbol{v} \in \mathbb{R}^6$ and $\hat{\boldsymbol{v}} \in \mathbb{R}^6$. However, we can see a serious problem with (4.73) because of the Coriolis matrix, i.e., $(\boldsymbol{v} - \hat{\boldsymbol{v}})^{\boldsymbol{T}}\big(\boldsymbol{C}(\boldsymbol{v})\boldsymbol{v} - \boldsymbol{C}(\hat{\boldsymbol{v}})\hat{\boldsymbol{v}}\big)$ is not nonnegative for all $\boldsymbol{v} \in \mathbb{R}^6$ and $\hat{\boldsymbol{v}} \in \mathbb{R}^6$. Therefore, only a local or semiglobal observer can be obtained.

The second approach involves a nontrivial coordinate transformation to transform the vessel's equations of motion to a new set of differential equations that are linear in unmeasured states. Then the output-injection method is used to design an observer. This approach usually results in a global observer if the nontrivial coordinate transformation can be found. Unfortunately, this coordinate transformation depends heavily on a solution of a set of partial differential equations, which in general are hard to solve. The main idea of this approach is to find a coordinate transformation

$$\boldsymbol{X} = \boldsymbol{Q}(\boldsymbol{\eta})\boldsymbol{v}, \tag{4.74}$$

where $\boldsymbol{Q}(\boldsymbol{\eta})$ is an invertible. This matrix is to be determined later. Substituting (4.74) into (4.26) results in

$$\begin{aligned}\dot{\boldsymbol{\eta}} &= \boldsymbol{J}(\boldsymbol{\eta})\boldsymbol{Q}^{-1}(\boldsymbol{\eta})\boldsymbol{X},\\ \dot{\boldsymbol{X}} &= \left[\dot{\boldsymbol{Q}}(\boldsymbol{\eta})\boldsymbol{v} - \boldsymbol{Q}(\boldsymbol{\eta})\boldsymbol{M}^{-1}\boldsymbol{C}(\boldsymbol{v})\boldsymbol{v}\right] - \boldsymbol{Q}(\boldsymbol{\eta})\boldsymbol{M}^{-1}\boldsymbol{D}\boldsymbol{Q}^{-1}\boldsymbol{X} + \boldsymbol{Q}(\boldsymbol{\eta})\boldsymbol{M}^{-1}\boldsymbol{\tau}.\end{aligned} \tag{4.75}$$

The goal is to determine the matrix $\boldsymbol{Q}(\boldsymbol{\eta})$ such that

$$\dot{\boldsymbol{Q}}(\boldsymbol{\eta})\boldsymbol{v} - \boldsymbol{Q}(\boldsymbol{\eta})\boldsymbol{M}^{-1}\boldsymbol{C}(\boldsymbol{v})\boldsymbol{v} = 0, \tag{4.76}$$

for all $\boldsymbol{\eta} \in \mathbb{R}^3$ and $\boldsymbol{v} \in \mathbb{R}^3$. With (4.76), we can write (4.75) as

$$\begin{aligned}\dot{\boldsymbol{\eta}} &= \boldsymbol{J}(\boldsymbol{\eta})\boldsymbol{Q}^{-1}(\boldsymbol{\eta})\boldsymbol{X},\\ \dot{\boldsymbol{X}} &= -\boldsymbol{Q}(\boldsymbol{\eta})\boldsymbol{M}^{-1}\boldsymbol{D}\boldsymbol{Q}^{-1}\boldsymbol{X} + \boldsymbol{Q}(\boldsymbol{\eta})\boldsymbol{M}^{-1}\boldsymbol{\tau}.\end{aligned} \tag{4.77}$$

It is seen that the transformed system (4.77) is linear in the unmeasured state $\boldsymbol{X}$. This allows us to design an exponential/asymptotical observer to estimate $\boldsymbol{X}$. After that an estimate, $\hat{\boldsymbol{v}}$, of $\boldsymbol{v}$ can be found from (4.74), i.e.,

$$\hat{\boldsymbol{v}} = \boldsymbol{Q}^{-1}(\boldsymbol{\eta})\hat{\boldsymbol{X}}, \tag{4.78}$$

where $\hat{\boldsymbol{X}}$ denotes an estimate of $\boldsymbol{X}$. It is noted that combining the first equation of (4.26) and (4.76) results in a set of partial differential equations. Finding a solution to this set of partial differential equations is a hard task. A simple application of the above idea gives the results in [101–104] for some single degree of freedom Lagrange systems. It is noted that the method of solving the set of partial differential equations in [101–104] is not applicable for systems of more than one degree of freedom. For more complicated Lagrange systems, it is hard to find a result in this approach. However, the reader is referred to [105] where an output eedback control solution for simultaneous stabilization and tracking control of an underactuated ODIN is given.

Remark 4.7. The main difficulty in designing an observer-based output feedback for surface ships and Lagrange systems in general is because of the Coriolis matrix, which results in cross terms of unmeasured velocities. In addition, the underactuation of surface ships makes the output feedback problem much more challenging. For example, many solutions proposed for robot control, see [14] and references therein, cannot directly be applied. The reader will find that a set of special coordinate transformations is derived in this book to transform the ship dynamics to a system that is linear in unmeasured velocities, and another set of coordinate transformations that makes it possible to design global output feedback control controllers for underactuated ships.

4.3 Conclusions

This chapter presented the main control properties of ocean vessels. The literature on the control of underactuated ocean vessels including ships and underwater vehicles was then reviewed. Through this review, several challenging questions were raised. These questions motivate contributions of the coming chapters of this book.

Part III
Control of Underactuated Ships

Chapter 5
Trajectory-tracking Control of Underactuated Ships

This chapter addresses the problem of trajectory-tracking control of underactuated surface ships. In particular, we present a method to design a controller for underactuated surface ships with only surge force and yaw moment available to globally asymptotically track a reference trajectory generated by a suitable virtual ship. The reference yaw velocity does not have to satisfy a persistently exciting condition as was often required in previous literature. Hence, the reference trajectory is allowed to be a curve including a straight line and a circle. In addition, a new solution to global K-exponential tracking as given in previous work is obtained. The control development is based on Lyapunov's direct method and the backstepping technique, and utilizes passive properties of ship dynamics and their interconnected structure.

5.1 Control Objective

We consider an underactuated ship with simplified dynamics meaning that all off-diagonal terms of the linear and nonlinear damping matrices, and environmental disturbances induced by waves, wind and ocean currents are ignored. For the reader's convenience the mathematical model of an underactuated ship moving in surge, sway, and yaw directions, see Section 3.4.1.2, is rewritten as

$$\begin{aligned}
\dot{x} &= u\cos(\psi) - v\sin(\psi),\\
\dot{y} &= u\sin(\psi) + v\cos(\psi),\\
\dot{\psi} &= r,\\
\dot{u} &= \frac{m_{22}}{m_{11}}vr - \frac{d_{11}}{m_{11}}u + \frac{1}{m_{11}}\tau_u,\\
\dot{v} &= -\frac{m_{11}}{m_{22}}ur - \frac{d_{22}}{m_{22}}v,\\
\dot{r} &= \frac{m_{11}-m_{22}}{m_{33}}uv - \frac{d_{33}}{m_{33}}r + \frac{1}{m_{33}}\tau_r,
\end{aligned} \tag{5.1}$$

where all symbols in (5.1) are defined as in Section 3.4.1.2. The available control inputs are the surge force τ_u and the yaw moment τ_r. Since the sway control force is not available, the ship model (5.1) is underactuated. The tracking control problem is to force the underactuated ship to track a reference trajectory generated by a virtual ship as

$$\begin{aligned}
\dot{x}_d &= u_d \cos(\psi_d) - v_d \sin(\psi_d),\\
\dot{y}_d &= u_d \sin(\psi_d) + v_d \cos(\psi_d),\\
\dot{\psi}_d &= r_d,\\
\dot{u}_d &= \frac{m_{22}}{m_{11}} v_d r_d - \frac{d_{11}}{m_{11}} u_d + \frac{1}{m_{11}} \tau_{ud},\\
\dot{v}_d &= -\frac{m_{11}}{m_{22}} u_d r_d - \frac{d_{22}}{m_{22}} v_d,\\
\dot{r}_d &= \frac{(m_{11}-m_{22})}{m_{33}} u_d v_d - \frac{d_{33}}{m_{33}} r_d + \frac{1}{m_{33}} \tau_{rd},
\end{aligned} \tag{5.2}$$

where all variables have similar meanings as in system (5.1) for the virtual reference ship. In this chapter, we propose a novel method to design a controller such that it forces the ship model (5.1) to globally asymptotically track a reference trajectory generated by a virtual ship described by (5.2) under the following assumption.

Assumption 5.1.

1. *The reference signals u_d, r_d, $\dot{u}_d$, $\ddot{u}_d$ and $\dot{r}_d$ are bounded.*
2. *One of the following conditions holds:*
   (1) There exists a positive constant σ_r such that, for any pair of (t_0, t), $0 \le t_0 \le t < \infty$,

$$\int_{t_0}^{t} r_d^2(\tau)\mathrm{d}\tau \ge \sigma_r (t - t_0). \tag{5.3}$$

   (2) There exist constants $\sigma_u^{\min} > 0$, $\chi_1 \ge 0$ and $\chi_2 > 0$ such that

$$\sigma_u^{\min} \le |u_d(t)|,\ |r_d(t)| \le \chi_1 e^{-\chi_2 (t-t_0)},\ \ \forall t \ge t_0 \ge 0. \tag{5.4}$$

Remark 5.1. A persistently exciting condition similar to (5.3) is also required in [19, 71, 106], and (5.4) implies that the sign of $u_d(t)$ remains unchanged. However, we only require either (5.3) or (5.4) to be satisfied. Assumption 5.1 is quite realistic from a practical point of view. Roughly speaking, either r_d or u_d is allowed to approach zero. When both r_d and u_d are equal to zero, the tracking problem becomes one of stabilizing (5.1) and cannot be solved by any time-invariant smooth feedback. It is also noted that the exponential vanishing of r_d in (5.4) can be easily replaced by $\int_0^{\infty} |r_d(t)|\,\mathrm{d}t \le \vartheta,\ 0 \le \vartheta < \infty$.

In [19, 71, 106], a coordinate transformation of the form

$$\begin{aligned} z_1 &= \cos(\psi)x + \sin(\psi)y, \\ z_2 &= -\sin(\psi)x + \cos(\psi)y, \\ z_3 &= \psi \end{aligned} \tag{5.5}$$

was used for both (5.1) and (5.2), then different controllers were designed for the resulting tracking error system. Consequently, the reference yaw velocity r_d must satisfy persistently exciting conditions of various kinds. Hence, the way-point tracking is excluded from consideration. In addition, the physical meaning of the tracking errors is less clear. To overcome the above-mentioned drawbacks, we introduce the position and orientation errors

$$\begin{aligned} &x - x_d, \\ &y - y_d, \\ &\psi - \psi_d \end{aligned} \tag{5.6}$$

in a frame attached to the ship body, see Figure 5.1. In this figure, $O_E X_E Y_E$ is the earth-fixed frame, $O_b X_b Y_b$ is the body-fixed frame, and CG is the center of gravity of the ship. With the above definition in mind, we have the error coordinates

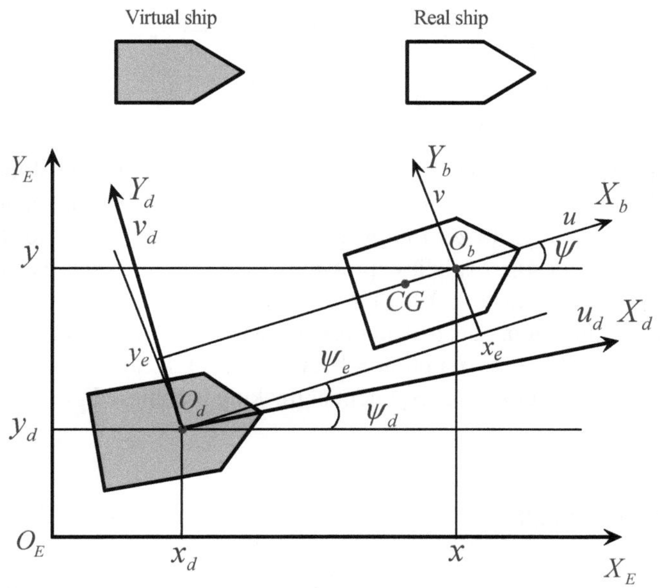

Figure 5.1 Definition of tracking errors

$$\begin{bmatrix} x_e \\ y_e \\ \psi_e \end{bmatrix} = \begin{bmatrix} \cos(\psi) & \sin(\psi) & 0 \\ -\sin(\psi) & \cos(\psi) & 0 \\ 0 & 0 & 1 \end{bmatrix} \begin{bmatrix} x - x_d \\ y - y_d \\ \psi - \psi_d \end{bmatrix}. \tag{5.7}$$

Indeed, convergence of $[x_e\ y_e\ \psi_e]^T$ to zero implies that of $[x - x_d\ y - y_d\ \psi - \psi_d]^T$ since the matrix

$$\begin{bmatrix} \cos(\psi) & \sin(\psi) & 0 \\ -\sin(\psi) & \cos(\psi) & 0 \\ 0 & 0 & 1 \end{bmatrix}$$

is nonsingular for all $\psi \in \mathbb{R}$. We also define the velocity tracking errors as

$$\begin{aligned} u_e &= u - u_d, \\ v_e &= v - v_d, \\ r_e &= r - r_d. \end{aligned} \tag{5.8}$$

Then, differentiating both sides of (5.7) and (5.8) along the solutions of (5.1) and (5.2) results in

$$\begin{aligned} \dot{x}_e &= u_e - u_d(\cos(\psi_e) - 1) - v_d \sin(\psi_e) + r_e y_e + r_d y_e, \\ \dot{y}_e &= v_e - v_d(\cos(\psi_e) - 1) + u_d \sin(\psi_e) - r_e x_e - r_d x_e, \\ \dot{\psi}_e &= r_e, \\ \dot{u}_e &= \frac{m_{22}}{m_{11}} v r - \frac{d_{11}}{m_{11}} u + \frac{1}{m_{11}} \tau_u - \dot{u}_d, \\ \dot{v}_e &= -\frac{m_{11}}{m_{22}} u_e r_d - \frac{m_{11}}{m_{22}} (u_e + u_d) r_e - \frac{d_{22}}{m_{22}} v_e, \\ \dot{r}_e &= \frac{(m_{11} - m_{22})}{m_{33}} u v - \frac{d_{33}}{m_{33}} r + \frac{1}{m_{33}} \tau_r - \dot{r}_d. \end{aligned} \tag{5.9}$$

It is now clear that the tracking problem of underactuated surface ships becomes one of stabilizing (5.9) at the origin since, as has already been mentioned convergence of x_e, y_e, and ψ_e to zero implies that of $x - x_d$, $y - y_d$, and $\psi - \psi_d$. In particular, we will design explicit expressions for τ_u and τ_r such that $\|X_e(t)\| \leq \gamma(\|X_e(t_0)\|) e^{-\mu(t-t_0)}$, $t_0 \geq 0$, $X_e(t_0) \in \mathbb{R}^6$, with $X_e = [x_e, y_e, \psi_e, u_e, v_e, r_e]^T$, γ being a class-K function, and μ being a nonnegative continuous decreasing function of $\|(X_e(t_0)\|$. When Assumption 5.1 satisfies (5.3), it can be shown that μ is a positive constant, i.e., global K-exponential tracking is achieved.

5.2 Control Design

In this section, a procedure to design a global asymptotic stabilizer for the tracking error system (5.9) is presented. The triangular structure of (5.9) suggests that we design the actual controls τ_u and τ_r in two stages. First, we design the virtual velocity controls u_e and r_e to globally asymptotically stabilize x_e, y_e, ψ_e, and v_e at the origin. Based on the backstepping technique, the controls τ_u and τ_r will then be designed.

By looking at (5.9), we can directly see that x_e and ψ_e can be stabilized at the origin by u_e and r_e. There are several options to stabilize y_e at the origin. We can either use r_e, v_e or ψ_e. If r_e is directly used, the control design will be extremely complicated since r_e enters all of the three equations of (5.9). In addition, r_e couples with x_e and y_e in the first and second equations of (5.9), respectively. This causes difficulties in obtaining a global solution. On the other hand, the use of v_e to stabilize y_e at the origin will result in an undesired feature of the vehicle control practice, namely the vehicle will slide in the sway direction. Hence we will choose ψ_e to stabilize the sway error y_e at the origin. This also coincides with the ship control practice as described in [107]: A good helmsman will use the ship course angle rather than use the ship sway velocity to steer the ship. Toward this end, we define the following coordinate

$$z_e = \psi_e + \arcsin\left(\frac{k u_d y_e}{\sqrt{1+x_e^2+y_e^2}}\right), \tag{5.10}$$

where k is a positive constant satisfying

$$k u_d^{\max} \leq k_* \tag{5.11}$$

for some positive constant $k_* < 1$ to be chosen later in the stability analysis. $u_d^{\max}$ denotes the maximum value of $|u_d|$. It is seen that (5.10) is well defined and convergence of x_e, z_e, and y_e to zero implies that of ψ_e. Using (5.10) instead of $z_e = \psi_e + k y_e$, we avoid the ship whirling around for large y_e. This equation is also different from the one in [108] and [18], which resulted in a local tracking result for mobile robots. With (5.10), the ship error dynamics (5.9) are rewritten as

$$\begin{aligned}
\dot{x}_e &= u_e + u_d\,(1-\varpi_e) + k u_d v_d y_e \varpi_{e1} + r_e y_e + r_d y_e + p_x,\\
\dot{y}_e &= v_e + v_d\,(1-\varpi_e) - k u_d^2 y_e \varpi_{e1} - r_e x_e - r_d x_e + p_y,\\
\dot{v}_e &= -\alpha\, v_e - \beta\, u_e r_d - \beta\,(u_e + u_d) r_e,\\
\dot{z}_e &= \left(1 - k u_d \varpi_{e2}^{-1} x_e\right) r_e + \varpi_{e2}^{-1}\,(p_z + p_u) - k u_d x_e y_e \varpi_{e2}^{-1} \varpi_{e1}^2 \tilde{u}_e,\\
\dot{u}_e &= \frac{m_{22}}{m_{11}} v r - \frac{d_{11}}{m_{11}} (u_e + u_d) + \frac{1}{m_{11}} \tau_u - \dot{u}_d,\\
\dot{r}_e &= \frac{m_{11}-m_{22}}{m_{33}} u v - \frac{d_{33}}{m_{33}}\,(r_e + r_d) + \frac{1}{m_{33}} \tau_r - \dot{r}_d,
\end{aligned} \tag{5.12}$$

where, for notational simplicity, we have defined

$$\begin{aligned}
\alpha &= \frac{d_{22}}{m_{22}},\ \beta = \frac{m_{11}}{m_{22}},\\
p_x &= -\left((\cos(z_e)-1)u_d + \sin(z_e) v_d\right) \varpi_e - \\
&\quad \left(\sin(z_e) u_d - (\cos(z_e)-1) v_d\right) k u_d y_e \varpi_{e1},\\
p_y &= -\left((\cos(z_e)-1) v_d - \sin(z_e) u_d\right) \varpi_e - \\
&\quad \left(\sin(z_e) v_d + (\cos(z_e)-1) u_d\right) k u_d y_e \varpi_{e1},
\end{aligned}$$

$$\begin{aligned}
p_z = \;& k u_d p_y - k u_d y_e \left(x_e p_x + y_e p_y\right) \varpi_{e1}^2, \\
p_u = \;& k u_d v_e + k u_d v_d (1-\varpi_e) - k^2 u_d^3 y_e \varpi_{e1} - k u_d r_d x_e + k \dot{u}_d y_e - \\
& k u_d \varpi_{e1}^2 y_e \big(x_e u_e^d + x_e u_d (1-\varpi_e) + k u_d v_d \varpi_{e1} x_e y_e + y_e v_e + \\
& y_e v_d (1-\varpi_e) - k u_d^2 \varpi_{e1} y_e^2\big), \\
\varpi_{e1} = \;& \sqrt{\frac{1}{1+x_e^2+y_e^2}}, \\
\varpi_{e2} = \;& \sqrt{1+x_e^2+(1-(k u_d)^2) y_e^2}, \\
\varpi_e = \;& \varpi_{e1} \varpi_{e2},
\end{aligned} \tag{5.13}$$

with u_e^d and $\tilde{u}_e$ being defined in (5.15). Before designing the control laws, τ_u and τ_r, we note from (5.11) that

$$0 < 1 - k_* \leq 1 - \frac{x_e}{\varpi_{e2}} k u_d. \tag{5.14}$$

This allows us to design global control laws τ_u and τ_r to asymptotically stabilize (5.12) at the origin. The control design consists of two steps as follows.

Step 1

We define the following virtual control errors

$$\begin{aligned}
\tilde{u}_e &= u_e - u_e^d, \\
\tilde{r}_e &= r_e - r_e^d,
\end{aligned} \tag{5.15}$$

where u_e^d and r_e^d are the virtual velocity controls of u_e and r_e, respectively.

The virtual surge and yaw velocity controls are chosen as

$$\begin{aligned}
u_e^d &= -k_1 x_e + k_2 r_d y_e, \\
r_e^d &= r_{1e}^d + r_{2e}^d
\end{aligned} \tag{5.16}$$

where

$$r_{1e}^d = -\frac{1}{\varpi_{e2} - k u_d x_e} p_u, \tag{5.17}$$

$$r_{2e}^d = -\frac{1}{\varpi_{e2} - k u_d x_e} \left(k_3 \varpi_{e2} z_e + p_z\right), \tag{5.18}$$

and $k_i, i = 1, 2, 3$, are positive constants to be selected later. We have written r_e^d as a sum of r_{1e}^d and r_{2e}^d to simplify notation in the stability analysis later.

Remark 5.2. Unlike the standard application of the backstepping technique, in order to reduce complexity of the controller expressions, we have chosen a simple virtual control law u_e^d without canceling the known terms. The first term in u_e^d is used to stabilize the x_e-dynamics while the second term plays the role of stabilizing the y_e-dynamics when (5.3) holds. From (5.17) and (5.13), we observe that r_{1e}^d is Lipschitz in (x_e, y_e, v_e) and with (5.18), r_{2e}^d exponentially vanishes when z_e does. This observation plays a crucial role in the stability analysis of the closed loop system.

Step 2

Differentiating (5.15) along the solutions of (5.12) and (5.16) yields

$$\begin{aligned}\dot{\tilde{u}}_e &= \frac{m_{22}}{m_{11}}vr - \frac{d_{11}}{m_{11}}u + \frac{1}{m_{11}}\tau_u - \dot{u}_d - \dot{u}_e^d,\\ \dot{\tilde{r}}_e &= \frac{(m_{11}-m_{22})}{m_{33}}uv - \frac{d_{33}}{m_{33}}r + \frac{1}{m_{33}}\tau_r - \dot{r}_d - \dot{r}_e^d,\end{aligned} \tag{5.19}$$

where $\dot{u}_e^d$ and $\dot{r}_e^d$ are the first time derivatives of u_e^d and r_e^d along the solutions of (5.12).

From (5.19) we choose the actual controls τ_u and τ_r without canceling the useful damping terms as

$$\begin{aligned}\tau_u &= m_{11}\left(-\delta_1\tilde{u}_e - \frac{m_{22}}{m_{11}}vr + \frac{d_{11}}{m_{11}}(u_e^d + u_d) + \dot{u}_d + \dot{u}_e^d\right),\\ \tau_r &= m_{33}\left(-\delta_2\tilde{r}_e - \frac{(m_{11}-m_{22})}{m_{33}}uv + \frac{d_{33}}{m_{33}}(r_e^d + r_d) + \dot{r}_d + \dot{r}_e^d\right),\end{aligned} \tag{5.20}$$

where δ_1 and δ_2 are positive constants to be chosen later. Substituting (5.16) and (5.20) into (5.12) yields the closed loop system

$$\begin{aligned}\dot{X}_{1e} &= f_1(t, X_{1e}) + g_1(t, X_{1e}, X_{2e}),\\ \dot{X}_{2e} &= f_2(t, X_{1e}, X_{2e}),\end{aligned} \tag{5.21}$$

where $X_{1e} = [x_e\ y_e\ v_e]^T$, $X_{2e} = [z_e\ \tilde{u}_e\ \tilde{r}_e]^T$,

$$f_1(t, X_{1e}) = \begin{bmatrix} -k_1x_e + k_2r_dy_e + u_d\,(1-\varpi_e) + (ku_dv_d\varpi_{e1} + r_{1e}^d + r_d)y_e \\ v_e + v_d\,(1-\varpi_e) - ku_d^2\varpi_{e1}y_e - r_{1e}^dx_e - r_dx_e \\ -\alpha\,v_e - \beta\,u_e^dr_d - \beta\,(u_e^d + u_d)r_{1e}^d \end{bmatrix}, \tag{5.22}$$

$$g_1(t, X_{1e}, X_{2e}) = \begin{bmatrix} r_{2e}^dy_e + \tilde{r}_ey_e + p_x + \tilde{u}_e \\ -r_{2e}^dx_e - \tilde{r}_ex_e + p_y \\ -\beta\left(\tilde{u}_er_d + \tilde{u}_e(r_{1e}^d + r_{2e}^d + \tilde{r}_e) + (u_e^d + u_d)(r_{2e}^d + \tilde{r}_e)\right) \end{bmatrix}, \tag{5.23}$$

$$f_2(t, X_{1e}, X_{2e}) = \begin{bmatrix} -k_3z_e + (1 - ku_dx_e/\varpi_{e2})\,\tilde{r}_e - ku_dx_ey_e\varpi_{e1}^2\tilde{u}_e/\varpi_{e2} \\ -(\delta_1 + d_{11}/m_{11})\,\tilde{u}_e \\ -(\delta_2 + d_{33}/m_{33})\,\tilde{r}_e \end{bmatrix}. \tag{5.24}$$

The time dependence in $f_1(t, X_{1e})$, $g_1(t, X_{1e}, X_{2e})$ and $f_2(t, X_{1e}, X_{2e})$ results from the time-varying reference velocities. We now state our main result, the proof of which is given in the next section.

Theorem 5.1. *Assume that the reference signals (x_d, y_d, ψ_d, v_d) are generated by the virtual ship model (5.2) and that the reference velocities (u_d, r_d) satisfy Assumption 5.1. If the state feedback control law (5.20) is applied to the ship system (5.1) then the tracking errors $x(t) - x_d(t)$, $y(t) - y_d(t)$, $\psi(t) - \psi_d(t)$, and $v(t) - v_d(t)$ converge to zero asymptotically from arbitrary initial values with an appropriate choice of the design constants k_i, $1 \leq i \leq 3$ and δ_j, $j = 1,2$, i.e., the closed loop system (5.21) is GAS at the origin. Furthermore if Assumption 5.1 holds with (5.3), the closed loop system (5.21) is globally K-exponentially stable at the origin.*

5.3 Stability Analysis

To prove the global asymptotic stability at the origin of the closed loop system (5.21), we note that (5.21) consists of two subsystems, X_{1e} and X_{2e}, in an interconnected structure. If X_{1e} does not appear in the second equation of (5.21), we can use the stability analysis approaches for a cascaded system in [17]. One might claim that if the system

$$
\begin{aligned}
\dot{X}_{1e} &= f_1(t, X_{1e}), \\
\dot{X}_{2e} &= f_2(t, X_{1e}, X_{2e})
\end{aligned} \tag{5.25}
$$

is uniformly globally asymptotically stable (UGAS), and the term $g_1(t, X_{1e}, X_{2e})$ somehow is of linear growth in X_{1e}, then the closed loop system (5.21) is UGAS. However, we are interested in proving stronger stability properties of (5.21). Hence we present the following lemma.

Lemma 5.1. *Consider the following nonlinear system:*

$$
\dot{x} = f(t, x) + g(t, x, \xi(t)) \tag{5.26}
$$

where $x \in \mathbb{R}^n$, $\xi(t) \in \mathbb{R}^m$, $f(t, x)$ is piecewise continuous in t and locally Lipschitz in x. Assume the following:

C1. There exists a proper function $V(t, x)$ satisfying:

$$
\begin{aligned}
& c_1 \|x\|^2 \leq V(t, x) \leq c_2 \|x\|^2, \\
& \left\| \frac{\partial V}{\partial x}(t, x) \right\| \leq c_3 \|x\|, \\
& \frac{\partial V}{\partial t} + \frac{\partial V}{\partial x} f(t, x) \leq -l_1(t) \|x\|^2 - l_2(t) \frac{\|x\|^2}{\sqrt{1 + \|x\|^2}},
\end{aligned} \tag{5.27}
$$

where $c_i > 0, i = 1,2,3$, *and the pair* $l_1(t)$ *and* $l_2(t)$ *satisfy one of the conditions:*

1. $\int_{t_0}^{t} l_1(\tau)\mathrm{d}\tau \geq \eta_1(t-t_0)$, $\eta_1 > 0$ *and* $l_2(t) \geq 0$, $\forall t \geq t_0 \geq 0$,
2. $0 < \eta_2 \leq l_2(t) < \infty$, *and* $|l_1(t)| \leq \rho_1 e^{-\rho_2(t-t_0)}$, $\rho_1 \geq 0, \rho_2 > 0$, $\forall t \geq t_0 \geq 0$.

C2. $g(t,x,\xi(t))$ *grows linearly in x:*

$$\|g(t,x,\xi(t))\| \leq (\lambda_1 + \lambda_2 \|x\|)\|\xi(t)\|, \ \lambda_i \geq 0, \ i = 1,2.$$

C3. $\xi(t)$ *vanishes exponentially:* $\|\xi(t)\| \leq \gamma_0(\|\xi(t_0)\|)e^{-\sigma_0(t-t_0)}$, $\sigma_0 > 0$, $\forall t \geq t_0 \geq 0$, *with* γ_0 *being a class-K function.*

Then, the solution $x(t)$ *of (5.26) is GAS in the sense that*

$$\|x(t)\| \leq \gamma(\|(x(t_0),\xi(t_0))\|)e^{-(\sigma_1+\sigma_2)(t-t_0)}, \ \forall t \geq t_0 \geq 0, \ \sigma_1 \geq 0, \tag{5.28}$$

where γ *is a class-K function,* σ_2 *is a nonnegative continuous decreasing function of* $\|(x(t_0),\xi(t_0))\|$, *and* $\sigma_1 > 0$ *if condition C1 holds with 1. It can be understood that* $\xi(t)$ *is generated from a globally K-exponentially stable dynamic system.*

Proof. We first consider the case of condition C1 satisfying 1. From conditions C1, C2, and C3, we have

$$\dot{V} \leq -\frac{l_1(t)}{c_2}V + c_3\left(\frac{\lambda_1}{\sqrt{c_1}}\sqrt{V} + \frac{\lambda_2}{c_1}V\right)\gamma_0(\|\xi(t_0)\|)e^{-\sigma_0(t-t_0)}. \tag{5.29}$$

By defining $\kappa(t) = \sqrt{V(t)}$, we can rewrite (5.29) as

$$\begin{aligned}\dot{\kappa}(t) \leq &-\left(\frac{l_1(t)}{2c_2} - \frac{c_3\lambda_2}{2c_1}\gamma_0(\|\xi(t_0)\|)e^{-\sigma_0(t-t_0)}\right)\kappa(t) + \frac{c_3\lambda_1}{2\sqrt{c_1}} \times \\ &\gamma_0(\|\xi(t_0)\|)e^{-\sigma_0(t-t_0)}.\end{aligned} \tag{5.30}$$

Solving the differential inequality (5.30) readily yields (5.28). It is of interest to note that in this case, the system (5.26) is GES due to independence of σ_1 in (5.28) on $\|(x(t_0),\xi(t_0))\|$. We now move to the case of condition C1 satisfying 2. We first show that $x(t)$ is bounded for all $t \geq t_0 \geq 0$. From conditions C1, C2, and C3, we have

$$\begin{aligned}\dot{V} \leq &\rho_1 e^{-\rho_2(t-t_0)}\|x\|^2 - l_2(t)\frac{\|x\|^2}{\sqrt{1+\|x\|^2}} + \\ &c_3\|x\|(\lambda_1 + \lambda_2\|x\|)\gamma_0(\|\xi(t_0)\|)e^{-\sigma_0(t-t_0)} \\ \leq &c_3\left(\frac{\lambda_1}{\sqrt{c_1}}\sqrt{V} + \frac{\lambda_2}{c_1}V\right)\gamma_0(\|\xi(t_0)\|)e^{-\sigma_0(t-t_0)} + \frac{\rho_1}{c_1}e^{-\rho_2(t-t_0)}V.\end{aligned} \tag{5.31}$$

From (5.31), it is not hard to show that the solution $x(t)$ is bounded by $\|x(t)\| \leq b(\|(x(t_0),\xi(t_0))\|)$ with b being a class-K function. By utilizing this bound and condition C1, we can rewrite the first line in (5.31) as

$$\dot{V} \leq -\frac{\eta_2}{c_2\sqrt{1+b^2}}V + c_3\left(\frac{\lambda_1}{\sqrt{c_1}}\sqrt{V} + \frac{\lambda_2}{c_1}V\right)\gamma_0\left(\|\xi(t_0)\|\right)e^{-\sigma_0(t-t_0)} + \frac{\rho_1}{c_1}\times$$
$$e^{-\rho_2(t-t_0)}V. \tag{5.32}$$

By defining $\kappa(t) = \sqrt{V(t)}$, we rewrite (5.32) as follows:

$$\dot{\kappa}(t) \leq -\left(\frac{\eta_2}{2c_2\sqrt{1+b^2}} - \frac{c_3\lambda_2}{2c_1}\gamma_0\left(\|\xi(t_0)\|\right)e^{-\sigma_0(t-t_0)} - \frac{\rho_1}{2c_1}e^{-\rho_2(t-t_0)}\right)\times$$
$$\kappa(t) + \frac{c_3\lambda_1}{2\sqrt{c_1}}\gamma_0\left(\|\xi(t_0)\|\right)e^{-\sigma_0(t-t_0)}. \tag{5.33}$$

Again, solving the differential inequality (5.33) readily yields (5.28). □

We now view our closed loop system (5.21) as a form of the system studied in Lemma 5.1 with $X_{1e}(t)$ as $x(t)$, and $X_{2e}(t)$ as $\xi(t)$. In the following, we will show that the closed loop system (5.21) satisfies all the conditions C1, C2, and C3.

Verifying Condition C1. To verify condition C1 of Lemma 5.1 for the system

$$\dot{X}_{1e} = f_1(t, X_{1e}), \tag{5.34}$$

the first time derivative of the following Lyapunov function

$$V_1 = \frac{1}{2}x_e^2 + \frac{1}{2}y_e^2 + \frac{k_4}{2}(v_e + k_5 y_e)^2, \tag{5.35}$$

where k_4 and k_5 are positive constants, along the solutions of (5.34) satisfies

$$\dot{V}_1 = -k_1 x_e^2 - \beta k_2 k_4 k_5 r_d^2 y_e^2 - k u_d^2(1 + k_4 k_5^2)\varpi_{e1} y_e^2 - k_4(\alpha - k_5)v_e^2 + M, \tag{5.36}$$

where

$$M = k_2 r_d x_e y_e + (1 - \varpi_e)(u_d x_e + k_4 k_5 v_d(v_e + k_5 y_e) + v_d y_e) + k u_d v_d \varpi_{e1}\times$$
$$x_e y_e - k_2 k_4 \beta r_d^2 y_e v_e - k_4 k_5 k u_d^2 \varpi_{e1} y_e v_e - \beta k_4(v_e + k_5 y_e)(k_2 r_d y_e + u_d) r_{1e}^d, \tag{5.37}$$

and we have chosen

$$\beta k_1 - k_5 = 0,\ 1 - k_4(\alpha - k_5)k_5 = 0 \tag{5.38}$$

to cancel some common terms. In addition, we note that

$$1 - \varpi_e = \frac{(k u_d)^2 y_e^2}{\sqrt{1 + x_e^2 + y_e^2}\left(\sqrt{1 + x_e^2 + y_e^2} + \sqrt{1 + x_e^2 + (1 - (k u_d)^2)y_e^2}\right)}. \tag{5.39}$$

We now write $M = M_1 + \sum_{i=1}^{4} N_i$, where

$$\begin{aligned}
M_1 &= k_2 r_d x_e y_e + (1-\varpi_e)(u_d x_e + k_4 k_5 v_d (v_e + k_5 y_e) + v_d y_e) + \\
&\quad k u_d v_d \varpi_{e1} x_e y_e - k_2 k_4 \beta r_d^2 y_e v_e - k_4 k_5 k u_d^2 \varpi_{e1} y_e v_e, \\
N_1 &= -\beta k_2 k_4 r_d y_e v_e r_{1e}^d,\ N_2 = -\beta k_4 u_d v_e r_{1e}^d, \\
N_3 &= -\beta k_2 k_4 k_5 r_d y_e^2 r_{1e}^d,\ N_4 = -\beta k_4 k_5 r_d y_e r_{1e}^d.
\end{aligned} \tag{5.40}$$

Applying (5.39) and the completing square to (5.40) yields

$$\begin{aligned}
M_1 &\le \frac{1}{4\varepsilon_1}\left(k_2 + k u_d^2 + k v_d^2\right) x_e^2 + (k_2\varepsilon_1 + \beta k_2 k_4 \varepsilon_2) r_d^2 y_e^2 + \\
&\quad (2k\varepsilon_1 + 2k k_4 k_5 \varepsilon_3)\frac{u_d^2 y_e^2}{\sqrt{1+x_e^2+y_e^2}} + k(1+k_4 k_5^2)|u_d v_d| \times \\
&\quad \frac{y_e^2}{\sqrt{1+x_e^2+y_e^2}} + \left(\frac{k k_4 k_5 v_d^2}{4\varepsilon_3} + \frac{\beta k_2 k_4 r_d^2}{4\varepsilon_2} + \frac{k k_4 k_5 u_d^2}{4\varepsilon_3}\right) v_e^2,
\end{aligned} \tag{5.41}$$

$$\begin{aligned}
N_1 &\le \frac{\beta k_2 k_4 k}{1-k_*}\left(\frac{|u_d r_d|}{\sqrt{1-k_*^2}} + |u_d r_d| + \frac{1}{4\varepsilon_2}\left(\left(u_d^2\left(v_d^2 + r_d^2\right) + 2k\left|u_d^3\right| + \right.\right.\right. \\
&\quad \left.\left.\left.\frac{\dot{u}_d^2}{\sqrt{1-k_*^2}} + (k_1 + k_2 r_d^2) u_d^2 + k_*^2 u_d^4 + k_*^2 v_d^2 + |u_d v_d|\right)\right)\right) v_e^2 + \frac{\beta k_2 k_4}{1-k_*} \\
&\quad \times \left(k\varepsilon_2\left(3 + k_1 + k_2 + 2k\left|u_d^3\right| + \frac{1}{\sqrt{1-k_*^2}} + |u_d v_d|\right) + k_*^2\varepsilon_2\right) r_d^2 y_e^2,
\end{aligned} \tag{5.42}$$

$$\begin{aligned}
N_2 &\le \frac{\beta k_4 k}{1-k_*}\left(2u_d^2 + u_d^2 |r_d| \varepsilon_1 + \frac{1}{4\varepsilon_3}\left(k_*^2 u_d^2 v_d^2 + 2k u_d^4 + \dot{u}_d^2 + k_1 u_d^2 + \right.\right. \\
&\quad \left.\left. k_2 u_d^2 r_d^2 + k_*^2 u_d^4 + k u_d^4 v_d^2 + k_*^2 u_d^2 v_d^2\right)\right) v_e^2 + \frac{\beta k k_4 u_d^2 |r_d|}{(1-k_*) 4\varepsilon_1} x_e^2 + \\
&\quad \frac{\beta k_4 \varepsilon_3 k}{1-k_*}\left(3k_*^2 + 2k u_d^2 + \frac{1}{\sqrt{1-k_*^2}} + k_1 + k_2 + k\right)\frac{u_d^2 y_e^2}{\sqrt{1+x_e^2+y_e^2}},
\end{aligned} \tag{5.43}$$

$$\begin{aligned}
N_3 &\le \frac{\beta k_2 k_4 k_5 k}{1-k_*}\left(\frac{2\varepsilon_2}{\sqrt{1-k_*^2}} + |u_d| + k_2 |u_d|\right) r_d^2 y_e^2 + \frac{\beta k_2 k_4 k_5 k}{1-k_*} \times \\
&\quad \left(\frac{|\dot{u}_d|}{\sqrt{1-k_*^2}} + k_1 |u_d|\right)|r_d| y_e^2 + \frac{\beta k_4 k_5 k_2 k}{1-k_*}\left(\frac{2k|u_d r_d|}{\sqrt{1-k_*^2}} + k^2 |r_d v_d| + \right. \\
&\quad \left. k_*^2 + k|r_d v_d| + k^2 |u_d r_d v_d|\right)\frac{u_d^2 y_e^2}{\sqrt{1+x_e^2+y_e^2}} + \frac{\beta k_2 k_4 k_5 k u_d^2 v_e^2}{(1-k_*) 2\varepsilon_2 \sqrt{1-k_*^2}},
\end{aligned} \tag{5.44}$$

$$
\begin{aligned}
N_4 \leq & \frac{\beta k_4 k_5}{1-k_*}\frac{k u_d^2}{2\varepsilon_3} v_e^2 + \frac{\beta k_4 k_5 k u_d^2}{1-k_*}\frac{1}{4\varepsilon_1} x_e^2 + \frac{\beta k_4 k_5 k u_d^2 \varepsilon_1}{1-k_*} r_d^2 y_e^2 + \\
& \frac{\beta k_4 k_5 k\,|\dot{u}_d|}{(1-k_*)\sqrt{1-k_*^2}} \frac{|u_d|\, y_e^2}{\sqrt{1+x_e^2+y_e^2}} + \frac{\beta k_4 k_5 k}{1-k_*}\Bigg(\frac{\varepsilon_3}{\sqrt{1-k_*^2}} + 2k_*^2 |v_d| + \\
& 2k u_d^2 + k_1 + k_2 |r_d| + k_*^2 |u_d| + k\,|u_d v_d| + 1\Bigg) \frac{u_d^2 y_e^2}{\sqrt{1+x_e^2+y_e^2}}.
\end{aligned} \tag{5.45}
$$

Substituting (5.37) with (5.41)–(5.45) into (5.36) yields

$$
\dot{V}_1 \leq -\mu_1(t) x_e^2 - \mu_{21}(t) y_e^2 - \frac{\mu_{22}(t) y_e^2}{\sqrt{1+x_e^2+y_e^2}} - \mu_3(t) v_e^2, \tag{5.46}
$$

where

$$
\begin{aligned}
\mu_1(t) =& k_1 - \frac{1}{4\varepsilon_1}\left(k_2 + k u_d^2 + k v_d^2 + \frac{\beta k_4 k u_d^2 |r_d|}{1-k_*} + \frac{\beta k_4 k_5 k u_d^2}{1-k_*}\right), \\
\mu_{21}(t) =& \mu_{211}(t) r_d^2 - \mu_{212}(t)\,|r_d|, \\
\mu_{211}(t) =& \Bigg[\beta k_2 k_4 (k_5 - \varepsilon_2) - k_2 \varepsilon_1 - \frac{\beta k_2 k_4}{1-k_*}\Bigg(k\varepsilon_2\Bigg(3 + k_1 + k_2 + \\
& 2k\,|u_d^3| + \frac{1}{\sqrt{1-k_*^2}} + |u_d v_d|\Bigg) + k_*^2 \varepsilon_2\Bigg) - \frac{\beta k_2 k_4 k_5}{1-k_*} \times \\
& \left(\frac{2k\varepsilon_2}{\sqrt{1-k_*^2}} + k\,|u_d| + k k_2\,|u_d|\right) - \frac{\beta k_4 k_5 k u_d^2 \varepsilon_1}{1-k_*}\Bigg], \\
\mu_{212}(t) =& \frac{\beta k_2 k_4 k_5}{1-k_*}\left(\frac{k\,|\dot{u}_d|}{\sqrt{1-k_*^2}} + k k_1\,|u_d|\right), \\
\mu_{22}(t) =& \mu_{221}(t) u_d^2 - \mu_{222}(t)\,|u_d|, \\
\mu_{221}(t) =& k(1 + k_4 k_5^2) - 2k\varepsilon_1 - 2k_4 k_5 k \varepsilon_3 - \frac{\beta k_4 \varepsilon_3 k}{1-k_*} \times \\
& \left(3k_*^2 + 2k u_d^2 + \frac{1}{\sqrt{1-k_*^2}} + k_1 + k_2 + k\right) - \frac{\beta k_4 k_5 k_2 k}{1-k_*} \times \\
& \left(\frac{2k\,|u_d r_d|}{\sqrt{1-k_*^2}} + k^2\,|r_d v_d| + k_*^2 + k\,|r_d v_d| + k^2\,|u_d r_d v_d|\right) - \\
& \frac{\beta k_4 k_5 k}{1-k_*}\Bigg(\frac{\varepsilon_3}{\sqrt{1-k_*^2}} + 2k_*^2 |v_d| + 2k u_d^2 + k_1 + k_2 |r_d| + \\
& k_*^2 |u_d| + k\,|u_d v_d| + 1\Bigg), \\
\mu_{222}(t) =& k(1 + k_4 k_5^2)\,|v_d| + \frac{\beta k_4 k_5 k\,|\dot{u}_d|}{(1-k_*)\sqrt{1-k_*^2}},
\end{aligned}
$$

$$\mu_3(t) = k_4(\alpha - k_5) - \frac{kk_4k_5v_d^2}{4\varepsilon_3} - \frac{\beta k_2k_4r_d^2}{4\varepsilon_2} - \frac{kk_4k_5u_d^2}{4\varepsilon_3} - \frac{\beta k_2k_4k_5ku_d^2}{(1-k_*)2\varepsilon_2\sqrt{1-k_*^2}} - \frac{\beta k_4k_5}{1-k_*}\frac{ku_d^2}{2\varepsilon_3} - \frac{\beta k_2k_4k}{1-k_*}\times \left(\frac{|u_dr_d|}{\sqrt{1-k_*^2}} + |u_dr_d| + \frac{1}{4\varepsilon_2}\left(u_d^2(v_d^2+r_d^2) + 2k\,|u_d^3| + \frac{\dot{u}_d^2}{\sqrt{1-k_*^2}} + (k_1+k_2r_d^2)u_d^2 + k_*^2u_d^4 + k_*^2v_d^2 + |u_dv_d|\right)\right) - \frac{\beta k_4k}{1-k_*}\left(2u_d^2 + u_d^2\,|r_d|\,\varepsilon_1\frac{1}{4\varepsilon_3}\left(k_*^2u_d^2v_d^2 + 2ku_d^4 + \dot{u}_d^2 + k_1u_d^2 + k_2u_d^2r_d^2 + k_*^2u_d^4 + ku_d^4v_d^2 + k_*^2u_d^2v_d^2\right)\right) \tag{5.47}$$

for some positive constants $\varepsilon_i, i = 1,2,3$. The time dependence of μ_1, μ_{21}, μ_{22}, and μ_3 is due to the time-varying reference velocities. Hence condition C1 is satisfied if there exist positive constants k_4 and k_5, and the design constants k, k_1, and k_2 such that the following conditions are satisfied:

1. $ku_d^{\max} \le k_* < 1$, $\beta k_1 - k_5 = 0$, $1 - k_4(\alpha - k_5)k_5 = 0$, $\mu_1(t) \ge \mu_1^*$, $\mu_3(t) \ge \mu_3^*$, $\forall t \ge 0$,
2. $\mu_{221}(t) \ge 0$, $\mu_{211}^{\sigma_r} - \mu_{212}^m r_d^{\max} - \mu_{222}^m u_d^{\max} \ge \mu_{21}^*$, if (5.3) holds, see Assumption 5.1,
3. $\mu_{22}(t) \ge \mu_{22}^*$, if (5.4) holds, see Assumption 5.1,

where μ_1^*, μ_{21}^*, μ_{22}^*, and μ_3^* are some positive constants, μ_{212}^m and μ_{222}^m are the values of $\mu_{211}(t)$ and $\mu_{212}(t)$ with the maximum values of $|u_d(t)|$, $|\dot{u}_d(t)|$ and $|r_d(t)|$ substituted in, and $r_d^{\max}$ denotes the maximum value of $|r_d|$. We now need to show that the above constants always exist. From (5.38), we have

$$k_1 = \frac{k_5}{\beta},\; k_5 = \frac{\alpha}{2} - \sqrt{\frac{\alpha^2}{4} - \frac{1}{k_4}}. \tag{5.48}$$

It is noted that we take $k_5 = \frac{\alpha}{2} - \sqrt{\frac{\alpha^2}{4} - \frac{1}{k_4}}$ instead of $k_5 = \frac{\alpha}{2} + \sqrt{\frac{\alpha^2}{4} - \frac{1}{k_4}}$ since we want to have a small k_5 for the condition $\mu_3(t) \ge \mu_3^* > 0$. We choose $k_4 \ge 4/\alpha^2$ to make k_5 real and positive, and set $k_2 = \delta k$ with δ being a small positive constant, say $\delta \le \max(u_d^{\max}, \dot{u}_d^{\max}, r_d^{\max})$, with $\dot{u}_d^{\max}$ being the maximum of $|\dot{u}_d|$. By observing from all functions $\mu_i(t)$, $i = 1,21,22,3$ that their negative parts have k as a factor, and that the mass including added mass in the sway dynamics, m_{22}, is always larger than that in the surge dynamics, m_{11}, for surface ships, i.e., $\beta < 1$, therefore we can always find a positive constant k or even $k = 0$ such that the conditions 1 and 2 hold for some small $\varepsilon_i > 0$, $i = 1,2,3$. For simplicity, one can take $\varepsilon_i = \beta$. In fact, $k = 0$ results in the controllers proposed in [19, 106] (with nonzero k_2 selected). However if $k = 0$ the condition 3 cannot be satisfied. Hence we should choose k to be a small positive constant. Note that a small k automatically implies a small k_*. The value of this constant should be reduced if the reference yaw, r_d, and surge, u_d, velocities and surge acceleration, $\dot{u}_d$, are large. The physical meaning of this interpretation is that the distance from the ship to the point it aims to track should be increased if the velocities are large, otherwise the ship will miss that point due to high velocities. Furthermore if α is small, k_5 is automatically small. This physically means that if the damping in the sway is small, the control gain in the surge dynamics should also be small, otherwise the ship will slide in the sway direction. This can be mathematically seen from the expression of $\mu_3(t)$. Due to complicated expressions of $\mu_i(t)$, we give a simple procedure to choose the design constants k_1 and k_2 rather than present their complex explicit expressions.

1. Pick a small positive constant k, say $k << \max(u_d^{\max}, r_d^{\max}, \dot{u}_d^{\max})$, δ such that $\delta \leq \max(u_d^{\max}, r_d^{\max})$ and set $k_2 = \delta k$, $k_4 = 4/\alpha^2$.
2. Substitute (5.48) into $\mu_i(t)$and slowly increase k and k_4 until all conditions 1, 2, and 3 hold with arbitrarily positive constants μ_1^*, μ_{21}^*, μ_{22}^*, and μ_3^*.

Once conditions 1, 2, and 3 hold, the design constant k_1 is calculated from (5.48) and $k_2 = \delta k$. Based on the above choice of design constants, it directly follows that
• if (5.3) holds:

$$\dot{V}_1 \leq -\mu_1^* x_e^2 - \left(\mu_{211}^m r_d^2 - \mu_{212}^m r_d^{\max} - \mu_{222}^m u_d^{\max}\right) y_e^2 - \mu_3^* v_e^2, \tag{5.49}$$

• if (5.4) holds:

$$\begin{aligned} \dot{V}_1 \leq & -\mu_1^* x_e^2 - \frac{\mu_{22}^* y_e^2}{\sqrt{1 + x_e^2 + y_e^2}} - \mu_3^* v_e^2 + \\ & (|\mu_{211}(t)| \chi_1 + |\mu_{212}(t)|) \chi_1 e^{-\chi_2(t-t_0)} y_e^2. \end{aligned} \tag{5.50}$$

It is now seen that condition C1 of Lemma 5.1 follows readily from (5.50) since the last term in the second line of (5.50) is linear in y_e^2 and its coefficient exponentially vanishes, see also proof of Lemma 5.1.

Remark 5.3. When the reference surge and sway velocities and surge acceleration are very large, the above procedure will result in very small control gains k, k_1, and k_2, which will give very slow convergence. Hence it is suggested that for large u_d, r_d, and $\dot{u}_d$, the control gains k, k_1, and k_2 should be chosen such that either conditions 1 and 2 or 1 and 3 hold to improve the convergence. The trade-off is that some "switches" in the control gains for different reference trajectories will occur.

Verifying Condition C2. We need to show that $g_1(t, X_{1e}, X_{2e})$ satisfies condition C2. It can be seen that

$$\begin{aligned} & \varpi_e \leq 1, \\ & k u_d \varpi_{e1} y_e \leq k^*, \\ & |\sin(z_e)/z_e| \leq 1, \\ & |(\cos(z_e) - 1)/z_e| \leq 1. \end{aligned} \tag{5.51}$$

From (5.13) and (5.51), it is not hard to show that:

$$\begin{aligned} |p_x| &\le (1+k^*)(|u_d|+|v_d|)|z_e|, \\ |p_y| &\le (1+k^*)(|u_d|+|v_d|)|z_e|, \\ |p_z| &\le k^*(2|p_y|+|p_x|). \end{aligned} \tag{5.52}$$

Applying (5.52) to (5.21) and noting Remark 5.3 yields the condition C2 readily.

Verifying Condition C3. This condition is verified by showing that the X_{2e}-subsystem is GES. Take the following quadratic function:

$$V_2 = \frac{1}{2}\left(z_e^2 + \tilde{u}_e^2 + \tilde{r}_e^2\right). \tag{5.53}$$

Differentiating (5.53) along the solutions of the X_{2e}-subsystem satisfies

$$\begin{aligned} \dot{V}_2 &\le -(k_3-1.5)z_e^2 - \left(\delta_1 + \frac{d_{11}}{m_{11}} - 0.5\right)\tilde{u}_e^2 - \left(\delta_2 + \frac{d_{33}}{m_{33}} - 1\right)\tilde{r}_e^2 \\ &\le -\rho_2 V_2, \end{aligned} \tag{5.54}$$

where

$$\rho_2 = \min\left(2(k_3-1.5), 2\left(\delta_1 + \frac{d_{11}}{m_{11}} - 0.5\right), 2\left(\delta_2 + \frac{d_{33}}{m_{33}} - 1\right)\right).$$

Thus it suffices to choose the design constants k_3, δ_1, and δ_2 such that

$$\begin{aligned} k_3 &> 1.5, \\ \delta_1 &\ge 0.5, \\ \delta_2 &\ge 1. \end{aligned} \tag{5.55}$$

Hence condition C3 follows from (5.54). We have thus verified all the conditions of Lemma 5.1. Therefore the closed loop system (5.21) is GAS at the origin.

Remark 5.4. From the proof of Theorem 5.1 and Lemma 5.1, it is seen that when the yaw reference velocity r_d satisfies the persistently exciting condition as stated in Assumption 5.1, the tracking error system is GES. This is the case reported in [19, 71, 106] based on different approaches. Therefore the results in those papers are a special case of the result in Theorem 5.1.

5.4 Simulations

This section validates the control laws (5.20) by simulating them on a monohull ship with the length of 32 m, mass of 118×10^3 kg, and other parameters are calculated by using MARINTEK Ship Motion program version 3.18, a program for calculating the added mass and damping matrices of the ship, as

$$\begin{aligned}
&m_{11} = 120\times10^3\text{kg},\ m_{22} = 177.9\times10^3\text{kg},\ m_{33} = 636\times10^5\text{kgm}^2,\\
&d_{11} = 215\times10^2\text{kgs}^{-1},\ d_{u2} = 43\times10^2\text{kgm}^{-1},\ d_{u3} = 21.5\times10^2\text{kgsm}^{-2},\\
&d_{22} = 117\times10^3\text{kgs}^{-1},\ d_{v2} = 23.4\times10^3\text{kgm}^{-1},\ d_{v3} = 11.7\times10^3\text{kgsm}^{-2},\\
&d_{33} = 802\times10^4\text{kgm}^2 s^{-1},\ d_{r2} = 160.4\times10^4\text{kgm}^2,\ d_{r3} = 80.2\times10^4\text{kgm}^2 s,\\
&d_{ui} = 0,\ d_{vi} = 0,\ d_{ri} = 0,\ \forall i > 3.
\end{aligned}$$

This ship has a minimum turning circle with the radius of 150 m, a maximum surge force of 5.2×10^9 N, and a maximum yaw moment of 8.5×10^8 Nm. From the ship parameters, we have $\alpha = 0.54$ and $\beta = 0.55$. The reference trajectory is generated by the virtual ship (5.2) where the reference surge force τ_{ud} is taken as $\tau_{ud} = d_{11} - m_{22}v_d r_d$, and the reference yaw moment τ_{rd} is taken as $\tau_{rd} = -(m_{11}-m_{22})u_d r_d$ for the first 800 seconds and $\tau_{rd} = -(m_{11}-m_{22})u_d r_d + 0.01d_{33}$ for the last 1000 seconds. This choice means that the reference surge velocity u_d is 1ms^{-1} over the entire simulation time, and that the reference trajectory is a straight line for the first 800 seconds and a circle with a radius of 200 m for the last 1000 seconds. We first pick $k = 0.02$ and $k_4 = 13.9$, then increase these constants until conditions a, b, and c hold. We have

$$\begin{aligned}
&k = 0.1,\ k_1 = 0.34,\ k_2 = 0.1,\ k_3 = 2,\\
&k_4 = 15.26,\ k_5 = 0.19,\ \delta_1 = 5,\ \delta_2 = 5.
\end{aligned}$$

The initial conditions are chosen as

$$\begin{aligned}
&[x(0), y(0), \psi(0), u(0), v(0), r(0)]\\
&= \left[15\text{m}, 50\text{m}, 0.7\text{rad}, 0\text{ms}^{-1}, 0.5\text{ms}^{-1}, 0\text{rads}^{-1}\right].
\end{aligned}$$

The reference trajectory is generated by a virtual ship with the initial conditions of

$$\begin{aligned}
&[x_d(0), y_d(0), \psi_d(0), u_d(0), v_d(0), r_d(0)]\\
&= \left[0.5\text{m}, 0.5\text{m}, 0.2\text{rad}, 1\text{ms}^{-1}, 0\text{ms}^{-1}, 0\text{rads}^{-1}\right].
\end{aligned}$$

The tracking errors are plotted in Figure 5.2a and control inputs are plotted in Figure 5.2b. In addition, the real and reference trajectories in the (x, y)-plane are plotted in Figure 5.3. It can be seen from Figure 5.2a that the tracking errors asymptotically converge to zero as proven in Theorem 5.1. From Figure 5.2b, we can see that the control inputs τ_u and τ_r are quite large in the first few seconds because of a short transient response time. Indeed, we can reduce the control effort by tuning the control design gains such as reducing the control gains δ_1 and δ_2. However, this will increase the transient response time.

For a comparison, we also simulate the controller proposed in [19] with the same initial conditions and reference trajectory in Figures 5.4 and 5.5. It is clearly seen that the controller proposed in [19] cannot track the straight line as discussed before. It should be noted that the controllers proposed in [71, 106] also result in similar nonzero errors although they were designed based on different approaches.

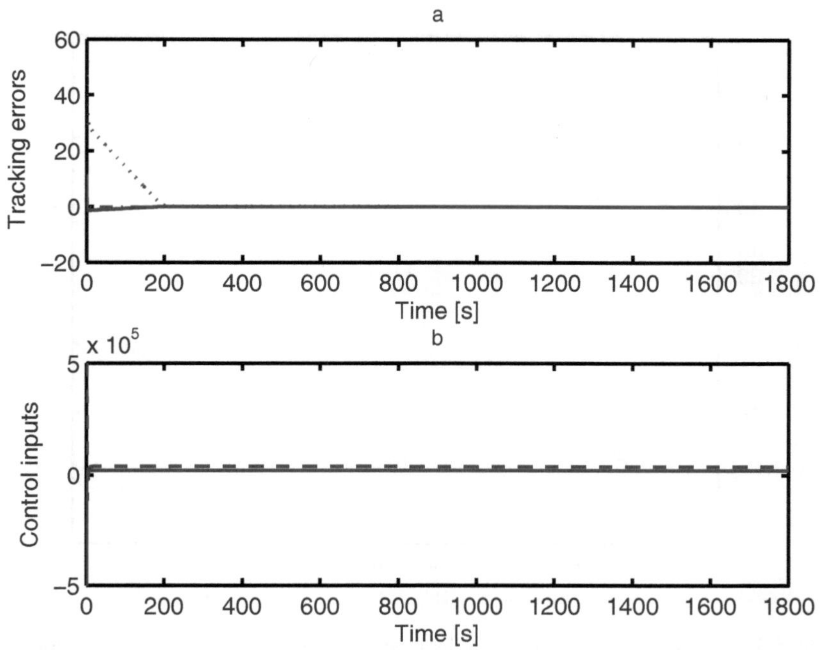

Figure 5.2 Simulation results of the controller proposed in this chapter: **a**. Convergence of tracking errors ($x - x_d$ [m]: *solid line*, $y - y_d$ [m]: *dashed line*, $\psi - \psi_d$ [rad]: *dash-dotted line*); and **b**. Control inputs (τ_u [N]: *solid line*, τ_r [Nm]: *dashed line*)

5.5 Conclusions

The key to control development is the coordinate change (5.10) to transform the tracking error system, in which the tracking errors are interpreted in the frame attached to the ship's body, to a triangular form, to which the popular backstepping technique can be applied. This coordinate change made it possible to relax the severe restrictions in [19,71,106] on the reference trajectories generated by the virtual ships in the sense that the developed controllers can force the vehicles to track a straight line (way-point tracking), a curve and a combination of both. Several assumptions on the ship dynamics have been made such as all off-diagonal terms of the damping matrix, nonlinear damping matrix, and environmental disturbances induced by waves, wind, and ocean currents being ignored. For a justification, the reader is referred to Section 3.4.1.2. It is seen from Assumption 5.1 that the surge and yaw reference velocities cannot be zero, i.e., the problem of stabilization and/or parking is not considered in this chapter. In the next chapter, this limitation will be removed when a simultaneous stabilization and tracking control is addressed. This chapter is based on [109–111].

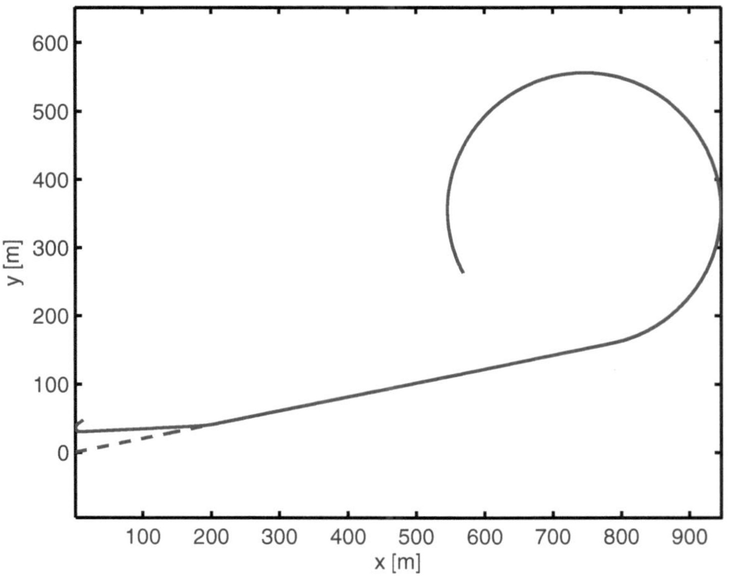

Figure 5.3 Simulation results of the controller proposed in this chapter: Tracking trajectory in the (x, y)-plane ((x, y): *solid line*, (x_d, y_d): *dashed line*)

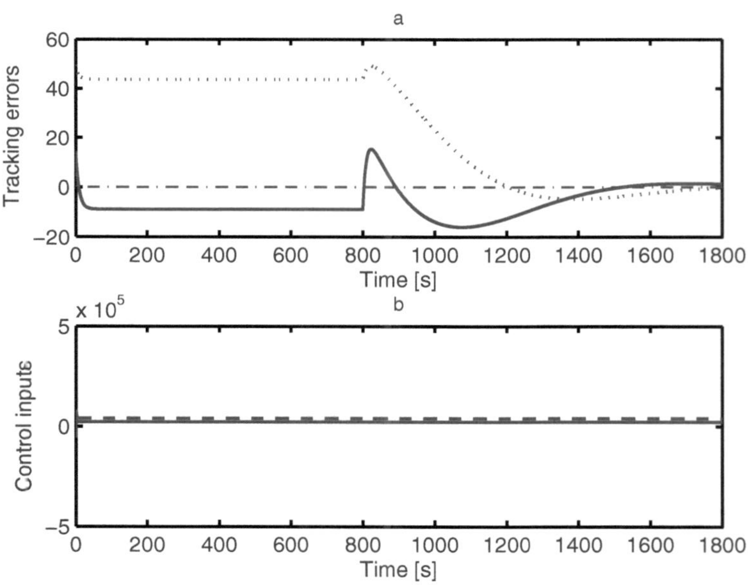

Figure 5.4 Simulation results of the controller proposed in [19]: **a**. Convergence of tracking errors ($x - x_d$ [m]: *solid line*, $y - y_d$ [m]: *dashed line*, $\psi - \psi_d$ [rad]: *dash-dotted line*); **b**. Control inputs (τ_u [N]: *solid line*, τ_r [Nm]: *dashed line*)

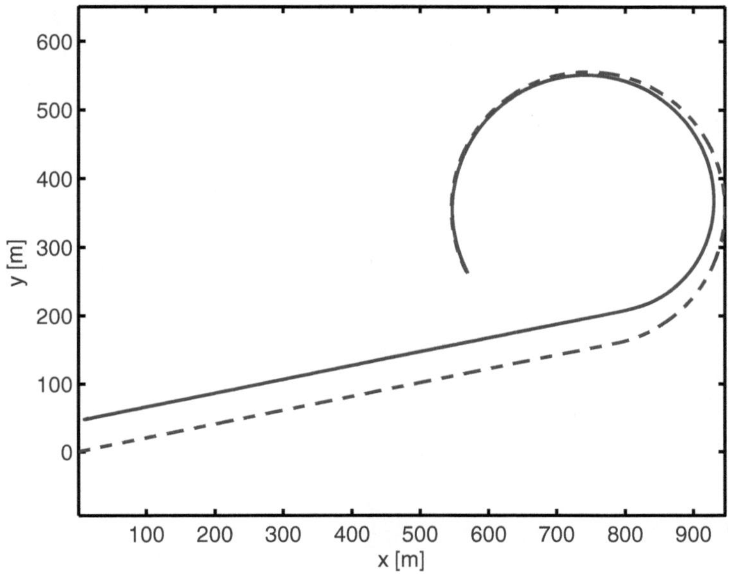

Figure 5.5 Simulation results of the controller proposed in [19]: Tracking trajectory in the (x, y)-plane ((x, y): *solid line*, (x_d, y_d): *dashed line*)

Chapter 6
Simultaneous Stabilization and Trajectory-tracking Control of Underactuated Ships

This chapter examines the problem of designing a single controller that achieves both stabilization and tracking simultaneously for underactuated surface ships without an independent sway actuator and with simplified dynamics. The proposed controller guarantees that stabilization and tracking errors converge to zero asymptotically from any initial values. In comparison with the preceding chapter, a path approaching the origin and a set-point can also be included in the reference trajectory, i.e., stabilization/regulation is also considered. The control development is based on some special coordinate transformations, Lyapunov's direct method and the backstepping technique, and utilizes passive properties of ship dynamics and their interconnected structure.

6.1 Control Objective

For the reader's convenience, the mathematical model of the underactuated ship moving in surge, sway, and yaw, see Section 3.4.1.2, is recaptured as

$$\begin{aligned}
\dot{x} &= u\cos(\psi) - v\sin(\psi),\\
\dot{y} &= u\sin(\psi) + v\cos(\psi),\\
\dot{\psi} &= r,\\
\dot{u} &= \frac{m_{22}}{m_{11}}vr - \frac{d_{11}}{m_{11}}u + \frac{1}{m_{11}}\tau_u,\\
\dot{v} &= -\frac{m_{11}}{m_{22}}ur - \frac{d_{22}}{m_{22}}v,\\
\dot{r} &= \frac{(m_{11}-m_{22})}{m_{33}}uv - \frac{d_{33}}{m_{33}}r + \frac{1}{m_{33}}\tau_r,
\end{aligned} \tag{6.1}$$

where all symbols in (6.1) are defined as in Section 3.4.1.2. The available controls are the surge force τ_u and the yaw moment τ_r. Since the sway control force is not available, the ship model (6.1) is underactuated. Similar to the preceding chapter, it is assumed that the reference trajectory is generated by a virtual ship as

$$\begin{aligned}
\dot{x}_d &= u_d \cos(\psi_d) - v_d \sin(\psi_d), \\
\dot{y}_d &= u_d \sin(\psi_d) + v_d \cos(\psi_d), \\
\dot{\psi}_d &= r_d, \\
\dot{u}_d &= \frac{m_{22}}{m_{11}} v_d r_d - \frac{d_{11}}{m_{11}} u_d + \frac{1}{m_{11}} \tau_{ud}, \\
\dot{v}_d &= -\frac{m_{11}}{m_{22}} u_d r_d - \frac{d_{22}}{m_{22}} v_d, \\
\dot{r}_d &= \frac{(m_{11} - m_{22})}{m_{33}} u_d v_d - \frac{d_{33}}{m_{33}} r_d + \frac{1}{m_{33}} \tau_{rd},
\end{aligned} \tag{6.2}$$

where all variables have similar meanings as in the system (6.1) for the virtual reference ship. In this chapter, a single controller that simultaneously solves stabilization and tracking problems of underactuated surface ships is proposed under the following assumptions.

Assumption 6.1. *The reference velocities, u_d and r_d, are bounded and differentiable with bounded derivatives $\dot{u}_d$, $\ddot{u}_d$ and $\dot{r}_d$.*

Assumption 6.2. *One of the following conditions holds:*

C1. $u_d = r_d = 0$.

C2. $\int_{t_0}^{t} r_d^2(\tau)\mathrm{d}\tau \geq \sigma_r (t - t_0),\ \sigma_r > 0, \forall\ 0 \leq t_0 \leq t < \infty$.

C3. $|u_d(t)| \geq \sigma_u > 0$ and $\int_0^{\infty} |r_d(t)|\mathrm{d}t \leq \mu_1,\ 0 \leq \mu_1 < \infty$.

C4. $\int_0^{\infty} (|r_d(t)| + |u_d(t)| + |\dot{u}_d(t)|)\,\mathrm{d}t \leq \mu_2,\ 0 \leq \mu_2 < \infty$.

Remark 6.1. The problems of regulation, stabilization, or dynamic positioning are included in condition C1. Tracking a circular path belongs to the case when condition C2 holds. Condition C3 covers the case of straight-line or way-point tracking. By Barbalat's lemma together with Assumption 6.2, Condition C4 implies that $\lim_{t\to\infty} r_d(t) = \lim_{t\to\infty} u_d(t) = 0$. Hence the parking problem is captured by condition C4. Indeed, one can see that Condition C4 covers Condition C1. However we study these two conditions separately because Condition C1 has been studied in the literature but Condition C4 has not.

Remark 6.2. The work in [19, 53–56, 71, 106] can deal with either Condition C1 or Condition C2. In the preceding chapter of this book, we allowed for either Condition C2 or Condition C3.

Similar to the preceding chapter, we introduce position and orientation errors $x - x_d$, $y - y_d$, and $\psi - \psi_d$ in a frame attached to the ship body. This results in the error coordinates as

$$\begin{bmatrix} x_e \\ y_e \\ \psi_e \end{bmatrix} = \begin{bmatrix} \cos(\psi) & \sin(\psi) & 0 \\ -\sin(\psi) & \cos(\psi) & 0 \\ 0 & 0 & 1 \end{bmatrix} \begin{bmatrix} x - x_d \\ y - y_d \\ \psi - \psi_d \end{bmatrix}. \tag{6.3}$$

One can see that the convergence of x_e, y_e, and ψ_e to the origin implies that of $x - x_d$, $y - y_d$ and $\psi - \psi_d$. We also define the velocity tracking errors as

$$u_e = u - u_d, \; v_e = v - v_d, \; r_e = r - r_d. \tag{6.4}$$

Differentiating both sides of (6.3) along the solutions of (6.1) and (6.2) yields:

$$\begin{aligned}
\dot{x}_e &= u_e - u_d(\cos(\psi_e) - 1) - v_d \sin(\psi_e) + r_e y_e + r_d y_e, \\
\dot{y}_e &= v_e - v_d(\cos(\psi_e) - 1) + u_d \sin(\psi_e) - r_e x_e - r_d x_e, \\
\dot{\psi}_e &= r_e, \\
\dot{u}_e &= \frac{m_{22}}{m_{11}} v r - \frac{d_{11}}{m_{11}} u + \frac{1}{m_{11}} \tau_u - \dot{u}_d, \\
\dot{v}_e &= -\frac{m_{11}}{m_{22}} u_e r_d - \frac{m_{11}}{m_{22}} (u_e + u_d) r_e - \frac{d_{22}}{m_{22}} v_e, \\
\dot{r}_e &= \frac{(m_{11} - m_{22})}{m_{33}} u v - \frac{d_{33}}{m_{33}} r + \frac{1}{m_{33}} \tau_r - \dot{r}_d.
\end{aligned} \tag{6.5}$$

The tracking and regulation problem of underactuated surface ships is therefore equivalent to stabilizing system (6.5) at the origin. In particular, we are interested in designing explicit expressions for τ_u and τ_r such that $\lim_{t \to \infty} \|X_e(t)\| = 0$, for all $t \geq t_0 \geq 0$ and $X_e(t_0) \in \mathbb{R}^6$ with $X_e = [x_e, y_e, \psi_e, u_e, v_e, r_e]^T$. In case of Condition C2, it will be shown that $\|X_e(t)\| \leq \gamma(\|X_e(t_0)\|) e^{-\mu(t-t_0)}$ with γ being a class-K function, and μ a positive constant, i.e., global K-exponential tracking is achieved.

6.2 Control Design

Similar to the preceding chapter, we observe from (6.5) that x_e and ψ_e can be stabilized by u_e and r_e. On the other hand, ψ_e should be chosen as a virtual control to stabilize the sway error y_e. However when the surge reference velocity is zero or approaches zero, i.e., the case of stabilization and parking, we cannot use ψ_e to stabilize y_e since $u_d \sin(\psi_e)$ is zero or approaches zero. In this case, we need some persistently exciting signal in r_e to stabilize y_e. As such, we choose the coordinate transformation

$$z_e = \psi_e + \arcsin\left(\frac{k(t) y_e}{\sqrt{1 + x_e^2 + y_e^2}} \right), \tag{6.6}$$

with

$$k(t) = \lambda_1 (u_d + \lambda_2 \cos(\lambda_3 t)),$$

where the constants λ_i, $1 \le i \le 3$ are such that

$$\sup_{t\ge 0} |k(t)| < 1. \tag{6.7}$$

They will be specified in the stability analysis later. Notice that (6.6) is well defined and that convergence of z_e and y_e to the origin implies that of ψ_e.

Remark 6.3. Using the nonlinear coordinate transformation (6.6) instead of $z_e = \psi_e + k(t)y_e$, we avoid the ship moving left and right largely when y_e is large. If one uses $z_e = \psi_e + \arcsin\left(k(t)y_e/\sqrt{1+y_e^2}\right)$, the problem of the ship whirling around is avoided. Indeed using these coordinate transformations will result in a much simpler tracking error system than using (6.6). However, it is readily shown that these coordinate transformations result in a local control solution because of the term $r_e x_e$ in the y_e-dynamics. The term $\lambda_1 u_d$ plays the role of stabilizing y_e when u_d is not zero and does not approach to zero while the term $\lambda_1\lambda_2\cos(\lambda_3 t)$ is used to introduce the persistently exciting signal in r_e to stabilize y_e when u_d is zero or approaches zero.

Remark 6.4. The function arcsin in (6.6) can be replaced by several other smooth functions such as arctan and tanh, or the identity function. We have chosen the function arcsin because of its simplicity.

Using the new coordinate (6.6), the ship error dynamics (6.5) are rewritten as

$$\begin{aligned}
\dot{x}_e &= u_e + u_d \frac{(\varpi_2 - \varpi_1)}{\varpi_2} + \frac{k(t) v_d y_e}{\varpi_2} + r_e y_e + r_d y_e + p_x, \\
\dot{y}_e &= v_e + v_d \frac{(\varpi_2 - \varpi_1)}{\varpi_2} - \frac{k(t) u_d y_e}{\varpi_2} - r_e x_e - r_d x_e + p_y, \\
\dot{z}_e &= \left(1 - \frac{k(t) x_e}{\varpi_1}\right) r_e + \frac{1}{\varpi_1}\left(\dot{k}(t) y_e + k(t)\left(v_e + v_d \frac{(\varpi_2 - \varpi_1)}{\varpi_2} - \right.\right. \\
&\quad \left.\left. \frac{k(t) u_d y_e}{\varpi_2} - r_d x_e\right) - \frac{k(t) y_e}{\varpi_2^2} (x_e u_e + y_e v_e + (x_e u_d + y_e v_d) \times \right. \\
&\quad \left. \frac{(\varpi_2 - \varpi_1)}{\varpi_2} + (x_e v_d - y_e u_d) \frac{k(t) y_e}{\varpi_2}\right) + p_z, \\
\dot{u}_e &= \frac{m_{22}}{m_{11}} v r - \frac{d_{11}}{m_{11}} (u_e + u_d) + \frac{1}{m_{11}} \tau_u - \dot{u}_d, \\
\dot{v}_e &= -\frac{d_{22}}{m_{22}} v_e - \frac{m_{11}}{m_{22}} u_e r_d - \frac{m_{11}}{m_{22}} (u_e + u_d) r_e, \\
\dot{r}_e &= \frac{m_{11} - m_{22}}{m_{33}} u v - \frac{d_{33}}{m_{33}} (r_e + r_d) + \frac{1}{m_{33}} \tau_r - \dot{r}_d,
\end{aligned} \tag{6.8}$$

where, for notational simplicity, we have used $k(t)$ instead of using its explicit expression given in (6.6), and defined as

$$
\begin{aligned}
\varpi_1 &= \sqrt{1+x_e^2+(1-k^2(t))y_e^2},\ \varpi_2 = \sqrt{1+x_e^2+y_e^2},\\
p_x &= -u_d\left((\cos(z_e)-1)\frac{\varpi_1}{\varpi_2}+\sin(z_e)\frac{k(t)y_e}{\varpi_2}\right)-\\
&\quad v_d\left(\sin(z_e)\frac{\varpi_1}{\varpi_2}-(\cos(z_e)-1)\frac{k(t)y_e}{\varpi_2}\right),\\
p_y &= -v_d\left((\cos(z_e)-1)\frac{\varpi_1}{\varpi_2}+\sin(z_e)k(t)\varpi_2^{-1}y_e\right)+\\
&\quad u_d\left(\sin(z_e)\frac{\varpi_1}{\varpi_2}-(\cos(z_e)-1)\frac{k(t)y_e}{\varpi_2}\right),\\
p_z &= \frac{1}{\varpi_1}\left(k(t)p_y-\frac{k(t)y_e}{\varpi_2}\left(x_e p_x+y_e p_y\right)\right).
\end{aligned}
\tag{6.9}
$$

The effort that we have made so far is to put the tracking error dynamics in the triangular form of (6.8), and to have the term $-k(t)u_d y_e/\varpi_2$ in the y_e-dynamics. This term plays an important role in stabilizing the y_e-dynamics. In this section, a procedure to design a universal controller for the tracking error system (6.8) is presented in detail. The triangular structure of (6.8) suggests that we design the actual controls τ_u and τ_r in two stages. First, we design virtual velocity controls u_e and r_e to globally asymptotically stabilize x_e, y_e, z_e and v_e at the origin. Based on the backstepping technique, the controls τ_u and τ_r are then designed to force the error between the virtual velocity controls and their actual values exponentially to zero. Before designing the control laws, τ_u and τ_r, we note if the constants λ_i, $1 \leq i \leq 3$ are chosen such that $\sup_{t\geq 0}|k(t)| < 1$, then (6.6) implies that

$$
0 < 1-\sup_{t\geq 0}|k(t)| \leq 1-\frac{k(t)x_e}{\varpi_1} < 2. \tag{6.10}
$$

This allows us to design control laws for τ_u and τ_r to globally asymptotically stabilize (6.8). The control design consists of two steps as follows.

Step 1

In this step, we define the following virtual control errors

$$
\begin{aligned}
\tilde{u}_e &= u_e - u_e^d,\\
\tilde{r}_e &= r_e - r_e^d,
\end{aligned}
\tag{6.11}
$$

where u_e^d and r_e^d are the virtual velocity controls of u_e and r_e. Unlike the standard application of backstepping, in order to reduce complexity of the controller expressions, we will choose a simpler virtual control law u_e^d without canceling the known terms. The virtual surge and yaw velocity controls are chosen as

$$\begin{aligned} u_e^d &= -k_1 x_e + k_2 r_d y_e, \\ r_e^d &= r_{1e}^d + r_{2e}^d \end{aligned} \tag{6.12}$$

where

$$\begin{aligned} r_{1e}^d = & -\frac{1}{(1-k(t)x_e/\varpi_1)}\frac{1}{\varpi_1}\left(\dot{k}(t)y_e + k(t)\left(v_e + v_d\frac{\varpi_2-\varpi_1}{\varpi_2} - \right.\right. \\ & \left.\frac{k(t)u_d y_e}{\varpi_2} - r_d x_e\right) - \frac{k(t)y_e}{\varpi_2^2}(x_e(-k_1x_e + k_2 r_d y_e) + y_e v_e + \\ & \left.\left.(x_e u_d + y_e v_d)\frac{\varpi_2-\varpi_1}{\varpi_2} + (x_e v_d - y_e u_d)\frac{k(t)y_e}{\varpi_2}\right)\right), \end{aligned} \tag{6.13}$$

$$r_{2e}^d = -\left[(1-k(t)x_e/\varpi_1)\right]^{-1}(k_3 z_e + p_z), \tag{6.14}$$

and k_i, $i = 1, 2, 3$, are positive constants to be selected later.

Step 2

Differentiating (6.11) along the solutions of (6.8) and (6.12) yields

$$\begin{aligned} \dot{\tilde{u}}_e &= \frac{m_{22}}{m_{11}}vr - \frac{d_{11}}{m_{11}}u + \frac{1}{m_{11}}\tau_u - \dot{u}_d - \dot{u}_e^d, \\ \dot{\tilde{r}}_e &= \frac{(m_{11}-m_{22})}{m_{33}}uv - \frac{d_{33}}{m_{33}}r + \frac{1}{m_{33}}\tau_r - \dot{r}_d - \dot{r}_e^d, \end{aligned} \tag{6.15}$$

where $\dot{u}_e^d$ and $\dot{r}_e^d$ are the first time derivatives of u_e^d and r_e^d along the solutions of (6.8). From (6.15) we choose the actual controls τ_u and τ_r as

$$\begin{aligned} \tau_u = & \; m_{11}\left(-c_1\tilde{u}_e - \frac{m_{22}}{m_{11}}vr + \frac{d_{11}}{m_{11}}(u_e^d + u_d) + \dot{u}_d + \dot{u}_e^d + \frac{k(t)x_e y_e z_e}{\varpi_1\varpi_2^2}\right), \\ \tau_r = & \; m_{33}\left(-c_2\tilde{r}_e - \frac{(m_{11}-m_{22})}{m_{33}}uv + \frac{d_{33}}{m_{33}}(r_e^d + r_d) + \right. \\ & \left.\dot{r}_d + \dot{r}_e^d - \left(1 - \frac{k(t)x_e}{\varpi_1}\right)z_e\right), \end{aligned} \tag{6.16}$$

where c_1 and c_2 are positive constants. Substituting (6.16) and (6.12) into (6.8) yields the closed loop system

$$\begin{aligned} \dot{x}_e = & -k_1 x_e + k_2 r_d y_e + u_d(\varpi_2 - \varpi_1)\varpi_2^{-1} + k(t)v_d\varpi_2^{-1}y_e + \\ & r_{1e}^d y_e + r_d y_e + r_{2e}^d y_e + \tilde{u}_e + \tilde{r}_e y_e + p_x, \\ \dot{y}_e = & \; v_e + v_d(\varpi_2 - \varpi_1)\varpi_2^{-1} - k(t)u_d\varpi_2^{-1}y_e - r_{1e}^d x_e - r_d x_e - \\ & r_{2e}^d x_e - \tilde{r}_e x_e + p_y, \end{aligned}$$

$$\begin{aligned}
\dot{v}_e &= -\frac{d_{22}}{m_{22}} v_e - \frac{m_{11}}{m_{22}} u_e^d r_d - \frac{m_{11}}{m_{22}} (u_e + u_d) r_{1e}^d - \frac{m_{11}}{m_{22}} (r_d + r_{1e}^d)\tilde{u}_e - \\
&\quad \frac{m_{11}}{m_{22}} (\tilde{u}_e + u_e^d + u_d)(r_{2e}^d + \tilde{r}_e), \\
\dot{z}_e &= -k_3 z_e + \left(1 - \frac{k(t)x_e}{\varpi_1}\right)\tilde{r}_e - \frac{k(t)x_e y_e \tilde{u}_e}{\varpi_1 \varpi_2^2}, \\
\dot{\tilde{u}}_e &= -\left(c_1 + \frac{d_{11}}{m_{11}}\right)\tilde{u}_e + \frac{k(t)x_e y_e z_e}{\varpi_1 \varpi_2^2}, \\
\dot{\tilde{r}}_e &= -\left(c_2 + \frac{d_{33}}{m_{33}}\right)\tilde{r}_e - \left(1 - \frac{k(t)x_e}{\varpi_1}\right) z_e.
\end{aligned} \tag{6.17}$$

We now state our main result, the proof of which is given in the next section.

Theorem 6.1. *Assume that the reference signals (x_d, y_d, ψ_d, v_d) are generated by the virtual ship model (6.2) and that the reference velocities (u_d, r_d) satisfy Assumptions 6.1 and 6.2. If the universal state feedback control law (6.16) is applied to the ship system (6.1) then the tracking errors $(x(t) - x_d(t), y(t) - y_d(t), \psi(t) - \psi_d(t)$, and $v(t) - v_d(t))$ converge to zero asymptotically from any initial values, i.e., the closed loop system (6.17) is GAS at the origin, with an appropriate choice of the design constants λ_i and k_i, $1 \leq i \leq 3$. Furthermore, if Assumption 6.2 holds with C2, the closed loop system (6.17) is globally K-exponentially stable at the origin.*

6.3 Stability Analysis

We only need to show that the transformed tracking errors, (x_e, y_e, z_e, v_e), converge to zero from any initial values. Under Assumption 6.1, boundedness of the controls τ_u and τ_r follows readily. We note that the closed loop system (6.17) consists of two subsystems, $(z_e, \tilde{u}_e, \tilde{r}_e)$ and (x_e, y_e, v_e), in an interconnected structure. To prove Theorem 6.1, we take the following Lyapunov function:

$$V_1(X_{1e}) = \frac{1}{2}\left(z_e^2 + \tilde{u}_e^2 + \tilde{r}_e^2\right) \tag{6.18}$$

for the $(z_e, \tilde{u}_e, \tilde{r}_e)$-subsystem, and

$$V_2(X_{2e}) = \frac{1}{2}(x_e^2 + y_e^2 + k_4(y_e + k_5 v_e)^2) \tag{6.19}$$

for the (x_e, y_e, v_e)-subsystem, where k_4 and k_5 are positive constants to be selected later and

$$X_{1e} = \begin{bmatrix} \tilde{z}_e \\ \tilde{u}_e \\ \tilde{r}_e \end{bmatrix}, \; X_{2e} = \begin{bmatrix} x_e \\ y_e \\ v_e \end{bmatrix}.$$

From (6.18) and (6.19), if we consider $V = V_1 + V_2$ as a Lyapunov function for the closed loop system (6.17) it can then be seen that this function has a connection with the energy of the ship system as follows. We write the function V as

$$V = X^T AX, \tag{6.20}$$

where $X = \left[x_e, y_e, z_e, u-(u_d+u_e^d), v-v_d, r-(r_d+r_e^d)\right]^T$ and A is a diagonal positive definite matrix. It is now seen that the function V consists of two positive definite parts. The first part related to (x_e, y_e, z_e) can be regarded as the "potential" energy. The second part related to $\left(u-(u_d+u_e^d), v-v_d, r-(r_d+r_e^d)\right)$ is referred to as the "kinetic" energy of the ship, which is defined with respect to a coordinate that moves at a speed of $\left((u_d+u_e^d), v_d, (r_d+r_e^d)\right)$. Hence we wish to bring the energy function V to zero. However considering this function together with the closed loop (6.17) is difficult, we consider the subsystems $(z_e, \tilde{u}_e, \tilde{r}_e)$ and (x_e, y_e, v_e) separately.

$(z_e, \tilde{u}_e, \tilde{r}_e)$-subsystem. From the last three equations of (6.17), it can be seen that the $(z_e, \tilde{u}_e, \tilde{r}_e)$-subsystem is GES at the origin by considering the Lyapunov function $V_1(X_{1e})$, see (6.18), whose time derivative along the solutions of the last three equations of (6.17) is

$$\dot{V}_1 = -k_3 z_e^2 - \left(c_1 + \frac{d_{11}}{m_{11}}\right)\tilde{u}_e^2 - \left(c_2 + \frac{d_{33}}{m_{33}}\right)\tilde{r}_e^2. \tag{6.21}$$

Therefore we have

$$\|X_{1e}(t)\| \leq \|X_{1e}(t_0)\|\, e^{-\sigma_1(t-t_0)}, \tag{6.22}$$

where $\sigma_1 = \min\{k_3, (c_1 + d_{11}/m_{11}), (c_2 + d_{33}/m_{33})\}$.

(x_e, y_e, v_e)-subsystem. To investigate stability of this subsystem, we take the Lyapunov function $V_2(X_{2e})$, see (6.19), whose first time derivative, after some manipulation, along the solutions of the first three equations of (6.17) is

$$\begin{aligned}\dot{V}_2 = \; & -k_1 x_e^2 - \frac{k(t)u_d(1+k_4)y_e^2}{\sqrt{1+x_e^2+y_e^2}} - \beta k_2 k_4 k_5 r_d^2 y_e^2 - \\ & k_4 k_5(\alpha k_5 - 1)v_e^2 + M + \Omega + \Phi,\end{aligned} \tag{6.23}$$

where for simplicity, we have defined $\alpha = d_{22}/m_{22}$, $\beta = m_{11}/m_{22}$,

$$\begin{aligned}M = \; & k_2 r_d x_e y_e + (u_d x_e + v_d(k_4+1)y_e)\frac{\varpi_2 - \varpi_1}{\varpi_2} + \\ & \frac{k(t)y_e(v_d x_e - k_4 k_5 u_d v_e)}{\varpi_2} + k_4 k_5 v_d \frac{\varpi_2 - \varpi_1}{\varpi_2} v_e - \\ & \beta k_2 k_4 k_5^2 r_d^2 y_e v_e - \beta k_4 k_5 (y_e + k_5 v_e)(k_2 r_d y_e + u_d) r_{1e}^d,\end{aligned} \tag{6.24}$$

$$\begin{aligned}\Omega = \; & x_e(p_x+\tilde{u}_e)+y_e p_y+k_4(y_e+k_5 v_e)\big(p_y-\tilde{r}_e x_e- \\ & r_{2e}^d x_e-\beta k_5\left((r_d+r_{1e}^d)\tilde{u}_e+(\tilde{u}_e+u_e^d+u_d)(\tilde{r}_e+r_{2e}^d)\right)\big),\end{aligned} \tag{6.25}$$

$$\Phi = k_4(y_e+k_5 v_e)(-1+\beta k_1 k_5)(r_{1e}^d+r_d)x_e+(1-k_4(\alpha k_5-1))y_e v_e. \tag{6.26}$$

We first choose the design constants such that

$$\begin{aligned}& 1-k_4(\alpha k_5-1)=0, \\ & \beta k_1 k_5-1=0\end{aligned} \tag{6.27}$$

to cancel the common term Φ. Noticing that

$$\begin{aligned}& \left|\frac{\sin(z_e)}{z_e}\right|\le 1,\ \left|\frac{\cos(z_e)-1}{z_e}\right|\le 1, \\ & \frac{\varpi_2-\varpi_1}{\varpi_2}=\frac{k^2(t)y_e^2}{\varpi_2(\varpi_2+\varpi_1)}, \\ & \frac{1}{\varpi_1}\le\frac{1}{\varpi_2\sqrt{1-k^2(t)}}\end{aligned} \tag{6.28}$$

after a lengthy but simple calculation of upper bounds of M and Ω by completing the squares, we arrive at

$$\begin{aligned}\dot{V}_2 \le & -p_1(t)x_e^2-p_{21}(t)y_e^2-\frac{p_{22}(t)y_e^2}{\sqrt{1+x_e^2+y_e^2}}+\frac{p_{23}(t)y_e^2}{\sqrt{1+x_e^2+y_e^2}}- \\ & p_3(t)v_e^2+(\chi_1 V_2+\chi_2)e^{-\sigma_1(t-t_0)},\end{aligned} \tag{6.29}$$

where

$$\begin{aligned}p_1(t)= \; & k_1-\frac{1}{4\rho_1}\left(k_2+|k(t)|\,|v_d|+\frac{\beta k_4 k_5\,|k(t)|\,|u_d|\,|r_d|}{1-|k(t)|}+\right. \\ & \left.\frac{\beta k_1 k_2 k_4 k_5\,|k(t)|}{1-|k(t)|}\right),\end{aligned} \tag{6.30}$$

$$\begin{aligned}p_{21}(t)= \; & p_{211}(t)r_d^2-p_{212}(t)\,|r_d|, \\ p_{211}(t)= \; & \beta k_2 k_4 k_5-k_2\rho_1-\frac{\beta k_2 k_4 k_5^2}{4\rho_2}-\frac{\beta k_2 k_4 k_5}{1-|k(t)|}\left(\frac{|k(t)|}{2\rho_2\sqrt{1-k^2(t)}}+\right. \\ & \left.(k_2+1+k_1\rho_1)\,|k(t)|\right)-\frac{\beta k_2 k_4 k_5^2}{4\rho_2\,(1-|k(t)|)}\left(\frac{\left|\dot{k}(t)\right|+|k(t)|\,k^2(t)}{\sqrt{1-k^2(t)}}+\right. \\ & \big(k^2(t)\,|u_d|+(1+k_1+k_2)\,|k(t)|+|k(t)|\,k^2(t)\,|u_d|+\end{aligned}$$

$$|k(t)|k^2(t)|v_d| + k^2(t)|u_d| + k^2(t)|v_d|\Big) - \frac{\beta k_4 k_5^2 |k(t)||u_d|}{4\rho_2(1-|k(t)|)},$$

$$p_{212}(t) = \frac{\beta k_2 k_4 k_5 \left|\dot{k}(t)\right|}{(1-|k(t)|)\sqrt{1-k^2(t)}}, \tag{6.31}$$

$$\begin{aligned}
p_{22}(t) &= (1+k_4)k(t)u_d - p_{221}(t),\\
p_{221}(t) &= k_4 k_5 \rho_3 |k(t)||u_d| + \frac{\beta k_2 k_4 k_5}{1-|k(t)|}\left(|k(t)| + 1 + \frac{1}{\sqrt{1-k^2(t)}}\right)\times\\
&\quad k^2(t)|u_d||r_d| + \frac{\beta k_4 k_5}{1-|k(t)|}\left(\frac{\left|\dot{k}(t)\right||u_d|}{\sqrt{1-k^2(t)}} + \frac{|k(t)||u_d|}{4\rho_2\sqrt{1-k^2(t)}} + k^2(t)\times\right.\\
&\quad \left(|k(t)||u_d||v_d| + u_d^2\right) + \frac{\rho_1 |k(t)||u_d||r_d|}{\sqrt{1-k^2(t)}} + k_1|k(t)||u_d| + k_2|k(t)|\times\\
&\quad |u_d||r_d| + \frac{|k(t)||u_d|}{4\rho_2\sqrt{1-k^2(t)}} + |k(t)|k^2(t)u_d^2 + |k(t)|k^2(t)|u_d||v_d| +\\
&\quad \left. k^2(t)(u_d^2 + |u_d||v_d|)\right) + \frac{\beta k_4 k_5^2}{4\rho_2(1-|k(t)|)}\left(\frac{\left|\dot{k}(t)\right||u_d| + |k^3(t)||u_d v_d|}{\sqrt{1-k^2(t)}} + \right.\\
&\quad k^2(t)u_d^2 + (k_1 + k_2|r_d|)|k(t)||u_d| + |k(t)|k^2(t)u_d^2 + k^2(t)(|k(t)|+1)\times\\
&\quad \left.|u_d||v_d| + k^2(t)u_d^2\right) + k^2(t)|u_d|,
\end{aligned} \tag{6.32}$$

$$\begin{aligned}
p_{23}(t) = &\left(k^2(t)(1+k_4) + \rho_1|k(t)| + \rho_3 k^2(t)k_4k_5\right)|v_d| + \frac{\beta k_4 k_2 k_5}{1-|k(t)|}\times\\
&\left(\frac{|k(t)|k^2(t)|v_d||r_d|}{\sqrt{1-k^2(t)}} + k^2(t)(|k(t)|+1)|v_d||r_d|\right),
\end{aligned} \tag{6.33}$$

$$\begin{aligned}
p_3(t) &= k_4k_5(k_5\alpha - 1) - \left(\frac{k_4k_5|k(t)||u_d|}{4\rho_3} + \frac{k_4k_5k^2(t)|v_d|}{4\rho_3} + \right.\\
&\quad \left.\rho_2\beta k_2k_4k_5^2r_d^2\right) - \frac{2\rho_2\beta k_2k_4k_5|k(t)|}{(1-|k(t)|)\sqrt{1-k^2(t)}} -\\
&\quad \frac{2\beta\rho_2k_4k_5|k(t)||u_d|}{1-|k(t)|} - \frac{\beta k_2k_4k_5^2}{1-|k(t)|}\left(\rho_2\left(\frac{\left|\dot{k}(t)\right|}{\sqrt{1-k^2(t)}} + \right.\right.\\
&\quad |k(t)|k^2(t)v_d^2 + k^2(t)|u_d| + |k(t)|r_d^2 + \left(k_1 + k_2r_d^2\right)\times\\
&\quad \left.|k(t)| + (|k(t)|+1)k^2(t)(|v_d| + |u_d|)\right) + \frac{2|k(t)||r_d|}{\sqrt{1-k^2(t)}}\Bigg) -
\end{aligned}$$

$$\frac{\beta k_4 k_5^2}{1-|k(t)|}\left(\rho_2\left(\left|\dot{k}(t)u_d\right| + |k(t)u_d|\left(|v_d|+3\right) + \left(k_1 + k_2|r_d|\right) \times \right.\right.$$
$$\left.\left. |k(t)u_d| + (|k(t)|+2)k^2(t)u_d^2 + k^2(t)|u_d v_d|(|k(t)|+1)\right)\right), \tag{6.34}$$

where χ_1, χ_2 are nondecreasing functions of $\|X_{1e}(t_0)\|$, σ_1 is given in (6.22), and $\rho_i > 0,\ i = 1,2,3$.

We are now in a position to select the design parameters, λ_i, $1 \le i \le 3$ and k_j, $j = 1,2,4,5$ so that the closed loop system (6.17) is GAS at the origin. We proceed to choose these parameters by investigating all the cases of Assumption 6.2. For each case, we choose a subset of λ_i and k_j. As discussed before, we firstly choose

$$0 < \lambda_1 < 1/(u_d^{\max} + \lambda_2),\ \lambda_2 > 0,\ \lambda_3 > 0, \tag{6.35}$$

so that $\sup_{t \ge 0}|k(t)| \le k_* < 1$, where $u_d^{\max}$ is the maximum value of $|u_d(t)|$. Note that this primary choice guarantees $\sup_{t \ge 0}|k(t)| < 1$.

Secondly, we choose λ_i, $1 \le i \le 3$ and k_j such that

$$p_1(t) \ge p_1^* > 0,\ p_3(t) \ge p_3^* > 0. \tag{6.36}$$

Before going further to choose λ_i and k_j, let us discuss each case of Assumption 6.2.

Case C1. From the fourth equation of (6.2), we have

$$v_d(t) = v_d(t_0)e^{-\alpha_1(t-t_0)},\ \forall t \ge t_0 \ge 0,\ \alpha_1 > 0.$$

Therefore there exists a nondecreasing function π_1 of $|v_d(t_0)|$ such that

$$p_{23}(t) \le \pi_1 e^{-\alpha_1(t-t_0)}. \tag{6.37}$$

It is noted that in this case

$$p_{21}(t) = p_{22}(t) = 0. \tag{6.38}$$

From (6.35)–(6.38), we can write (6.29) as

$$\dot{V}_2 \le -p_1^* x_e^2 - p_3^* v_e^2 + \pi_1 e^{-\alpha_1(t-t_0)}\frac{y_e^2}{\sqrt{1+x_e^2+y_e^2}} + (\chi_1 V_2 + \chi_2)e^{-\sigma_1(t-t_0)}, \tag{6.39}$$

which yields

$$\dot{V}_2 \le -p_1^* x_e^2 - p_3^* v_e^2 + (\eta_1 V_2 + \eta_2)e^{-\sigma_2(t-t_0)},\ \ \sigma_2 = \min(\alpha_1, \sigma_1) \tag{6.40}$$

for some nondecreasing functions η_1 and η_2 of $\|(X_{1e}(t_0), v_d(t_0))\|$. We now present a technical lemma to investigate stability of the (x_e, y_e, v_e)-subsystem, in this case based on (6.40).

Lemma 6.1. *Consider the following first-order scalar differential equation*

$$\dot{x} = (ax + b)e^{-c(t-t_0)} + \delta(t)x, \tag{6.41}$$

where $a \geq 0, b \geq 0$ and $c > 0$ are constants. If $\delta(t)$ enjoys the property that there exists a constant $\sigma \geq 0$ such that

$$\int_0^\infty |\delta(t)|\,\mathrm{d}t \leq \sigma < \infty \tag{6.42}$$

then the solution of (6.41) satisfies

$$|x(t)| \leq |x(t_0)|\,e^{\sigma} e^{a/c} + \frac{b}{a} e^{\sigma}\left(e^{a/c} - 1\right) := \pi\left(|x(t_0)|\right). \tag{6.43}$$

Proof. Proof of this lemma is followed directly by solving the differential equation (6.41). □

By applying Lemma 6.1 and the comparison principle found in [6] to (6.40), we have

$$V_2(t) \leq V_2(t_0)e^{\chi_1/\sigma_1} + \frac{\chi_2}{\chi_1}\left(e^{\chi_1/\sigma_1} - 1\right) := \eta_3\left(\|(X_{1e}(t_0), X_{2e}(t_0))\|\right). \tag{6.44}$$

Therefore we now rewrite (6.40) as

$$\dot{V}_2 \leq -W_2(x_e, v_e) + (\eta_1\eta_3 + \eta_2)e^{-\sigma_2(t-t_0)} \tag{6.45}$$

where $W_2(x_e, v_e) = p_1^* x_e^2 + p_3^* v_e^2$. Integrating both sides of (6.45) yields

$$\int_{t_0}^{t} W_2(x_e(\tau), v_e(\tau))\mathrm{d}\tau \leq V_2(t_0) + \int_{t_0}^{t} (\eta_1\eta_3 + \eta_2)e^{-\sigma_2(\tau-t_0)}\mathrm{d}\tau. \tag{6.46}$$

It is seen that the right-hand side of (6.46) exists and is bounded. On the other hand, $W_2(x_e(t), v_e(t))$ is uniformly continuous because its time derivative is bounded. Hence from Barbalat's lemma, we have $W_2(x_e(t), v_e(t)) \to 0$ as $t \to \infty$. Therefore $(x_e(t), v_e(t)) \to 0$ as $t \to \infty$. To prove that $y_e(t) \to 0$ as $t \to \infty$, applying Lemma 2.6 to the x_e-dynamic equation of (6.17) yields

$$\lim_{t\to\infty}\left(k_2 r_d y_e + u_d\frac{\varpi_2 - \varpi_1}{\varpi_2} + \frac{k(t)v_d y_e}{\varpi_2} + r_{1e}^d y_e + r_d y_e + \right.$$
$$\left. r_{2e}^d y_e + \tilde{u}_e + \tilde{r}_e y_e + p_x\right) = 0. \tag{6.47}$$

Since $u_d = 0$ and $(X_{1e}(t), x_e(t), v_e(t), v_d(t)) \to 0$ as $t \to \infty$, (6.47) is equivalent to

$$\lim_{t\to\infty}\left(r_{1e}^d y_e\right) = 0. \tag{6.48}$$

From the expression of r_{1e}^d, it is directly shown that (6.48) is equivalent to

$$\lim_{t\to\infty}\left(\frac{y_e^2}{\sqrt{1+(1-k^2(t))y_e^2}}\dot{k}(t)\right)=0. \tag{6.49}$$

On the other hand from (6.45), we have

$$\frac{\mathrm{d}}{\mathrm{d}t}\left(V_2+\sigma_2^{-1}(\eta_1\eta_3+\eta_2)e^{-\sigma_2(t-t_0)}\right)\leq 0$$

which implies that

$$V_2+\sigma_2^{-1}(\eta_1\eta_3+\eta_2)e^{-\sigma_2(t-t_0)}$$

is nonincreasing. Since V_2 is bounded from below by zero, V_2 tends to a finite nonnegative constant depending on $\|X_e(t_0)\|$. This implies that the limit of $|y_e(t)|$ exists and is finite, say l_{y_e}. If l_{y_e} were not zero, there would exist a sequence of increasing time instants $\{t_i\}_{i=1}^{\infty}$ with $t_i\to\infty$, such that both of the limits of $\dot{k}(t_i)$ and $\dot{k}(t_i)y_e^2(t_i)$ were not zero, which is impossible because of (6.49). Hence l_{y_e} must be zero. Therefore we conclude from (6.49) that $y_e(t)\to 0$ as $t\to\infty$. Hence we have proven $\lim_{t\to\infty}X_{2e}(t)=0$ for this case. We define the subset of the design parameters that satisfy the conditions (6.27), (6.35), and (6.36) as $\Xi_1^{\lambda k}$.

Case C2. In this case, we choose the design parameters λ_i and k_j such that

$$p_{211}(t)\geq p_{211}^*>0. \tag{6.50}$$

Under Assumption 6.1, there exists a choice of λ_i and k_j such that (6.27), (6.35), (6.36), and (6.50) hold. Substituting (6.35), (6.36), and (6.50) into (6.29) yields

$$\begin{aligned}\dot{V}_2\leq\; & -p_1^*x_e^2-\left(p_{211}^*r_d^2-|p_{22}(t)|-p_{212}(t)|r_d|-p_{23}(t)\right)y_e^2-\\ & p_3^*v_e^2+(\chi_1V_2+\chi_2)e^{-\sigma_1(t-t_0)}.\end{aligned} \tag{6.51}$$

On the other hand, it is noted that

$$V_2\leq\frac{1}{2}\left(x_e^2+(1+k_4+k_4k_5)y_e^2+k_4k_5(1+k_5)v_e^2\right). \tag{6.52}$$

From (6.52) and (6.51), we have

$$\begin{aligned}\dot{V}_2\leq\; & -2\min\left(p_1^*,\frac{p_{211}^*r_d^2-|p_{22}(t)|-p_{212}(t)|r_d|-p_{23}(t)}{1+k_4+k_4k_5},\frac{p_3^*}{k_4k_5(1+k_5)}\right)\times\\ & V_2+(\chi_1V_2+\chi_2)e^{-\sigma_1(t-t_0)}.\end{aligned} \tag{6.53}$$

By applying Lemma 2.6 to (6.53), there exist a positive constant σ_3 independent of initial conditions and a nondecreasing function π_3 of $\|(X_{1e}(t_0),X_{2e}(t_0))\|$ such that

$$\|X_{2e}(t)\|\leq\pi_3\left(\|(X_{1e}(t_0),X_{2e}(t_0))\|\right)e^{-\sigma_3(t-t_0)}, \tag{6.54}$$

as long as

$$\int_{t_0}^{t}\left(p_{211}^{*}r_d^2(\tau)-p_{212}(\tau)\left|r_d(\tau)\right|-\left|p_{22}(\tau)\right|-p_{23}(\tau)\right)\mathrm{d}\tau\geq p_{21}^{*}(t-t_0), \quad (6.55)$$

where $p_{21}^{*}>0$. For simplicity of calculation, we can replace the last three terms in the left-hand side of (6.55) by their maximum values. Note that (6.54) implies that $X_{2e}(t)$ is globally K-exponentially stable at the origin but not GES in the sense of Lyapunov, i.e., π_3 in (6.54) must be a linear function of $\|(X_{1e}(t_0),X_{2e}(t_0))\|$. However, it can be shown that the local exponential stability (in the sense of Lyapunov) of the closed loop system (6.17) is guaranteed.

We define the subset of λ_i and k_j satisfying the conditions (6.27), (6.35), (6.36), (6.50), and (6.55) as $\Xi_2^{\lambda k}$.

Case C3. Under Assumption 6.2 and $\int\limits_0^\infty |r_d(t)|\mathrm{d}t\leq\mu_1<\infty$, there exists

$$0\leq\sigma_{p21}<\infty$$

such that

$$\int\limits_0^\infty |p_{21}(t)|\mathrm{d}t\leq\sigma_{p21}. \quad (6.56)$$

In this case, we choose λ_i and k_j such that

$$p_{22}(t)-p_{23}(t)\geq p_{22}^{*}>0. \quad (6.57)$$

With (6.36) and (6.57), we rewrite (6.29) as

$$\begin{aligned}\dot V_2\leq\ & -p_1^{*}x_e^2-p_{22}^{*}\frac{y_e^2}{\sqrt{1+x_e^2+y_e^2}}-p_3^{*}v_e^2+\\ & |p_{21}(t)|\,y_e^2+(\chi_1V_2+\chi_2)\,e^{-\sigma_1(t-t_0)}.\end{aligned} \quad (6.58)$$

Again by applying Lemma 6.1 to (6.58), we have

$$\begin{aligned}V_2(t)\leq\ & V_2(t_0)e^{2\sigma_{p21}}e^{\chi_1/\sigma_1}+\frac{\chi_2}{\chi_1}e^{2\sigma_{p21}}\left(e^{\chi_1/\sigma_1}-1\right)\\ :=\ & \eta_3\left(\|(X_{1e}(t_0),X_{2e}(t_0))\|\right).\end{aligned} \quad (6.59)$$

Substituting (6.59) and (6.52) into (6.58) yields

$$\dot V_2\leq-\rho_4V_2+2\left|p_{21}(t)\right|V_2+(\chi_1V_2+\chi_2)\,e^{-\sigma_1(t-t_0)}, \quad (6.60)$$

where

$$\rho_4=2\min\left(p_1^{*},\frac{p_{22}^{*}}{\sqrt{1+2\eta_3}\,(1+k_4+k_4k_5)},\frac{p_3^{*}}{k_4k_5\,(1+k_5)}\right). \quad (6.61)$$

From (6.52) and (6.60), it can be shown that

$$\|X_{2e}(t)\| \leq \pi_4 \left(\|(X_{1e}(t_0), X_{2e}(t_0))\|\right) e^{-\sigma_4 (t-t_0)}, \tag{6.62}$$

where

$$\begin{aligned}
&\text{if } \rho_4 \neq \sigma_1 \text{ then} \\
&\quad \sigma_4 = 0.5 \min(\rho_4, \sigma_1), \\
&\quad \pi_4(\bullet) = \sqrt{4\lambda_4^{-1} e^{2\sigma_{p21}} e^{\chi_1/\sigma_1} \left(V_2(t_0) + \chi_2 |\sigma_1 - \rho_4|^{-1}\right)}, \\
&\text{if } \rho_4 = \sigma_1 \text{ then} \\
&\quad \sigma_4 = 0.5(\sigma_1 - d_4), \\
&\quad \pi_4(\bullet) = \sqrt{4\lambda_4^{-1} e^{2\sigma_{p21}} e^{\chi_1/\sigma_1} \left(V_2(t_0) + \chi_2 \phi_4\right)},
\end{aligned} \tag{6.63}$$

with

$$\lambda_4 = \min\left(2, 1 + k_4 + k_4 k_5^2 - \sqrt{\left(1 + k_4 - k_4 k_5^2\right)^2 + 4(k_4 k_5)^2}\right),$$

$$0 < d_4 < \sigma_1,$$
$$\phi_4 \geq (t - t_0) e^{-d_4 (t-t_0)}.$$

Note that ϕ_4 is finite for an arbitrarily small d_4. It can be seen from (6.59), (6.61) and (6.63) that the rate $\sigma_4 > 0$ depends on the initial conditions $\|(X_{1e}(t_0), X_{2e}(t_0))\|$. Hence the closed loop system (6.17) is GAS but not exponentially stable at the origin. In other words, global asymptotic tracking is achieved in this case. We define the subset of λ_i and k_j satisfying the conditions (6.27), (6.35), (6.36), and (6.57) as $\Xi_3^{\lambda k}$.

Case C4. We first show that V_2 is bounded. Substituting (6.35) and (6.36) into (6.29) yields

$$\dot{V}_2 \leq -p_1^* x_e^2 - p_3^* v_e^2 + p_4(t) y_e^2 + (\chi_1 V_2 + \chi_2) e^{-\sigma_1 (t-t_0)}, \tag{6.64}$$

where

$$p_4(t) = |p_{21}(t)| + |p_{22}(t)| + p_{23}(t). \tag{6.65}$$

Note that in this case, there exists $0 \leq \sigma_{p4} < \infty$ such that $\int_0^\infty p_4(t) \mathrm{d}t \leq \sigma_{p4}$. From (6.19) and (6.65), we have

$$\dot{V}_2 \leq 2 p_4(t) V_2 + (\chi_1 V_2 + \chi_2) e^{-\sigma_1 (t-t_0)}. \tag{6.66}$$

Applying Lemma 6.1 to (6.66) yields

$$\begin{aligned} V_2(t) \leq & \ V_2(t_0)e^{2\sigma_{p4}}e^{\chi_1/\sigma_1} + \frac{\chi_2}{\chi_1}e^{2\sigma_{p4}}\left(e^{\chi_1/\sigma_1} - 1\right) \\ & := \eta_4\left(\|(X_{1e}(t_0), X_{2e}(t_0))\|\right). \end{aligned} \tag{6.67}$$

Hence y_e is also bounded. Substituting this upper bound into (6.64) yields

$$\dot{V}_2 \leq -p_1^* x_e^2 - p_3^* v_e^2 + 2\eta_4 p_4(t) + (\chi_1\eta_4 + \chi_2)e^{-\sigma_1(t-t_0)}. \tag{6.68}$$

From (6.68), we have

$$\frac{\mathrm{d}}{\mathrm{d}t}\left(V_2 - 2\eta_4\int_0^t p_4(\tau)\mathrm{d}\tau + \sigma_1^{-1}(\chi_1\eta_4 + \chi_2)e^{-\sigma_1(t-t_0)}\right) \leq 0, \tag{6.69}$$

which implies that $(x_e, v_e) \to 0$ as $t \to \infty$ and $\lim_{t\to\infty} y_e = 0$. We define the subset of λ_i and k_j satisfying the conditions (6.27), (6.35), and (6.36) as $\Xi_4^{\lambda k}$.

6.4 Selection of Design Constants

In this section, we show that $\bigcap_{1\leq l\leq 4} \Xi_l^{\lambda k} \neq \emptyset$. From the above section, it can be seen that these design constants must satisfy (6.27), (6.35), (6.36), (6.50), (6.55), and (6.57). From (6.27), we have

$$k_1 = \frac{1}{\beta k_5},\ k_4 = \frac{1}{\alpha k_5 - 1}. \tag{6.70}$$

Hence we first choose $k_5 > 1/\alpha$ and set $k_2 = \delta k_*$ with δ being a small positive constant, say $\delta \leq \max(u_d^{\max}, \dot{u}_d^{\max}, r_d^{\max})$, with $\dot{u}_d^{\max}$ being the maximum value of $|\dot{u}_d|$. Next we replace k in all negative terms of $p_i(t)$, $i = 1, 21, 22, 23, 3$ by k_*. It is now observed that the negative parts of $p_i(t)$ have k_* as a factor. The mass including added mass in the sway dynamics, m_{22}, is always larger than that in the surge dynamics, m_{11}, for surface ships, i.e., $\beta < 1$. Therefore we can always find positive constants k_* and k_5 such that the conditions (6.36), (6.50), (6.55), and (6.57) hold with some small $\rho_i > 0$, $i = 1, 2, 3$, and λ_i being chosen based on (6.35). The value of constant k_* should be reduced if the reference yaw, r_d, and surge, u_d, velocities and surge acceleration, $\dot{u}_d$, are large. This physically means that the distance from the ship to the point it aims to track should be increased if the velocities and surge acceleration are large, otherwise the ship will miss that point. In fact, setting $\lambda_i = 0$ and picking $k_2 > 0$ results in the tracking controllers proposed in [19,71,106] with a restrictive assumption of the yaw reference velocity satisfying a persistently exciting condition. It can also be seen that setting $\lambda_2 = 0$ yields the tracking controller in the preceding chapter, which does not require the yaw reference velocity to be persistently exciting. Furthermore, small α automatically results in large k_5 and small k_1, see (6.70). This can be physically interpreted as follows: If the damping in the sway dynamics is small, the control gain in the surge dynamics should be small, otherwise the ship will slide in the sway direction. Due to complicated expressions of $p_i(t)$, we provide some guidelines to choose the design constants rather than present their extremely complex explicit expressions.

1. Pick positive constants λ_i such that the constant k_* is small, say $k_* << \max(u_d^{\max}, r_d^{\max}, \dot{u}_d^{\max})$, δ such that $\delta \leq \max(u_d^{\max}, r_d^{\max})$ and set $k_2 = \delta k_*$, $k_5 = 1/\alpha + \varepsilon$ with ε being a small positive constant.
2. Substitute (6.70) into $p_i(t)$ and slowly increase λ_i and k_5 until all the conditions (6.36), (6.50), (6.55), and (6.57) hold with some positive constants p_1^*, p_{211}^*, p_{21}^*, p_{22}^*, and p_3^*.

Remark 6.5. When the reference surge and sway velocities and surge acceleration are very large, the above procedure will result in very small control gains k, k_1, and k_2, which will give slow convergence. Hence it is suggested that for large u_d, r_d, and $\dot{u}_d$, the control gains k, k_1, and k_2 should be chosen such that either conditions ((6.27), (6.35), (6.36)) or ((6.27), (6.35), (6.36), (6.50), (6.55)), or ((6.27), (6.35), (6.36), (6.57)) hold to improve the convergence. The trade-off is that the control gains will be different for different reference trajectories.

6.5 Sensitivity Analysis

The control law (6.16) has been designed under the assumptions that the system parameters are precisely known and there are no environmental disturbances. Indeed, these assumptions are unrealistic in practice. The aim of this section is to discuss the sensitivity of our proposed controller in relation to the inaccurate knowledge of the ship parameters. A discussion related to environmental disturbances can be carried out similarly.

The control law (6.16) can be easily modified to account for the inaccurate knowledge of the ship parameters entering (6.15), by adding some robustifying terms or can be changed into an adaptive version such that the $(z_e, \tilde{u}_e, \tilde{r}_e)$-subsystem is globally exponentially/asymptotically stable at the origin. However, this control law cannot overcome the imprecise knowledge of the ship parameters, which enter the last equation of (6.2) due to the fact that the reference trajectory is generated by the virtual ship model (6.2). Therefore, we only focus on this issue to simplify the analysis, i.e., we are interested in the question: Assuming that the control law (6.16) has been modified as above, how do the ship parameters affect the tracking errors when the virtual ship model (6.2) is formed by the nominal values of the ship parameters?

Defining

$$\Delta\alpha = \alpha - \alpha_c,\ \Delta\beta = \beta - \beta_c, \tag{6.71}$$

where α_c and β_c are the known parts of α and β, we rewrite the v_e-dynamics of (6.8) as

$$\dot{v}_e = -\alpha\, v_e - \beta\, u_e r_d - \beta\,(u_e + u_d) r_e + \Delta\alpha v_d + \Delta\beta u_d r_d. \tag{6.72}$$

Notice that we cannot choose the design constants such that $1-k_4(\alpha k_5-1)=0$ as in the previous sections to cancel the term $y_e v_e(1-k_4(\alpha k_5-1))$, see (6.26), due to the inaccurately known ship parameters. If one replaces α in the above expression by $\alpha_c+\Delta\alpha$, the term $y_e v_e(1-k_4(\alpha_c k_5-1))$ will be canceled. However the remaining term $\Delta\alpha k_4 k_5 y_e v_e$ cannot be canceled in the case of C3 because this term does not contain any reference velocities. Therefore, we replace k_5 in the Lyapunov function (6.19) by k_5^*/α where k_5^* is a positive constant to be determined, i.e.,

$$V_2^*(X_{2e})=\frac{1}{2}\left(x_e^2+y_e^2+k_4(y_e+k_5^*\alpha^{-1}v_e)^2\right). \tag{6.73}$$

The condition (6.27) is replaced by

$$1-k_4(k_5^*-1)=0,\ \beta_c\alpha_c^{-1}k_1k_5^*-1=0. \tag{6.74}$$

Note that the term Φ is now written as

$$\Phi=\Delta\beta^*k_1k_4\frac{k_5^*}{\alpha}(y_e+\frac{k_5^*}{\alpha}v_e)(r_{1e}^d+r_d)x_e, \tag{6.75}$$

where $\Delta\beta^*=\beta/\alpha-\beta_c/\alpha_c$. With (6.72) and (6.75), we have

$$\begin{aligned}\dot{V}_2^* \leq\ & -q_1(t)x_e^2-q_{21}(t)y_e^2-\frac{q_{22}(t)y_e^2}{\sqrt{1+x_e^2+y_e^2}}+\frac{q_{23}(t)y_e^2}{\sqrt{1+x_e^2+y_e^2}}-q_3(t)v_e^2+\\ & (\chi_1V_2+\chi_2)e^{-\sigma_1(t-t_0)}+\left(\frac{1}{4\phi_1}\frac{k_4k_5^*}{\alpha}+\frac{1}{4\phi_2}\frac{k_4k_5^{*2}}{\alpha^2}\right)|\Delta\alpha v_d+\Delta\beta u_d r_d|,\end{aligned} \tag{6.76}$$

where

$$\begin{aligned}q_1(t)=\ & \hat{p}_1(t)-\frac{k_1k_4k_5^*|\Delta\beta^*|}{\alpha}\left(\frac{1}{4\rho_1}+\frac{k_5^*}{\alpha}+\frac{|k(t)|}{1-|k(t)|}\left(\frac{1+4\rho_1|r_d|}{4\rho_1\sqrt{1-k^2(t)}}+\right.\right.\\ & \left.k_1+k_2|r_d|+k^2(t)|u_d|+|k(t)v_d|\right)+\frac{|k(t)|k_5^*}{4\rho_1\alpha(1-|k(t)|)}\times\\ & \left.\left(k_1+|r_d|+k_2|r_d|+k^2(t)|u_d|+|k(t)v_d|+|k(t)u_d|\right)\right),\end{aligned} \tag{6.77}$$

$$q_{21}(t)=\hat{p}_{21}(t)-\left(k_4\rho_1|\Delta\beta^*|r_d^2+\frac{k_4k_5^*}{\alpha}|\Delta\alpha v_d+\Delta\beta u_d r_d|\right), \tag{6.78}$$

$$\begin{aligned}q_{22}(t)=\ & \hat{p}_{22}(t)-\frac{k_1k_4k_5^*|\Delta\beta^*|}{\alpha(1-|k(t)|)}\left(\frac{|\dot{k}(t)|+|k(t)|\rho_1}{\sqrt{1-k^2(t)}}+\frac{k_5^*}{4\rho_2\alpha}\times\right.\\ & \left.\frac{|\dot{k}(t)|}{\sqrt{1-k^2(t)}}+2\left(k^2(t)|u_d|+|k^3(t)v_d|\right)\left(1+\frac{k_5^*}{4\rho_2\alpha}\right)\right),\end{aligned} \tag{6.79}$$

$$q_{23}(t)=\hat{p}_{23}(t), \tag{6.80}$$

$$
\begin{aligned}
q_3(t) = \;& \hat{p}_3(t) - \left(\frac{k_4 k_5^{*2}}{\alpha^2} \left| \Delta\alpha v_d + \Delta\beta u_d r_d \right| \phi_2 + \frac{k_4 k_5^* \left| \Delta\beta^* \right| \rho_1}{\alpha} + \right. \\
& \frac{2\left|k(t)\right| k_4 \left|\Delta\beta^*\right| \rho_1}{\left(1-\left|k(t)\right|\right)\sqrt{1-k^2(t)}} + \frac{k_1 k_4 k_5^{*2} \left|\Delta\beta^*\right|}{\alpha^2 \left(1-\left|k(t)\right|\right)} \left(\frac{\left|\dot{k}(t)\right| \rho_2}{\sqrt{1-k^2(t)}} + \right. \\
& 2\left|k(t)\right| + (\rho_1+\rho_2)\left|k^3(t)\right| \left(|v_d|+|u_d|\right) + \rho_1 \left|k(t)\right| (k_1 + \\
& \left.\left. k_2 + |r_d|) + \rho_1 k^2(t)\left(|v_d|+|u_d|\right)\right)\right), \qquad (6.81)
\end{aligned}
$$

$\hat{p}_1$, $\hat{p}_{21}$, $\hat{p}_{22}$, $\hat{p}_{23}$, and $\hat{p}_3$ are p_1, p_{21}, p_{22}, p_{23}, and p_3 respectively, with k_5 being replaced by k_5^*/α, and ϕ_1 and ϕ_2 are some positive constants. From (6.76), the stability analysis of the (x_e, y_e, v_e)-subsystem can be carried out using the same arguments as in the previous section. It can be shown that there always exist the design constants k_1, k_3, k_4, and k_5^* such that the required conditions hold if $\Delta\alpha$ and $\Delta\beta$ are not too large, say $\Delta\alpha << \alpha$, $\Delta\beta << \beta$ and $\Delta\beta^* << \min(\beta,\alpha)$. It is noted that when Assumption 6.2 holds with either cases C1, C3, or C4, the (x_e, y_e, v_e)-subsystem is still GAS since the term $|\Delta\alpha v_d + \Delta\beta u_d r_d|$ in (6.76) globally exponentially/asymptotically vanishes at the origin. However, only global practical stability can be achieved in the case of C2. This coincides with the ship control practice in the sense that a big ship, in general, cannot accurately track the reference trajectory generated by a small ship because of underactuated configuration in the sway dynamics.

Remark 6.6. Since the inaccurate knowledge of the ship parameters directly affects the tracking errors when Assumption 6.2 holds with case C2 and the v_e-dynamics is globally ISS with respect to u_e and r_e as inputs, see (6.8), we can also treat the v_e-dynamics as unmodeled dynamics. In this case, one can apply the methodology proposed in this chapter and the work on controlling nonlinear systems with unmodeled dynamics [112].

6.6 Simulations

The same monohull ship for the trajectory tracking simulations in Section 5.4 is used in this section to validate the control law (6.16) where all the four conditions in Assumption 6.2 are considered. The reference trajectory is generated by the virtual ship with $[x_d(0), y_d(0), \psi_d(0), u_d(0), v_d(0), r_d(0)] = \left[0\,\text{m}, 0\,\text{m}, 0\,\text{ms}^{-1}, 0\,\text{rads}^{-1}\right]$ and

- for the case C1: $\tau_{ud} = -m_{22} v_d r_d$, $\tau_{rd} = -(m_{11} - m_{22}) u_d r_d$, i.e., the reference trajectory is a point at the origin,
- for the case C2: $\tau_{ud} = -m_{22} v_d r_d + d_{11}$, $\tau_{rd} = -(m_{11} - m_{22}) u_d r_d + 0.01 d_{33}$, i.e., the reference trajectory is a circle with a radius of 200 m and the reference velocity u_d is 1 ms^{-1},
- for the case C3: $\tau_{ud} = -m_{22} v_d r_d + d_{11}$, $\tau_{rd} = -(m_{11} - m_{22}) u_d r_d$, i.e., the reference trajectory is a straight line and the reference velocity u_d is 1 ms^{-1},

- for the case C4: $\tau_{ud} = -m_{22}v_d r_d + 0.1d_{11}e^{-0.3t}$, $\tau_{rd} = -(m_{11}-m_{22})u_d r_d + 0.1d_{33}e^{-5t}$, i.e., the reference trajectory is a path approaching to the origin.

We first choose $k_3 = 0.8$, $\lambda_3 = 3.5$, and $c_1 = c_2 = 5$. Note that these constants can be any arbitrarily positive real numbers. The bigger these constants, the faster the transient response will be, but the control effort becomes larger. Next, we initialize $k_* = 0.01$ and $k_5 = 1.8$, then increase these values until all of the conditions (6.36), (6.50), (6.55), and (6.57) hold. We get $k_* = 0.02$ and $k_5 = 2.14$, which result in $k_1 = 0.85$, $k_2 = 0.54$, $k_4 = 6.4$, $\lambda_1 = 0.015$, and $\lambda_2 = 0.005$, see (6.35) and (6.70). With the above design constant selection, one can verify that all conditions (6.27), (6.35), (6.36), (6.50), (6.55), and (6.57) hold with $p_1^* = 0.47$, $p_{211}^* = 0.15$, $p_{21}^* = 0.07$, $p_{22}^* = 0.13$, $p_3^* = 0.16$, $\rho_1 = 0.2$, $\rho_2 = 2$, and $\rho_3 = 0.5$. The initial conditions of the ship are chosen as $[x(0), y(0), \psi(0), u(0), v(0), r(0)] = [5\,\text{m}, -40\,\text{m}, 0.7\,\text{rad}, 0\,\text{ms}^{-1}, 0.5\,\text{ms}^{-1}, 0\,\text{rads}^{-1}]$.

Figures 6.1–6.4 plot the tracking errors, control inputs, and tracking trajectory in the (x, y)-plane for cases C1–C4. It can be seen from Figures 6.1a–6.4a that the tracking errors asymptotically (K-exponentially for the case of C2) converge to zero as proven in Theorem 6.1. It is noted that the transient response for cases C1 and C4 is much slower and is more oscillatory than cases C2 and C3 because the reference velocities u_d and r_d are zero or approach zero. Hence the term $\lambda_1 u_d$ does not help in the control gain $k(t)$, see (6.6). The observation about the transient response for cases C1 and C4 should not be surprising due to the sinusoidal signal introduced in the control gain $k(t)$, and is also true for any time-varying controllers that stabilize an underactuated system at the origin [15].

To test the robustness of our proposed controller with respect to the ship parameters and small environmental disturbances induced by wave, wind, and ocean current, we simulate the control law (6.16) with the same design constants selected as above. Figures 6.5-6.8 plot the tracking errors, control inputs, and tracking trajectory in (x, y)-plane for cases C1–C4 with 8% variation in all of the system parameters in the sense that the ship parameters are taken as $0.92m_{11}$, $1.08m_{22}$, $0.92m_{33}$, $0.92d_{11}$, $1.08d_{22}$, $1.08d_{33}$, and with the environmental disturbances acting on the surge, sway and yaw dynamics as $\tau_{wu}(t) = 0.1m_{11}\text{rand}(\cdot)$, $\tau_{wv}(t) = 0.1m_{22}\text{rand}(\cdot)$, $\tau_{wv}(t) = 0.1m_{33}\text{rand}(\cdot)$, where rand$(\cdot)$ is the zero-mean random noise with the uniform distribution on the interval $[-0.5\;0.5]$. The above choice of the variation in the ship parameters is only for illustrating robustness properties of the controller designed in this chapter with respect to the system parameter uncertainties. Note that this variation directly affects the tracking errors in Case C2 since the reference sway velocity has to be generated by the virtual ship. It is worth mentioning that under arbitrarily small nonvanishing environmental disturbances, the discontinuous stabilization controller proposed in [55] results in an unstable closed loop system in the sense that the closed loop trajectory goes to infinity exponentially fast.

With the above disturbances, the last three equations of (6.1) are of the form

$$\dot{u} = \frac{m_{22}}{m_{11}}vr - \frac{d_{11}}{m_{11}}u + \frac{1}{m_{11}}\tau_u + \frac{1}{m_{11}}\tau_{wu}(t),$$

$$\dot{v} = -\frac{m_{11}}{m_{22}}ur - \frac{d_{22}}{m_{22}}v + \frac{1}{m_{22}}\tau_{wv}(t),$$
$$\dot{r} = \frac{(m_{11} - m_{22})}{m_{33}}uv - \frac{d_{33}}{m_{33}}r + \frac{1}{m_{33}}\tau_r + \frac{1}{m_{33}}\tau_{wr}(t). \quad (6.82)$$

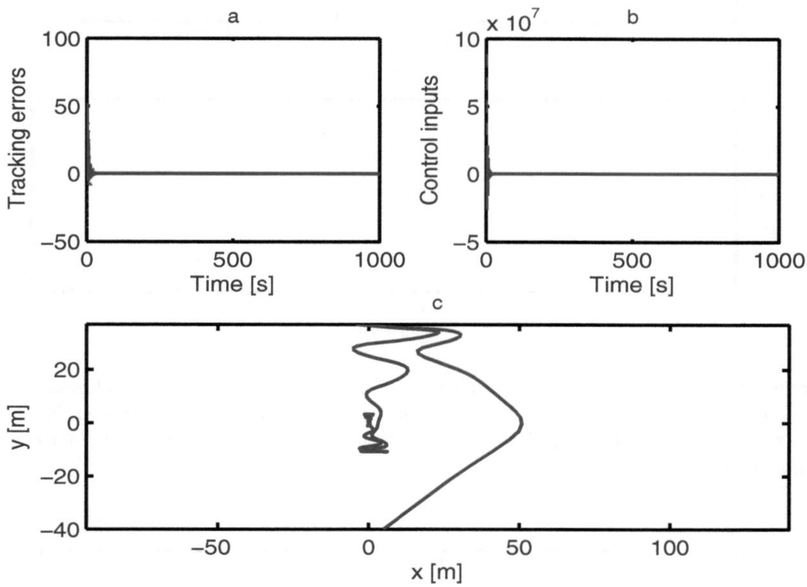

Figure 6.1 Simulation results without disturbances for case C1: **a**. Convergence of tracking errors ($x - x_d$[m]: *solid line*, $y - y_d$[m]: *dot-dotted line*, $\psi - \psi_d$[rad]: *dash-dotted line*); **b**. Control inputs (τ_u[N]: *solid line*, τ_r[Nm]: *dashed line*); **c**. Tracking trajectory in the (x, y)-plane ((x, y): *solid line*, (x_d, y_d): *dashed line*)

6.7 Conclusions

Restrictive assumptions on reference velocities required in the literature and Chapter 5 have been removed thanks to the coordinate transformation (6.6). The proposed methodology in this chapter can be readily extended to design a single controller for simultaneous stabilization and tracking of underactuated underwater vehicles. The work presented in this chapter is based on [105, 113–115].

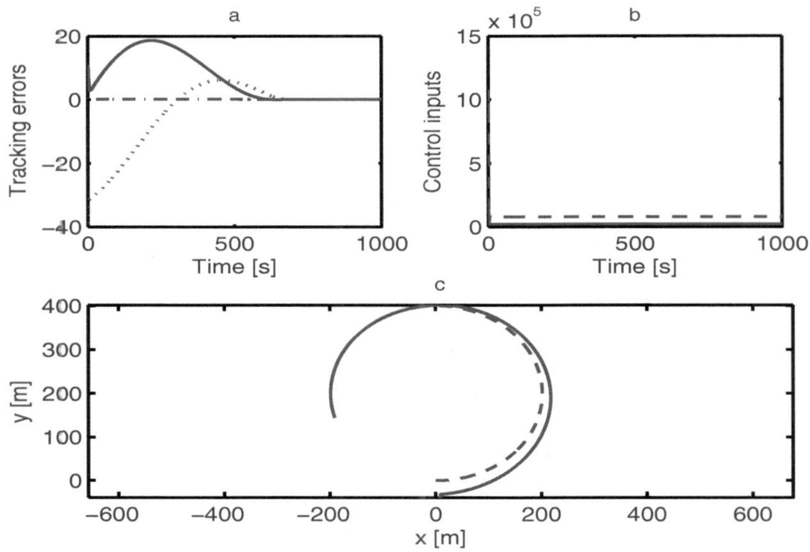

Figure 6.2 Simulation results without disturbances for case C2: **a**. Convergence of tracking errors ($x - x_d$ [m]: *solid line*, $y - y_d$ [m]: *dot-dotted line*, $\psi - \psi_d$ [rad]: *dash-dotted line*); **b**. Control inputs (τ_u[N]: *solid line*, τ_r [Nm]: *dashed line*); **c**. Tracking trajectory in the (x, y)-plane ((x, y): *solid line*, (x_d, y_d): *dashed line*)

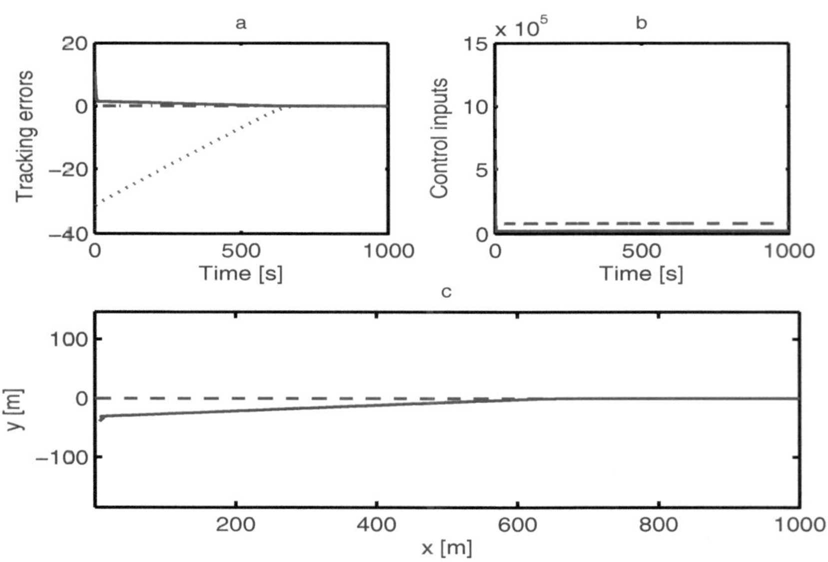

Figure 6.3 Simulation results without disturbances for case C3: **a**. Convergence of tracking errors ($x - x_d$ [m]: *solid line*, $y - y_d$ [m]: *dot-dotted line*, $\psi - \psi_d$ [rad]: *dash-dotted line*); **b**. Control inputs (τ_u[N]: *solid line*, τ_r [Nm]: *dashed line*); **c**. Tracking trajectory in the (x, y)-plane ((x, y): *solid line*, (x_d, y_d): *dashed line*)

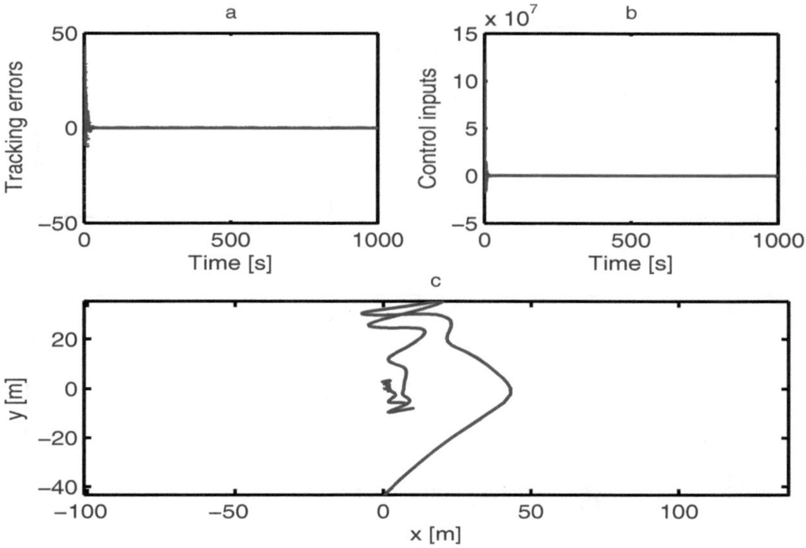

Figure 6.4 Simulation results without disturbances for case C4: **a**. Convergence of tracking errors ($x - x_d$[m]: *solid line*, $y - y_d$[m]: *dot-dotted line*, $\psi - \psi_d$[rad]: *dash-dotted line*); **b**. Control inputs (τ_u[N]: *solid line*, τ_r[Nm]: *dashed line*); **c**. Tracking trajectory in the (x, y)-plane ((x, y): *solid line*, (x_d, y_d): *dashed line*)

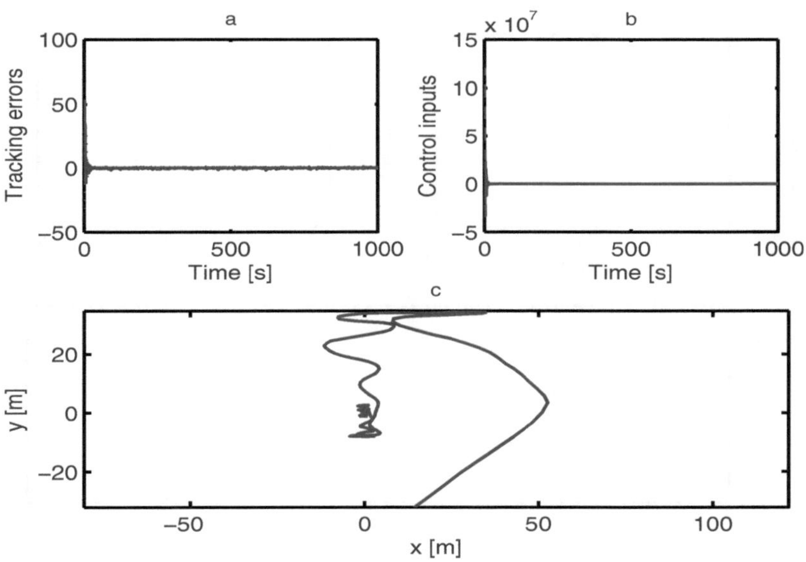

Figure 6.5 Simulation results with disturbances for case C1: **a**. Convergence of tracking errors ($x - x_d$[m]: *solid line*, $y - y_d$[m]: *dot-dotted line*, $\psi - \psi_d$[rad]: *dash-dotted line*); **b**. Control inputs (τ_u[N]: *solid line*, τ_r[Nm]: *dashed line*); **c**. Tracking trajectory in the (x, y)-plane ((x, y): *solid line*, (x_d, y_d): *dashed line*)

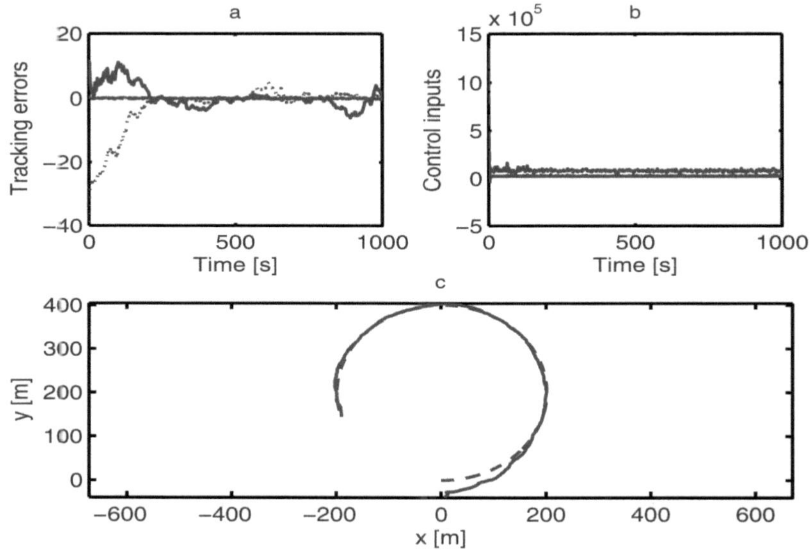

Figure 6.6 Simulation results with disturbances for case C2: **a**. Convergence of tracking errors ($x - x_d$ [m]: *solid line*, $y - y_d$ [m]: *dot-dotted line*, $\psi - \psi_d$ [rad]: *dash-dotted line*); **b**. Control inputs (τ_u[N]: *solid line*, τ_r [Nm]: *dashed line*); **c**. Tracking trajectory in the (x, y)-plane ((x, y): *solid line*, (x_d, y_d): *dashed line*)

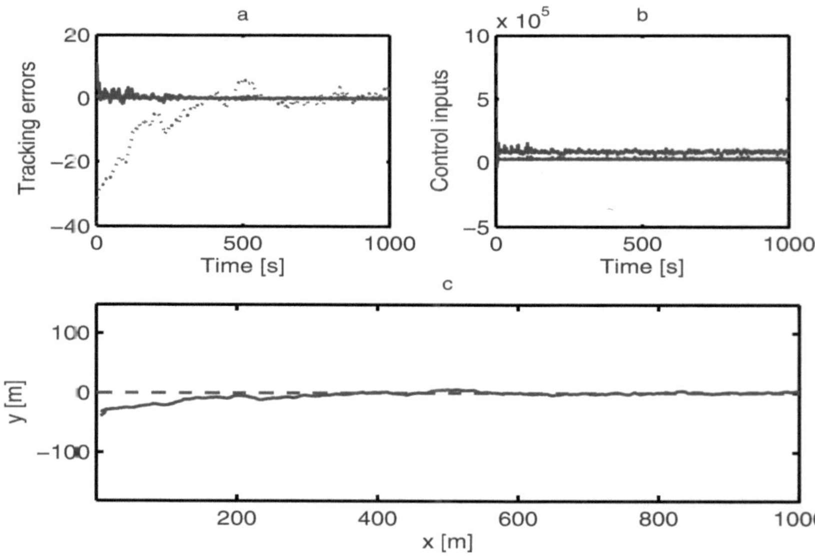

Figure 6.7 Simulation results with disturbances for case C3: **a**. Convergence of tracking errors ($x - x_d$ [m]: *solid line*, $y - y_d$ [m]: *dot-dotted line*, $\psi - \psi_d$ [rad]: *dash-dotted line*); **b**. Control inputs (τ_u[N]: *solid line*, τ_r [Nm]: *dashed line*); **c**. Tracking trajectory in the (x, y)-plane ((x, y): *solid line*, (x_d, y_d): *dashed line*)

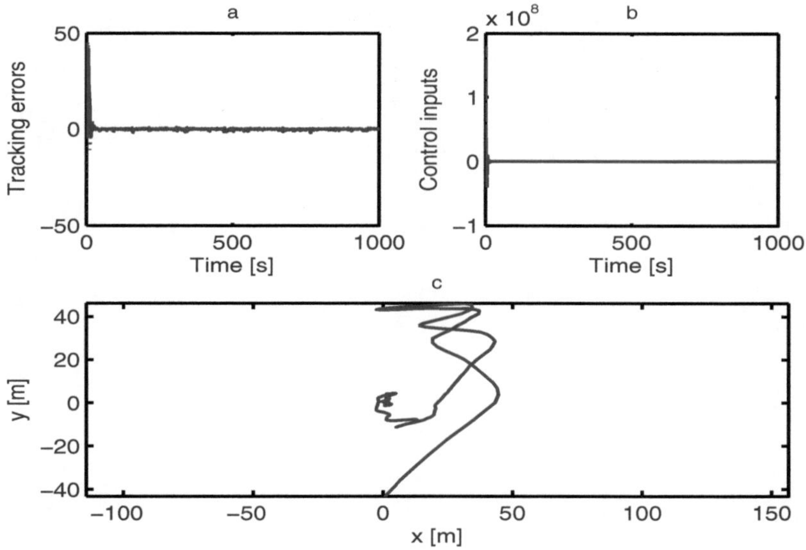

Figure 6.8 Simulation results with disturbances for case C4: **a**. Convergence of tracking errors ($x - x_d$[m]: *solid line*, $y - y_d$[m]: *dot-dotted line*, $\psi - \psi_d$[rad]: *dash-dotted line*); **b**. Control inputs (τ_u[N]: *solid line*, τ_r[Nm]: *dashed line*); **c**. Tracking trajectory in the (x, y)-plane ((x, y): *solid line*, (x_d, y_d): *dashed line*)

Chapter 7
Partial-state and Output Feedback Trajectory-tracking Control of Underactuated Ships

Global partial-state feedback and output feedback control schemes are discussed in this chapter for tracking control of an underactuated surface ship without sway force. For the case of partial-state feedback, we do not require measurements of the ship sway and surge velocities, while for the case of output feedback, none of the ship velocities are required for feedback. The reference trajectory to be tracked can be a curve including a straight line. Global nonlinear coordinate changes are introduced to transform the ship dynamics to a system affine in the ship velocities to design observers to globally exponentially estimate unmeasured velocities. These observers plus the techniques in the previous chapter facilitate the development of controllers in the following sections.

7.1 Control Objective

For the convenience of the reader, the mathematical model of the underactuated ship moving in surge, sway and yaw, see Section 3.4.1.2 (i.e., (3.45) and (3.46)), is once again presented:

$$\begin{aligned}\dot{\boldsymbol{\eta}} &= \boldsymbol{J}(\boldsymbol{\eta})\boldsymbol{v},\\ \boldsymbol{M}\dot{\boldsymbol{v}} &= -\boldsymbol{C}(\boldsymbol{v})\boldsymbol{v} - \boldsymbol{D}\boldsymbol{v} + \boldsymbol{\tau},\end{aligned} \tag{7.1}$$

where the matrices $\boldsymbol{J}(\boldsymbol{\eta})$, $\boldsymbol{M}$, $\boldsymbol{C}(\boldsymbol{v})$, and $\boldsymbol{D}$ are given by

$$\begin{aligned}\boldsymbol{J}(\boldsymbol{\eta}) &= \begin{bmatrix}\cos(\psi) & -\sin(\psi) & 0\\ \sin(\psi) & \cos(\psi) & 0\\ 0 & 0 & 1\end{bmatrix}, \quad \boldsymbol{M} = \begin{bmatrix} m_{11} & 0 & 0\\ 0 & m_{22} & 0\\ 0 & 0 & m_{33}\end{bmatrix},\\ \boldsymbol{C}(\boldsymbol{v}) &= \begin{bmatrix}0 & 0 & -m_{22}v\\ 0 & 0 & m_{11}u\\ m_{22}v & -m_{11}u & 0\end{bmatrix}, \quad \boldsymbol{D} = \begin{bmatrix} d_{11} & 0 & 0\\ 0 & d_{22} & 0\\ 0 & 0 & d_{33}\end{bmatrix},\end{aligned} \tag{7.2}$$

with

$$\begin{aligned} m_{11} &= m - X_{\dot{u}},\ m_{22} = m - Y_{\dot{v}},\ m_{33} = I_z - N_{\dot{r}}, \\ d_{11} &= -X_u,\ d_{22} = -Y_v,\ d_{33} = -N_r. \end{aligned} \tag{7.3}$$

The propulsion force and moment vector $\boldsymbol{\tau}$ is still given by (3.43), i.e.,

$$\boldsymbol{\tau} = \begin{bmatrix} \tau_u \\ 0 \\ \tau_r \end{bmatrix}. \tag{7.4}$$

We assume that the reference trajectory is generated by a virtual ship as follows:

$$\begin{aligned} \dot{\boldsymbol{\eta}}_d &= \boldsymbol{J}(\boldsymbol{\eta}_d)\boldsymbol{v}_d, \\ \dot{v}_d &= -\frac{m_{11}}{m_{22}} u_d r_d - \frac{d_{22}}{m_{22}} v_d, \end{aligned} \tag{7.5}$$

where all the variables have similar meanings as in system (7.1). It is noted that we do not require the reference surge and yaw velocities to be generated by the virtual ship. In this chapter we impose the following assumptions on the reference model (7.5):

Assumption 7.1. *The reference signals $x_d, y_d, u_d, r_d, \dot{u}_d, \ddot{u}_d$ and $\dot{r}_d$ are bounded. There exists a strictly positive constant $u_{d\min}$, such that $|u_d(t)| \geq u_{d\min}$, $\forall t \geq 0$. The reference sway velocity satisfies $|v_d(t)| < |u_d(t)|$, $\forall t \geq 0$.*

Assumption 7.2. *One of the following conditions holds:*

C1. The surge and sway displacements (x, y), yaw angle, ψ, and yaw velocity, r, are measurable but the surge and sway velocities, u and v, are not.
C2. The surge and sway displacements (x, y) and yaw angle ψ are measurable but none of the velocities u, v, and r are measurable.

Remark 7.1. Condition $|u_d(t)| \geq u_{d\min}$, $\forall t \geq 0$ implies that the reference surge velocity is always nonzero but can be either positive or negative. This means that we consider both forward and backward tracking. From a practical control viewpoint of surface ships, the condition $|u_d(t)| \geq u_{d\min}$, $\forall t \geq 0$ is much less restrictive than a persistently exciting condition on the yaw reference velocity in the references [19, 71, 106] in the sense that tracking of a straight line is included. Surface ships are often equipped with a rudder or a pair of propellers or water jets. The yaw moment to steer the ship is generated by changing the rudder angle or the speed of each propeller or water jet. These facts imply that the tracking control is carried out only when the surge speed is nonzero. The condition $|v_d(t)| < |u_d(t)|$, $\forall t \geq 0$ implies that the ship cannot track a circle with arbitrarily small radius due to the ship's high inertia and underactuation in the sway direction.

Remark 7.2. Assumption 7.2.C1 means that we need to solve a partial-state feedback control problem. Although the yaw velocity is measurable, there is still a cross term uv in the yaw velocity dynamics, see the last equation of (7.1). Assumption 7.2.C2 implies that we need to solve an output feedback control problem. Indeed, Assumption 7.2.C2 covers Assumption 7.2.C1. We will, however, show later that design of an output feedback tracking controller is much more involved than that of the partial-state feedback controller.

7.2 Partial-state Feedback

7.2.1 Observer Design

As discussed above, since the term $\boldsymbol{C}(\boldsymbol{v})\boldsymbol{v}$ in (7.1) causes difficulties in observer design, we first remove this term by proposing the following coordinate transformation:

$$\boldsymbol{X} = e^{\boldsymbol{Q}(t,\boldsymbol{\eta})\boldsymbol{v}}, \tag{7.6}$$

where $\boldsymbol{Q}(t,\boldsymbol{\eta})$ is a matrix to be determined. Differentiating both sides of (7.6) along the solutions of the second equation of (7.1) yields

$$\dot{\boldsymbol{X}} = e^{\boldsymbol{Q}(t,\boldsymbol{\eta})}[\dot{\boldsymbol{Q}}(t,\boldsymbol{\eta}) - \boldsymbol{M}^{-1}\boldsymbol{C}(\boldsymbol{v})]\boldsymbol{v} + e^{\boldsymbol{Q}(t,\boldsymbol{\eta})}(-\boldsymbol{D}\boldsymbol{v} + \boldsymbol{\tau}). \tag{7.7}$$

It can be seen that the square bracket on the right-hand side of (7.7) is zero, if the matrix $\boldsymbol{Q}(t,\boldsymbol{\eta})$ is chosen such that

$$\dot{\boldsymbol{Q}}(t,\boldsymbol{\eta}) - \boldsymbol{M}^{-1}\boldsymbol{C}(\boldsymbol{v}) = 0. \tag{7.8}$$

By using the first equation of (7.1), a particular solution of (7.8) is

$$\boldsymbol{Q}(t,\boldsymbol{\eta}) = \boldsymbol{M}^{-1}\begin{bmatrix} 0 & 0 & -m_{22}q_{13} \\ 0 & 0 & m_{11}q_{23} \\ m_{22}q_{13} & -m_{11}q_{23} & 0 \end{bmatrix}, \tag{7.9}$$

with

$$\begin{aligned} q_{13} &= y\cos(\psi) - x\sin(\psi) + p_{13}(t), \\ q_{23} &= y\sin(\psi) + x\cos(\psi) + p_{23}(t), \\ \dot{p}_{13} &= y\sin(\psi)r + x\cos(\psi)r, \\ \dot{p}_{23} &= -y\cos(\psi)r + x\sin(\psi)r. \end{aligned} \tag{7.10}$$

Note that the matrix $\boldsymbol{Q}(t,\boldsymbol{\eta})$ contains only the available signals since we assume that x, y, ψ, and r are measurable. Using the Taylor expansion the matrix $e^{\boldsymbol{Q}(t,\boldsymbol{\eta})}$ can be expanded as

$$e^{\mathbf{Q}(t,\eta)} = \begin{bmatrix} \frac{1}{2a_5}\left(2a_2a_4 + a_1a_3\left(e^{-\sqrt{a_5}} + e^{\sqrt{a_5}}\right)\right) & \frac{1}{2a_5}\left(a_1a_4\left(e^{-\sqrt{a_5}} + e^{\sqrt{a_5}} - 2\right)\right) & \frac{1}{2\sqrt{a_5}}\left(a_1\left(-e^{-\sqrt{a_5}} + e^{\sqrt{a_5}}\right)\right) \\ \frac{1}{2a_5}\left(a_2a_3\left(e^{-\sqrt{a_5}} + e^{\sqrt{a_5}} - 2\right)\right) & \frac{1}{2a_5}\left(2a_1a_3 + a_2a_4\left(e^{-\sqrt{a_5}} + e^{\sqrt{a_5}}\right)\right) & \frac{1}{2\sqrt{a_5}}\left(a_2\left(-e^{-\sqrt{a_5}} + e^{\sqrt{a_5}}\right)\right) \\ \frac{1}{2\sqrt{a_5}}\left(a_3\left(-e^{-\sqrt{a_5}} + e^{\sqrt{a_5}}\right)\right) & \frac{1}{2\sqrt{a_5}}\left(a_4\left(-e^{-\sqrt{a_5}} + e^{\sqrt{a_5}}\right)\right) & \frac{1}{2}\left(e^{-\sqrt{a_5}} + e^{\sqrt{a_5}}\right) \end{bmatrix}, \tag{7.11}$$

where

$$\begin{aligned} a_1 &= -\frac{m_{22}}{m_{11}} q_{13}, \\ a_2 &= \frac{m_{11}}{m_{22}} q_{23}, \\ a_3 &= \frac{m_{22}}{m_{33}} q_{13}, \\ a_4 &= -\frac{m_{11}}{m_{33}} q_{23}, \\ a_5 &= a_1a_3 + a_2a_4. \end{aligned} \tag{7.12}$$

Similarly, $e^{-\boldsymbol{Q}(t,\eta)}$ has the same form as (7.11), but all of the terms $a_i, 1 \le i \le 4$ have an opposite sign to those defined in (7.12). From (7.11) and noticing that $a_5 \le 0, \forall (q_{13}, q_{23}) \in \mathbb{R}^2$, it is easily seen that all elements of $e^{\boldsymbol{Q}(t,\eta)}$ or $e^{-\boldsymbol{Q}(t,\eta)}$ are bounded by some constants, which depend only on the ship parameters, m_{11}, m_{22}, and m_{33}. Using the coordinate change (7.6), the ship system (7.1) is written in $(\eta, \boldsymbol{X})$ coordinates as

$$\begin{aligned} \dot{\eta} &= \boldsymbol{J}(\eta) e^{-\boldsymbol{Q}(t,\eta)} \boldsymbol{X}, \\ \dot{\boldsymbol{X}} &= -e^{\boldsymbol{Q}(t,\eta)} \boldsymbol{M}^{-1} \boldsymbol{D} e^{-\boldsymbol{Q}(t,\eta)} \boldsymbol{X} + e^{\boldsymbol{Q}(t,\eta)} \boldsymbol{M}^{-1} \boldsymbol{\tau}. \end{aligned} \tag{7.13}$$

The system (7.13) has a very nice structure, namely linear in the unmeasured states. Of course, a reduced-order observer can be designed but it is often noise-sensitive. Here we use the following nonlinear observer to construct the unmeasured surge and sway velocities:

$$\begin{aligned} \dot{\hat{\eta}} &= \boldsymbol{J}(\eta) e^{-\boldsymbol{Q}(t,\eta)} \hat{\boldsymbol{X}} + \boldsymbol{K_0}(\eta - \hat{\eta}), \\ \dot{\hat{\boldsymbol{X}}} &= -e^{\boldsymbol{Q}(t,\eta)} \boldsymbol{M}^{-1} \boldsymbol{D} e^{-\boldsymbol{Q}(t,\eta)} \hat{\boldsymbol{X}} + e^{\boldsymbol{Q}(t,\eta)} \boldsymbol{M}^{-1} \boldsymbol{\tau} + \\ &\quad (\boldsymbol{J}(\eta) e^{-\boldsymbol{Q}(t,\eta)})^T (\eta - \hat{\eta}), \end{aligned} \tag{7.14}$$

where $\hat{\eta}$ and $\hat{X}$ are the estimates of η and X, respectively; $K_0 = K_0^T$ is the positive diagonal observer gain matrix. From (7.13) and (7.14), we have

$$\begin{aligned}\dot{\tilde{\eta}} &= J(\eta)e^{-Q(t,\eta)}\tilde{X} - K_0\tilde{\eta},\\ \dot{\tilde{X}} &= -e^{Q(t,\eta)}M^{-1}De^{-Q(t,\eta)}\tilde{X} - (J(\eta)e^{-Q(t,\eta)})^T\tilde{\eta},\end{aligned} \tag{7.15}$$

where $\tilde{\eta} = \eta - \hat{\eta}$ and $\tilde{X} = X - \hat{X}$. From (7.15), one can show that

$$\left\|(\tilde{\eta}(t), \tilde{X}(t))\right\| \le \left\|(\tilde{\eta}(t_0), \tilde{X}(t_0))\right\| e^{-\sigma_0(t-t_0)}, \ \forall\, 0 \le t_0 \le t < \infty, \tag{7.16}$$

with $\sigma_0 = \min\left(\lambda_{\min}(K_0), \lambda_{\min}(M^{-1}D)\right)$, which in turn implies that (7.14) is a global exponential observer of (7.13). We define $\hat{v} = [\hat{u},\ \hat{v},\ \hat{r}]^T$ being an estimate of the velocity vector v as

$$\hat{v} = e^{-Q(t,\eta)}\hat{X}. \tag{7.17}$$

Using (7.17) and (7.14), we rewrite (7.1) in $(\eta, \hat{v})$ coordinates as

$$\begin{bmatrix}\dot{\eta}\\ \dot{\hat{v}}\end{bmatrix} = \begin{bmatrix}J(\eta)\hat{v}\\ -M^{-1}C(\hat{v})\hat{v} - M^{-1}D\hat{v} + M^{-1}\tau\end{bmatrix} + \begin{bmatrix}0_{3\times 3} & H_{12}\\ H_{21} & H_{22}\end{bmatrix}\begin{bmatrix}\tilde{\eta}\\ \tilde{X}\end{bmatrix}, \tag{7.18}$$

where

$$\begin{aligned}H_{12} &= J(\eta)e^{-Q(t,\eta)},\\ H_{21} &= e^{-Q(t,\eta)}(J(\eta)e^{-Q(t,\eta)})^T,\\ H_{22} &= M^{-1}C(\hat{v} + e^{-Q(t,\eta)}\tilde{X})e^{-Q(t,\eta)} - M^{-1}C^*(\hat{v})e^{-Q(t,\eta)} +\\ &\quad e^{-Q(t,\eta)}M^{-1}C(\hat{v} + e^{-Q(t,\eta)}\tilde{X}),\end{aligned} \tag{7.19}$$

with $C^*(\hat{v})$ being defined such that $C^*(\hat{v})\tilde{v} = C(\tilde{v})\hat{v}$. It is now observed that the systems (7.15) and (7.18) are in a cascaded structure. It is also observed that the system $(\tilde{\eta}, \tilde{X})$ is GES at the origin and that the connected terms H_{12}, H_{21}, and H_{22} are Lipschitz in η and $\hat{v}$. Furthermore from (7.17) and (7.6), the velocity estimate error vector, $\tilde{v} = [\tilde{u},\ \tilde{v},\ \tilde{r}]^T = v - \hat{v}$, satisfies

$$\tilde{v} = e^{-Q(t,\eta)}\tilde{X}. \tag{7.20}$$

Since all elements of $e^{-Q(t,\eta)}$ are bounded, (7.16) and (7.20) imply that there exists a positive constant γ_0 such that

$$\|(\tilde{\eta}(t), \tilde{v}(t))\| \le \gamma_0 \|(\tilde{\eta}(t_0), \tilde{v}(t_0))\| e^{-\sigma_0(t-t_0)}, \ \forall 0 \le t_0 \le t < \infty, \tag{7.21}$$

which means that the estimation errors $\tilde{\eta}(t)$ and $\tilde{v}(t)$ globally exponentially converge to the origin.

7.2.2 Coordinate Transformations

We now interpret the position and orientation errors $x - x_d$, $y - y_d$, and $\psi - \psi_d$ in a frame attached to the ship body. That is, we consider the error coordinates

$$\begin{bmatrix} x_e \\ y_e \\ \psi_e \end{bmatrix} = \boldsymbol{J}^{-1}(\eta) \begin{bmatrix} x - x_d \\ y - y_d \\ \psi - \psi_d \end{bmatrix}, \quad \begin{bmatrix} u_e \\ v_e \\ r_e \end{bmatrix} = \begin{bmatrix} \hat{u} - u_d \\ \hat{v} - v_d \\ \hat{r} - r_d \end{bmatrix}. \tag{7.22}$$

Differentiating both sides of (7.22) along the solutions of (7.18) and (7.5) yields the error dynamics of the "kinematic part" in the transformed coordinates:

$$\begin{aligned} \dot{x}_e &= u_e - u_d(\cos(\psi_e) - 1) - v_d \sin(\psi_e) + r_e y_e + r_d y_e + h_x, \\ \dot{y}_e &= v_e - v_d(\cos(\psi_e) - 1) + u_d \sin(\psi_e) - r_e x_e - r_d x_e + h_y, \\ \dot{\psi}_e &= r_e + h_\psi \end{aligned} \tag{7.23}$$

where h_x, h_y, and h_ψ are the first, second, and third rows of $\boldsymbol{J}^{-1}(\eta) H_{12} \tilde{\boldsymbol{X}} + \tilde{r} \begin{bmatrix} y_e \\ -x_e \\ 0 \end{bmatrix}$, respectively.

By looking at (7.23), we see that x_e and ψ_e can be stabilized by u_e and r_e. There are several options to stabilize y_e, namely r_e, v_e, or ψ_e. If r_e is used, the control design will be extremely complicated since r_e enters all of the three equations of (7.23). On the other hand, the use of v_e to stabilize y_e will result in an undesired feature of ship control practice, namely the ship will slide in the sway direction. Hence we will use ψ_e to stabilize the sway error y_e. As such, we define the following coordinate transformation

$$z_e = \psi_e + \arcsin\left(\frac{k y_e}{\sqrt{c^2 + x_e^2 + y_e^2 + v_e^2}} \right), \tag{7.24}$$

where the constants k and c are such that $|k| < 1$ and $c \geq 1$ and will be specified later. It is seen that (7.24) is well defined and that convergence of z_e and y_e implies that of ψ_e. By using the nonlinear coordinate transformation (7.24) instead of a linear one like $z_e = \psi_e + k y_e$, we avoid the ship whirling around when y_e is large. The coordinate (7.24) is slightly different from the one in the preceding chapter. This will result in bounded virtual velocity controls. Using the nonlinear coordinate (7.24) together with (7.23), the ship error dynamics are rewritten as

$$\begin{aligned} \dot{x}_e = {} & u_e + u_d \varpi_2^{-1}(\varpi_2 - \varpi_1) + k v_d \varpi_2^{-1} y_e + r_e y_e + r_d y_e + p_x + h_x, \\ \dot{y}_e = {} & v_e + v_d \varpi_2^{-1}(\varpi_2 - \varpi_1) - k u_d \varpi_2^{-1} y_e - r_e x_e - r_d x_e + p_y + h_y, \end{aligned}$$

$$
\begin{aligned}
\dot{z}_e =&\ \left(1-k\varpi_1^{-1}x_e+k\beta\varpi_1^{-1}\varpi_2^{-2}y_e v_e(u_e+u_d)\right) r_e + \\
&k\varpi_1^{-1}\left(v_e+v_d\varpi_2^{-1}(\varpi_2-\varpi_1)-ku_d\varpi_2^{-1}y_e-r_d x_e- y_e\varpi_2^{-2}\times\right.\\
&x_e u_e+y_e v_e+(x_e u_d+y_e v_d)\varpi_2^{-1}(\varpi_2-\varpi_1)+(x_e v_d-y_e u_d))\\
&\left.\left(k\varpi_2^{-1}y_e-v_e(\alpha v_e+\beta u_e r_d)\right)\right)+p_z+h_z,\\
\dot{u}_e =&\ \frac{m_{22}}{m_{11}}\hat{v}\hat{r}-\frac{d_{11}}{m_{11}}\hat{u}+\frac{1}{m_{11}}\tau_u-\dot{u}_d+h_u,\\
\dot{v}_e =&\ -\frac{m_{11}}{m_{22}}u_e r_d-\frac{m_{11}}{m_{22}}(u_e+u_d)r_e-\frac{d_{22}}{m_{22}}v_e+h_v,\\
\dot{r}_e =&\ \frac{(m_{11}-m_{22})}{m_{33}}\hat{u}\hat{v}-\frac{d_{33}}{m_{33}}\hat{r}+\frac{1}{m_{33}}\tau_r-\dot{r}_d+h_r,
\end{aligned}
\tag{7.25}
$$

where h_u, h_v, and h_r are the first, second, and third rows of $H_{21}\tilde{\eta}+H_{22}\tilde{X}$, respectively. Also, for notational simplicity, we have defined

$$
\begin{aligned}
\varpi_1 =&\ \sqrt{c^2+x_e^2+(1-k^2)y_e^2+v_e^2},\\
\varpi_2 =&\ \sqrt{c^2+x_e^2+y_e^2+v_e^2},\\
\alpha =&\ \frac{d_{22}}{m_{22}},\quad \beta=\frac{m_{11}}{m_{22}},\\
p_x =&\ -u_d\left((\cos(z_e)-1)\varpi_1\varpi_2^{-1}+\sin(z_e)k\varpi_2^{-1}y_e\right)-\\
&v_d\left(\sin(z_e)\varpi_1\varpi_2^{-1}-(\cos(z_e)-1)k\varpi_2^{-1}y_e\right),\\
p_y =&\ -v_d\left((\cos(z_e)-1)\varpi_1\varpi_2^{-1}+\sin(z_e)k\varpi_2^{-1}y_e\right)+\\
&u_d\left(\sin(z_e)\varpi_1\varpi_2^{-1}-(\cos(z_e)-1)k\varpi_2^{-1}y_e\right),\\
p_z =&\ k\varpi_1^{-1}\left(p_y-y_e\varpi_2^{-2}\left(x_e p_x+y_e p_y\right)\right),\\
h_z =&\ k\varpi_1^{-1}\left(h_y-y_e\varpi_2^{-1}\left(x_e h_x+y_e h_y+v_e h_v\right)\right)+h_\psi.
\end{aligned}
\tag{7.26}
$$

It is now clear that the problem of forcing the underactuated ship (7.1) to track the virtual ship (7.5) becomes one of stabilizing the system (7.25) at the origin.

7.2.3 Control Design

The triangular structure of (7.25) suggests that we design the actual controls τ_u and τ_r in two stages. First, we design the virtual velocity controls for u_e and r_e to globally asymptotically stabilize x_e, y_e, z_e and v_e at the origin. Based on the backstepping technique, the controls τ_u and τ_r will be then designed. It is noted that the term $\left(1-kx_e/\varpi_1+k\beta y_e v_e(u_e+u_d)/(\varpi_1\varpi_2^2)\right)$ in the z_e-dynamics may vanish and therefore might prevent a global design. This problem can be fixed by decomposing u_e and r_e as

$$\begin{aligned} u_e &= u_e^d + \tilde{u}_e, \\ r_e &= r_e^d + \tilde{r}_e, \end{aligned} \tag{7.27}$$

where u_e^d and r_e^d are the virtual velocity controls of u_e and r_e; $\tilde{u}_e$ and $\tilde{r}_e$ are the virtual control errors.

Step 1

In this step, the virtual surge and yaw velocity controls are chosen as

$$u_e^d = -\frac{k_1 x_e}{\varpi_2},\ r_e^d = r_{1e}^d + r_{2e}^d, \tag{7.28}$$

where

$$\begin{aligned} r_{1e}^d &= -\frac{1}{1 - kx_e/\varpi_1 + k\beta y_e v_e(u_e^d + u_d)/(\varpi_1 \varpi_2^2)} \Bigg(\frac{k}{\varpi_1} \Bigg(v_e + \frac{v_d}{\varpi_2} \times \\ &\quad (\varpi_2 - \varpi_1)) - \frac{k u_d y_e}{\varpi_2} - r_d x_e - \Bigg(\frac{y_e}{\varpi_2} \Big(x_e u_e^d + y_e v_e + (x_e u_d + \\ &\quad y_e v_d) \frac{(\varpi_2 - \varpi_1)}{\varpi_2^{-1}} + (x_e v_d - y_e u_d) \frac{k y_e}{\varpi_2} - v_e(\alpha v_e + \beta u_e^d r_d) \Big) \Bigg) \Bigg), \\ r_{2e}^d &= -\frac{1}{1 - kx_e/\varpi_1 + k\beta y_e v_e(u_e^d + u_d)/(\varpi_1 \varpi_2^2)} \left(\frac{k_2 z_e}{\sqrt{1 + z_e^2}} + p_z \right), \end{aligned} \tag{7.29}$$

and k_i, $i = 1, 2$, are positive constants to be selected later. We have written r_e^d as a sum of r_{1e}^d and r_{2e}^d to simplify notation in the stability analysis later. Notice that

$$1 - \frac{kx_e}{\varpi_1} + \frac{k\beta y_e v_e(u_e^d + u_d)}{\varpi_1 \varpi_2^2} \geq 1 - |k| \left(1 + \frac{0.5\beta(k_1 + |u_d|)}{c} \right), \tag{7.30}$$

therefore r_{1e}^d and r_{2e}^d are well defined if the design constants k, c and k_1 are chosen such that

$$1 - |k|(1 + 0.5\beta(k_1 + |u_d|)/c) \geq k^* > 0. \tag{7.31}$$

By noting

$$\begin{aligned} |p_x| &\leq |u_d|(2 + |k|) + |v_d|(1 + 2|k|), \\ |p_y| &\leq |v_d|(2 + |k|) + |u_d|(1 + 2|k|), \\ |p_z| &\leq |k|(2|p_y| + |p_x|), \end{aligned} \tag{7.32}$$

we can show from (7.28) and (7.29) that u_e^d and r_e^d are bounded by some constants.

Remark 7.3. Unlike the standard application of the backstepping technique, in order to reduce complexity of the controller expressions, we have chosen a simple virtual control law u_e^d without canceling the known terms. From (7.29) and (7.28), we observe that r_{1e}^d is Lipschitz in (x_e, y_e, v_e) and r_{2e}^d vanishes when z_e does. This observation plays a crucial role in the stability analysis of the closed loop system.

Step 2

By differentiating (7.27) along the solutions of (7.25) and (7.29), the actual controls τ_u and τ_r without canceling the useful damping terms are chosen as

$$\begin{aligned}
\tau_u &= -m_{11}\Big(c_1\tilde{u}_e + m_{22}m_{11}^{-1}\hat{v}\hat{r} - d_{11}m_{11}^{-1}(u_e^d + u_d) - \dot{u}_d - \\
&\quad \frac{\partial u_e^d}{\partial x_e}\left(u_e + u_d\,\varpi_2^{-1}(\varpi_2 - \varpi_1) + kv_d\varpi_2^{-1}y_e + r_e y_e + r_d y_e\right) - \\
&\quad \frac{\partial u_e^d}{\partial y_e}\left(v_e + v_d\,\varpi_2^{-1}(\varpi_2 - \varpi_1) - ku_d\varpi_2^{-1}y_e - r_e x_e - r_d x_e\right) - \\
&\quad \frac{\partial u_e^d}{\partial v_e}m_{22}^{-1}\left(-m_{11}u_e r_d - m_{11}(u_e + u_d)r_e - d_{22}v_e\right) + \\
&\quad \varpi_1^{-1}\varpi_2^{-2}\left(k\beta y_e v_e r - k y_e(x_e - \beta r_d v_e)\right)z_e\Big), \\
\tau_r &= -m_{33}\Big(c_2\tilde{r}_e + (m_{11} - m_{22})m_{33}^{-1}\hat{u}\hat{v} - d_{33}m_{33}^{-1}(r_e^d + r_d) - \dot{r}_d - \\
&\quad \frac{\partial r_e^d}{\partial u_d}\dot{u}_d - \frac{\partial r_e^d}{\partial v_d}\dot{v}_d - \frac{\partial r_e^d}{\partial r_d}\dot{r}_d - \frac{\partial r_e^d}{\partial x_e}\big(u_e + u_d\,\varpi_2^{-1}(\varpi_2 - \varpi_1) + \\
&\quad kv_d\varpi_2^{-1}y_e + r_e y_e + r_d y_e\big) - \frac{\partial r_e^d}{\partial y_e}\big(v_e + v_d\,\varpi_2^{-1}(\varpi_2 - \varpi_1) - \\
&\quad ku_d\varpi_2^{-1}y_e - r_e x_e - r_d x_e\big) - \frac{\partial r_e^d}{\partial v_e}m_{22}^{-1}\big(-m_{11}u_e r_d - m_{11}(u_e + \\
&\quad u_d)r_e - d_{22}v_e\big) - \frac{\partial r_e^d}{\partial \psi_e}r_e + \big(1 - k\varpi_1^{-1}x_e + k\beta\varpi_1^{-1}\varpi_2^{-2} \times \\
&\quad y_e v_e(u_e^d + u_d)\big)z_e\Big),
\end{aligned} \tag{7.33}$$

where c_i, $i = 1, 2$, are positive constants. We now state the first main result of this chapter, the proof of which is given in the next section.

Theorem 7.1. *Under Assumption 7.1 assume the following:*

1. *There are no environmental disturbances*
2. *The ship parameters are known*
3. *The reference signals are generated by the virtual ship model (7.5) and the reference velocities satisfy Assumption 7.1.*

If the partial state feedback control law (7.33) together with the observer (7.18) are applied to the ship system (7.1), then the tracking errors $x(t) - x_d(t)$, $y(t) - y_d(t)$, $\psi(t) - \psi_d(t)$, *and* $v(t) - v_d(t)$ *globally asymptotically and locally exponentially* *converge to zero with an appropriate choice of the design constants c, k and k_i, $i = 1,2$. Furthermore, the virtual velocity controls, u_e^d and r_e^d, are bounded by some computable positive constants.*

7.2.4 Stability Analysis

Substituting (7.33) and (7.28) into (7.25) results in the following closed loop system:

$$\begin{aligned}\dot{\boldsymbol{X}}_{1e} &= \boldsymbol{f_1}(t,\boldsymbol{X_e})+\boldsymbol{g_1}(t,\boldsymbol{X_e})+\boldsymbol{\phi_1}(t,\hat{\boldsymbol{v}},\boldsymbol{X_e},\boldsymbol{\eta},\tilde{\boldsymbol{\eta}},\tilde{\boldsymbol{X}}),\\ \dot{\boldsymbol{X}}_{2e} &= \boldsymbol{f_2}(t,\boldsymbol{X_e})+\boldsymbol{\phi_2}(t,\hat{\boldsymbol{v}},\boldsymbol{X_e},\boldsymbol{\eta},\tilde{\boldsymbol{\eta}},\tilde{\boldsymbol{X}}),\end{aligned} \tag{7.34}$$

where

$$\boldsymbol{X}_{1e} = \begin{bmatrix} x_e \\ y_e \\ v_e \end{bmatrix},\ \boldsymbol{X}_{2e} = \begin{bmatrix} z_e \\ \tilde{u}_e \\ \tilde{r}_e \end{bmatrix},\ \boldsymbol{X_e} = \begin{bmatrix} \boldsymbol{X}_{1e} \\ \boldsymbol{X}_{2e} \end{bmatrix},$$

$$\boldsymbol{f_1}(t,\boldsymbol{X_e}) = \begin{bmatrix} f_{11} \\ f_{12} \\ f_{13} \end{bmatrix},\ \boldsymbol{f_2}(t,\boldsymbol{X_e}) = \begin{bmatrix} f_{21} \\ f_{22} \\ f_{23} \end{bmatrix},\ \boldsymbol{g_1}(t,\boldsymbol{X_e}) = \begin{bmatrix} p_x+\tilde{u}_e \\ p_y \\ g_{13} \end{bmatrix},$$

$$\boldsymbol{\phi_1}(t,\hat{\boldsymbol{v}},\boldsymbol{X_e},\boldsymbol{\eta},\tilde{\boldsymbol{\eta}},\tilde{\boldsymbol{X}}) = \begin{bmatrix} h_x^* \\ h_y^* \\ h_v \end{bmatrix},\ \boldsymbol{\phi_2}(t,\hat{\boldsymbol{v}},\boldsymbol{X_e},\boldsymbol{\eta},\tilde{\boldsymbol{\eta}},\tilde{\boldsymbol{X}}) = \begin{bmatrix} h_z \\ \varphi_{22} \\ \varphi_{23} \end{bmatrix},$$

$$\begin{aligned}
f_{11} &= -\frac{k_1 x_e}{\varpi_2}+u_d\frac{\varpi_2-\varpi_1}{\varpi_2}+\frac{k v_d y_e}{\varpi_2}+y_e(r_e+r_d+\tilde{r}),\\
f_{12} &= v_e+v_d\frac{\varpi_2-\varpi_1}{\varpi_2}-\frac{k u_d y_e}{\varpi_2}-x_e(r_e+r_d+\tilde{r}),\\
f_{13} &= -\alpha v_e-\beta u_e^d r_d-\beta(u_e^d+u_d)r_{1e}^d,\\
g_{13} &= -\beta\left(\tilde{u}_e(r_e^d+\tilde{r}_e+r_d)+(u_e^d+u_d)(r_{2e}^d+\tilde{r}_e)\right),\\
f_{21} &= -\frac{k_2 z_e}{\sqrt{1+z_e^2}}+\left(1-\frac{k x_e}{\varpi_1}+\frac{k\beta y_e v_e(u_e^d+u_d)}{\varpi_1\varpi_2^2}\right)\tilde{r}_e+\\
&\quad\left(\frac{k\beta y_e v_e r_e}{\varpi_1\varpi_2^2}-\frac{k y_e}{\varpi_1\varpi_2^2}(x_e-\beta r_d v_e)\right)\tilde{u}_e,\\
f_{22} &= -\left(c_1+\frac{d_{11}}{m_{11}}\right)\tilde{u}_e-\left(\frac{k\beta y_e v_e r_e}{\varpi_1\varpi_2^2}-\frac{k y_e}{\varpi_1\varpi_2^2}(x_e-\beta r_d v_e)\right)z_e,\\
f_{23} &= -\left(c_2+\frac{d_{33}}{m_{33}}\right)\tilde{r}_e-\left(1-\frac{k x_e}{\varpi_1}+\frac{k\beta y_e v_e(u_e^d+u_d)}{\varpi_1\varpi_2^2}\right)z_e,
\end{aligned}$$

$$\varphi_{22} = h_u - \frac{\partial u_e^d}{\partial x_e}(p_x + h_x) - \frac{\partial u_e^d}{\partial y_e}(p_y + h_y) - \frac{\partial u_e^d}{\partial v_e} h_v,$$
$$\varphi_{23} = h_r - \frac{\partial r_e^d}{\partial x_e}(p_x + h_x) - \frac{\partial r_e^d}{\partial y_e}(p_y + h_y) - \frac{\partial r_e^d}{\partial v_e} h_v - \frac{\partial r_e^d}{\partial \psi_e} h_\psi, \quad (7.35)$$

with h_x^* and h_y^* being the first and second rows of $\boldsymbol{J}^{-1}(\boldsymbol{\eta}) H_{12} \tilde{\boldsymbol{X}}$. The time dependence of $\boldsymbol{f_i}(t, \boldsymbol{X_e})$, $\boldsymbol{g_1}(t, \boldsymbol{X_e})$, and $\boldsymbol{\phi_i}(t, \hat{\boldsymbol{v}}, \boldsymbol{X_e}, \boldsymbol{\eta}, \tilde{\boldsymbol{\eta}}, \tilde{\boldsymbol{X}})$, $i = 1,2$, is due to the time-varying reference velocities. Observe that the closed loop system (7.34) consists of the $(\boldsymbol{X_{1e}}, \boldsymbol{X_{2e}})$-subsystem and $(\tilde{\boldsymbol{\eta}}, \tilde{\boldsymbol{X}})$-subsystem (see (7.15)) in a cascaded structure. From (7.35) it can be readily shown that the connected terms $\boldsymbol{\phi_i}(t, \hat{\boldsymbol{v}}, \boldsymbol{X_e}, \boldsymbol{\eta}, \tilde{\boldsymbol{\eta}}, \tilde{\boldsymbol{X}})$, $i = 1,2$, satisfy

$$\left\| \boldsymbol{\phi_i}(t, \hat{\boldsymbol{v}}, \boldsymbol{X_e}, \boldsymbol{\eta}, \tilde{\boldsymbol{\eta}}, \tilde{\boldsymbol{X}}) \right\| \leq \phi_i(\hat{\boldsymbol{v}}, \boldsymbol{\eta}) \left\| (\tilde{\boldsymbol{\eta}}, \tilde{\boldsymbol{X}}) \right\|, \quad (7.36)$$

where the functions $\phi_i(\hat{\boldsymbol{v}}, \boldsymbol{\eta})$ are Lipschitz in $\hat{\boldsymbol{v}}$ and are bounded with respect to any $\boldsymbol{\eta}$. Also we note that by definition $\hat{\mathbf{v}} = [\hat{u}, \hat{v}, \hat{r}]^T = [u_e^d + u_d + \tilde{u}_e, v_e + v_d, r_e^d + r_d + \tilde{r}_e]^T$, the $(\tilde{\boldsymbol{\eta}}, \tilde{\boldsymbol{X}})$-subsystem is GES at the origin, and the reference velocities u_d, v_d and r_d are bounded. On the other hand, the virtual velocity controls, u_e^d and r_e^d are bounded. Hence, using the recent stability results for cascade systems given in [17,69], we need to show that there exist the design constants c, k, k_1, and k_2 such that the $(\boldsymbol{X_{1e}}, \boldsymbol{X_{2e}})$-subsystem without the connected terms $\boldsymbol{\phi_i}(t, \hat{\boldsymbol{v}}, \boldsymbol{\eta}, \tilde{\boldsymbol{\eta}}, \tilde{\boldsymbol{X}})$, $i = 1,2$, is GAS at the origin. That is why we did not include some nonlinear damping terms in the control law (7.33). From the above discussions, we will study the system given by

$$\begin{aligned} \dot{\boldsymbol{X}}_{1e} &= \boldsymbol{f_1}(t, \boldsymbol{X_e}) + \boldsymbol{g_1}(t, \boldsymbol{X_e}), \\ \dot{\boldsymbol{X}}_{2e} &= \boldsymbol{f_2}(t, \boldsymbol{X_e}). \end{aligned} \quad (7.37)$$

To further simplify the investigation of stability of the system (7.37), we note that this system also consists of the $\boldsymbol{X_{1e}}$-subsystem and the $\boldsymbol{X_{2e}}$-subsystem in a cascaded structure. From (7.35), it is not hard to show that the connected term $\boldsymbol{g_1}(t, \boldsymbol{X_e})$ satisfies $\|\boldsymbol{g_1}(t, \boldsymbol{X_e})\| \leq \kappa_1 \|\boldsymbol{X_{2e}}\|$ with κ_1 being some positive constant. Therefore global stability of $\dot{\boldsymbol{X}}_{1e} = \boldsymbol{f_1}(t, \boldsymbol{X_e})$ and $\dot{\boldsymbol{X}}_{2e} = \boldsymbol{f_2}(t, \boldsymbol{X_e})$ implies that of (7.37). We will first study stability of the subsystem $\dot{\boldsymbol{X}}_{2e} = \boldsymbol{f_2}(t, \boldsymbol{X_e})$ then move to the subsystem $\dot{\boldsymbol{X}}_{1e} = \boldsymbol{f_1}(t, \boldsymbol{X_e})$.

Subsystem $\dot{\boldsymbol{X}}_{2e} = \boldsymbol{f_2}(t, \boldsymbol{X_e})$. By differentiating $V_1 = 0.5\left(z_e^2 + \tilde{u}_e^2 + \tilde{r}_e^2\right)$ along the solutions of $\dot{\boldsymbol{X}}_{2e} = \boldsymbol{f_2}(t, \boldsymbol{X_e})$, one can show that this system is globally asymptotically and locally exponentially stable at the origin for any constants $k_2 > 0$, $c_1 \geq 0$, and $c_2 \geq 0$.

Subsystem $\dot{\boldsymbol{X}}_{1e} = \boldsymbol{f_1}(t, \boldsymbol{X_e})$. To investigate the stability of this subsystem, we take the Lyapunov function

$$V_2 = \sqrt{c^2 + x_e^2 + y_e^2 + v_e^2} + \frac{1}{2}k_3 v_e^2 - c, \tag{7.38}$$

where k_3 is a positive constant to be selected later. After some lengthy but simple calculation using the completing squares, the time derivative of (7.38) along the solutions of $\dot{\boldsymbol{X}}_{1e} = \boldsymbol{f}_1(t, \boldsymbol{X}_e)$ satisfies

$$\dot{V}_2 \leq -p_1(t) x_e^2 \varpi_2^{-2} - p_2(t) y_e^2 \varpi_2^{-2} - p_3(t) v_e^2, \tag{7.39}$$

with

$$\begin{aligned}
p_1(t) &= k_1 - \beta(k_3 + 1/c)k_1 |r_d|/(4\varepsilon_3) - [|k v_d|/(4\varepsilon_1) + \beta(k_3 + 1/c) \times \\
&\quad (k_1 + |u_d|)|k|(|r_d|/(4\varepsilon_3(1-k^2)) + k_1/(2c\sqrt{1-k^2}))/k^*\big], \\
p_2(t) &= k u_d - 1/(4\varepsilon_2) - \big[k^2(0.5|u_d| + |v_d|)/c + \varepsilon_1 |k v_d| + \beta(k_3 + 1/c) \times \\
&\quad (k_1 + |u_d|)|k|\left(k^2|v_d| + (k^2 + |k|)(|u_d| + |v_d|) + |k u_d|/(c4\varepsilon_3)\right)/k^*\big], \\
p_3(t) &= \alpha k_3 - \varepsilon_2 - \beta(k_3 + 1/c)k_1 |r_d| \varepsilon_3 - [\beta(k_3 + 1/c)(k_1 + |u_d|)|k| \times \\
&\quad (2/c + |k u_d|\varepsilon_3/c + \varepsilon_3 |r_d| + 0.5\alpha/c + \beta k_1 |r_d|/c^2)/k^*\big],
\end{aligned} \tag{7.40}$$

where ε_i, $1 \leq i \leq 3$ are positive constants. Hence the subsystem $\dot{\boldsymbol{X}}_{1e} = \boldsymbol{f}_1(t, \boldsymbol{X}_e)$ is GAS at the origin if the design constants are chosen such that

$$p_i(t) \geq p_i^* \tag{7.41}$$

for some positive constants $p_i^*, i = 1, 2, 3$. In summary, we need to choose the constants c, k, k_1, and k_3 such that they satisfy (7.31) and (7.41). Note that the condition (7.31) automatically implies that $|k| < 1$ is required in (7.24). In the next section, we will show that under Assumption 7.1, there always exist the constants c, k, k_1, and k_3 such that (7.31) and (7.41) hold.

7.2.5 Selection of Design Constants

To choose the design constants c, k, k_1, and k_3, we observe the following: First, it is noted from the expression of $p_2(t)$ that the sign of constant k must have the same sign as that of the reference surge velocity, u_d (this sign does not change under Assumption 7.1). Second, it is observed that the condition (7.31) can be rewritten in the form of

$$1 - |k|\left(1 + 0.5\beta(k_1 + u_d^{\max})/c\right) \geq k^* > 0, \tag{7.42}$$

which implies that for each fixed $k^* < 1$, $k_1 > 0$, $u_d^{\max}$, $|k| < k^*$, we can always pick a large enough constant c such that (7.42) holds. Third, under Assumption 7.1, the magnitude of the reference sway velocity is always less than that of the reference surge velocity. Fourth, the mass including added mass in the sway dynamics, m_{22}, is always larger than that in the surge dynamics, m_{11}, for surface ships, i.e., $\beta < 1$.

Finally, all of the terms in the square brackets in $p_i, i = 1,2,3$ have the constant k as a factor. These terms also decrease when the constant c increases. Looking closely at p_i with the above observations, if the constant k is chosen small enough and the constant c is selected large enough, then we can pick a positive constant k_1 such that (7.31) and (7.41) hold with some large enough k_3. It is noted that $|k|$ should be decreased and c should be increased if u_d is large. This physically means that the distance from the ship to the point it aims to track should be increased if the velocities and surge acceleration are large, otherwise the ship will miss that point. Furthermore when α is small, $|k|$ and k_1 should be decreased, and c should be increased. This can be physically interpreted as follows: If the damping in the sway dynamics is small, the control gain in the surge dynamics should also be small otherwise the ship will slide in the sway direction. Due to complicated expressions of $p_i(t)$, we provide some general guidelines to choose the design constants rather than present their extremely complex explicit expressions: Pick $k^* < 1$, small values for $|k|$ and k_1, larger values for c and k_3. Then increase c and k_3 until (7.31) and (7.41) hold.

7.3 Output Feedback

7.3.1 Observer Design

We now introduce a more general coordinate change than (7.6) to cancel the term $\boldsymbol{C}(\boldsymbol{v})\boldsymbol{v}$ in (7.1) as follows:

$$\boldsymbol{X} = \boldsymbol{P}(\boldsymbol{\eta})\boldsymbol{v}, \tag{7.43}$$

where $\boldsymbol{P}(\boldsymbol{\eta}) \in \mathbb{R}^{3\times 3}$ is a global invertible matrix to be determined. With (7.43), the second equation of (7.1) is written as

$$\dot{\boldsymbol{X}} = \Big[\dot{\boldsymbol{P}}(\boldsymbol{\eta})\boldsymbol{v} - \boldsymbol{P}(\boldsymbol{\eta})\boldsymbol{M}^{-1}\boldsymbol{C}(\boldsymbol{v})\boldsymbol{v}\Big] - \boldsymbol{P}(\boldsymbol{\eta})\boldsymbol{M}^{-1}\boldsymbol{D}\boldsymbol{P}^{-1}(\boldsymbol{\eta})\boldsymbol{X} + \boldsymbol{P}(\boldsymbol{\eta})\boldsymbol{M}^{-1}\boldsymbol{\tau}. \tag{7.44}$$

Our goal is to cancel the terms in the square bracket on the right-hand side of (7.44). Assuming that the elements of $\boldsymbol{P}(\boldsymbol{\eta})$ are $p_{ij}(\boldsymbol{\eta}), i = 1,2,3,\ j = 1,2,3$, the first bracket in the right-hand side of (7.44) is zero if

$$\begin{aligned} &\dot{p}_{i1}u + \dot{p}_{i2}v + \dot{p}_{i3}r + \frac{m_{22}}{m_{11}}p_{i1}vr - \frac{m_{11}}{m_{22}}p_{i2}ur + \frac{m_{11}-m_{22}}{m_{33}}p_{i3}uv = 0, \\ &i = 1,2,3,\ \forall(\boldsymbol{\eta}, u, v, r) \in \mathbb{R}^6, \end{aligned} \tag{7.45}$$

where for brevity, we omit the argument $\boldsymbol{\eta}$ of $p_{ij}(\boldsymbol{\eta})$. With the first equation of (7.1), we expand (7.45) as

$$\left(\frac{\partial p_{i1}}{\partial x}\cos(\psi)+\frac{\partial p_{i1}}{\partial y}\sin(\psi)\right)u^2+\left(-\frac{\partial p_{i2}}{\partial x}\sin(\psi)+\frac{\partial p_{i2}}{\partial y}\cos(\psi)\right)v^2+$$
$$\frac{\partial p_{i3}}{\partial \psi}r^2+\left(-\frac{\partial p_{i1}}{\partial x}\sin(\psi)+\frac{\partial p_{i1}}{\partial y}\cos(\psi)+\frac{\partial p_{i2}}{\partial x}\cos(\psi)+\frac{\partial p_{i2}}{\partial y}\sin(\psi)+\right.$$
$$\left.\frac{m_{11}-m_{22}}{m_{33}}p_{i3}\right)uv+\left(\frac{\partial p_{i1}}{\partial \psi}+\frac{\partial p_{i3}}{\partial x}\cos(\psi)+\frac{\partial p_{i3}}{\partial y}\sin(\psi)-\frac{m_{11}}{m_{22}}p_{i2}\right)ur+$$
$$\left(\frac{\partial p_{i2}}{\partial \psi}-\frac{\partial p_{i3}}{\partial x}\sin(\psi)+\frac{\partial p_{i3}}{\partial y}\cos(\psi)+\frac{m_{22}}{m_{11}}p_{i1}\right)vr=0. \tag{7.46}$$

Therefore (7.46) holds for all $(\boldsymbol{\eta}, u, v, r)\in\mathbb{R}^6$ if

$$\begin{aligned}
&\frac{\partial p_{i1}}{\partial x}\cos(\psi)+\frac{\partial p_{i1}}{\partial y}\sin(\psi)=0,\\
&-\frac{\partial p_{i2}}{\partial x}\sin(\psi)+\frac{\partial p_{i2}}{\partial y}\cos(\psi)=0,\\
&\frac{\partial p_{i3}}{\partial \psi}=0,\\
&\left(\frac{\partial p_{i2}}{\partial y}-\frac{\partial p_{i1}}{\partial x}\right)\sin(\psi)+\left(\frac{\partial p_{i1}}{\partial y}+\frac{\partial p_{i2}}{\partial x}\right)\cos(\psi)+\frac{m_{11}-m_{22}}{m_{33}}p_{i3}=0,\\
&\frac{\partial p_{i1}}{\partial \psi}+\frac{\partial p_{i3}}{\partial x}\cos(\psi)+\frac{\partial p_{i3}}{\partial y}\sin(\psi)-\frac{m_{11}}{m_{22}}p_{i2}=0,\\
&\frac{\partial p_{i2}}{\partial \psi}-\frac{\partial p_{i3}}{\partial x}\sin(\psi)+\frac{\partial p_{i3}}{\partial y}\cos(\psi)+\frac{m_{22}}{m_{11}}p_{i1}=0.
\end{aligned} \tag{7.47}$$

A family of solutions of the above set of six partial differential equations is

$$\begin{aligned}
p_{i1}&=\frac{((m_{11}C_{i3}x+m_{33}C_{i1})\sin(\psi)-(m_{11}C_{i3}y-m_{33}C_{i2})\cos(\psi))}{m_{33}},\\
p_{i2}&=\frac{m_{22}\left((m_{11}C_{i3}x+m_{33}C_{i1})\cos(\psi)+(m_{11}C_{i3}y-m_{33}C_{i2})\sin(\psi)\right)}{m_{11}m_{33}},\\
p_{i3}&=C_{i3},
\end{aligned} \tag{7.48}$$

where C_{i1}, C_{i2} and C_{i3} are arbitrary constants. It is noted that the above solutions can be obtained by the following Maple™ code:

```
>PDE1:=diff(p11(x,y,\psi),x)*cos(\psi)+
   diff(p11(x,y,\psi),y)*sin(\psi)=0,
   -diff(p12(x,y,\psi),x)*sin(\psi)+
   diff(p12(x,y,\psi),y)*cos(\psi)=0,
   diff(p13(x,y,\psi),\psi)=0,
   (diff(p12(x,y,\psi),y)-diff(p11(x,y,\psi),x))*
   sin(\psi)+(diff(p11(x,y,\psi),y)+diff(p12(x,y,
   \psi),x))*cos(\psi)+(m11-m22)/m33*p13(x,y,\psi)=0,
   diff(p11(x,y,\psi),\psi)+diff(p13(x,y,\psi),x)*
```

```
   cos(\psi)+diff(p13(x,y,\psi),y)*sin(\psi)-
   m11/m22*p12(x,y,\psi)=0,
   diff(p12(x,y,\psi),\psi)-diff(p13(x,y,\psi),x)*
   sin(\psi)+diff(p13(x,y,\psi),y)*cos(\psi)+
   m22/m11*p11(x,y,\psi)=0;


>PDE2:=diff(p21(x,y,\psi),x)*cos(\psi)+
   diff(p21(x,y,\psi),y)*sin(\psi)=0,
   -diff(p22(x,y,\psi),x)*sin(\psi)+
   diff(p22(x,y,\psi),y)*cos(\psi)=0,
   diff(p23(x,y,\psi),\psi)=0,
   (diff(p22(x,y,\psi),y)-diff(p21(x,y,\psi),x))*
   sin(\psi)+(diff(p21(x,y,\psi),y)+diff(p22(x,y,
   \psi),x))*cos(\psi)+(m11-m22)/m33*p23(x,y,\psi)=0,
   diff(p21(x,y,\psi),\psi)+diff(p23(x,y,\psi),x)*
   cos(\psi)+diff(p23(x,y,\psi),y)*sin(\psi)-
   m11/m22*p22(x,y,\psi)=0,
   diff(p22(x,y,\psi),\psi)-diff(p23(x,y,\psi),x)*
   sin(\psi)+diff(p23(x,y,\psi),y)*cos(\psi)+
   m22/m11*p21(x,y,\psi)=0;


>PDE3:=diff(p31(x,y,\psi),x)*cos(\psi)+
   diff(p31(x,y,\psi),y)*sin(\psi)=0,
   -diff(p32(x,y,\psi),x)*sin(\psi)+
   diff(p32(x,y,\psi),y)*cos(\psi)=0,
   diff(p33(x,y,\psi),\psi)=0,
   (diff(p32(x,y,\psi),y)-diff(p31(x,y,\psi),x))*
   sin(\psi)+(diff(p31(x,y,\psi),y)+diff(p32(x,y,
   \psi),x))*cos(\psi)+(m11-m22)/m33*p33(x,y,\psi)=0,
   diff(p31(x,y,\psi),\psi)+diff(p33(x,y,\psi),x)*
   cos(\psi)+diff(p33(x,y,\psi),y)*sin(\psi)-
   m11/m22*p32(x,y,\psi)=0,
   diff(p32(x,y,\psi),\psi)-diff(p33(x,y,\psi),x)*
   sin(\psi)+diff(p33(x,y,\psi),y)*cos(\psi)+
   m22/m11*p31(x,y,\psi)=0;

>solutions:=pdsolve([PDE1, PDE2, PDE3]);
```

We now choose the constants C_{i1}, C_{i2}, and C_{i3} such that the matrix $\boldsymbol{P}(\boldsymbol{\eta})$ is invertible. A choice of $C_{13} = C_{11} = 0$, $C_{12} = 1$, $C_{23} = C_{22} = 0$, $C_{21} = 1$, $C_{31} = C_{32} = 0$, and $C_{33} = 1$ results in

$$\boldsymbol{P}(\boldsymbol{\eta}) = \begin{bmatrix} \cos(\psi) & -\dfrac{m_{22}\sin(\psi)}{m_{11}} & 0 \\ \sin(\psi) & \dfrac{m_{22}\cos(\psi)}{m_{11}} & 0 \\ \dfrac{m_{11}(\sin(\psi)x - \cos(\psi)y)}{m_{33}} & \dfrac{m_{22}(\cos(\psi)x + \sin(\psi)y)}{m_{33}} & 1 \end{bmatrix}. \tag{7.49}$$

Substituting (7.49) into (7.44) and using (7.43) and the first equation of (7.1), we have

$$\begin{aligned} \dot{\boldsymbol{\eta}} &= \boldsymbol{J}(\boldsymbol{\eta})\boldsymbol{P}^{-1}(\boldsymbol{\eta})\boldsymbol{X}, \\ \dot{\boldsymbol{X}} &= -\boldsymbol{D}_{\eta}(\boldsymbol{\eta})\boldsymbol{X} + \boldsymbol{P}(\boldsymbol{\eta})\boldsymbol{M}^{-1}\boldsymbol{\tau}, \end{aligned} \tag{7.50}$$

with $\boldsymbol{D}_{\eta}(\boldsymbol{\eta}) = \boldsymbol{P}(\boldsymbol{\eta})\boldsymbol{M}^{-1}\boldsymbol{D}\boldsymbol{P}^{-1}(\boldsymbol{\eta})$. It can be seen that the matrix $\boldsymbol{P}(\boldsymbol{\eta})$ given in (7.49) does not use any ship velocities. This feature results in the main difference between the partial-state feedback design in Section 7.2 and the output feedback design in this section. From (7.50), we use the following full-order nonlinear observer to construct the unmeasured ship velocities:

$$\begin{aligned} \dot{\hat{\boldsymbol{\eta}}} &= \boldsymbol{J}(\boldsymbol{\eta})\boldsymbol{P}^{-1}(\boldsymbol{\eta})\hat{\boldsymbol{X}} + \boldsymbol{K}_{01}(\boldsymbol{\eta} - \hat{\boldsymbol{\eta}}), \\ \dot{\hat{\boldsymbol{X}}} &= -\boldsymbol{D}_{\eta}(\boldsymbol{\eta})\hat{\boldsymbol{X}} + \boldsymbol{P}(\boldsymbol{\eta})\boldsymbol{M}^{-1}\boldsymbol{\tau} + \boldsymbol{K}_{02}(\boldsymbol{\eta} - \hat{\boldsymbol{\eta}}), \end{aligned} \tag{7.51}$$

where $\hat{\boldsymbol{\eta}}$ and $\hat{\boldsymbol{X}}$ are the estimates of $\boldsymbol{\eta}$ and $\boldsymbol{X}$, respectively. The observer gain matrices $\boldsymbol{K}_{01}$ and $\boldsymbol{K}_{02}$ are chosen such that

$$\begin{aligned} \boldsymbol{Q}_{01} &= \boldsymbol{K}_{01}^{T}\boldsymbol{P}_{01} + \boldsymbol{P}_{01}\boldsymbol{K}_{01}, \\ \boldsymbol{Q}_{02} &= \boldsymbol{D}_{\eta}^{T}(\boldsymbol{\eta})\boldsymbol{P}_{02} + \boldsymbol{P}_{02}\boldsymbol{D}_{\eta}(\boldsymbol{\eta}) \end{aligned}$$

are positive definite, and that

$$(\boldsymbol{J}(\boldsymbol{\eta})\boldsymbol{P}^{-1}(\boldsymbol{\eta}))^{T}\boldsymbol{P}_{01} - \boldsymbol{P}_{02}\boldsymbol{K}_{02} = 0, \tag{7.52}$$

with $\boldsymbol{P}_{01}$ and $\boldsymbol{P}_{02}$ being positive definite matrices. It is straightforward to show that $\boldsymbol{K}_{01}$ and $\boldsymbol{K}_{02}$ always exist since $\boldsymbol{D}_{\eta}(\boldsymbol{\eta})$ is positive definite. From (7.50) and (7.51), we have

$$\begin{aligned} \dot{\tilde{\boldsymbol{\eta}}} &= \boldsymbol{J}(\boldsymbol{\eta})\boldsymbol{P}^{-1}(\boldsymbol{\eta})\tilde{\boldsymbol{X}} - \boldsymbol{K}_{01}\tilde{\boldsymbol{\eta}}, \\ \dot{\tilde{\boldsymbol{X}}} &= -\boldsymbol{D}_{\eta}(\boldsymbol{\eta})\tilde{\boldsymbol{X}} - \boldsymbol{K}_{02}\tilde{\boldsymbol{\eta}}, \end{aligned} \tag{7.53}$$

where $\tilde{\boldsymbol{\eta}} := (\tilde{x}, \tilde{y}, \tilde{\psi})^{T} = \boldsymbol{\eta} - \hat{\boldsymbol{\eta}}$ and $\tilde{\boldsymbol{X}} := (\tilde{x}_1, \tilde{x}_2, \tilde{x}_3)^{T} = \boldsymbol{X} - \hat{\boldsymbol{X}}$. From (7.53) and (7.52), we can show that there exist strictly positive constants γ_0 and σ_0 such that

$$\left\|(\tilde{\boldsymbol{\eta}}(t), \tilde{\boldsymbol{X}}(t))\right\| \leq \gamma_0 \left\|(\tilde{\boldsymbol{\eta}}(t_0), \tilde{\boldsymbol{X}}(t_0))\right\| e^{-\sigma_0(t-t_0)}, \ \forall\, t \geq t_0 \geq 0. \tag{7.54}$$

Define $\hat{\boldsymbol{v}} = [\hat{u}, \hat{v}, \hat{r}]^T$ being an estimate of the velocity vector $\boldsymbol{v}$ as

$$\hat{\boldsymbol{v}} = \boldsymbol{P}^{-1}(\boldsymbol{\eta})\hat{\boldsymbol{X}}. \tag{7.55}$$

Then the velocity estimate error vector, $\tilde{\boldsymbol{v}} := [\tilde{u}, \tilde{v}, \tilde{r}]^T = \boldsymbol{v} - \hat{\boldsymbol{v}}$, satisfies

$$\tilde{\boldsymbol{v}} = \boldsymbol{P}^{-1}(\boldsymbol{\eta})\tilde{\boldsymbol{X}}. \tag{7.56}$$

Based on (7.56), we cannot conclude anything about the convergence of the velocity estimate errors since some elements of the matrix $\boldsymbol{P}(\boldsymbol{\eta})$, see (7.49), depend linearly on x and y. However, our controller will guarantee that (x, y) are globally bounded. Then (7.56) implies that the velocity estimate errors globally exponentially converge to zero. Indeed, the linear dependence of $\boldsymbol{P}(\boldsymbol{\eta})$ on x and y will result in a challenging problem, which the control design will have to take care of. To prepare for the control design, using (7.55), we rewrite (7.51) as

$$\begin{aligned}
\dot{x} &= \hat{u}\cos(\psi) - \hat{v}\sin(\psi) + \left(\cos^2(\psi) + m_{11}m_{22}^{-1}\sin^2(\psi)\right)\tilde{x}_1 + \\
&\quad 0.5(m_{22} - m_{11})m_{22}^{-1}\sin(2\psi)\tilde{x}_2, \\
\dot{y} &= \hat{u}\sin(\psi) + \hat{v}\cos(\psi) + 0.5(m_{22} - m_{11})m_{22}^{-1}\sin(2\psi)\tilde{x}_1 + \\
&\quad \left(\sin^2(\psi) + m_{11}m_{22}^{-1}\cos^2(\psi)\right)\tilde{x}_2, \\
\dot{\psi} &= \hat{r} + m_{11}m_{33}^{-1}\left(y\tilde{x}_1 - x\tilde{x}_2\right) + \tilde{x}_3, \\
\dot{\hat{u}} &= m_{22}m_{11}^{-1}\hat{v}\hat{r} - d_{11}m_{11}^{-1}\hat{u} + m_{11}^{-1}\tau_u + \hat{v}\left(m_{22}m_{33}^{-1}\left(y\tilde{x}_1 - x\tilde{x}_2\right) + \right. \\
&\quad \left. m_{22}m_{11}^{-1}\tilde{x}_3\right) + \cos(\psi)\tilde{x} + \sin(\psi)\tilde{y} - m_{11}m_{33}^{-1}\left(\sin(\psi)x - \cos(\psi)y\right)\tilde{\psi}, \\
\dot{\hat{v}} &= -m_{11}m_{22}^{-1}\hat{u}\hat{r} - d_{22}m_{22}^{-1}\hat{v} - \hat{u}m_{11}m_{22}^{-1}\left(m_{11}m_{33}^{-1}\left(y\tilde{x}_1 - x\tilde{x}_2\right) + \tilde{x}_3\right) - \\
&\quad m_{11}^2m_{22}^{-2}\left(\sin(\psi)\tilde{x} - \cos(\psi)\tilde{y}\right) - m_{11}^2m_{22}^{-1}m_{33}^{-1}\left(\cos(\psi)x + \sin(\psi)y\right)\tilde{\psi}, \\
\dot{\hat{r}} &= (m_{11} - m_{22})m_{33}^{-1}\hat{u}\hat{v} - d_{33}m_{33}^{-1}\hat{r} + m_{33}^{-1}\tau_r + m_{11}^2m_{22}^{-1}m_{33}^{-1}\hat{u} \times \\
&\quad (-\sin(\psi)\tilde{x}_1 + \cos(\psi)\tilde{x}_2) - m_{22}m_{33}^{-1}\hat{v} \times (\cos(\psi)\tilde{x}_1 + \\
&\quad \sin(\psi)\tilde{x}_2) + (m_{11}^2 + m_{33}^2)m_{33}^{-2}\left(x^2 + y^2\right)\tilde{\psi} + m_{22}^{-1}m_{33}^{-1} \times \\
&\quad \left(\left(0.5(m_{11}^2 - m_{11}m_{22})\sin(2\psi)x + \left(m_{11}m_{22}\cos^2(\psi) + \right.\right.\right. \\
&\quad \left.\left. m_{11}^2\sin^2(\psi)\right)y\right)\tilde{x} + \left(0.5(m_{11}^2 - m_{11}m_{22})\sin(2\psi)y - \right. \\
&\quad \left.\left.(m_{11}m_{22}\sin^2(\psi) + m_{11}^2\cos^2(\psi))x\right)\tilde{y}\right),
\end{aligned} \tag{7.57}$$

where for simplicity, we have taken $\boldsymbol{K}_{02} = (\boldsymbol{J}(\boldsymbol{\eta})\boldsymbol{P}^{-1}(\boldsymbol{\eta}))^{\boldsymbol{T}}$.

7.3.2 Coordinate Transformations

If one applies the coordinate change (7.22) to (7.57), it will result in a very complicated system, namely some quadratic terms of (x_e, y_e, ψ_e) multiplied by the observer errors appearing in the kinematic part of the transformed system due to linear dependence of x and y on some elements of the matrix $\boldsymbol{P}(\eta)$. This makes the control design extremely difficult and might result in a finite escape. To avoid the said difficulty, we propose the following coordinate transformation:

$$\begin{bmatrix} x_e \\ y_e \\ \psi_e \end{bmatrix} = \boldsymbol{J}^{-1}(\boldsymbol{\eta}_d) \begin{bmatrix} x - x_d \\ y - y_d \\ \psi - \psi_d \end{bmatrix}, \quad \begin{cases} u_e = \hat{u} - u_d, \\ v_e = \hat{v} - v_d, \\ r_e = \hat{r} - r_d. \end{cases} \tag{7.58}$$

Indeed, convergence to zero of (x_e, y_e, ψ_e) implies that of $(x - x_d, y - y_d, \psi - \psi_d)$. Differentiating both sides of (7.58) along the solutions of (7.57) and (7.5) yields the error dynamics of the "kinematic part":

$$\begin{aligned} \dot{x}_e &= u_e + (u_e + u_d)(\cos(\psi_e) - 1) - (v_e + v_d)\sin(\psi_e) + r_d y_e + h_x, \\ \dot{y}_e &= v_e + (v_e + v_d)(\cos(\psi_e) - 1) + (u_e + u_d)\sin(\psi_e) - r_d x_e + h_y, \\ \dot{\psi}_e &= r_e + \Omega_\psi + h_\psi, \end{aligned} \tag{7.59}$$

where, for notational simplicity, we have defined

$$\begin{aligned} h_x &= \cos(\psi_d)\Delta_x + \sin(\psi_d)\Delta_y, \\ h_y &= -\sin(\psi_d)\Delta_x + \cos(\psi_d)\Delta_y, \\ h_\psi &= m_{11}m_{33}^{-1}(y_d\tilde{x}_1 - x_d\tilde{x}_2) + \tilde{x}_3, \\ \Omega_\psi &= m_{11}m_{33}^{-1}((\sin(\psi_d)x_e + \cos(\psi_d)y_e)\tilde{x}_1 - (\cos(\psi_d)x_e - \text{sins}(\psi_d)y_e)\tilde{x}_2), \\ \Delta_x &= (\cos^2(\psi) + \sin^2(\psi)m_{11}m_{22}^{-1})\tilde{x}_1 + 0.5\sin(2\psi)\tilde{x}_2(m_{11} - m_{22})m_{22}^{-1}, \\ \Delta_y &= 0.5\sin(2\psi)\tilde{x}_1(m_{11} - m_{22})m_{22}^{-1} + (\sin^2(\psi) + \cos^2(\psi)m_{11}m_{22}^{-1})\tilde{x}_2. \end{aligned} \tag{7.60}$$

We define the following coordinate transformation, which is slightly different from (7.24):

$$z_e = \psi_e + \arcsin\left(\frac{k u_d y_e}{\sqrt{1 + x_e^2 + y_e^2}}\right), \tag{7.61}$$

where the constant k is such that $|k u_d(t)| < 1, \forall t \geq 0$. This constant will be specified later. Using the nonlinear coordinate transformation (7.61) together with (7.59), the ship error dynamics are rewritten as

$$\begin{aligned} \dot{x}_e &= u_e + (u_e + u_d)p_x - (v_e + v_d)p_y + (v_e + v_d)k u_d \varpi_2^{-1} y_e + \\ &\quad r_d y_e + h_x, \\ \dot{y}_e &= v_e + (v_e + v_d)p_x + (u_e + u_d)p_y - k u_d^2 \varpi_2^{-1} y_e - \\ &\quad k u_d \varpi_2^{-1} u_e y_e - r_d x_e + h_y, \\ \dot{z}_e &= r_e + f_z + g_z u_e + \Omega_z + h_z, \\ \dot{u}_e &= m_{22}m_{11}^{-1}\hat{v}\hat{r} - d_{11}m_{11}^{-1}\hat{u} + m_{11}^{-1}\tau_u - \dot{u}_d + \Omega_u + h_u, \\ \dot{v}_e &= -m_{11}m_{22}^{-1}(u_e r_e + u_d r_e + u_e r_d) - d_{22}m_{22}^{-1}v_e + \Omega_v + h_v, \\ \dot{r}_e &= (m_{11} - m_{22})m_{33}^{-1}\hat{u}\hat{v} - d_{33}m_{33}^{-1}\hat{r} + m_{33}^{-1}\tau_r - \dot{r}_d + \Omega_r + h_r, \end{aligned} \tag{7.62}$$

where, for notational simplicity and convenience of the control design, we have defined the following:

The terms $\omega_1, \omega_2, p_x, p_y, f_z$, and g_z are defined as

$$\begin{aligned}
\varpi_1 &= \sqrt{1+x_e^2+(1-k^2u_d^2)y_e^2},\ \varpi_2 = \sqrt{1+x_e^2+y_e^2},\\
p_x &= \varpi_2^{-1}\left((\cos(z_e)-1)\varpi_1+(\varpi_1-\varpi_2)+\sin(z_e)ku_d y_e\right),\\
p_y &= \varpi_2^{-1}\left(\sin(z_e)\varpi_1-(\cos(z_e)-1)ku_d y_e\right),\\
f_z &= k\varpi_1^{-1}\left(\dot{u}_d+u_d\left((1+x_e^2)\varpi_2^{-2}\left(v_e-ku_d^2\varpi_2^{-1}y_e+(v_e+v_d)\times\right.\right.\right.\\
&\quad p_x+u_d p_y)-r_d x_e-\varpi_2^{-2}x_e y_e(u_d p_x-(v_e+v_d)(p_y-\\
&\quad ku_d\varpi_2^{-1}y_e)))),\\
g_z &= ku_d\varpi_1^{-1}((1+x_e^2)\varpi_2^{-2}(-ku_d\varpi_2^{-1}y_e+p_y)-\varpi_2^{-2}y_e x_e(1+p_x)).
\end{aligned} \tag{7.63}$$

The terms $\Omega_z, \Omega_u, \Omega_v$, and Ω_r containing states multiplied by the observer errors are defined as

$$\begin{aligned}
\Omega_z &= \Omega_\psi,\\
\Omega_u &= \hat{v}(m_{22}m_{33}^{-1}(y\tilde{x}_1-x\tilde{x}_2)+m_{22}m_{11}^{-1}\tilde{x}_3)-\\
&\quad m_{11}m_{33}^{-1}(\sin(\psi_e)x_e-\cos(\psi_e)y_e)\tilde{\psi},\\
\Omega_v &= -(u_e+u_d)m_{11}m_{22}^{-1}(m_{11}m_{33}^{-1}(y\tilde{x}_1-x\tilde{x}_2)+\tilde{x}_3)-\\
&\quad m_{11}^2m_{22}^{-1}m_{33}^{-1}(\cos(\psi_e)x_e+\sin(\psi_e)y_e)\tilde{\psi},\\
\Omega_r &= m_{11}^2m_{22}^{-1}m_{33}^{-1}\hat{u}(-\sin(\psi)\tilde{x}_1+\cos(\psi)\tilde{x}_2)-\\
&\quad m_{22}m_{33}^{-1}\hat{v}(\cos(\psi)\tilde{x}_1+\sin(\psi)\tilde{x}_2)+\\
&\quad m_{22}^{-1}m_{33}^{-1}(0.5(m_{11}^2-m_{11}m_{22})\sin(2\psi)\Delta_{xd}+\\
&\quad (m_{11}m_{22}\cos^2(\psi)+m_{11}^2\sin^2(\psi))\Delta_{yd})\tilde{x}+\\
&\quad m_{22}^{-1}m_{33}^{-1}(0.5(m_{11}^2-m_{11}m_{22})\sin(2\psi)\Delta_{yd}-\\
&\quad (m_{11}m_{22}\sin^2(\psi)+m_{11}^2\cos^2(\psi))\Delta_{xd})\tilde{y}+\\
&\quad (m_{11}^2m_{33}^{-2}+1)(x_e^2+y_e^2+2\sin(\psi_d)(x_e y_d-\\
&\quad y_e x_d)+2\cos(\psi_d)(x_e x_d+y_e y_d))\tilde{\psi},
\end{aligned} \tag{7.64}$$

with $\Delta_{xd}=\cos(\psi_d)x_e-\sin(\psi_d)y_e$, and $\Delta_{yd}=\sin(\psi_d)x_e+\cos(\psi_d)y_e$.

The terms h_z, h_u, h_v, and h_r containing the observer errors multiplied by bounded terms are defined as

$$\begin{aligned}
h_z &= ku_d\varpi_1^{-1}((1+x_e^2)\varpi_2^{-2}h_y-\varpi_2^{-2}y_e x_e h_x)+h_\psi,\\
h_u &= \cos(\psi)\tilde{x}+\sin(\psi)\tilde{y}-m_{11}m_{33}^{-1}(\sin(\psi)x_d-\cos(\psi)y_d)\tilde{\psi},\\
h_v &= -m_{11}^2m_{22}^{-1}(\sin(\psi)\tilde{x}-\cos(\psi)\tilde{y})-m_{11}^2m_{22}^{-1}m_{33}^{-1}(\cos(\psi)x_d+\sin(\psi)y_d)\tilde{\psi},\\
h_r &= m_{22}^{-1}m_{33}^{-1}(0.5(m_{11}^2-m_{11}m_{22})\sin(2\psi)(x_d\tilde{x}-y_d\tilde{y})+(m_{11}m_{22}\cos^2(\psi)+\\
&\quad m_{11}^2\sin^2(\psi))y_d\tilde{x}-(m_{11}m_{22}\sin^2(\psi)+m_{11}^2\cos^2(\psi))x_d\tilde{y})+m_{33}^{-2}(m_{11}^2+\\
&\quad m_{33}^2)(x_d^2+y_d^2)\tilde{\psi}.
\end{aligned} \tag{7.65}$$

The problem of forcing the underactuated ship (7.1) to track the virtual ship (7.5) becomes one of stabilizing the system (7.62) at the origin. The effort, we have made so far is to obtain the stabilizing term $-ku_d^2 y_e/\varpi_2$ in the y_e-dynamics.

7.3.3 Control Design

Before designing the control inputs, it is important to note that the terms Ω_u and Ω_r can be dominated by adding some nonlinear damping terms in the control inputs τ_u and τ_r. However the term Ω_v cannot be dominated by any nonlinear damping terms in τ_u and τ_r. Since Ω_v contains $u_e x_e$ and $u_e y_e$ multiplied by the observer errors, with x and y being substituted in from (7.58), if one designs a virtual control of u_e, which is linear in x_e and y_e, the sway velocity dynamics might have a finite escape time due to the fact that separation principle does not hold for the nonlinear system in question. The coordinate change (7.58) allows us to design a virtual control of u_e such that it is bounded for all x_e and y_e and stabilizes the x_e-dynamics at the origin. We design the controls τ_u and τ_r in two steps.

Step 1

Define the virtual control errors as

$$\begin{aligned}\tilde{u}_e &= u_e - u_e^d,\\ \tilde{r}_e &= r_e - r_e^d,\end{aligned} \tag{7.66}$$

where u_e^d and r_e^d are the virtual velocity controls of u_e and r_e, respectively. The virtual controls u_e^d and r_e^d are chosen as follows:

$$\begin{aligned}u_e^d &= -\frac{k_1 x_e}{\varpi_2},\\ r_e^d &= -k_2 z_e - f_z - g_z u_e^d,\end{aligned} \tag{7.67}$$

where k_1 and k_2 are positive design constants to be specified later.

Step 2

By differentiating (7.66) along the solutions of (7.62) and (7.67), the actual controls τ_u and τ_r with some nonlinear damping terms to overcome the effect of observer errors, and without canceling the useful damping terms, are chosen as

$$\tau_u = m_{11}\Bigg(-m_{22}m_{11}^{-1}\hat{v}\hat{r} + d_{11}m_{11}^{-1}(u_e^d + u_d) + \dot{u}_d + \frac{\partial u_e^d}{\partial x_e}(u_e + (u_e + u_d)\times$$

$$
\begin{aligned}
&p_x + (v_e + v_d)\left(-p_y + k u_d \varpi_2^{-1} y_e\right) + r_d y_e\right) \frac{\partial u_e^d}{\partial y_e}(v_e + (v_e + v_d) p_x + \\
&(u_e + u_d) p_y - k u_d y_e \varpi_2^{-1}(u_e + u_d) - r_d x_e\right) - c_1 \tilde{u}_e - g_z z_e + \\
&k_3 k_4^{-1} m_{11} m_{22}^{-1} \hat{r} v_e - \delta_1 \tilde{u}_e \tau_{u\mathrm{dam}}\right),
\end{aligned}
$$

$$
\begin{aligned}
\tau_r &= m_{33}\Big(-(m_{11} - m_{22}) m_{33}^{-1} \hat{u}\hat{v} + d_{33} m_{33}^{-1}(r_e^d + r_d) + \dot{r}_d + \\
&\frac{\partial r_e^d}{\partial x_e}\left(u_e + (u_e + u_d) p_x + (v_e + v_d)\left(-p_y + k u_d \varpi_2^{-1} y_e\right) + r_d y_e\right) + \\
&\frac{\partial r_e^d}{\partial y_e}\left(v_e + (v_e + v_d) p_x + (u_e + u_d) p_y - k u_d y_e \varpi_2^{-1}(u_e + u_d) - \right. \\
&\left. r_d x_e\right) + \frac{\partial r_e^d}{\partial z_e}(r_e + f_z + g_z u_e) - \frac{\partial r_e^d}{\partial v_e}\left(d_{22} m_{22}^{-1} v_e + m_{11} m_{22}^{-1}(u_e r_e + \right. \\
&\left. u_d r_e + u_e r_d)\right) + \frac{\partial r_e^d}{\partial u_d}\dot{u}_d + \frac{\partial r_e^d}{\partial \dot{u}_d}\ddot{u}_d + \frac{\partial r_e^d}{\partial v_d}\dot{v}_d + \frac{\partial r_e^d}{\partial r_d}\dot{r}_d - c_2 \tilde{r}_e - z_e + \\
&k_3 k_4^{-1} m_{11} m_{22}^{-1}(u_e^d + u_d) v_e - \delta_2 \tilde{r}_e \tau_{r\mathrm{dam}}\Big),
\end{aligned} \tag{7.68}
$$

where c_1, c_2, k_3, and k_4 are positive constants to be specified later, δ_1 and δ_2 are arbitrarily positive constants. We introduced the ratio k_3/k_4 to enhance the feasibility of design constants. The nonlinear damping terms $\tau_{u\mathrm{dam}}$ and $\tau_{r\mathrm{dam}}$ are defined as:

$$
\begin{aligned}
\tau_{u\mathrm{dam}} &= (\hat{v}^2 + 1)(x_e^2 + y_e^2) + \hat{v}^2, \\
\tau_{r\mathrm{dam}} &= \hat{u}^2 + \hat{v}^2 + (x_e^2 + y_e^2)^2 + (x_e^2 + y_e^2)(\hat{u}^2 + 1).
\end{aligned} \tag{7.69}
$$

We now state the second main result of this chapter, the proof of which is given in the next section.

Theorem 7.2. *Under Assumption 7.2, assume that (a) there are no environmental disturbances; (b) the ship parameters are known; (c) the reference signals* (x_d, y_d, ψ_d, v_d) *are generated by the virtual ship model (7.5), and Assumption 7.1 holds. If the output feedback control law (7.68) together with the observer (7.51) are applied to the ship system (7.1), then the tracking errors* $(x(t) - x_d(t), y(t) - y_d(t), \psi(t) - \psi_d(t), v(t) - v_d(t))$ *globally asymptotically and locally exponentially converge to zero with an appropriate choice of the design constants* k, c_1, c_2, *and* k_i, $i = 1, \ldots, 4$.

Remark 7.4. The main differences between the partial-state feedback and output feedback designs are the nonlinear coordinate transformations (7.6), (7.24), (7.43), (7.58), and (7.61). Furthermore, the partial-state feedback controller can allow the reference trajectory (x_d, y_d) to exponentially grow but the output feedback controller cannot. This is because the observer errors of the partial-state feedback design do not depend on x and y while those of the output feedback design depend linearly on x and y. Indeed, the output feedback control design can directly yield a controller for the partial-state feedback case but not vice versa. We have, however, presented both control designs for the sake of completeness.

7.3.4 Stability Analysis

Substituting (7.68), (7.67), and (7.66) into (7.62) results in the following closed loop system:

$$\begin{aligned}
\dot{x}_e &= -k_1\varpi_2^{-1}x_e + (-k_1\varpi_2^{-1}x_e + u_d)p_x - (v_e+v_d)p_y + \\
&\quad (v_e+v_d)ku_d\varpi_2^{-1}y_e + r_d y_e + h_x + \tilde{u}_e(1+p_x),\\
\dot{y}_e &= v_e + (v_e+v_d)p_x + (-k_1\varpi_2^{-1}x_e + u_d)p_y - ku_d^2\varpi_2^{-1}y_e + \\
&\quad k_1ku_d\varpi_2^{-1}x_e y_e - r_d x_e + h_y + \tilde{u}_e(1-ku_d\varpi_2^{-1}y_e),\\
\dot{z}_e &= -k_2 z_e + \Omega_z + h_z + g_z\tilde{u}_e + \tilde{r}_e,\\
\dot{\tilde{u}}_e &= -\left(c_1 + d_{11}m_{11}^{-1}\right)\tilde{u}_e - g_z z_e + \frac{k_3}{k_4}\frac{m_{11}}{m_{22}}\hat{r}v_e - \delta_1\tilde{u}_e\tau_{u\text{dam}} + \\
&\quad \Omega_u + h_u - \frac{\partial u_e^d}{\partial x_e}h_x - \frac{\partial u_e^d}{\partial y_e}h_y,\\
\dot{v}_e &= -d_{22}m_{22}^{-1}v_e - m_{11}m_{22}^{-1}(u_e^d r_e^d + u_d r_e^d + u_e^d r_d) - m_{11}m_{22}^{-1}\times\\
&\quad (u_e^d + u_d)\tilde{r}_e - m_{11}m_{22}^{-1}\hat{r}\tilde{u}_e + \Omega_v + h_v,\\
\dot{\tilde{r}}_e &= -\left(c_2 + d_{33}m_{33}^{-1}\right)\tilde{r}_e - z_e - \frac{\partial r_e^d}{\partial x_e}h_x - \frac{\partial r_e^d}{\partial y_e}h_y - \frac{\partial r_e^d}{\partial z_e}(\Omega_z + h_z) - \\
&\quad \frac{\partial r_e^d}{\partial v_e}(\Omega_v + h_v) + k_3k_4^{-1}m_{11}m_{22}^{-1}(u_e^d + u_d)v_e - \delta_2\tilde{r}_e\tau_{r\text{dam}} + \Omega_r + h_r,
\end{aligned} \tag{7.70}$$

where, for brevity, we did not substitute the expressions of u_e^d and r_e^d into the sway dynamics. To prove Theorem 7.2, we just need to show that the closed loop system (7.70) is globally asymptotically and locally exponentially stable at the origin. It is noted that Ω_z contains x_e and y_e multiplied by the observer errors, see (7.64) and (7.60). On the other hand, the x_e and y_e-dynamics are stabilized by the terms $-k_1x_e/\varpi_2$ and $-ku_d^2y_e/\varpi_2$, respectively. This makes the stability analysis of (7.70) difficult, i.e., we cannot consider the (x_e, y_e, v_e) and $(z_e, \tilde{u}_e, \tilde{r}_e)$-subsystems separately as is often done in applying stability results for cascade systems. To illustrate our idea of proving asymptotic stability of (7.70), we first give a simple example. For any initial conditions $(\xi_1(t_0), \xi_2(t_0))$, the system

$$\begin{aligned}
\dot{\xi}_1 &= -\frac{\xi_1}{\sqrt{1+\xi_1^2}} + \phi(\xi_2),\\
\dot{\xi}_2 &= -\xi_2 + \xi_1 e^{-\sigma_{\xi 2}(t-t_0)}
\end{aligned} \tag{7.71}$$

is GAS at the origin for $t \geq t_0 \geq 0$, any $\sigma_{\xi 2} > 0$, and $|\phi(\xi_2(t))| \leq A|\xi_2(t)|$ with A being any positive constant. In the first step, we show that there exists a positive constant $\sigma_{\xi 1} < \sigma_{\xi 2}$ such that $|\xi_1(t)| \leq \lambda_1(\cdot)e^{\sigma_{\xi 1}(t-t_0)}$, with $\lambda_1(\cdot)$ being a nondecreasing function of $\|(\xi_1(t_0), \xi_2(t_0))\|$. Take the Lyapunov function $W = 0.5\left(\xi_1^2 + K\xi_2^2\right)$, with K being a positive constant, whose derivative along the solution of (7.71) satisfies

$$\dot{W} \leq \sigma_{\xi 1}\xi_1^2 + \left(0.25A^2/\sigma_{\xi 1} - K\right)\xi_2^2 + K\xi_1\xi_2 e^{-\sigma_{\xi 2}(t-t_0)}. \tag{7.72}$$

Pick K such that $A^2/(4\sigma_{\xi 1}) - K < 0$, then (7.72) implies that

$$\dot{W} \leq 2\sigma_{\xi 1}W + K\max(1, K)We^{-\sigma_{\xi 2}(t-t_0)}, \tag{7.73}$$

which in turn yields $|\xi_1(t)| \leq \lambda_1(\cdot)e^{\sigma_{\xi 1}(t-t_0)}$. Therefore the term $\xi_1 e^{-\sigma_{\xi 2}(t-t_0)}$ in (7.71) globally exponentially converges to zero. The second step of proving asymptotic stability can be carried out easily by taking the Lyapunov function $W_1 = \sqrt{1+\xi_1^2} - 1 + K_1\xi_2^2$, $K_1 > 0$. We now present the proof of asymptotic stability of the closed loop system (7.70) in two parts.

Part 1. In this part, we show that there exists a positive constant $\sigma_1 < \sigma_0$ such that

$$\|(x_e(t), y_e(t))\| \leq \gamma_{11}(\cdot)e^{\sigma_1(t-t_0)} + \gamma_{10}(\cdot), \tag{7.74}$$

where $\gamma_{11}(\cdot)$ and $\gamma_{10}(\cdot)$ are nondecreasing functions of $\|\varXi(t_0)\|$ with

$$\begin{aligned}
\varXi(t_0) &:= [\tilde{\eta}(t_0), \tilde{X}(t_0), X_e(t_0)]^T, \\
X_e &:= [x_e, y_e, z_e, v_e, \tilde{u}_e, \tilde{r}_e]^T.
\end{aligned}$$

Consider the following Lyapunov function:

$$V_1 = \frac{1}{2}\left(x_e^2 + y_e^2 + k_3 v_e^2 + k_4(z_e^2 + \tilde{u}_e^2 + \tilde{r}_e^2)\right), \tag{7.75}$$

with k_4 being a positive constant, whose time derivative along the solutions of (7.70), after some lengthy but simple calculation by completing squares satisfies

$$\begin{aligned}
\dot{V}_1 \leq{}& -k_1\varpi_2^{-1}x_e^2 - ku_d^2\varpi_2^{-1}y_e^2 + 7\varepsilon_1(x_e^2 + y_e^2) - a_z z_e^2 - a_v v_e^2 - a_u\tilde{u}_e^2 - \\
& a_r\tilde{r}_e^2 + (\chi_{11}(\cdot)V_1 + \chi_{10}(\cdot))e^{-\sigma_0(t-t_0)} + a_0,
\end{aligned} \tag{7.76}$$

where

$$\begin{aligned}
a_z &= k_2k_4 - 0.25\varepsilon_1^{-1}(k_2k_3m_{11}m_{22}^{-1}(k_1 + |u_d|))^2 - \varepsilon_1^{-1}(k_1 + |u_d|)^2, \\
a_u &= k_4(c_1 + d_{11}m_{11}^{-1}) - 5\varepsilon_1^{-1}, \\
a_v &= k_3d_{22}m_{22}^{-1} - 6.75\varepsilon_1^{-1} - k_3m_{11}m_{22}^{-1}(k_1 + |u_d|)(8 + |ku_d|)|ku_d|, \\
a_r &= k_4(c_2 + d_{33}m_{33}^{-1}),
\end{aligned} \tag{7.77}$$

with ε_1 and a_0 being positive constants. We choose the design constants k, c_1, c_2, and k_i, $i = 1,\ldots,4$ such that

$$\varepsilon_1 < \sigma_0/7,\ a_z > 0,\ a_v > 0,\ a_u > 0,\ a_r > 0. \tag{7.78}$$

It is not hard to show that there always exist k, c_1, c_2 and k_i, $i = 1,\ldots,4$ such that (7.78) holds for arbitrarily small ε_1. We will discuss more detail of (7.78) later. With the choice of (7.78), we can write (7.76) as

$$\begin{aligned}\dot{V}_1 \leq\ & 14\varepsilon_1 (V_1 + (a_0 + \chi_{10}(\cdot))/(14\varepsilon_1)) + \\ & \chi_{11}(\cdot)(V_1 + (a_0 + \chi_{10}(\cdot))/(14\varepsilon_1))e^{-\sigma_0(t-t_0)},\end{aligned} \tag{7.79}$$

which, together with (7.75), yields (7.74).

Part 2. We now prove that the closed loop system (7.70) is GAS at the origin by taking the Lyapunov function

$$V_2 = \sqrt{1 + x_e^2 + y_e^2} - 1 + \frac{1}{2}\left(k_3 v_e^2 + k_4(z_e^2 + \tilde{u}_e^2 + \tilde{r}_e^2)\right) \tag{7.80}$$

whose time derivative along the solutions of (7.70), after some lengthy but simple calculation by completing squares, and using (7.74), satisfies

$$\begin{aligned}\dot{V}_2 \leq\ & -\mu_x \varpi_2^{-2} x_e^2 - \mu_y(t)\varpi_2^{-2} y_e^2 - \mu_z(t) z_e^2 - \mu_v(t) v_e^2 - \mu_u \tilde{u}_e^2 - \mu_r \tilde{r}_e^2 + \\ & \chi_{21}(\cdot) V_2 e^{-\sigma_2(t-t_0)} + \chi_{20}(\cdot) e^{-\sigma_2(t-t_0)},\end{aligned} \tag{7.81}$$

where

$$\begin{aligned}
\mu_x =\ & k_1 - 6\rho_1, \\
\mu_y(t) =\ & k u_d^2 - (k u_d)^2 (k_1 + |u_d| + |v_d| + 0.5) - |k u_d|(k_1 + \rho_1) - \\
& 0.25\rho_1^{-1}(k u_d v_d)^2 - 10\rho_1, \\
\mu_z(t) =\ & k_2 k_4 - \rho_1^{-1}(0.5(k_1 + |u_d|)^2 + 5(1 + v_d^2) + 0.25(k_3 m_{11} m_{22}^{-1} \times \\
& (k_1 + |u_d|))^2 (k_2^2 + (k u_d)^2 (k_1 + |u_d| + v_d^2) + 4 + 4 v_d^2)), \\
\mu_v(t) =\ & k_3 d_{22} m_{22}^{-1} - 7\rho_1 - 0.25\rho_1^{-1}(|k u_d| + 1) - 0.5(k u_d)^2 - \\
& k_3 m_{11} m_{22}^{-1}(k_1 + |u_d|)\,|k u_d|\,(8 + |k u_d|) - 0.25\rho_1^{-1}(k_3 m_{11} m_{22}^{-1} \times \\
& (k_1 + |u_d|))^2 ((k\dot{u}_d)^2/(1 - (k u_d)^2) + (k u_d)^4 (u_d^2 + k_1^2 + \\
& (k u_d v_d)^2 + r_d^2 + 2) + (k u_d)^2), \\
\mu_u =\ & k_4(c_1 + d_{11}/m_{11}) - 29/(4\rho_1), \\
\mu_r =\ & k_4(c_2 + d_{33}/m_{33}),
\end{aligned} \tag{7.82}$$

where $\sigma_2 = \sigma_0 - \sigma_1$, ρ_1 is a positive constant, and $\chi_{21}(\cdot)$ and $\chi_{20}(\cdot)$ are nondecreasing functions of $\|(\tilde{\eta}(t_0), \tilde{\boldsymbol{X}}(t_0), \boldsymbol{X}_e(t_0))\|$. We now choose the constants k, c_1, c_2, and k_i, $i = 1,\ldots,4$, such that

$$\begin{aligned}
&\mu_x \geq \mu_x^*,\\
&\mu_y(t) \geq \mu_y^*,\\
&\mu_z(t) \geq \mu_z^*,\\
&\mu_v(t) \geq \mu_v^*,\\
&\mu_u \geq \mu_u^*,\\
&\mu_r \geq \mu_r^*
\end{aligned} \tag{7.83}$$

for all $t \geq 0$, where μ_x^*, μ_y^*, μ_z^*, μ_v^*, μ_u^*, and μ_r^* are positive constants. Substituting (7.83) into (7.81) yields

$$\begin{aligned}
\dot{V}_2 \leq& -\mu_x^* \varpi_2^{-2} x_e^2 - \mu_y^* \varpi_2^{-2} y_e^2 - \mu_z^* z_e^2 - \mu_v^* v_e^2 - \mu_u^* \tilde{u}_e^2 - \mu_r^* \tilde{r}_e^2 + \\
&\chi_{21}(\cdot) V_2 e^{-\sigma_2(t-t_0)} + \chi_{20}(\cdot) e^{-\sigma_2(t-t_0)}.
\end{aligned} \tag{7.84}$$

From (7.84) we have $\dot{V}_2 \leq \chi_{21}(\cdot) V_2 e^{-\sigma_2(t-t_0)} + \chi_{20}(\cdot) e^{-\sigma_2(t-t_0)}$, which implies that $V_2 \leq \chi_{22}(\cdot)$, with $\chi_{22}(\cdot)$ being a nondecreasing function of $\|\Xi(t_0)\|$. With $V_2 \leq \chi_{22}(\cdot)$ in mind, one can show from (7.84) that there exists $\sigma_3 > 0$ depending on $\|\Xi(t_0)\|$ such that $\|\boldsymbol{X_e}(t)\| \leq \gamma_2(\cdot) e^{-\sigma_3(t-t_0)}$, where $\gamma_2(\cdot)$ is a nondecreasing function of $\|\Xi(t_0)\|$, which in turn implies that the closed loop system (7.70) is asymptotically stable at the origin. However one can straightforwardly show that (7.70) is also locally exponentially stable at the origin. By carrying out a similar arguments in Section 7.2.5, one can show that there always exist the design constants k, c_1, and c_2 and k_i, $i = 1, 2, 3, 4$, such that $|k u_d(t)| < 1$ and that the conditions (7.78) and (7.83) hold.

7.4 Robustness Discussion

In this section, we discuss robustness of the output feedback tracking controller. A discussion for the partial-state feedback can be carried out in a similar way. The control law (7.68) has been designed under the assumption that there are no environmental disturbances. Indeed, this assumption is unrealistic in practice. The aim of this section is to discuss the robustness property of our proposed controller in relation to environmental disturbances. Under additive environmental disturbances, it is not hard to show that the observer (7.51) guarantees that the observer errors $(\tilde{\eta}(t), \tilde{X}(t))$ globally exponentially converge to a ball centered at the origin. Moreover, one can prove that the tracking error vector $\boldsymbol{X_e}(t)$ also globally asymptotically converges to a ball centered at the origin. The radius of this ball can be adjusted by changing the control gains if the environmental disturbances are not too large. When the environmental disturbances are large enough, the observer (7.51) cannot provide a sufficiently accurate estimate of unmeasured velocities, and the control law (7.68) cannot compensate for considerable environmental disturbances acting on the sway axis. These will result in an unstable closed loop system, especially at a low forward speed. This phenomenon should not be surprising since the vessel in question is not actuated in the sway axis and does not have velocities available for feedback. One can see this phenomenon by observing the simple example system (7.71) with some additive disturbance in the first equation. It is easy to show that when this additive disturbance has magnitude larger than 1, then the system (7.71) will be unstable. The robustness issue is still a challenging problem in control of underactuated ocean vehicles and underactuated systems in general.

7.5 Simulations

This section illustrates the soundness of the control laws (7.68) by simulating them on the same monohull ship in the previous two chapters. Details of the ship parameters are listed in Section 5.4.

The initial conditions of the reference trajectories are chosen as $(x_d(0), y_d(0), \psi_d(0), \upsilon_d(0)) = (0\,\text{m}, 0\,\text{m}, 0\,\text{rad}, 0\,\text{ms}^{-1})$. The reference velocities are $u_d = 4\text{ms}^{-1}$ and $r_d = 0\text{rads}^{-1}$ for the first 300 seconds, and $u_d = 4\,\text{ms}^{-1}$ and $r_d = 0.02\,\text{rads}^{-1}$ for the rest of the simulation time. This choice means that the reference trajectory is a straight line for the first 300 seconds and then followed by a circle with a radius of 200 m. Indeed the above choice of reference velocities satisfies Assumption 7.1. All of the initial conditions of $\hat{\boldsymbol{\eta}}$ and $\hat{\boldsymbol{v}}$ are chosen to be zero. We first choose $k_2 = 5$, $c_1 = 1$, $c_2 = 2$, $\delta_1 = \delta_2 = 0.1$, $\boldsymbol{K}_{01} = 10\,\text{diag}(1,1,1)$, $\boldsymbol{P}_{01} = \boldsymbol{P}_{02} = 0.5\,\text{diag}(1,1,1)$, and $\boldsymbol{K}_{02} = \left(\boldsymbol{J}(\eta)\boldsymbol{P}^{-1}(\eta)\right)^T$. The other design constants are chosen as: $k = 0.05$, $k_1 = 1$, $k_3 = 5$, and $k_4 = 100$. Simulation results are plotted in Figures 7.1 and 7.2 with the initial conditions of the ship: $x = -10\,\text{m}, y = 10\,\text{m}, \psi = 0.1\,\text{rad}, u = 0\,\text{ms}^{-1}, \upsilon = 0\,\text{ms}^{-1}$, and $r = 0\,\text{rads}^{-1}$. It is seen from Figures 7.1a and 7.1b that all of the tracking and observer errors converge to zero. The control inputs, τ_u and τ_r, are within their limits. As always, the magnitude of τ_u and τ_r can be reduced by adjusting the control gains such as c_1 and c_2. However, the trade-off is a longer transient response time. For clarity, we only plot the tracking errors for the first 180 seconds, and the observer errors for the first 10 seconds. To illustrate robustness of our proposed controller, we also simulate with the same control gains and initial conditions chosen as above, and the environmental disturbance vector $\boldsymbol{\tau_w}(t) = 0.5\boldsymbol{M}([\sin(t), \cos(t), \sin(t)]^T + 1.5)$, i.e., the last equation of (7.1) is in the form of $\boldsymbol{M}\dot{\boldsymbol{v}} = -\boldsymbol{C}(\boldsymbol{v})\boldsymbol{v} - \boldsymbol{D}\boldsymbol{v} + \boldsymbol{\tau} + \boldsymbol{\tau_w}(t)$. Simulation results are plotted in Figures 7.3 and 7.4. For clarity, we only plot the tracking errors for the first 300 seconds and the observer errors for the first 10 seconds. It can be seen that the environmental disturbances deteriorate the performance of the controlled loop system in the sense that the tracking errors do not converge to zero but to a ball centered at the origin. This shows an important property of robustness of the controlled system with respect to the environmental disturbances.

7.6 Conclusions

The key to the control developments is an introduction of the global nonlinear coordinate transformations (7.6), (7.24), (7.43), (7.58), and (7.61) to obtain an exponential observer and to transform the tracking error dynamics to a suitable nonlinear system, to which Lyapunov's direct method and the backstepping technique can be applied. The work presented in this chapter is based on [94, 114, 116, 117].

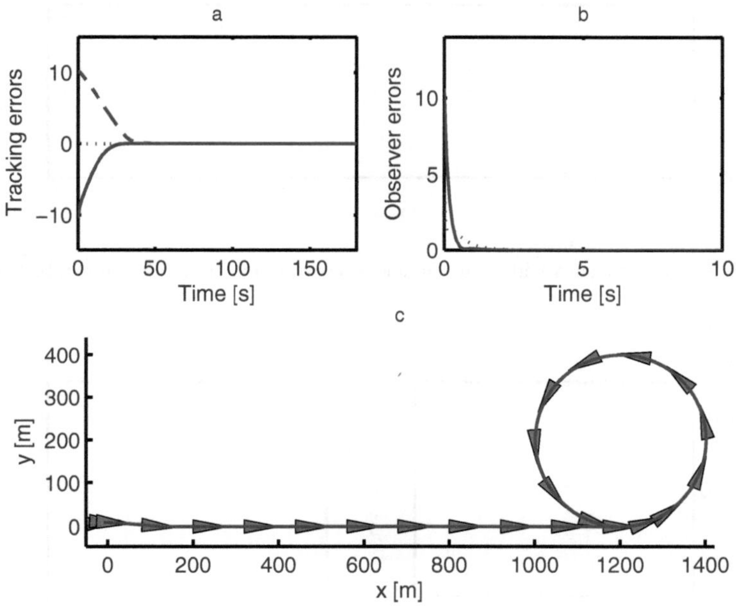

Figure 7.1 Simulation results without disturbances: **a**. Tracking errors ($x - x_d$[m]: *solid line*, $y - y_d$ [m]: *dashed line*, $\psi - \psi_d$ [rad]]: *dotted line*); **b**. Observer errors ($\sqrt{(x-\hat{x})^2+(y-\hat{y})^2+(\psi-\hat{\psi})^2}$: *solid line*, $\sqrt{(u-\hat{u})^2+(v-\hat{v})^2+(r-\hat{r})^2}$): *dotted line*; **c**. Ship position and orientation in the (x, y)-plane

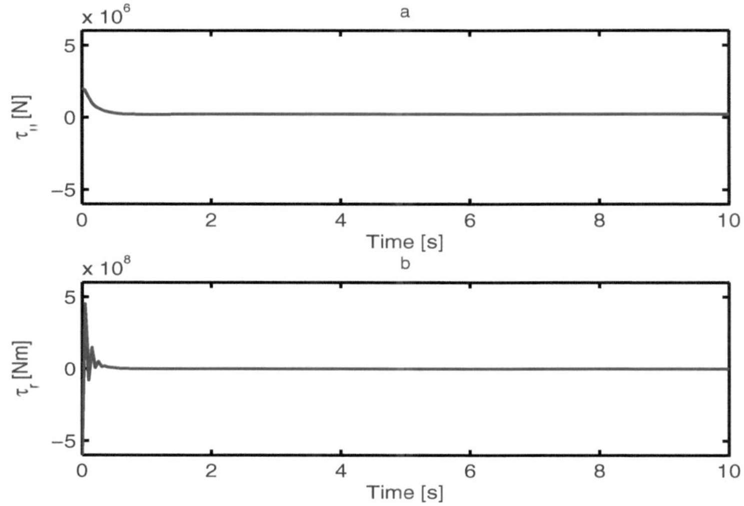

Figure 7.2 Simulation results without disturbances (control inputs): **a**. Surge force; **b**. Yaw moment

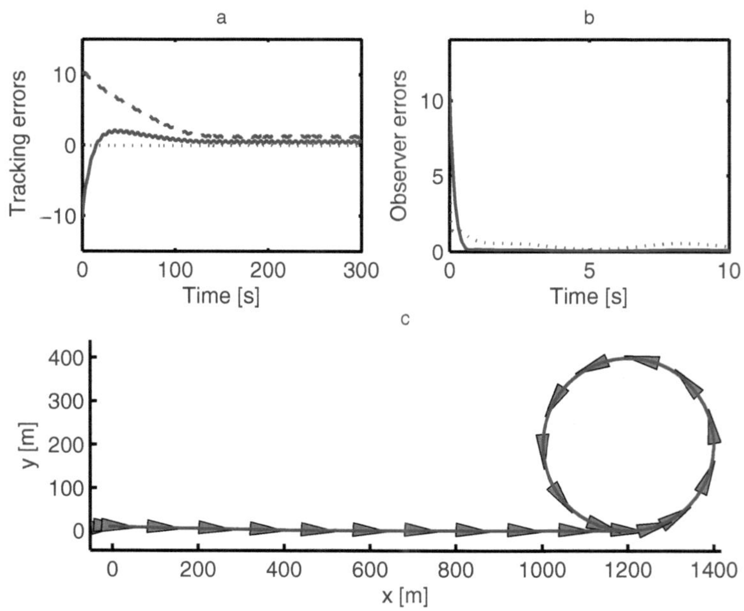

Figure 7.3 Simulation results with disturbances: **a**. Tracking errors ($x - x_d$[m]: *solid line*, $y - y_d$ [m]: *dashed line*, $\psi - \psi_d$ [rad]]: *dotted line*); **b**. Observer errors ($\sqrt{(x-\hat{x})^2+(y-\hat{y})^2+(\psi-\hat{\psi})^2}$: *solid line*, $\sqrt{(u-\hat{u})^2+(v-\hat{v})^2+(r-\hat{r})^2}$): *dotted line*; **c**. Ship position and orientation in the (x, y)-plane

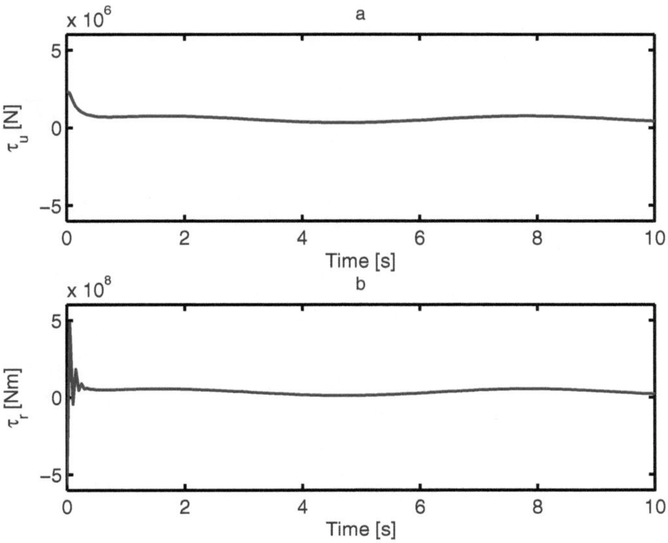

Figure 7.4 Simulation results with disturbances (control inputs): **a**. Surge force; **b**. Yaw moment

Chapter 8
Path-tracking Control of Underactuated Ships

This chapter deals with the problem of designing controllers to force an underactuated surface ship with standard dynamics to track a reference path. Both full-state feedback and output feedback cases are considered. In comparison with the preceding three chapters, the requirement that the reference trajectories be generated by virtual ships is relaxed, and the control design is simpler and more amenable for implementation in practice. The control development is based on a series of coordinate transformations, the backstepping technique, and utilizing the dynamic structure of the ship.

8.1 Full-State Feedback

8.1.1 Control Objective

In addition to the assumptions made in Section 3.4.1.1, we ignore the nonlinear damping terms. The case with the nonlinear damping terms requires more effort, and is treated in Section 8.2 where an output feedback control design is addressed. As such, the resulting mathematical model of the underactuated ship moving in surge, sway, and yaw is rewritten as

$$\begin{aligned}\dot{\boldsymbol{\eta}} &= \boldsymbol{J}(\psi)\boldsymbol{v},\\ \boldsymbol{M}\dot{\boldsymbol{v}} &= -\boldsymbol{C}(\boldsymbol{v})\boldsymbol{v} - \boldsymbol{D}\boldsymbol{v} + \boldsymbol{\tau},\end{aligned} \tag{8.1}$$

with

$$\boldsymbol{\eta} = \begin{bmatrix} x \\ y \\ \psi \end{bmatrix}, \quad \boldsymbol{v} = \begin{bmatrix} u \\ v \\ r \end{bmatrix}, \quad \boldsymbol{\tau} = \begin{bmatrix} \tau_u \\ 0 \\ \tau_r \end{bmatrix},$$

$$\boldsymbol{J}(\psi) = \begin{bmatrix} \cos(\psi) & -\sin(\psi) & 0 \\ \sin(\psi) & \cos(\psi) & 0 \\ 0 & 0 & 1 \end{bmatrix},$$

$$\boldsymbol{M} = \begin{bmatrix} m_{11} & 0 & 0 \\ 0 & m_{22} & m_{23} \\ 0 & m_{32} & m_{33} \end{bmatrix}, \quad \boldsymbol{D} = \begin{bmatrix} d_{11} & 0 & 0 \\ 0 & d_{22} & d_{23} \\ 0 & d_{32} & d_{33} \end{bmatrix},$$

$$\boldsymbol{C}(\boldsymbol{v}) = \begin{bmatrix} 0 & 0 & -m_{22}v - \frac{1}{2}(m_{23}+m_{32})r \\ 0 & 0 & m_{11}u \\ m_{22}v + \frac{1}{2}(m_{23}+m_{32})r & -m_{11}u & 0 \end{bmatrix},$$

where a definition of all symbols is given in Section 3.3.

Moreover, in order to clearly illustrate our idea we did not include the disturbances in (8.1) but will address this issue in Section 8.1.5 by a simple modification of the control design. In this section, we consider the following control objective.

Control Objective. Design the controls τ_u and τ_r such that the position (x, y) of the ship (8.1) globally tracks a reference path $\boldsymbol{\Omega}$ parameterized by $(x_d(s), y_d(s))$ with s being the path parameter, the ship total linear velocity is tangential to the reference trajectory $\boldsymbol{\Omega}$, and the desired surge velocity can be adjusted on-line.

Since the ship system (8.1) is underactuated, it is not expected to force the ship to track an arbitrary path $\boldsymbol{\Omega}$. We here impose the following sufficient conditions on the path $\boldsymbol{\Omega}$:

Assumption 8.1. *There exist strictly positive constants ε_i, $1 \leq i \leq 4$, such that*

$$x_d'^2(s) + y_d'^2(s) \geq \varepsilon_1, \ \forall s \in \mathbb{R}, \tag{8.2}$$

$$\frac{\sqrt{2}}{2}\alpha \bar{u}_d(t) - |\gamma| \geq \varepsilon_2, \ \forall t \geq 0, \tag{8.3}$$

$$\beta \bar{u}_d(t) + \dot{\bar{u}}_d(t) \geq \varepsilon_3, \ \forall t \geq 0, \tag{8.4}$$

and the solutions of the following differential equation

$$(\bar{u}_d \cos(\delta)(1-\alpha) + \gamma)\dot{\delta} = -(\beta \bar{u}_d + \dot{\bar{u}}_d)\sin(\delta) - (\alpha \bar{u}_d \cos(\delta) - \gamma)\bar{r}_d,$$
$$\delta(t_0) = 0 \tag{8.5}$$

satisfy

$$|\delta(t)| \leq \frac{\pi}{4} - \varepsilon_4, \ \forall t \geq t_0 \geq 0, \tag{8.6}$$

where $x_d'(s)$, $y_d'(s)$, $\bar{u}_d$ and $\bar{r}_d$ are defined as

$$x_d'(s) = \frac{\partial x_d}{\partial s},\; y_d'(s) = \frac{\partial y_d}{\partial s},$$
$$\bar{u}_d = \sqrt{x_d'^2(s) + y_d'^2(s)}\dot{s}, \tag{8.7}$$
$$\bar{r}_d = \frac{(x_d'(s)y_d''(s) - x_d''(s)y_d'(s))}{(x_d'^2(s) + y_d'^2(s))}\dot{s},$$

and the constants α, β, and γ are

$$\alpha = \frac{m_{11}}{m_{22}},\; \beta = \frac{d_{22}}{m_{22}},\; \gamma = \frac{d_{22}m_{23}}{m_{22}^2 - d_{23}/m_{22}}. \tag{8.8}$$

Remark 8.1.

1. Condition (8.2) implies that the path $\boldsymbol{\Omega}$ is regular with respect to the path parameter s. If the path is not regular, one can break it into different regular paths and consider each path separately.
2. Conditions (8.3) and (8.4) mean that the path $\boldsymbol{\Omega}$ does not contain or approach a set-point, i.e., we do not consider stabilization/regulation problems. These conditions also imply that $\bar{u}_d(t)$ should not vary too fast. For clarity we only consider the forward tracking case, i.e. the reference velocity $\bar{u}_d(t)$ is always larger than a positive constant. The backward tracking case (i.e. $\bar{u}_d(t)$ is always less than a negative constant) can be done similarly.
3. Conditions (8.5) and (8.6) imply that the desired surge velocity is always larger than the desired sway velocity. These conditions are reasonable since the ship in question is underactuated in the sway axis, one should not expect to force it to track a path with an arbitrarily large curvature. These conditions place certain restrictions on the path $\boldsymbol{\Omega}$ but allow many types of trajectories such as a straight line, an arc, and a sinusoidal path. For vessels with three planes of symmetry (like spherical underwater vehicles see Section 3.4.1.3, i.e., $\alpha = 1, \gamma = 0$), the differential equation (8.5) reduces to an algebraic equation, and condition (8.6) holds if
$$\frac{\bar{u}_d(t)\,|\bar{r}_d(t)|}{\beta\bar{u}_d(t) + \dot{\bar{u}}_d(t)} \leq 1 - \varepsilon_5,\; \forall t \geq 0, \tag{8.9}$$
where ε_5 is an arbitrarily small positive constant. For other cases, an application of Lemma 9.3 in [6] to (8.5) readily shows that sufficient conditions such that (8.6) holds can be expressed as
$$0.5\sqrt{2}\bar{u}_d(t)(1-\alpha) - |\gamma| > \varepsilon_6,\; \forall t \geq 0, \tag{8.10}$$
$$\frac{\alpha\bar{u}_d(t) + |\gamma|}{\beta\bar{u}_d(t) + \dot{\bar{u}}_d(t)}|\bar{r}_d(t)| \leq \frac{\sqrt{2}}{2} - \varepsilon_7, \forall t \geq 0, \tag{8.11}$$
where ε_6 and ε_7 are arbitrarily small positive constants.
4. The variable $\dot{s}$ can be used to specify the desired forward speed.

The main ideas to solve the control objective are as follows:

1. Introduce a coordinate to change the ship position (x, y) such that the ship model (8.1) can be transformed to a diagonal form to overcome difficulties caused by nonzero off-diagonal terms in the system matrices.
2. Interpret tracking errors in a frame attached to the path Ω such that the tracking error dynamics are of a triangular form to which the backstepping technique can be applied.
3. Use the orientation tracking error as a virtual control to stabilize the cross-track error.

8.1.2 Coordinate Transformations

8.1.2.1 Transform Ship Dynamics to a "Diagonal Form"

Introduce the following coordinate transformation (changing the ship position, see Figure 8.1):

$$\begin{aligned}\bar{x} &= x + \varepsilon \cos(\psi),\\ \bar{y} &= y + \varepsilon \sin(\psi),\\ \bar{v} &= v + \varepsilon r,\end{aligned} \tag{8.12}$$

where $\varepsilon = m_{23}/m_{22}$. Using the above change of coordinates, the ship dynamics (8.1) can be written as

$$\begin{aligned}\dot{\bar{x}} &= u\cos(\psi) - \bar{v}\sin(\psi),\\ \dot{\bar{y}} &= u\sin(\psi) + \bar{v}\cos(\psi),\\ \dot{\psi} &= r,\\ \dot{u} &= \bar{\tau}_u,\\ \dot{\bar{v}} &= -\alpha u r - \beta \bar{v} + \gamma r,\\ \dot{r} &= \bar{\tau}_r,\end{aligned} \tag{8.13}$$

where we have chosen the primary controls τ_u and τ_r as:

$$\begin{aligned}\tau_u &= m_{11}\bar{\tau}_u - m_{22}vr - \frac{m_{23}+m_{32}}{2}r^2 + d_{11}u,\\ \tau_r &= \frac{m_{22}m_{33}-m_{23}m_{32}}{m_{22}}\bar{\tau}_r - \frac{1}{m_{22}}\Big((m_{11}m_{22}-m_{22}^2)uv + (m_{11}m_{32}-\\ &\quad m_{22}\frac{m_{23}+m_{32}}{2}ur + (m_{32}d_{22}-m_{22}d_{32})v - (m_{22}d_{33}-m_{32}d_{23})r\Big),\end{aligned} \tag{8.14}$$

with $\bar{\tau}_u$ and $\bar{\tau}_r$ being considered as new controls to be designed later.

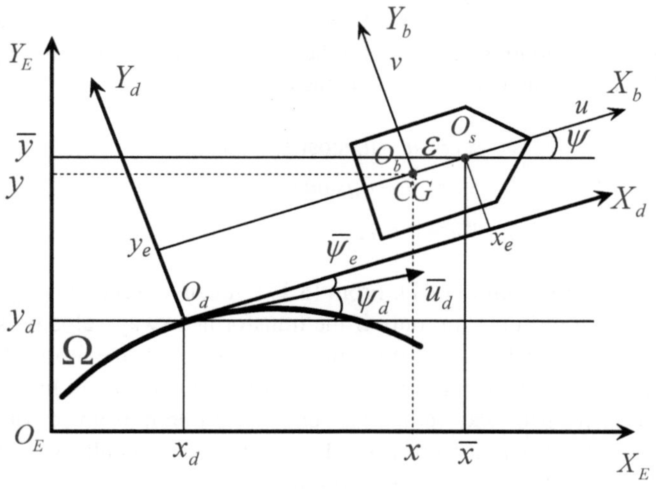

Figure 8.1 Interpretation of tracking errors

Remark 8.2. After the ship system (8.1) is transformed to the diagonal form (8.13), the methods mentioned in the preceding three chapters can be used to obtain controllers for specific tasks such as stabilization and model reference trajectory-tracking if $(\bar{x}, \bar{y})$ are considered as the ship position instead of (x, y).

8.1.2.2 Transform Path-tracking Errors

We now interpret the path-tracking errors in a frame attached to the reference path $\boldsymbol{\Omega}$ as follows (see Figure 8.1):

$$\begin{bmatrix} x_e \\ y_e \\ \bar{\psi}_e \end{bmatrix} = \boldsymbol{J}^T(\psi) \begin{bmatrix} \bar{x} - x_d \\ \bar{y} - y_d \\ \psi - \psi_d \end{bmatrix}, \tag{8.15}$$

where ψ_d is the angle between the path and the X-axis defined by

$$\psi_d = \arctan\left(\frac{y_d'(s)}{x_d'(s)}\right), \tag{8.16}$$

with $x_d'(s)$ and $y_d'(s)$ being defined in (8.7).

In Figure 8.1, $O_E X_E Y_E$ is the earth-fixed frame, $O_d X_d Y_d$ is a frame attached to the path $\boldsymbol{\Omega}$ such that $O_d X_d$ and $O_d Y_d$ are parallel to the surge and sway axes of the ship, $\bar{u}_d$ is tangential to the path, $CG \equiv O_b$ is the center of gravity, and O_s is referred to as the center of oscillation of the ship. Therefore x_e, y_e, and $\bar{\psi}_e$ can be referred to as the tangential tracking error, cross-tracking error, and heading error, respectively. Differentiating (8.15) along the solutions of the first three equations of (8.13) results in the kinematic error dynamics:

$$\begin{aligned}\dot{x}_e &= u - \bar{u}_d \cos(\bar{\psi}_e) + r y_e, \\ \dot{y}_e &= \bar{v} + \bar{u}_d \sin(\bar{\psi}_e) - r x_e, \\ \dot{\bar{\psi}}_e &= r - \bar{r}_d,\end{aligned} \tag{8.17}$$

where $\bar{u}_d$ and $\bar{r}_d$ are given in (8.7). From (8.17), it can be seen that $(x_e, y_e, \bar{\psi}_e) = (0,0,0)$ is the equilibrium point only if the transformed sway velocity is zero, i.e., $\bar{v} = 0$, which means that the ship must move on a straight line at the steady state. Since we allow the reference path $\boldsymbol{\Omega}$ to be different from a straight line, but also include it, the transformed sway velocity is generally different from zero at the steady state, i.e., $\bar{v}$ can be different from zero. To resolve this difficulty we introduce an angle δ to the orientation error $\bar{\psi}_e$ by defining

$$\psi_e = \bar{\psi}_e + \delta. \tag{8.18}$$

Substituting (8.18) into (8.17) yields

$$\begin{aligned}\dot{x}_e &= u - u_d \cos(\psi_e) - v_d \sin(\psi_e) + r y_e, \\ \dot{y}_e &= u_d \sin(\psi_e) + v_e - v_d(\cos(\psi_e) - 1) - r x_e, \\ \dot{\psi}_e &= r - r_d,\end{aligned} \tag{8.19}$$

where

$$\begin{aligned}v_e &= \bar{v} - v_d, \\ u_d &= \bar{u}_d \cos(\delta), \\ v_d &= \bar{u}_d \sin(\delta), \\ r_d &= \bar{r}_d - \dot{\delta}.\end{aligned} \tag{8.20}$$

We refer to u_d, v_d, and r_d as the desired surge, sway, and yaw reference velocities. It can now be seen that $(x_e, y_e, \psi_e) = (0,0,0)$ is the equilibrium point of (8.19) if $v_e = 0$. At the steady state, the surge and yaw velocities (u, r) should converge to their desired values (u_d, r_d) with the help of a control to be designed later. Therefore, to determine the angle δ that guarantees $v_e = 0$ at the steady state, from the fifth equation of (8.13), the desired sway velocity should be chosen as

$$\dot{v}_d = -\alpha u_d r_d - \beta v_d + \gamma r_d. \tag{8.21}$$

Substituting (8.20) into (8.21) results in (8.5). We make the following important observations:

Remark 8.3. With $|\delta(t)| < \pi/4$, $\forall t \geq t_0 \geq 0$, we have $|v_d(t)| < |u_d(t)|$ since $u_d = \bar{u}_d \cos(\delta)$ and $v_d = \bar{u}_d \sin(\delta)$. In addition, with the help of a control (to be designed later), convergence of (x_e, y_e, ψ_e) to zero implies that $\bar{\psi}_e$ converges to $-\delta(t)$. Since $|\delta(t)| < \pi/4$, the ship will not turn around.

To prepare for the control design, we rewrite the path-tracking error dynamics as follows:

$$\begin{aligned}
\dot{x}_e &= u - u_d \cos(\psi_e) - v_d \sin(\psi_e) + r y_e, \\
\dot{y}_e &= u_d \sin(\psi_e) + v_e - v_d(\cos(\psi_e) - 1) - r x_e, \\
\dot{\psi}_e &= r - r_d, \\
\dot{v}_e &= -\alpha(ur - u_d r_d) - \beta v_e + \gamma(r - r_d), \\
\dot{u} &= \bar{\tau}_u, \\
\dot{r} &= \bar{\tau}_r.
\end{aligned} \tag{8.22}$$

The control objective has been converted to one of stabilizing (8.22) at the origin. The tracking error dynamics (8.22) are similar to the ones in the preceding chapters. The techniques in these chapters can be applied to design the controls $\bar{\tau}_u$ and $\bar{\tau}_r$ but will result in an extremely complex procedure to determine the control gains. In this chapter we will propose a much simpler technique to design $\bar{\tau}_u$ and $\bar{\tau}_r$ to globally stabilize (8.22) at the origin.

8.1.3 *Control Design*

We divide the control design into two stages. At the first stage, we consider the first four equations of (8.22), with u and r being viewed as controls. At the second stage, the last two equations of (8.22) are considered to design the actual controls $\bar{\tau}_u$ and $\bar{\tau}_r$ using the backstepping technique.

8.1.3.1 Kinematic Control Design

Substep 1 (Stabilizing x_e-dynamics)

Define

$$u_e = u - \alpha_u, \tag{8.23}$$

where α_u is a virtual control of u. The first equation of (8.22) suggests that we choose

$$\alpha_u = -k_1 x_e + u_d, \tag{8.24}$$

where k_1 is a positive constant to be selected later. Substituting (8.23) and (8.24) into the first equation of (8.22) yields

$$\dot{x}_e = -k_1 x_e - u_d(\cos(\psi_e) - 1) - v_d \sin(\psi_e) + r y_e + u_e. \tag{8.25}$$

Note that we have chosen α_u not to cancel $-u_d(\cos(\psi_e)-1)-v_d\sin(\psi_e)+ry_e$ to simplify the control design.

Substep 2 (Stabilizing (y_e, v_e)-dynamics)

Introduce a coordinate change

$$z_1 = y_e + \frac{v_e}{\beta}. \tag{8.26}$$

Differentiating (8.26) along the solutions of the second and fourth equations of (8.22), and using (8.23) and (8.24) yields

$$\begin{aligned}\dot{z}_1 = \; & u_d\sin(\psi_e)-v_d(\cos(\psi_e)-1)-rx_e- \\ & \frac{\alpha}{\beta}((u_e-k_1x_e+u_d)r-u_dr_d)+\frac{\gamma}{\beta}(r-r_d).\end{aligned} \tag{8.27}$$

For simplicity, we now choose

$$k_1 = \frac{\beta}{\alpha} \tag{8.28}$$

to cancel the term $-rx_e$ on the right-hand side of (8.27) to have z_1-dynamics "independent" of x_e. Substituting (8.28) into (8.27) yields

$$\dot{z}_1 = u_d\sin(\psi_e)-v_d(\cos(\psi_e)-1)-\frac{1}{\beta}(\alpha u_d-\gamma)(r-r_d)-\frac{\alpha}{\beta}u_er. \tag{8.29}$$

It is now seen that the third equation of (8.22) and (8.29) are of a forward structure with r being considered as a control. One can apply forwarding control design techniques to design a control law for r. However, it is difficult to find a proper Lyapunov function to handle cross terms. The reader is referred to [7] for more details on control design techniques for forward systems. Here we use the backstepping technique by using ψ_e as a virtual control to stabilize z_1. Define

$$z_2 = \psi_e - \alpha_{\psi_e}, \tag{8.30}$$

where α_{ψ_e} is a virtual control of ψ_e. By noticing (8.3), (8.7) and $|\delta(t)| < \pi/4,\ \forall t \geq t_0 \geq 0$, we have $u_d(t) > 0$. Therefore, a simple control law for α_{ψ_e} can be chosen as

$$\alpha_{\psi_e} = -\arcsin\left(\frac{k_2z_1}{\Delta}\right),\ \Delta = \sqrt{1+(k_2z_1)^2}, \tag{8.31}$$

where k_z is a positive constant to be selected later. It is of interest to note that (8.31) is well defined for all $z_1 \in \mathbb{R}$.

Substep 3 (Stabilizing z_2-dynamics)

Define

$$r_e = r - r_d - \alpha_r, \tag{8.32}$$

where α_r is a virtual control of r. Differentiating both sides of (8.30) along the solutions of the third equation of (8.22) and (8.29) yields

$$\begin{aligned}\dot{z}_2 = & \left(1 - k_2 \frac{\alpha u_d - \gamma}{\beta \Delta^2}\right)(r_e + \alpha_r) + \\ & \frac{k_2}{\Delta^2}(u_d \sin(\psi_e) - v_d(\cos(\psi_e) - 1)) - \frac{k_2 \alpha}{\beta \Delta^2} u_e r.\end{aligned} \tag{8.33}$$

We choose the constant k_2 such that

$$1 - k_2 \frac{\alpha u_d - \gamma}{\beta \Delta^2} > 0 \Rightarrow 0 < k_2 \leq \frac{\beta}{\alpha u_d - |\gamma|}. \tag{8.34}$$

There always exists k_2 satisfying (8.34) under condition (8.3) since $|\delta(t)| < \pi/4$ and $u_d = \bar{u}_d \cos(\delta), \forall t \geq t_0 \geq 0$. We choose the virtual control α_r from (8.33) as

$$\alpha_r = \frac{1}{1 - \frac{k_2}{\beta \Delta^2}(\alpha u_d - \gamma)} \left(- \frac{c_1 z_2}{\sqrt{1 + z_2^2}} - \frac{k_2}{\Delta^2}(u_d \sin(\psi_e) - v_d(\cos(\psi_e) - 1)) \right), \tag{8.35}$$

where c_1 is a positive constant. It is noted that α_r is bounded by some constant for all $(z_1, z_2, \psi_e) \in \mathbb{R}^3$.

8.1.3.2 Kinetic Control Design

By differentiating both sides of (8.23) and (8.32) along the solutions of the first, third, fifth, and sixth equations of (8.22), the control $\bar{\tau}_u$ is chosen as

$$\bar{\tau}_u = -c_2 u_e + \frac{\partial \alpha_u}{\partial x_e} \dot{x}_e + \frac{\partial \alpha_u}{\partial u_d} \dot{u}_d + \frac{k_2 \alpha}{\beta \Delta^2} z_2 r, \tag{8.36}$$

and the control $\bar{\tau}_r$ is chosen as

$$\begin{aligned}\bar{\tau}_r = & \frac{\partial \alpha_r}{\partial z_1} \dot{z}_1 + \frac{\partial \alpha_r}{\partial z_2} \dot{z}_2 + \frac{\partial \alpha_r}{\partial \psi_e} \dot{\psi}_e + \frac{\partial \alpha_r}{\partial u_d} \dot{u}_d + \frac{\partial \alpha_r}{\partial v_d} \dot{v}_d - \\ & \left(1 - \frac{k_2}{\beta \Delta^2}(\alpha u_d - \gamma)\right) z_2 - c_3 r_e + \dot{r}_d,\end{aligned} \tag{8.37}$$

where c_2 and c_3 are positive constants. Substituting (8.30), (8.31), (8.35), (8.36), and (8.37) into (8.29) and (8.33), derivatives of u_e and r_e along the solutions of the last two equations of (8.22) gives the following closed loop system:

$$\begin{aligned}
\dot{x}_e &= -k_1 x_e - u_d(\cos(\psi_e)-1) - v_d\sin(\psi_e) + (r_d+\alpha_r+r_e)(z_1 - \\
&\quad \beta^{-1}v_e) + u_e, \\
\dot{v}_e &= -\alpha((u_e+\alpha_u)(r_d+\alpha_r+r_e) - u_d r_d) - \beta v_e + \gamma(\alpha_r+r_e), \\
\dot{z}_1 &= -k_2 u_d z_1/\Delta - v_d(\cos(\psi_e)-1) - \beta^{-1}(\alpha u_d - \gamma)(\alpha_r+r_e) + \\
&\quad u_d(\sin(z_2)\cos(\alpha_{\psi_e}) + (\cos(z_2)-1)\cos(\alpha_{\psi_e})) - \alpha\beta^{-1}u_e(r_d + \\
&\quad \alpha_r + r_e), \\
\dot{z}_2 &= -c_1 z_2/\sqrt{1+z_2^2} + (1 - k_2(\alpha u_d - \gamma)\beta^{-1}\Delta^{-2})r_e - k_2\alpha\beta^{-1}\Delta^{-2}u_e r, \\
\dot{u}_e &= -c_2 u_e + k_2\alpha\beta^{-1}\Delta^{-2}z_2 r, \\
\dot{r}_e &= -c_3 r_e - (1 - k_2(\alpha u_d - \gamma)\beta^{-1}\Delta^{-2})z_2.
\end{aligned} \tag{8.38}$$

We now present the first main result of this chapter, the proof of which is given in the next section.

Theorem 8.1. *Assume that the path* $\boldsymbol{\Omega}$ *satisfies all conditions specified in Assumption 8.1, then the controls* τ_u *and* τ_r *given by (8.14), (8.36), and (8.37) force the transformed tracking errors* (x_e, y_e, ψ_e) *to converge to zero globally asymptotically and locally exponentially with an appropriate choice of* k_2. *As a result, the actual position tracking errors* $(x - x_d, y - y_d)$ *and orientation tracking error* $(\psi - \psi_d)$ *globally asymptotically and locally exponentially converge to balls with radii of* $|m_{23}/m_{22}|$ *and* $|\delta(t)| \leq \arcsin((\alpha\bar{u}_d + |\gamma|)|\bar{r}_d|/(\beta\bar{u}_d + \bar{\dot{u}}_d)) < \pi/4$, *respectively. Furthermore, the desired forward speed of the ship on the path can be adjusted by adjusting* $\dot{s}$.

8.1.4 Stability Analysis

To prove Theorem 8.1, we first show that the closed loop system (8.38) is forward complete, i.e., there is no finite escape in the closed loop system. We then consider the (z_2, u_e, r_e)-, z_1- and (x_e, v_e)-subsystems separately. We first consider the Lyapunov function

$$V_1 = x_e^2 + z_1^2 + z_2^2 + v_e^2 + u_e^2 + r_e^2. \tag{8.39}$$

A simple calculation shows that the derivative of V_1 along the solutions of (8.38) satisfies

$$\begin{aligned}
&\dot{V}_1 \leq \rho_{11}V_1 + \rho_{12} \\
\Rightarrow\quad &V_1(t) \leq (V_1(t_0) + \rho_{12}/\rho_{11})e^{\rho_{11}(t-t_0)},
\end{aligned} \tag{8.40}$$

where ρ_{11} and ρ_{12} are some positive constants, which implies that the closed loop system (8.38) is forward complete.

To investigate stability of the (z_2, u_e, r_e)-subsystem, we consider the Lyapunov function

$$V_{21} = \frac{1}{2}(z_2^2 + u_e^2 + r_e^2), \tag{8.41}$$

whose derivative along the solutions of the last three equations of (8.38) satisfies

$$\dot{V}_{21} = -c_1 \frac{z_2^2}{\sqrt{1+z_2^2}} - c_2 u_e^2 - c_3 r_e^2 \leq 0. \tag{8.42}$$

Since $\dot{V}_{21} \leq 0$, we have $V_{21}(t) \leq \chi_{21}(\cdot)$ where $\chi_{21}(\cdot)$ is a class-K_∞ function of $X_1(t_0) = \|(z_2(t_0), u_e(t_0), r_e(t_0))\|$. With this upper bound of V_{21}, we have

$$\dot{V}_{21} = -c_1 \frac{z_2^2}{\sqrt{1+2\chi_{21}(\cdot)}} - c_2 u_e^2 - c_3 r_e^2, \tag{8.43}$$

which implies that the (z_2, u_e, r_e)-subsystem is globally asymptotically and locally exponentially stable at the origin, i.e. there exist a class-K function $\gamma_{21}(\cdot)$ and a constant σ_{21} depending on $X_1(t_0)$ such that

$$\|(z_2(t), u_e(t), r_e(t))\| \leq \gamma_{21}(\cdot) e^{-\sigma_{21}(t-t_0)}. \tag{8.44}$$

For the z_1-subsystem, we take the Lyapunov function

$$V_{22} = \frac{1}{2} z_1^2. \tag{8.45}$$

By substituting α_r, see (8.35), into the derivative of V_{22} along the solutions of the third equation of (8.38), we have

$$\begin{aligned}\dot{V}_{22} \leq\ & -k_2 \left(1 - k_2 \frac{|\alpha u_d - \gamma|}{(\beta - k_2\,|\alpha u_d - \gamma|)}\right) \frac{u_d - |v_d|}{\Delta} z_1^2 + \\ & \gamma_{21}(\cdot)(V_{22} + \gamma_{22}(\cdot)) e^{-\sigma_{21}(t-t_0)},\end{aligned} \tag{8.46}$$

where $\gamma_{21}(\cdot)$ and $\gamma_{22}(\cdot)$ are class-K_∞ functions of $X_1(t_0)$. We now choose k_2 such that

$$\begin{aligned} & 1 - \frac{k_2(\alpha u_d - |\gamma|)}{\beta - k_2(\alpha u_d - |\gamma|)} > 0 \\ \Rightarrow\quad & 0 < k_2 < 0.5 \frac{\beta}{\alpha u_d - |\gamma|}, \end{aligned} \tag{8.47}$$

then (8.46) together with conditions (8.3) and (8.10) in Assumption 8.1 imply that

$$\dot{V}_{22} \leq \gamma_{21}(\cdot)(V_{22} + \gamma_{22}(\cdot)) e^{-\sigma_{21}(t-t_0)}. \tag{8.48}$$

Solving the above differential inequality results in $V_{22}(t) \leq \gamma_{23}(\cdot)$ with $\gamma_{23}(\cdot)$ being a class-K_∞ function of $X_2(t_0) = \|(z_1(t_0), z_2(t_0), u_e(t_0), r_e(t_0))\|$. Substituting this upper bound of V_{22} into (8.46) yields

$$\begin{aligned}\dot{V}_{22} &\leq -k_2\left(1-k_2\frac{|\alpha u_d-\gamma|}{\beta-k_2|\alpha u_d-\gamma|}\right)\frac{u_d-|v_d|}{\Delta}z_1^2+\gamma_{24}(\cdot)e^{-\sigma_{21}(t-t_0)} \\ &\leq \gamma_{24}(\cdot)e^{-\sigma_{21}(t-t_0)}, \end{aligned} \tag{8.49}$$

where $\gamma_{24}(\cdot)=\gamma_{21}(\cdot)(\gamma_{23}(\cdot)+\gamma_{22}(\cdot))$. The second inequality of (8.49) implies that there exists a class-K_∞ function $\gamma_{25}(\cdot)$ of $X_2(t_0)$ such that $|z_1(t)|\leq\gamma_{25}(\cdot)$. Substituting this upper bound into the expression of Δ and using the first inequality of (8.49) yields global asymptotic and local exponential stability of the z_1-subsystem, i.e., there exist a class-K_∞ function of $X_2(t_0)$ and a constant $\sigma_{22}>0$ depending on $X_2(t_0)$ such that $|z_1(t)|\leq\gamma_{26}(\cdot)e^{-\sigma_{22}(t-t_0)}$. Note that condition (8.47) covers the condition (8.34).

For the (x_e,v_e)-subsystem, we consider the Lyapunov function

$$V_3=\frac{1}{2}(x_e^2+v_e^2). \tag{8.50}$$

By taking the derivative of this function along the solutions of the first two equations of (8.38), it can readily be shown that there exist class-K_∞ functions $\gamma_{31}(\cdot)$ and $\gamma_{32}(\cdot)$ of $X_2(t_0)$ such that

$$\begin{aligned}\dot{V}_3 \leq &-\frac{\beta}{\alpha}x_e^2-\beta v_e^2+\gamma_{31}(\cdot)(V_3+\gamma_{32}(\cdot))e^{-\sigma_{22}(t-t_0)} \\ \leq &\ \gamma_{31}(\cdot)(V_3+\gamma_{32}(\cdot))e^{-\sigma_{22}(t-t_0)}. \end{aligned} \tag{8.51}$$

The second inequality of (8.51) means that $V_3(t)\leq\gamma_{33}(\cdot)$ with $\gamma_{33}(\cdot)$ being a class-K_∞ function of $\|(x_e(t_0),v_e(t_0)z_1(t_0),z_2(t_0),u_e(t_0),r_e(t_0))\|$. Substituting this upper bound into the first line of (8.51) readily yields global asymptotic and local exponential stability of the (x_e,v_e)-subsystem. We have so far proven that the transformed tracking errors $(x_e,z_1,z_2,v_e,u_e,r_e)$ globally asymptotically and locally exponentially converge to zero. By (8.26), convergence of z_1 and v_e to zero implies that of y_e. Convergence of the actual position tracking error $(x-x_d,y-y_d)$ and orientation tracking error $(\psi-\psi_d)$ to balls with radii of $|m_{23}/m_{22}|$ and $|\delta(t)|\leq(\alpha\bar{u}_d+|\gamma|)|\bar{r}_d|/|\beta\bar{u}_d+\dot{\bar{u}}_d|<\pi/4$ follows from (8.15) and (8.18). Finally, since $v_d=\bar{u}_d\sin(\delta)$, $u_d=\bar{u}_d\cos(\delta)$ and $\bar{u}_d=\sqrt{x_d'^2+y_d'^2}\dot{s}$, we can see that the total linear velocity of the ship is tangential to the path and the desired forward speed u_d can be adjusted by adjusting $\dot{s}$.

8.1.5 Dealing with Environmental Disturbances

To clearly illustrate our idea, we have not included constant (or at least slowly time-varying) environmental disturbances induced by waves, wind, and ocean currents in the control design. To use the proposed controller in practice, the environmental disturbances should essentially be taken into account in the control design. The aim of this section is to discuss a simple way to modify the proposed control design in the previous section to handle these disturbances. When the constant disturbances are present, the last equation of (8.1) is of the form

$$\boldsymbol{M}\dot{\boldsymbol{v}} = -\boldsymbol{C}(\boldsymbol{v})\boldsymbol{v} - \boldsymbol{D}\boldsymbol{v} + \boldsymbol{\tau} + \boldsymbol{\tau}_{\boldsymbol{E}}, \tag{8.52}$$

where $\boldsymbol{\tau}_{\boldsymbol{E}} = [\tau_{uE}\ \tau_{vE}\ \tau_{rE}]^T$ with τ_{uE}, τ_{vE} and τ_{rE} being the environmental disturbances acting on the surge, sway, and yaw axes. Processing the same coordinate transformations as in Section 8.1.2, the last three equations of (8.13) are written as

$$\begin{aligned}
\dot{u} &= \bar{\tau}_u + \bar{\tau}_{uE}, \\
\dot{\bar{v}} &= -\alpha u r - \beta \bar{v} + \gamma r + \bar{\tau}_{vE}, \\
\dot{r} &= \bar{\tau}_r + \bar{\tau}_{rE},
\end{aligned} \tag{8.53}$$

with

$$\begin{aligned}
\bar{\tau}_{uE} &= \frac{1}{m_{11}} \tau_{uE}, \\
\bar{\tau}_{vE} &= \frac{1}{m_{22}} \tau_{vE}, \\
\bar{\tau}_{rE} &= \frac{(-m_{23}\tau_{vE} + m_{22}\tau_{rE})}{(m_{22}m_{33} - m_{23}m_{32})}.
\end{aligned} \tag{8.54}$$

We first deal with $\bar{\tau}_{vE}$. The idea to handle $\bar{\tau}_{vE}$ is to introduce an angle to the yaw angle. This angle together with the ship forward speed will compensate for $\bar{\tau}_{vE}$. We design an observer to estimate $\bar{\tau}_{vE}$ as

$$\begin{aligned}
\dot{\hat{\bar{v}}} &= k_{01}(\bar{v} - \hat{\bar{v}}) - \alpha u r - \beta \hat{\bar{v}} + \gamma r + \hat{\bar{\tau}}_{vE}, \\
\dot{\hat{\bar{\tau}}}_{vE} &= -\gamma_{01} \mathrm{proj}(\bar{v} - \hat{\bar{v}}, \hat{\bar{\tau}}_{vE}),
\end{aligned} \tag{8.55}$$

where k_{01} and γ_{01} are positive constants, $\hat{\bar{\tau}}_{vE}$ is an estimate of $\bar{\tau}_{vE}$, the operator proj represents the Lipschitz projection algorithm [118] as

$$\begin{aligned}
&\mathrm{proj}(\varpi, \hat{\omega}) = \varpi \text{ if } \Xi(\hat{\omega}) \le 0, \\
&\mathrm{proj}(\varpi, \hat{\omega}) = \varpi \text{ if } \Xi(\hat{\omega}) \ge 0 \text{ and } \Xi_{\hat{\omega}}(\hat{\omega})\varpi \le 0, \\
&\mathrm{proj}(\varpi, \hat{\omega}) = (1 - \Xi(\hat{\omega}))\varpi \text{ if } \Xi(\hat{\omega}) > 0 \text{ and } \Xi_{\hat{\omega}}(\hat{\omega})\varpi > 0,
\end{aligned}$$

where $\Xi(\hat{\omega}) = (\hat{\omega}^2 - \omega_M^2)/(\xi^2 + 2\xi\omega_M)$, $\Xi_{\hat{\omega}}(\hat{\omega}) = \partial \Xi(\hat{\omega})/\partial \hat{\omega}$, ξ is an arbitrarily small positive constant, $\hat{\omega}$ is an estimate of ω, and $|\omega| \le \omega_M$. The projection algorithm is such that if $\dot{\hat{\omega}} = \mathrm{proj}(\varpi, \hat{\omega})$ and $\hat{\omega}(t_0) \le \omega_M$, then

1. $\hat{\omega}(t) \le \omega_M + \xi,\ \forall 0 \le t_0 \le t < \infty$,
2. $\mathrm{proj}(\varpi, \hat{\omega})$ is Lipschitz continuous,
3. $|\mathrm{proj}(\varpi, \hat{\omega})| \le |\varpi|$,
4. $\tilde{\omega}\mathrm{proj}(\varpi, \hat{\omega}) \ge \tilde{\omega}\varpi$ with $\tilde{\omega} = \omega - \hat{\omega}$.

From (8.55), the second equation of (8.53), and property 4 of the proj operator, it is readily shown that $\lim_{t\to\infty} \tilde{\bar{\tau}}_{vE}(t) = 0$ with $\tilde{\bar{\tau}}_{vE}(t) = \bar{\tau}_{vE} - \hat{\bar{\tau}}_{vE}(t)$. Now, by comparing the transformed sway dynamics with disturbances (the second equation of (8.53)) and the one without disturbances (the fifth equation of (8.13)), the desired sway velocity v_d dynamics should now be modified from (8.21) as

$$\dot{v}_d = -\alpha u_d r_d - \beta v_d + \gamma r_d + \hat{\bar{\tau}}_{vE}. \tag{8.56}$$

With this choice, we have

$$\dot{v}_e = -\alpha(ur - u_d r_d) - \beta v_e + \gamma(r - r_d) + \tilde{\bar{\tau}}_{vE}. \tag{8.57}$$

Notice that (8.56) is equivalent to

$$\begin{aligned}(\bar{u}_d \cos(\delta)(1-\alpha) + \gamma)\dot{\delta} = -(\beta\bar{u}_d + \dot{\bar{u}}_d)\sin(\delta) - \\ (\alpha\bar{u}_d \cos(\delta) - \gamma)\bar{r}_d + \hat{\bar{\tau}}_{vE},\ \delta(t_0) = 0.\end{aligned} \tag{8.58}$$

Therefore the tracking objective is solvable if condition (8.11) is replaced by

$$\frac{(\alpha\bar{u}_d + |\gamma|)\,|\bar{r}_d| + \max(|\bar{\tau}_{vE}|) + \xi_{01}}{\beta\bar{u}_d + \dot{\bar{u}}_d} \le \frac{\sqrt{2}}{2} - \varepsilon_7, \tag{8.59}$$

where property 1 of the projection algorithm that guarantees $\left|\hat{\bar{\tau}}_{vE}\right| \le \max(|\bar{\tau}_{vE}|) + \xi_{01}$ has been utilized, with ξ_{01} an arbitrarily small positive constant. Hence, if the disturbances are not too large and the path $\boldsymbol{\Omega}$ is feasible, i.e., Assumption 8.1 with condition (8.11) being replaced by (8.59) holds, then the control objective is solvable under constant disturbances. Now processing the same coordinate changes as in Section 8.1.2 results in the same tracking error dynamics as (8.22) but the last three equations of (8.22) are replaced by

$$\begin{aligned}\dot{v}_e &= -\alpha(ur - u_d r_d) - \beta v_e + \gamma(r - r_d) + \tilde{\bar{\tau}}_{vE},\\ \dot{u} &= \bar{\tau}_u + \bar{\tau}_{uE},\\ \dot{r} &= \bar{\tau}_r + \bar{\tau}_{rE}.\end{aligned} \tag{8.60}$$

Since $\lim_{t\to\infty} \tilde{\bar{\tau}}_{vE}(t) = 0$, and $\bar{\tau}_{uE}$ and $\bar{\tau}_{rE}$ satisfy a matching condition, the control design is much the same as in Section 8.1.3 and is briefly described in the following.

Kinematic Control Design. The same as in Section 8.1.3.

Kinetic Control Design. The controls $\bar{\tau}_u$ and $\bar{\tau}_r$ are almost the same as in Section 8.1.3 but we cannot directly use $\dot{z}_1$ and $\dot{z}_2$ since they contain an unmeasured term $\tilde{\bar{\tau}}_{vE}(t)$ and we need to compensate $\bar{\tau}_{uE}$ and $\bar{\tau}_{rE}$. These controls are designed as follows:

$$
\begin{aligned}
\bar{\tau}_u &= -c_2 u_e + \frac{\partial \alpha_u}{\partial x_e}\dot{x}_e + \frac{\partial \alpha_u}{\partial u_d}\dot{u}_d + \frac{k_2 \alpha}{\beta \Delta^2} z_2 r - \hat{\bar{\tau}}_{uE}, \\
\bar{\tau}_r &= -c_3 r_e + \dot{r}_d + \frac{\partial \alpha_r}{\partial z_1}(u_d \sin(\psi_e) - v_d(\cos(\psi_e) - 1) - \\
& \frac{1}{\beta}(\alpha u_d - \gamma)(r - r_d) - \frac{\alpha}{\beta} u_e r + \frac{\partial \alpha_r}{\partial z_2}\left(-\frac{c_1 z_2}{\sqrt{1+z_2^2}} + \right. \\
& \left. \left(1 - \frac{k_2(\alpha u_d - \gamma)}{\beta \Delta^2}\right) r_e - \frac{k_2 \alpha}{\beta \Delta^2} u_e r \right) + \frac{\partial \alpha_r}{\partial \psi_e}\dot{\psi}_e + \frac{\partial \alpha_r}{\partial u_d}\dot{u}_d + \\
& \frac{\partial \alpha_r}{\partial v_d}\dot{v}_d - \frac{(1 - k_2(\alpha u_d - \gamma)}{\beta \Delta^2} z_2 - \hat{\bar{\tau}}_{rE},
\end{aligned} \tag{8.61}
$$

where $\hat{\bar{\tau}}_{uE}$ and $\hat{\bar{\tau}}_{rE}$ are estimates of $\bar{\tau}_{uE}$ and $\bar{\tau}_{rE}$ updated by

$$
\begin{aligned}
\dot{\hat{\bar{\tau}}}_{uE} &= -\gamma_{02} \text{proj}(u_e, \hat{\bar{\tau}}_{uE}), \\
\dot{\hat{\bar{\tau}}}_{rE} &= -\gamma_{03} \text{proj}(r_e, \hat{\bar{\tau}}_{rE}),
\end{aligned} \tag{8.62}
$$

where γ_{02} and γ_{03} are positive constants. From the above control design, we have a closed loop system as

$$
\begin{aligned}
\dot{x}_e =& -k_1 x_e - u_d(\cos(\psi_e) - 1) - v_d \sin(\psi_e) + (r_d + \alpha_r + r_e)(z_1 - \\
& \beta^{-1} v_e) + u_e, \\
\dot{v}_e =& -\alpha((u_e + \alpha_u)(r_d + \alpha_r + r_e) - u_d r_d) - \beta v_e + \gamma(\alpha_r + r_e) + \tilde{\bar{\tau}}_{vE}, \\
\dot{z}_1 =& -k_2 u_d z_1/\Delta - v_d(\cos(\psi_e) - 1) - \beta^{-1}(\alpha u_d - \gamma)(\alpha_r + r_e) + \\
& u_d(\sin(z_2)\cos(\alpha_{\psi_e}) + (\cos(z_2) - 1)\cos(\alpha_{\psi_e})) - \alpha \beta^{-1} u_e(r_d + \\
& \alpha_r + r_e) + \beta^{-1}\tilde{\bar{\tau}}_{vE}, \\
\dot{z}_2 =& -c_1 z_2/\sqrt{1+z_2^2} + (1 - k_2(\alpha u_d - \gamma)\beta^{-1}\Delta^{-2}) r_e - k_2 \alpha \beta^{-1} \Delta^{-2} u_e r + \\
& k_2 \beta^{-1} \Delta^{-2}\, \tilde{\bar{\tau}}_{vE}, \\
\dot{u}_e =& -c_2 u_e + k_2 \alpha \beta^{-1} \Delta^{-2} z_2 r + \tilde{\bar{\tau}}_{uE}, \\
\dot{r}_e =& -c_3 r_e - (1 - \frac{k_2}{\beta \Delta^2}(\alpha u_d - \gamma)) z_2 - \frac{\tilde{\bar{\tau}}_{vE}}{\beta}\frac{\partial \alpha_r}{\partial z_1} - \frac{k_2 \tilde{\bar{\tau}}_{vE}}{\beta \Delta^2}\frac{\partial \alpha_r}{\partial z_2} + \tilde{\bar{\tau}}_{rE},
\end{aligned} \tag{8.63}
$$

where $\tilde{\bar{\tau}}_{uE} = \bar{\tau}_{uE} - \hat{\bar{\tau}}_{uE}$ and $\tilde{\bar{\tau}}_{rE} = \bar{\tau}_{rE} - \hat{\bar{\tau}}_{rE}$. We now present a result for the case with constant disturbances.

Theorem 8.2. *Assume that the path* $\boldsymbol{\Omega}$ *satisfies all conditions specified in Assumption 8.1 except for (8.5) being replaced by (8.58) and that the disturbances are not too large, i.e., (8.59) holds, the controls* τ_u *and* τ_r *given by (8.14) and (8.61) force the transformed tracking errors* (x_e, y_e, ψ_e) *to globally asymptotically converge to zero with an appropriate choice of* k_2. *As a result, the ac-*

tual position tracking errors $(x - x_d, y - y_d)$ *and the orientation tracking error* $(\psi - \psi_d)$ *globally asymptotically converge to balls with radii of* $|m_{23}/m_{22}|$ *and* $|\delta(t)| \leq \arcsin((\alpha\bar{u}_d + |\gamma|)|\bar{r}_d| + \max(|\bar{\tau}_{vE}|) + \xi_{01})/(\beta\bar{u}_d + \dot{\bar{u}}_d)) < \pi/4$, *respectively.*

Note that convergence of the tracking errors in the case with disturbances is asymptotic (not locally exponential due to adaptation), and the magnitude of $\delta(t)$ is generally larger than the one in the case without disturbances. Since

$$\lim_{t\to\infty}(\tilde{\bar{\tau}}_{uE}(t), \tilde{\bar{\tau}}_{uE}(t), \tilde{\bar{\tau}}_{rE}(t)) = 0,$$

$\partial\alpha_r/\partial z_1$ and $\partial\alpha_r/\partial z_2$ are bounded by some constants, the structure of (8.63) is very similar to (8.38), and the proof of Theorem 8.2 follows the same lines as for Theorem 8.1 plus an application of Barbalat's lemma, we omit proof of Theorem 8.2.

8.1.6 Numerical Simulations

We perform some numerical simulations to illustrate the effectiveness of our proposed control laws given by (8.14), (8.36), and (8.37) on a supply vessel with the length of $L = 76.2$ m and a mass of $m = 226 \times 10^3$ kg. This vessel has a minimum turning circle with the radius of 250m, a maximum surge force of 6.5×10^{10} N, and a maximum yaw moment of 3.2×10^{12} Nm. The Bis-scale parameters of the vessel taken from [11], pp. 446, are as follows:

$$m_{11} = 1.1274,\ m_{22} = 1.8902,\ m_{23} = -0.0744,\ m_{33} = 0.1278,$$
$$d_{11} = 0.0358,\ d_{22} = 0.1183,\ d_{23} = -0.0124,\ d_{32} = -0.0041,\ d_{33} = 0.0308.$$

From these values, we have $\alpha = 0.5964$, $\beta = 0.0626$, and $\gamma = 0.0041$. The reference path $\boldsymbol{\Omega}$ is chosen as $x_d(s) = s$, $y_d(s) = 100\sin(0.01s)$ and $\dot{s} = 1$. Therefore, $\bar{u}_d = 2\sqrt{1+\cos(0.01s)^2}$ and $\bar{r}_d = -0.02\sin(0.01s)/(1+\cos(0.01s)^2)$. A simple calculation shows that all of conditions of Assumption 8.1 hold. In the simulations, the initial conditions are chosen as $x(0) = -20$, $y(0) = 50$, $\psi(0) = 0.5$, $u(0) = 0$, $v(0) = 0.1$, and $r(0) = 0$. The control gains are chosen as $k_1 = \beta/\alpha = 0.1049$, $k_2 = 0.025$, and $c_1 = c_2 = c_3 = 2$. A simple calculation shows that k_2 satisfies the condition specified in (8.47). Simulation results without disturbances are plotted in Figures 8.2 and 8.3. In these figures, all variables are converted back to their original values using the following table with g being the gravitational acceleration:

The tracking errors are plotted in Figure 8.2a, and the angle δ is plotted in Figure 8.2b. It is seen that the proposed controller nicely forces the ship to track the given reference path in the sense that the tracking errors asymptotically converge to zero. From (8.20) and the fact that the surge and sway velocities (u, v) converge to their desired values (u_d, v_d), the change of the angle δ reflects the change of the surge and sway velocities of the ship with the total ship velocity $\bar{u} = \sqrt{u^2 + v^2}$ as time

Table 8.1 Normalization variables used for the Bis system

Variable	Bis-system
Time	$\sqrt{L/g}$
Position	L
Angle	1
Linear velocity	$\sqrt{Lg}$
Angular velocity	$\sqrt{g/L}$
Force	mg
Moment	mgL

evolutes. The ship and reference trajectories in the (x,y)-plane, and the control inputs are plotted in Figure 8.3.

To illustrate how our controller in Section 8.1.5 can compensate for the disturbances, we also simulate with the disturbances as $\boldsymbol{\tau}_E = [0.15, 0.08, 0.05]^T$ in the Bis system. This choice of $\boldsymbol{\tau}_E$ implies that the disturbance forces on the surge and sway are 0.15 and 0.08, and the disturbance moment on the yaw is 0.05. The adaptation gains are chosen as $\gamma_{0i} = 1$, $k_{0i} = 1$, and $\xi_{0i} = 0.05$, $i = 1,2,3$. With these disturbances, one can verify that condition (8.59) holds, i.e., $((\alpha\bar{u}_d + |\gamma|)\,|\bar{r}_d| + \max(|\bar{\tau}_{vE}|) + \xi_{01})/(\beta\bar{u}_d + \dot{\bar{u}}_d)) \leq 0.71$. Simulation results are plotted in Figures 8.4 and 8.5. It is seen that our controller forces the heading angle to a value $\delta(t)$ to compensate for the disturbances.

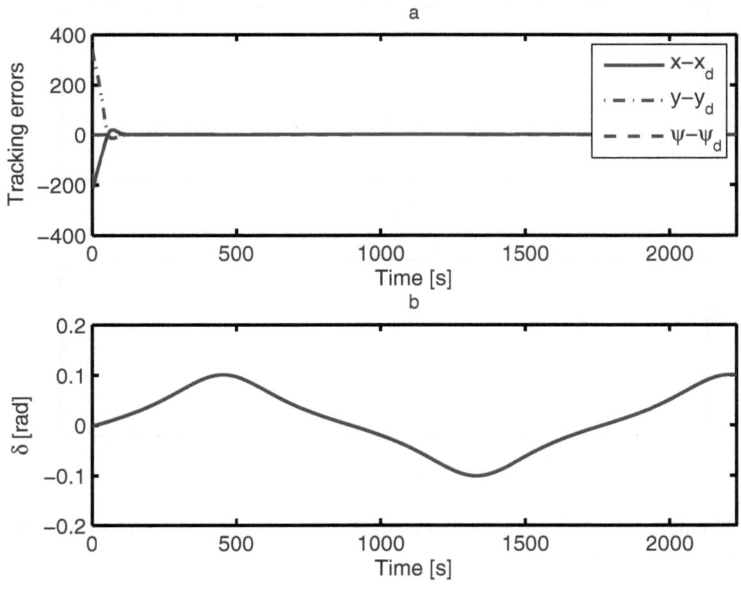

Figure 8.2 Simulation results without disturbances: **a**. Tracking errors $((x-x_d),(y-y_d),(\psi-\psi_d))$; **b**. Variation of angle δ

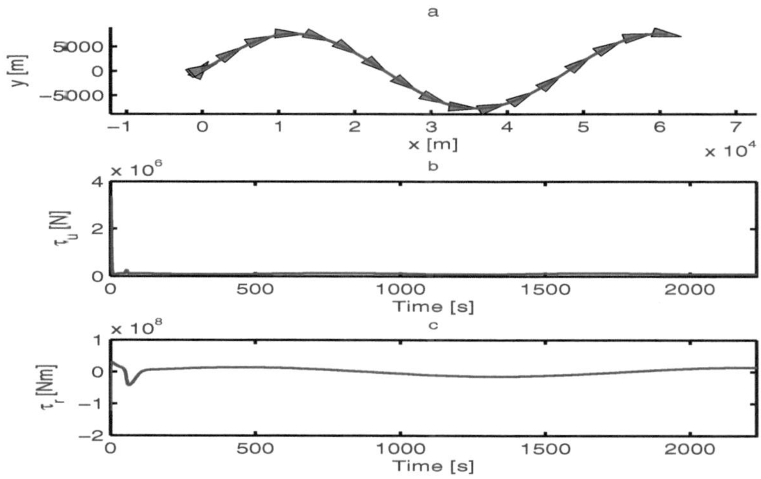

Figure 8.3 Simulation results without disturbances: **a**. Ship position and orientation in the (x, y)-plane; **b**. Control input τ_u; **c**. Control input τ_r

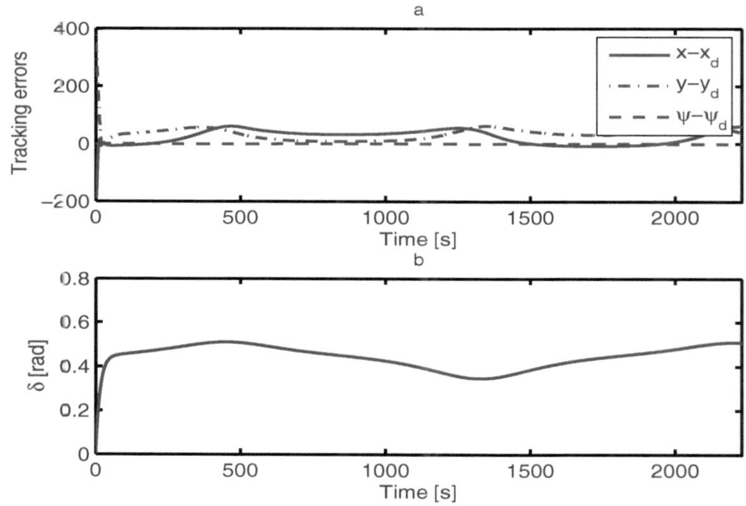

Figure 8.4 Simulation results with disturbances: **a**. Tracking errors $((x - x_d), (y - y_d), (\psi - \psi_d))$; **b**. Variation of angle δ

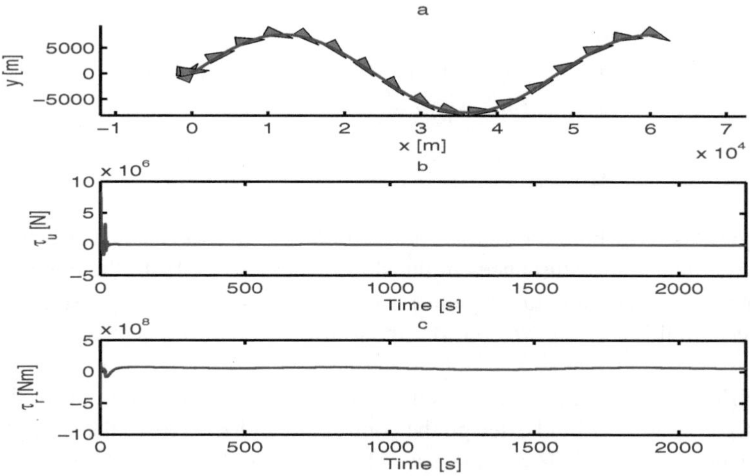

Figure 8.5 Simulation results with disturbances: **a**. Ship position and orientation in the (x, y) plane; **b**. Control input τ_u; **c**. Control input τ_r

8.2 Output Feedback

8.2.1 Control Objective

For the reader's convenience the mathematical model of an underactuated ship moving in a horizontal plane is rewritten here, see Section 3.4.1.1:

$$\begin{aligned} \dot{\boldsymbol{\eta}} &= \boldsymbol{J}(\psi)\boldsymbol{v}, \\ \boldsymbol{M}\dot{\boldsymbol{v}} &= -\boldsymbol{C}(\boldsymbol{v})\boldsymbol{v} - (\boldsymbol{D} + \boldsymbol{D_n}(\boldsymbol{v}))\boldsymbol{v} + \boldsymbol{\tau}, \end{aligned} \tag{8.64}$$

where

$$\boldsymbol{\eta} = \begin{bmatrix} x \\ y \\ \psi \end{bmatrix}, \quad \boldsymbol{v} = \begin{bmatrix} u \\ v \\ r \end{bmatrix}, \quad \boldsymbol{\tau} = \begin{bmatrix} \tau_u \\ 0 \\ \tau_r \end{bmatrix},$$

$$\boldsymbol{J}(\psi) = \begin{bmatrix} \cos(\psi) & -\sin(\psi) & 0 \\ \sin(\psi) & \cos(\psi) & 0 \\ 0 & 0 & 1 \end{bmatrix}, \boldsymbol{M} = \begin{bmatrix} m_{11} & 0 & 0 \\ 0 & m_{22} & m_{23} \\ 0 & m_{32} & m_{33} \end{bmatrix},$$

$$\boldsymbol{C}(\boldsymbol{v}) = \begin{bmatrix} 0 & 0 & C_{13} \\ 0 & 0 & C_{23} \\ C_{31} & C_{32} & 0 \end{bmatrix}, \boldsymbol{D} + \boldsymbol{D_n}(\boldsymbol{v}) = \begin{bmatrix} D_{11} & 0 & 0 \\ 0 & D_{22} & D_{23} \\ 0 & D_{32} & D_{33} \end{bmatrix},$$

with

$$\begin{aligned}
&m_{11} = m - X_{\dot{u}}, m_{22} = m - Y_{\dot{v}}, \; m_{23} = mx_g - Y_{\dot{r}}, \\
&m_{32} = mx_g - N_{\dot{v}}, m_{33} = I_z - N_{\dot{r}}, \\
&C_{13} = -C_{31} = -(m - Y_{\dot{v}})v - (mx_g - 0.5(Y_{\dot{r}} + N_{\dot{v}}))r, \\
&C_{23} = -C_{32} = (m - X_{\dot{u}})u, \\
&D_{11} = -(X_u + X_{|u|u}|u|), \; D_{22} = -(Y_v + Y_{|v|v}|v|), \\
&D_{23} = -Y_r, \; D_{32} = -N_v, \; D_{33} = -(N_r + N_{|r|r}|r|).
\end{aligned}$$

All the symbols used here are defined in Section 3.3. It is noted that we did not include environmental disturbances in the ship dynamics (8.64). In the control design, we will add integrators to the path-tracking error outputs to compensate for the disturbances. In this section, we consider the following control objective.

Control Objective. Assume that the ship velocities, u, v, and r are not available and that Assumption 8.2 holds, design the controls τ_u and τ_r to force the ship (8.64) to follow a prescribed path $\boldsymbol{\Omega}$ parameterized by $(x_d(s), y_d(s))$ with s being the path parameter in the sense that the position of the ship (8.64) globally tracks the path $\boldsymbol{\Omega}$, the ship total linear velocity is tangential to the path $\boldsymbol{\Omega}$, and let the desired surge speed, $u_{d0}(t)$, be adjusted on-line. The desired surge speed, $u_{d0}(t)$, is assumed to be bounded and twice differentiable.

Assumption 8.2. *There exist strictly positive constants ε_i, $0 \le i \le 5$ such that*

$$\varepsilon_0 \le x_d^{'2}(s) + y_d^{'2}(s) \le \varepsilon_1, \; \forall s \in \mathbb{R}, \tag{8.65}$$

$$u_{d0}(t) \ge \varepsilon_2, \; \forall t \ge 0, \tag{8.66}$$

$$N_{\dot{v}} = Y_{\dot{r}}, \tag{8.67}$$

$$1 - (\alpha(u_{d0}(t) + \varepsilon_3) + |\gamma|)/(u_{d0}(t) - \varepsilon_4) \ge \varepsilon_5, \; \forall t \ge 0, \tag{8.68}$$

where $x_d'(s)$, $y_d'(s)$, α and γ are defined as

$$\begin{aligned}
x_d'(s) &= \frac{\partial x_d}{\partial s}, \; y_d'(s) = \frac{\partial y_d}{\partial s}, \\
\alpha &= \frac{m_{11}}{m_{22}}, \; \gamma = \frac{Y_r}{m_{22}}.
\end{aligned} \tag{8.69}$$

Remark 8.4. The following observations are made on Assumption 8.2.

1. Condition (8.65) implies that the path $\boldsymbol{\Omega}$ is regular with respect to the path parameter s. If the path is not regular, one can break it into different regular paths and consider each path separately.
2. Condition (8.66) means that the path $\boldsymbol{\Omega}$ does not contain or approach a set-point, i.e., we do not consider stabilization/regulation problems. For clarity we only consider forward tracking. Backward tracking can be done similarly.
3. Condition (8.67) implies that the added mass matrix is symmetric. This condition is needed for designing an observer. If this condition does not hold, we can use

an acceleration feedback in the yaw dynamics [12] to reshape the added mass matrix.
4. Condition (8.68) is needed to guarantee stability of the sway dynamics under the proposed controller (to be designed later). Since α is strictly smaller than 1 (added mass, $-X_{\dot{u}}$, in surge is smaller than that, $-Y_{\dot{v}}$, in sway) and $|\gamma|$ is small for most ships, Condition (8.68) is satisfied under (8.66) with ε_2 being strictly larger than $|\gamma|$.

The main ideas in solving the above control objective are as follows:

1. Choose an appropriate body-fixed frame origin to avoid the yaw moment control τ_r acting directly on the sway dynamics. This choice will overcome the difficulty caused by off-diagonal terms in the mass matrix.
2. Transform the ship dynamics to a system, which is linear and monotonic in unmeasured velocities, so that an observer can be easily designed.
3. Interpret path-tracking errors in a frame attached to the path $\boldsymbol{\Omega}$ such that the error dynamics are of a triangular form to which the backstepping technique can be applied.
4. Use the orientation error as a virtual control to stabilize the cross-track error where a filter of the sway velocity is designed and used in the virtual control. This filter and the derivative of the path parameter used as an additional control allow a global controller and desired path with arbitrary curvature.
5. Add projection integrators to the path-tracking error dynamics' output to compensate for constant bias of the disturbances.

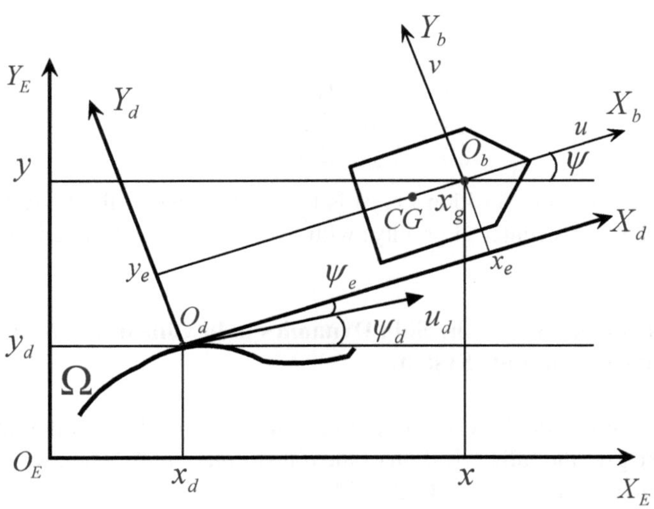

Figure 8.6 Interpretation of path-tracking errors

8.2.2 Coordinate Transformations

8.2.2.1 Choosing the Body-fixed Frame Origin

To avoid the yaw moment control τ_r acting directly on the sway dynamics, we choose the body-fixed frame origin such that it is on the center line of the ship (see Figure 8.6), i.e., $y_g = 0$ and that:

$$x_g = \frac{Y_{\dot{r}}}{m}. \tag{8.70}$$

With this choice, the ship model (8.64) can be written as:

$$\begin{aligned} \dot{\boldsymbol{\eta}} &= \boldsymbol{J}(\psi)\boldsymbol{v}, \\ \bar{\boldsymbol{M}}\dot{\boldsymbol{v}} &= -\bar{\boldsymbol{C}}(\boldsymbol{v})\boldsymbol{v} - (\bar{\boldsymbol{D}} + \bar{\boldsymbol{D}}_{\boldsymbol{n}}(\boldsymbol{v}))\boldsymbol{v} + \boldsymbol{\tau}, \end{aligned} \tag{8.71}$$

where

$$\bar{\boldsymbol{M}} = \begin{bmatrix} m_{11} & 0 & 0 \\ 0 & m_{22} & 0 \\ 0 & 0 & m_{33} \end{bmatrix}, \quad \bar{\boldsymbol{D}}_{\boldsymbol{n}}(\boldsymbol{v}) = \begin{bmatrix} d_{n1}|u| & 0 & 0 \\ 0 & d_{n2}|v| & 0 \\ 0 & 0 & d_{n3}|r| \end{bmatrix},$$

$$\bar{\boldsymbol{D}} = \begin{bmatrix} d_{11} & 0 & 0 \\ 0 & d_{22} & d_{23} \\ 0 & d_{32} & d_{33} \end{bmatrix}, \quad \bar{\boldsymbol{C}}(\boldsymbol{v}) = \begin{bmatrix} 0 & 0 & -m_{22}v \\ 0 & 0 & m_{11}u \\ m_{22}v & -m_{11}u & 0 \end{bmatrix}, \tag{8.72}$$

with

$$\begin{aligned} d_{11} &= -X_u, \ d_{22} = -Y_v, \\ d_{23} &= -Y_r, \ d_{32} = -N_v, \\ d_{33} &= -N_r, \ d_{n1} = -X_{|u|u}, \\ d_{n2} &= Y_{|v|v}, \ d_{n3} = N_{|r|r}. \end{aligned}$$

It is noted that the ship position (x, y) is the coordinates of the body-fixed frame origin (not the ship's center of gravity) with respect to the earth-fixed frame.

8.2.2.2 Transformation of the Ship Dynamics to be Linear and Monotonic in Unmeasured State System

The transformation here is very similar to that in Section 7.3.1. For the sake of completeness, we give the full derivation of the transformation. As such, we first remove the term $\bar{\boldsymbol{C}}(\boldsymbol{v})\boldsymbol{v}$ in the right-hand side of the second equation of (8.71), which causes difficulties in the observer design, by introducing the following coordinate:

$$\boldsymbol{X} = \boldsymbol{P}(\eta)\boldsymbol{v}, \tag{8.73}$$

where $\boldsymbol{P}(\boldsymbol{\eta}) \in \mathbb{R}^{3\times 3}$ is a global invertible matrix to be designed. With (8.73), the second equation of (8.71) is written as

$$\begin{aligned}\dot{\boldsymbol{X}} = & \left[\dot{\boldsymbol{P}}(\boldsymbol{\eta})\boldsymbol{v} - \boldsymbol{P}(\boldsymbol{\eta})\bar{\boldsymbol{M}}^{-1}\bar{\boldsymbol{C}}(\boldsymbol{v})\boldsymbol{v}\right] - \\ & \boldsymbol{P}(\boldsymbol{\eta})\bar{\boldsymbol{M}}^{-1}(\bar{\boldsymbol{D}} + \bar{\boldsymbol{D}}_{\boldsymbol{n}}(\boldsymbol{P}^{-1}(\boldsymbol{\eta})\boldsymbol{X}))\boldsymbol{P}^{-1}(\boldsymbol{\eta})\boldsymbol{X} + \boldsymbol{P}(\boldsymbol{\eta})\bar{\boldsymbol{M}}^{-1}\boldsymbol{\tau}. \qquad (8.74)\end{aligned}$$

It can be seen that in order to remove the term $\bar{\boldsymbol{C}}(\boldsymbol{v})\boldsymbol{v}$, we need to find the matrix $\boldsymbol{P}(\boldsymbol{\eta})$ such that

$$\left[\dot{\boldsymbol{P}}(\boldsymbol{\eta})\boldsymbol{v} - \boldsymbol{P}(\boldsymbol{\eta})\bar{\boldsymbol{M}}^{-1}\bar{\boldsymbol{C}}(\boldsymbol{v})\boldsymbol{v}\right] = 0, \qquad (8.75)$$

for all $(\boldsymbol{\eta}, \boldsymbol{v}) \in \mathbb{R}^6$. Assuming that the elements of $\boldsymbol{P}(\boldsymbol{\eta})$ are $p_{ij}(\boldsymbol{\eta}), i = 1,2,3, j = 1,2,3$, (8.75) is expanded to:

$$\begin{aligned}& \dot{p}_{i1}u + \dot{p}_{i2}v + \dot{p}_{i3}r + \frac{m_{22}}{m_{11}}p_{i1}vr - \frac{m_{11}}{m_{22}}p_{i2}ur + \frac{m_{11}-m_{22}}{m_{33}}p_{i3}uv = 0, \\ & i = 1,2,3, \ \forall (\boldsymbol{\eta}, u, v, r) \in \mathbb{R}^6, \qquad (8.76)\end{aligned}$$

where for brevity, we omit the argument $\boldsymbol{\eta}$ of $p_{ij}(\boldsymbol{\eta})$. With the first equation of (8.71), we expand (8.76) as

$$\begin{aligned}& \left(\frac{\partial p_{i1}}{\partial x}\cos(\psi) + \frac{\partial p_{i1}}{\partial y}\sin(\psi)\right)u^2 + \left(-\frac{\partial p_{i2}}{\partial x}\sin(\psi) + \frac{\partial p_{i2}}{\partial y}\cos(\psi)\right)v^2 + \\ & \frac{\partial p_{i3}}{\partial \psi}r^2 + \left(-\frac{\partial p_{i1}}{\partial x}\sin(\psi) + \frac{\partial p_{i1}}{\partial y}\cos(\psi) + \frac{\partial p_{i2}}{\partial x}\cos(\psi) + \frac{\partial p_{i2}}{\partial y}\sin(\psi) + \right. \\ & \left.\frac{m_{11}-m_{22}}{m_{33}}p_{i3}\right)uv + \left(\frac{\partial p_{i1}}{\partial \psi} + \frac{\partial p_{i3}}{\partial x}\cos(\psi) + \frac{\partial p_{i3}}{\partial y}\sin(\psi) - \frac{m_{11}}{m_{22}}p_{i2}\right)ur + \\ & \left(\frac{\partial p_{i2}}{\partial \psi} - \frac{\partial p_{i3}}{\partial x}\sin(\psi) + \frac{\partial p_{i3}}{\partial y}\cos(\psi) + \frac{m_{22}}{m_{11}}p_{i1}\right)vr = 0. \qquad (8.77)\end{aligned}$$

Therefore (8.77) holds for all $(x, y, \psi, u, v, r) \in \mathbb{R}^6$ if

$$\begin{aligned}& \frac{\partial p_{i1}}{\partial x}\cos(\psi) + \frac{\partial p_{i1}}{\partial y}\sin(\psi) = 0, \\ & -\frac{\partial p_{i2}}{\partial x}\sin(\psi) + \frac{\partial p_{i2}}{\partial y}\cos(\psi) = 0, \\ & \frac{\partial p_{i3}}{\partial \psi} = 0, \\ & \left(\frac{\partial p_{i2}}{\partial y} - \frac{\partial p_{i1}}{\partial x}\right)\sin(\psi) + \left(\frac{\partial p_{i1}}{\partial y} + \frac{\partial p_{i2}}{\partial x}\right)\cos(\psi) + \frac{m_{11}-m_{22}}{m_{33}}p_{i3} = 0, \\ & \frac{\partial p_{i1}}{\partial \psi} + \frac{\partial p_{i3}}{\partial x}\cos(\psi) + \frac{\partial p_{i3}}{\partial y}\sin(\psi) - \frac{m_{11}}{m_{22}}p_{i2} = 0, \\ & \frac{\partial p_{i2}}{\partial \psi} - \frac{\partial p_{i3}}{\partial x}\sin(\psi) + \frac{\partial p_{i3}}{\partial y}\cos(\psi) + \frac{m_{22}}{m_{11}}p_{i1} = 0. \qquad (8.78)\end{aligned}$$

A family of solutions of the above set of six partial differential equations is

$$\begin{aligned}
p_{i1} &= \frac{((m_{11}C_{i3}x + m_{33}C_{i1})\sin(\psi) - (m_{11}C_{i3}y - m_{33}C_{i2})\cos(\psi))}{m_{33}},\\
p_{i2} &= \frac{m_{22}((m_{11}C_{i3}x + m_{33}C_{i1})\cos(\psi) + (m_{11}C_{i3}y - m_{33}C_{i2})\sin(\psi))}{m_{11}m_{33}},\\
p_{i3} &= C_{i3},
\end{aligned} \tag{8.79}$$

where C_{i1}, C_{i2}, and C_{i3} are arbitrary constants.

We now choose the constants C_{i1}, C_{i2}, and C_{i3} such that the matrix $\boldsymbol{P}(\eta)$ is invertible for all $(x, y, \psi) \in \mathbb{R}^3$. A trivial choice of $C_{13} = C_{11} = 0$, $C_{12} = 1$, $C_{23} = C_{22} = 0$, $C_{21} = 1$, $C_{31} = C_{32} = 0$, and $C_{33} = 1$ results in

$$\boldsymbol{P}(\eta) = \begin{bmatrix} \cos(\psi) & -\sin(\psi)\frac{m_{22}}{m_{11}} & 0 \\ \sin(\psi) & \cos(\psi)\frac{m_{22}}{m_{11}} & 0 \\ (\sin(\psi)x - \cos(\psi)y)\frac{m_{11}}{m_{33}} & (\cos(\psi)x + \sin(\psi)y)\frac{m_{22}}{m_{33}} & 1 \end{bmatrix}. \tag{8.80}$$

Indeed this matrix is invertible for all $(x, y, \psi) \in \mathbb{R}^3$. Substituting (8.80) into (8.74) gives

$$\dot{\boldsymbol{X}} = -\boldsymbol{P}(\eta)\bar{\boldsymbol{M}}^{-1}(\bar{\boldsymbol{D}} + \bar{\boldsymbol{D}}_{\boldsymbol{n}}(\boldsymbol{P}^{-1}(\eta)\boldsymbol{X}))\boldsymbol{P}^{-1}(\eta)\boldsymbol{X} + \boldsymbol{P}(\eta)\bar{\boldsymbol{M}}^{-1}\boldsymbol{\tau}. \tag{8.81}$$

8.2.3 Observer Design

It can be seen that system (8.81) is linear and monotonic in the unmeasured state vector $\boldsymbol{X}$. We now design a simple observer to estimate $\boldsymbol{X}$ as:

$$\dot{\hat{\boldsymbol{X}}} = -\boldsymbol{P}(\eta)\bar{\boldsymbol{M}}^{-1}(\bar{\boldsymbol{D}} + \bar{\boldsymbol{D}}_{\boldsymbol{n}}(\boldsymbol{P}^{-1}(\eta)\hat{\boldsymbol{X}}))\boldsymbol{P}^{-1}(\eta)\hat{\boldsymbol{X}} + \boldsymbol{P}(\eta)\bar{\boldsymbol{M}}^{-1}\boldsymbol{\tau}, \tag{8.82}$$

where $\hat{\boldsymbol{X}}$ is an estimate of $\boldsymbol{X}$. From (8.81) and (8.82), we have

$$\begin{aligned}
\dot{\tilde{\boldsymbol{X}}} = &-\boldsymbol{P}(\eta)\bar{\boldsymbol{M}}^{-1}\bar{\boldsymbol{D}}\boldsymbol{P}^{-1}(\eta)\tilde{\boldsymbol{X}} - \boldsymbol{P}(\eta)\bar{\boldsymbol{M}}^{-1}\times \\
&[\bar{\boldsymbol{D}}_{\boldsymbol{n}}(\boldsymbol{P}^{-1}(\eta)\boldsymbol{X}))\boldsymbol{P}^{-1}(\eta)\boldsymbol{X} - \bar{\boldsymbol{D}}_{\boldsymbol{n}}(\boldsymbol{P}^{-1}(\eta)\hat{\boldsymbol{X}}))\boldsymbol{P}^{-1}(\eta)\hat{\boldsymbol{X}}],
\end{aligned} \tag{8.83}$$

where $\tilde{\boldsymbol{X}} := (\tilde{x}_1, \tilde{x}_2, \tilde{x}_3)^T = \boldsymbol{X} - \hat{\boldsymbol{X}}$. We will use (8.83) in stability analysis of the closed loop system later. Define $\hat{\boldsymbol{v}} = [\hat{u}, \hat{v}, \hat{r}]^T$ as being an estimate of the velocity vector $\boldsymbol{v}$ as:

$$\hat{\boldsymbol{v}} = \boldsymbol{P}^{-1}(\eta)\hat{\boldsymbol{X}}. \tag{8.84}$$

Then the velocity estimate error vector, $\tilde{\boldsymbol{v}} := [\tilde{u}, \tilde{v}, \tilde{r}]^T = \boldsymbol{v} - \hat{\boldsymbol{v}}$, satisfies:

$$\begin{bmatrix} \tilde{u} \\ \tilde{v} \\ \tilde{r} \end{bmatrix} = \begin{bmatrix} \tilde{x}_1 \cos(\psi) + \tilde{x}_2 \sin(\psi) \\ \frac{m_{11}}{m_{22}}(-\tilde{x}_1 \sin(\psi) + \tilde{x}_2 \cos(\psi)) \\ \frac{m_{11}}{m_{33}}(\tilde{x}_1 y - \tilde{x}_2 x) + \tilde{x}_3 \end{bmatrix}. \tag{8.85}$$

Using (8.84), we rewrite the first equation of (8.71) and (8.82) as:

$$\begin{aligned} \dot{x} &= (\hat{u} + \tilde{u})\cos(\psi) - (v + \tilde{v})\sin(\psi), \\ \dot{y} &= (\hat{u} + \tilde{u})\sin(\psi) + (v + \tilde{v})\cos(\psi), \\ \dot{\psi} &= \hat{r} + \tilde{r}, \\ \dot{\hat{u}} &= \frac{m_{22}}{m_{11}}\hat{v}\hat{r} - \frac{d_{11}}{m_{11}}\hat{u} - \frac{d_{n1}}{m_{11}}|\hat{u}|\,\hat{u} + \frac{1}{m_{11}}\tau_u + \frac{m_{22}}{m_{11}}\hat{v}\tilde{r}, \\ \dot{\hat{v}} &= -\frac{m_{11}}{m_{22}}\hat{u}\hat{r} - \frac{d_{22}}{m_{22}}\hat{v} - \frac{d_{23}}{m_{22}}\hat{r} - \frac{d_{n2}}{m_{22}}|\hat{v}|\,\hat{v} - \frac{m_{11}}{m_{22}}\hat{u}\tilde{r}, \\ \dot{\hat{r}} &= \frac{m_{11} - m_{22}}{m_{33}}\hat{u}\hat{v} - \frac{d_{22}}{m_{33}}\hat{v} - \frac{d_{33}}{m_{33}}\hat{r} - \frac{d_{n3}}{m_{33}}|\hat{r}|\,\hat{r} + \\ &\quad \frac{1}{m_{33}}\tau_r + \frac{m_{11}}{m_{33}}\hat{u}\tilde{v} - \frac{m_{22}}{m_{33}}\hat{v}\tilde{u}. \end{aligned} \tag{8.86}$$

Transformation of Path-tracking Errors

We now interpret the path-tracking errors in a frame attached to the reference path $\boldsymbol{\Omega}$ as follows (see Figure 8.6):

$$\begin{bmatrix} x_e \\ y_e \\ \psi_e \end{bmatrix} = \boldsymbol{J}^T(\psi) \begin{bmatrix} x - x_d \\ y - y_d \\ \psi - \psi_d \end{bmatrix}, \tag{8.87}$$

where ψ_d is the angle between the path and the X-axis defined by

$$\psi_d = \arctan\left(\frac{y_d'(s)}{x_d'(s)}\right), \tag{8.88}$$

with $x_d'(s)$ and $y_d'(s)$ being defined in (8.69).

In Figure 8.6, $O_E X_E Y_E$ is the earth-fixed frame; $O_p X_p Y_p$ is a frame attached to the path $\boldsymbol{\Omega}$ such that $O_p X_p$ and $O_p Y_p$ are parallel to the surge and sway axes of the ship, respectively, u_d is tangential to the path, CG is the center of gravity of the ship; and $O_b X_b Y_b$ is the body-fixed frame. Therefore x_e, y_e, and ψ_e can be referred to as tangential, cross and heading errors, respectively. Differentiating both sides of (8.87) along the solutions of the first three equations of (8.86) results in the kinematic path-tracking errors:

$$\begin{aligned}\dot{x}_e &= \hat{u} - u_d\cos(\psi_e) + (\hat{r} + \tilde{r})y_e + \tilde{u},\\ \dot{y}_e &= \hat{v} + u_d\sin(\psi_e) - (\hat{r} + \tilde{r})x_e + \tilde{v},\\ \dot{\psi}_e &= \hat{r} - r_d + \tilde{r},\end{aligned} \tag{8.89}$$

where u_d and r_d are given by:

$$\begin{aligned}u_d &= \sqrt{x_d'^2(s) + y_d'^2(s)}\dot{s},\\ r_d &= \frac{x_d'(s)y_d''(s) - x_d''(s)y_d'(s)}{x_d'^2(s) + y_d'^2(s)}\dot{s}.\end{aligned} \tag{8.90}$$

8.2.4 Control Design with Integral Action

There are often two methods to let a control system compensate for constant (or at least slowly-varying) environmental disturbances: (1) deriving an adaptive control law to estimate a constant bias (as the one in Section 8.1.5), and (2) adding extra integrators to the output of the path-tracking error system. Here we use the second method for simplicity since the first one will result in a complicated control law due to the observer developed in the previous section, see Section 8.2.6.1 for a discussion. We divide the control design procedure into two separate stages. At the first stage, we consider (8.89) with $\hat{u}$ and $\hat{r}$ being viewed as controls. At the second stage, the last three equations of (8.86) are considered to design the actual controls τ_u and τ_r using the backstepping technique.

8.2.4.1 Kinematic Control Design

Substep 1 (Stabilizing (x_e, y_e)-dynamics)

First we add two projection integrators to the position path-tracking error dynamics as

$$\begin{aligned}\dot{\sigma}_{xe} &= \text{proj}(\gamma_1 x_e/\Delta, \sigma_{xe}),\\ \dot{\sigma}_{ye} &= \text{proj}(\gamma_2 y_e/\Delta, \sigma_{ye}),\\ \dot{x}_e &= \hat{u} - u_d\cos(\psi_e) + (\hat{r} + \tilde{r})y_e + \tilde{u},\\ \dot{y}_e &= \hat{v} + u_d\sin(\psi_e) - (\hat{r} + \tilde{r})x_e + \tilde{v},\end{aligned} \tag{8.91}$$

where $\Delta = \sqrt{1 + x_e^2 + y_e^2}$, γ_1 and γ_2 are positive constants and the operator proj represents the Lipschitz projection algorithm (repeated here for the reader's convenience)

$$\begin{aligned}
&\text{proj}(\varpi,\hat{\omega}) = \varpi \text{ if } \Xi(\hat{\omega}) \le 0,\\
&\text{proj}(\varpi,\hat{\omega}) = \varpi \text{ if } \Xi(\hat{\omega}) \ge 0 \text{ and } \Xi_{\hat{\omega}}(\hat{\omega})\varpi \le 0,\\
&\text{proj}(\varpi,\hat{\omega}) = (1-\Xi(\hat{\omega}))\varpi \text{ if } \Xi(\hat{\omega}) > 0 \text{ and } \Xi_{\hat{\omega}}(\hat{\omega})\varpi > 0,
\end{aligned}$$

where $\Xi(\hat{\omega}) = (\hat{\omega}^2 - \omega_M^2)/(\xi^2 + 2\xi\omega_M)$, $\Xi_{\hat{\omega}}(\hat{\omega}) = \partial\Xi(\hat{\omega})/\partial\hat{\omega}$, ξ is an arbitrarily small positive constant and ω_M is the maximum value of the constant bias that $\hat{\omega}$ will compensate for. The projection algorithm (repeated here for convenience of the reader) is such that if $\dot{\hat{\omega}} = \text{proj}(\varpi,\hat{\omega})$ and $\hat{\omega}(t_0) \le \omega_M$ then:

1. $\hat{\omega}(t) \le \omega_M + \xi$, $\forall 0 \le t_0 \le t < \infty$,
2. $\text{proj}(\varpi,\hat{\omega})$ is Lipschitz continuous,
3. $|\text{proj}(\varpi,\hat{\omega})| \le |\varpi|$,
4. $\tilde{\omega}\text{proj}(\varpi,\hat{\omega}) \ge \tilde{\omega}\varpi$ with $\tilde{\omega} = \omega - \hat{\omega}$.

It is noted that we have added the integrators via the projection algorithm to guarantee that σ_{xe} and σ_{ye} are always bounded by some predefined constants. Define

$$\begin{aligned}
u_e &= \hat{u} - \alpha_u,\\
\bar{\psi}_e &= \psi_e - \alpha_{\psi_e},
\end{aligned} \tag{8.92}$$

where α_u and α_{ψ_e} are virtual controls of $\hat{u}$ and ψ_e, respectively. Substituting (8.92) into the last two equations of (8.91) gives

$$\begin{aligned}
\dot{x}_e = {} & \alpha_u + u_e - u_d\cos(\alpha_{\psi_e}) - u_d((\cos(\bar{\psi}_e) - 1)\cos(\alpha_{\psi_e}) - \\
& \sin(\bar{\psi}_e)\sin(\alpha_{\psi_e})) + (\hat{r} + \tilde{r})y_e + \tilde{u},\\
\dot{y}_e = {} & \hat{v} + u_d\sin(\alpha_{\psi_e}) + u_d(\sin(\bar{\psi}_e)\cos(\alpha_{\psi_e}) + (\cos(\bar{\psi}_e) - 1)\times\\
& \sin(\alpha_{\psi_e})) - (\hat{r} + \tilde{r})x_e + \tilde{v}.
\end{aligned} \tag{8.93}$$

From the first equation of (8.91) and first equation of (8.93), we can easily design a control law for α_u to stabilize the x_e-dynamics as follows

$$\alpha_u = -\varphi_1 - \sigma_{xe} + u_d\cos(\alpha_{\psi_e}), \tag{8.94}$$

with

$$\varphi_1 = k_1\frac{x_e}{\Delta},$$

where k_1 is a positive constant to be specified later. It is noted that we have chosen α_u not to cancel the known term,$-u_d((\cos(\bar{\psi}_e) - 1)\cos(\alpha_{\psi_e}) - \sin(\bar{\psi}_e)\sin(\alpha_{\psi_e})) + \hat{r}y_e$, to simplify the controller expression.

However, stabilizing the second equation of (8.91) and second equation of (8.93) is more difficult. We cannot directly use α_{ψ_e} to cancel $\hat{v}$ since the term $u_d\sin(\alpha_{\psi_e})$ cannot cancel $\hat{v}$ without imposing a restriction on the initial conditions, i.e., no global result can be obtained. To resolve this difficulty, the derivative of the path parameter is used as an additional control and is designed as:

$$\dot{s} = \frac{\sqrt{u_{d0}^2 + \sigma_{xe}^2 + (\varphi_2 + \sigma_{ye} + v_d)^2}}{\sqrt{x_d'^2(s) + y_d'^2(s)}}, \tag{8.95}$$

with $\varphi_2 = k_2 \frac{y_e}{\Delta}$, where from now we drop the argument t of u_{d0} and its derivatives, k_2 is a positive constant to be specified later, and v_d is a filtered value of $\hat{v}$ to be designed later. From (8.95) and (8.90), we have:

$$u_d = \sqrt{u_{d0}^2 + \sigma_{xe}^2 + (\varphi_2 + \sigma_{ye} + v_d)^2}. \tag{8.96}$$

From (8.96), the second equation of (8.93), and (8.90), we design a control law for α_{ψ_e} as:

$$\alpha_{\psi_e} = -\arctan\left(\frac{\varphi_2 + \sigma_{ye} + v_d}{\Upsilon_{u_{d0}}}\right), \tag{8.97}$$

with

$$\Upsilon_{u_{d0}} = \sqrt{u_{d0}^2 + \sigma_{xe}^2}. \tag{8.98}$$

Substituting (8.97), (8.95), and (8.94) into (8.93) results in

$$\begin{aligned}
\dot{x}_e = {} & -k_1 x_e/\Delta - \sigma_{xe} + u_e - u_d((\cos(\bar{\psi}_e) - 1)\cos(\alpha_{\psi_e}) - \\
& \sin(\bar{\psi}_e)\sin(\alpha_{\psi_e})) + (\hat{r} + \tilde{r})y_e + \tilde{u}, \\
\dot{y}_e = {} & -k_2 y_e/\Delta - \sigma_{ye} + v_e + u_d(\sin(\bar{\psi}_e)\cos(\alpha_{\psi_e}) + \\
& (\cos(\bar{\psi}_e) - 1)\sin(\alpha_{\psi_e})) - (\hat{r} + \tilde{r})x_e + \tilde{v},
\end{aligned} \tag{8.99}$$

where $v_e = \hat{v} - v_d$.

Substep 2 (Stabilizing $\bar{\psi}_e$ dynamics)

Define

$$r_e = \hat{r} - \alpha_r, \tag{8.100}$$

where α_r is a virtual control of $\hat{r}$. Differentiating both sides of the second equation of (8.92) along the solutions of (8.97), and adding a projection integrator gives

$$\begin{aligned}
& \dot{\sigma}_{\bar{\psi}_e} = \operatorname{proj}\left(\gamma_3 \frac{\bar{\psi}_e}{\sqrt{1 + \bar{\psi}_e^2}}, \sigma_{\bar{\psi}_e}\right), \\
& \dot{\bar{\psi}}_e = b_1(r_e + \alpha_r + \tilde{r}) - r_d + \frac{\Upsilon_{u_{d0}}(\Pi + \dot{v}_d + \dot{\sigma}_{ye})}{u_d^2} - \\
& \frac{(\varphi_2 + \sigma_{ye} + v_d)(\dot{u}_{d0} u_{d0} + \dot{\sigma}_{xe}\sigma_{xe})}{u_d^2 \Upsilon_{u_{d0}}} - \frac{k_2 \Upsilon_{u_{d0}}(x_e y_e(u_e + \tilde{u}) + (1 + x_e^2)(v_e + \tilde{v}))}{u_d^2 \Delta^3},
\end{aligned} \tag{8.101}$$

where γ_3 is a positive constant and

$$b_1 = 1 - k_2 \frac{\Upsilon_{udo} x_e}{u_d^2 \Delta^3},$$
$$\Pi = k_2 \left(\frac{v_d + u_d \sin(\psi_e)}{\Delta} - \frac{y_e(x_e(\alpha_u - u_d \cos(\psi_e)) + y_e(v_d + u_d \sin(\psi_e))}{\Delta^3} \right). \quad (8.102)$$

From the first equation of (8.102), we choose the design constant k_2 such that:

$$k_2 \le \min(u_{do}) - \varepsilon_6, \quad (8.103)$$

where ε_6 is a strictly positive constant. Then b_1 is always larger than some strictly positive constant. Now the virtual control α_r can be designed from (8.101) as

$$\alpha_r = \frac{1}{b_1} \Big(- \frac{k_3 \bar{\psi}_e}{\sqrt{1 + \bar{\psi}_e^2}} - \sigma_{\bar{\psi}e} + r_d - \frac{\Upsilon_{udo}}{u_d^2} (\Pi + \dot{v}_d + \dot{\sigma}_{ye}) + \frac{\varphi_2 + \sigma_{ye} + v_d}{u_d^2 \Upsilon_{udo}} \times$$
$$(\dot{u}_{do} u_{do} + \dot{\sigma}_{xe} \sigma_{xe}) \Big), \quad (8.104)$$

where k_3 is a positive constant. Substituting (8.104) into the second equation of (8.101) gives:

$$\dot{\bar{\psi}}_e = -k_3 \frac{\bar{\psi}_e}{\sqrt{1 + \bar{\psi}_e^2}} - \sigma_{\bar{\psi}e} - \frac{k_2 \Upsilon_{udo} x_e y_e}{u_d^2 \Delta^3} (u_e + \tilde{u}) + \frac{k_2 \Upsilon_{udo} (1 + x_e^2)}{u_d^2 \Delta^3} \times$$
$$(v_e + \tilde{v}) + b_1 (r_e + \tilde{r}). \quad (8.105)$$

To determine v_d, we differentiate $v_e = \hat{v} - v_d$ along the solutions of the fifth equation of (8.86) to obtain:

$$\dot{v}_e = -\frac{m_{11}}{m_{22}} (\alpha_u + u_e)(\alpha_r + r_e) - \frac{d_{22}}{m_{22}} (v_e + v_d) - \frac{d_{23}}{m_{22}} (\alpha_r + r_e) -$$
$$\frac{d_{n2}}{m_{22}} |\hat{v}| (v_e + v_d) - \frac{m_{11}}{m_{22}} \hat{u} \tilde{r} - \dot{v}_d \quad (8.106)$$

which suggests that we choose:

$$\dot{v}_d = -\frac{m_{11} \alpha_u + d_{23}}{m_{22}} \alpha_r - \frac{d_{22}}{m_{22}} v_d - \frac{d_{n2}}{m_{22}} |v_d| v_d + \frac{k_2 \Upsilon_{udo} (1 + x_e^2)}{u_d^2 \Delta^3} \frac{\bar{\psi}_e}{\sqrt{1 + \bar{\psi}_e^2}}, \quad (8.107)$$

where the last term on the right-hand side of (8.107) is added to take care of the second last term in the right-hand side of (8.105). Substituting (8.107) into (8.106) gives:

$$\dot{v}_e = -\frac{d_{22}}{m_{22}}v_e - \frac{d_{n2}}{m_{22}}|\hat{v}|\,v_e - \frac{m_{11}\hat{u}+d_{23}}{m_{22}}r_e - \frac{m_{11}}{m_{22}}\hat{r}u_e - \frac{d_{23}}{m_{22}}r_e - \frac{m_{11}}{m_{22}}(u_e+\alpha_u)\tilde{r} - \frac{k_2\Upsilon_{ud0}(1+x_e^2)}{u_d^2\Delta^3}\frac{\bar{\psi}_e}{\sqrt{1+\bar{\psi}_e^2}}. \tag{8.108}$$

Now notice that α_r depends on $\dot{v}_d$, see (8.104). Hence, we substitute (8.104) into (8.107) to obtain:

$$\dot{v}_d = \frac{1}{b_2}\Bigg(-\frac{d_{22}v_d+d_{n2}|v_d|\,v_d}{m_{22}} - \frac{m_{11}\alpha_u+d_{23}}{b_1 m_{22}}\Bigg(-\frac{k_3\bar{\psi}_e}{\sqrt{1+\bar{\psi}_e^2}} - \sigma_{\bar{\psi}e} + r_d - \frac{\Upsilon_{ud0}(\Pi+\dot{\sigma}_{ye})}{u_d^2} + \frac{\varphi_2+\sigma_{ye}+v_d}{u_d^2\Upsilon_{ud0}}(\dot{u}_{d0}u_{d0}+\dot{\sigma}_{xe}\sigma_{xe})\Bigg) - k_2\frac{\Upsilon_{ud0}(1+x_e^2)\bar{\psi}_e}{u_d^2\Delta^3\sqrt{1+\bar{\psi}_e^2}}\Bigg), \tag{8.109}$$

where

$$b_2 = 1 - \frac{m_{11}\alpha_u+d_{23}}{m_{22}}\frac{\Upsilon_{ud0}}{u_d^2 b_1} \quad\Rightarrow\quad b_2 \geq 1 - \frac{m_{11}(u_{d0}+k_1)+|d_{23}|}{m_{22}(u_{d0}-k_2)}. \tag{8.110}$$

From condition (8.68) in Assumption 8.2, if we pick k_1 and k_2 such that

$$k_1 \leq \varepsilon_3,\ k_2 \leq \varepsilon_4, \tag{8.111}$$

then we have $b_2 \geq \varepsilon_5$, i.e., there is no singularity in (8.109), see Assumption 8.2 for the constants ε_3, ε_4 and ε_5.

8.2.4.2 Kinetic Control Design

Before designing the actual controls τ_u and τ_r, we note that the virtual control α_u is a smooth function of x_e, y_e, u_{d0}, and σ_{xe}, and the virtual control α_r is a smooth function of $x_e, y_e, \psi_e, \sigma_{xe}, \sigma_{ye}, \sigma_{\bar{\psi}e}, u_{d0}, \dot{u}_{d0}, v_d$, and s. By differentiating both sides of (8.100) and the first equation of (8.92), and noting (8.101) and (8.108), we choose the actual controls τ_u and τ_r as:

$$\tau_u = -c_1 m_{11}u_e - m_{22}\hat{v}\hat{r} + (d_{11}-d_{n1}|\hat{u}|)\alpha_u + m_{11}\Bigg(\frac{\partial\alpha_u}{\partial x_e}(\hat{u}-u_d\cos(\psi_e)+\hat{r}y_e) + \frac{\partial\alpha_u}{\partial y_e}(\hat{v}+u_d\sin(\psi_e)-\hat{r}x_e) + \frac{\partial\alpha_u}{\partial\sigma_{xe}}\dot{\sigma}_{xe} + \frac{\partial\alpha_u}{\partial u_{d0}}\dot{u}_{d0} + \frac{k_2\Upsilon_{ud0}x_e y_e}{u_d^2\Delta^3}\frac{\bar{\psi}_e}{\sqrt{1+\bar{\psi}_e^2}} + \frac{m_{11}}{m_{22}}\hat{r}v_e\Bigg),$$

$$
\begin{aligned}
\tau_r = \; & -c_2 m_{33} r_e + (d_{33} + d_{n3}|\hat{r}|)\alpha_r + d_{32}\hat{v} - (m_{11} - m_{22})\hat{u}\hat{v} + \\
& m_{33}\left(\frac{\partial \alpha_r}{\partial x_e}(\hat{u} - u_d \cos(\psi_e) + \hat{r} y_e) + \frac{\partial \alpha_r}{\partial y_e}(\hat{v} + u_d \sin(\psi_e) - \hat{r} x_e) + \right. \\
& \frac{\partial \alpha_r}{\partial s}\dot{s} + \frac{\partial \alpha_r}{\partial \psi_e}(\hat{r} + r_d) + \frac{\partial \alpha_r}{\partial \sigma_{xe}}\dot{\sigma}_{xe} + \frac{\partial \alpha_r}{\partial \sigma_{ye}}\dot{\sigma}_{ye} + \frac{\partial \alpha_r}{\partial \sigma_{\bar{\psi}e}}\dot{\sigma}_{\bar{\psi}e} + \\
& \left. \frac{\partial \alpha_r}{\partial u_{d0}}\dot{u}_{d0} + \frac{\partial \alpha_r}{\partial \dot{u}_{d0}}\ddot{u}_{d0} + \frac{\partial \alpha_r}{\partial v_d}\dot{v}_d - \frac{b_1 \bar{\psi}_e}{\sqrt{1 + \bar{\psi}_e^2}} + \frac{(m_{11}\hat{u} + d_{23})v_e}{m_{22}} \right),
\end{aligned}
\tag{8.112}
$$

where c_1 and c_2 are positive constants. The closed loop system is:

$$
\dot{\sigma}_{xe} = \text{proj}\left(\frac{\gamma_1 x_e}{\Delta}, \sigma_{xe}\right), \; \dot{\sigma}_{ye} = \text{proj}\left(\frac{\gamma_2 y_e}{\Delta}, \sigma_{ye}\right), \; \dot{\sigma}_{\bar{\psi}e} = \text{proj}\left(\frac{\gamma_3 \bar{\psi}_e}{\sqrt{1 + \bar{\psi}_e^2}}, \sigma_{\bar{\psi}e}\right),
$$

$$
\begin{aligned}
\dot{x}_e = \; & -k_1 \frac{x_e}{\Delta} - \sigma_{xe} + u_e - u_d((\cos(\bar{\psi}_e) - 1)\cos(\alpha_{\psi_e}) - \\
& \sin(\bar{\psi}_e)\sin(\alpha_{\psi_e})) + (\hat{r} + \tilde{r})y_e + \tilde{u}, \\
\dot{y}_e = \; & -k_2 y_e/\Delta - \sigma_{ye} + v_e + u_d(\sin(\bar{\psi}_e)\cos(\alpha_{\psi_e}) + \\
& (\cos(\bar{\psi}_e) - 1)\sin(\alpha_{\psi_e})) - (\hat{r} + \tilde{r})x_e + \tilde{v}, \\
\dot{\bar{\psi}}_e = \; & -k_3 \frac{\bar{\psi}_e}{\sqrt{1 + \bar{\psi}_e^2}} - \sigma_{\bar{\psi}e} + b_1(r_e + \tilde{r}) - \frac{k_2 \Upsilon_{u_{d0}} x_e y_e}{u_d^2 \Delta^3}(u_e + \tilde{u}) + \\
& \frac{k_2 \Upsilon_{u_{d0}}(1 + x_e^2)}{u_d^2 \Delta^3}(v_e + \tilde{v}), \\
\dot{u}_e = \; & -(c_1 + \frac{d_{11} + d_{n1}|\hat{u}|}{m_{11}})u_e + \frac{m_{22}}{m_{11}}\hat{v}\tilde{r} - \frac{\partial \alpha_u}{\partial x_e}(\tilde{u} + y_e \tilde{r}) - \\
& \frac{\partial \alpha_u}{\partial y_e}(\tilde{v} - x_e \tilde{r}) + \frac{k_2 \Upsilon_{u_{d0}} x_e y_e}{u_d^2 \Delta^3}\frac{\bar{\psi}_e}{\sqrt{1 + \bar{\psi}_e^2}} + \frac{m_{11}}{m_{22}}\hat{r} v_e, \\
\dot{v}_e = \; & -\frac{d_{22}}{m_{22}}v_e - \frac{d_{n2}}{m_{22}}|\hat{v}|\, v_e - \frac{m_{11}\hat{u} + d_{23}}{m_{22}}r_e - \frac{m_{11}}{m_{22}}\hat{r} u_e - \frac{d_{23}}{m_{22}}r_e - \\
& \frac{m_{11}}{m_{22}}(u_e + \alpha_u)\tilde{r} - \frac{k_2 \Upsilon_{u_{d0}}(1 + x_e^2)}{u_d^2 \Delta^3}\frac{\bar{\psi}_e}{\sqrt{1 + \bar{\psi}_e^2}}, \\
\dot{r}_e = \; & -(c_2 + \frac{d_{33} + d_{n3}|\hat{r}|}{m_{33}})r_e + \frac{m_{11}}{m_{33}}\hat{u}\tilde{v} - \frac{m_{22}}{m_{33}}\hat{v}\tilde{u} - \frac{\partial \alpha_r}{\partial x_e}(\tilde{u} + y_e \tilde{r}) - \\
& \frac{\partial \alpha_r}{\partial y_e}(\tilde{v} - x_e \tilde{r}) - \frac{\partial \alpha_r}{\partial \psi_e}\tilde{r} + \frac{(m_{11}\hat{u} + d_{23})v_e}{m_{22}} - \frac{b_1 \bar{\psi}_e}{\sqrt{1 + \bar{\psi}_e^2}},
\end{aligned}
$$

$$\dot{v}_d = \frac{1}{b_2}\left(-\frac{d_{22}v_d + d_{n2}|v_d|v_d}{m_{22}} - \frac{m_{11}\alpha_u + d_{23}}{b_1 m_{22}}\left(-\frac{k_3\bar{\psi}_e}{\sqrt{1+\bar{\psi}_e^2}} - \sigma_{\bar{\psi}e} + \right.\right.$$

$$\left. r_d - \frac{\Upsilon_{ud0}(\Pi + \dot{\sigma}_{ye})}{u_d^2} + \frac{\varphi_2 + \sigma_{ye} + v_d}{u_d^2 \Upsilon_{ud0}}(\dot{u}_{d0}u_{d0} + \dot{\sigma}_{xe}\sigma_{xe})\right) -$$

$$\left. k_2 \frac{\Upsilon_{ud0}(1+x_e^2)\bar{\psi}_e}{u_d^2 \Delta^3 \sqrt{1+\bar{\psi}_e^2}}\right),$$

$$\dot{x}_d = \frac{\partial x_d}{\partial s}\frac{u_d}{\sqrt{x_d'^2(s) + y_d'^2(s)}}, \ \dot{y}_d = \frac{\partial y_d}{\partial s}\frac{u_d}{\sqrt{x_d'^2(s) + y_d'^2(s)}}. \tag{8.113}$$

We now present the second main result of this chapter, the proof of which is given in the next subsection.

Theorem 8.3. *Assume that Assumption 8.2 holds, the controls τ_u and τ_r given by (8.112) solve the control objective with an appropriate choice of k_1 and k_2 such that (8.103) and (8.111) hold. Particularly, the transformed path-tracking errors $(x_e, y_e, \bar{\psi}_e)$ globally asymptotically converge to zero. As a result, the actual position path-tracking errors $(x - x_d, y - y_d)$ and orientation path-tracking error $(\psi - \psi_d)$ globally asymptotically converge to zero and to a ball with a radius of smaller than 0.5π, respectively. Furthermore, the desired forward speed of the ship on the path can be adjusted by adjusting $u_{d0}(t)$ and the total linear velocity of the ship is tangential to the path.*

8.2.5 Stability Analysis

To prove Theorem 8.3, we first show that the closed loop system (8.113) is forward complete (i.e. no finite escape in the closed loop system), and there exist an arbitrarily small positive constant ϕ_0, and a class-K function Ξ_0 of $\|\chi_0(t_0)\|$, with

$$\chi_0(t_0) := \Big[x_e(t_0), y_e(t_0), \psi_e(t_0), x_d(t_0), y_d(t_0), \sigma_{xe}(t_0),$$
$$\sigma_{ye}(t_0), \sigma_{\bar{\psi}e}(t_0), u_e(t_0), v_e(t_0), r_e(t_0), v_d(t_0)\Big]^T$$

such that

$$\|(x_e(t), y_e(t), x_d(t), x_d(t))\| \le \Xi_0 e^{\phi_0(t-t_0)}, \ \forall t \ge t_0.$$

The reason for doing this is that the term, $\tilde{r}$, contains (x, y), see (8.85). We then consider the $(\sigma_{\bar{\psi}e}, \bar{\psi}_e, u_e, v_e, r_e)$-subsystem and prove that v_d is bounded by some constant. Finally we consider the $(\sigma_{xe}, \sigma_{ye}, x_e, y_e)$-subsystem.

To prove that the closed loop system (8.113) is forward complete, we consider the Lyapunov function

$$W = \frac{1}{2}\log(1+W_0) + \frac{1}{2}K_2\tilde{\boldsymbol{X}}^T\tilde{\boldsymbol{X}}, \tag{8.114}$$

where

$$W_0 = x_e^2 + y_e^2 + x_d^2 + y_d^2 + \sigma_{xe}^2 + \sigma_{ye}^2 + \sigma_{\bar{\psi}e}^2 + K_1(\psi_e^2 + u_e^2 + v_e^2 + r_e^2 + v_d^2),$$

K_1 and K_2 are positive constants to be picked later. Differentiating both sides of (8.114) along the solutions of (8.113) and (8.83), and using properties of the projection algorithm, after some calculations we obtain:

$$\dot{W} \le \frac{1}{2}\phi_0 + \frac{\phi_1}{1+W_0} + \phi_2\|\tilde{\boldsymbol{X}}\|^2, \tag{8.115}$$

with constants ϕ_i, $i = 0,1,2$ being defined as

$$\phi_0 = 2\left(5\varepsilon_{01} + 4\varepsilon_{02} + 16\varepsilon_{03} + \frac{1}{2K_1\varepsilon_{01}} - \left(c_{12} + \frac{d_{11}}{m_{11}} + \frac{d_{22}}{m_{22}} + \frac{d_{33}}{m_{33}}\right)\right),$$

$$\begin{aligned}
\phi_1 &= \frac{\bar{\sigma}_{xe}^2 + \bar{\sigma}_{ye}^2}{4\varepsilon_{01}} + \frac{9}{\varepsilon_{01}}(\bar{u}_{d0}^2 + \bar{\sigma}_{xe}^2 + 3\bar{\sigma}_{ye}^2 + 3k_2^2) + (\bar{\sigma}_{xe}^2 + \bar{\sigma}_{ye}^2 + \bar{\sigma}_{\bar{\psi}e}^2)\times \\
&\quad \max(\bar{\dot{\sigma}}_{xe}, \bar{\dot{\sigma}}_{ye}, \bar{\dot{\sigma}}_{\bar{\psi}e}) + K_1\left(\frac{0.75}{\varepsilon_{03}} + \frac{\varepsilon_{04}}{\underline{b}_2} + \frac{1}{4\varepsilon_{05}}\right)\left(\frac{|d_{23}|}{\underline{b}_1 m_{22}} + \right. \\
&\quad \left.\frac{m_{11}(k_1 + \bar{u}_{d0})}{\underline{b}_1 m_{22}}\left(k_3 + \bar{\sigma}_{\bar{\psi}e} + \frac{(2 + k_2 + \bar{\dot{\sigma}}_{xe} + \bar{\dot{\sigma}}_{ye} + \bar{\sigma}_{xe} + \bar{\sigma}_{ye})}{\underline{u}_{d0}}\right)\right) \\
\phi_2 &= -K_2\lambda_{\min}(\bar{\boldsymbol{M}}^{-1}\bar{\boldsymbol{D}}) + \frac{1}{2\varepsilon_{01}} + K_1\Big(\frac{2.25}{\varepsilon_{02}} + \frac{1}{\varepsilon_{03}}(0.5k_1^2 + 0.25(A_1^2 + \\
&\quad A_2^2) + 8.25\frac{m_{22}^2}{m_{33}^2} + (2k_1^2 + 2k_2^2 + 3(1 + k_1 + \bar{u}_{d0})^2 + 2A_1^2 + \\
&\quad 2(A_2^2 + A_3^2)),
\end{aligned} \tag{8.116}$$

where $c_{12} = c_1 + c_2$, $\varepsilon_{0i}, i = 1,...,5$ are positive constants. The constants A_i, $i = 1,2,3$ denote upper bounds of $|\partial\alpha_r/\partial x_e|$, $|\partial\alpha_r/\partial y_e|$, $|\partial\alpha_r/\partial\psi_e|$, respectively. These constants can be calculated by taking corresponding partial derivatives of α_r, see (8.104). Inequality (8.115) implies that the solution of the closed loop system (8.113) exists.

Now we pick the constants $\varepsilon_{0i}, i = 1,2,3$ and K_1 such that ϕ_0 is strictly less than $\lambda_{\min}(\bar{\boldsymbol{M}}^{-1}\bar{\boldsymbol{D}})$. Then picking the constant K_2 such that $\phi_2 \le 0$, we have

$$\dot{W} \le \frac{1}{2}\phi_0 + \frac{\phi_1}{1+W_0}. \tag{8.117}$$

On the other hand, it is noted from (8.114) that

$$\dot{W} \geq \frac{\dot{W}_0}{2(1+W_0)} - \lambda_{\max}(\bar{M}^{-1}\bar{D})\|\tilde{X}\|^2 \geq \frac{\dot{W}_0}{2(1+W_0)}, \tag{8.118}$$

where $\lambda_{\max}(\bar{\boldsymbol{M}}^{-1}\bar{\boldsymbol{D}}) > 0$ is the maximum eigenvalue of $\bar{\boldsymbol{M}}^{-1}\bar{\boldsymbol{D}}$. From (8.117) and (8.118), we have

$$\dot{W}_0 \leq \phi_0 W_0 + 2\phi_1 + \phi_0. \tag{8.119}$$

From (8.119) and the expression for W_0, we have

$$\|(x_e(t), y_e(t), x_d(t), y_d(t))\| \leq \Xi_0 e^{\phi_0(t-t_0)},\ \forall t \geq t_0.$$

To show that $\tilde{\boldsymbol{X}}(t)$ globally exponentially converges to zero, we take the Lyapunov function

$$V_0 = \frac{1}{2}\tilde{\boldsymbol{X}}^{\boldsymbol{T}}\tilde{\boldsymbol{X}},$$

whose derivative along the solutions of (8.83) satisfies

$$\begin{aligned} &\dot{V}_0 \leq -\lambda_{\min}(\bar{\boldsymbol{M}}^{-1}\bar{\boldsymbol{D}})\|\tilde{X}\|^2 \\ \Rightarrow \quad &\|\tilde{\boldsymbol{X}}(t)\| \leq \left\|\tilde{\boldsymbol{X}}(t_0)\right\| e^{-\lambda_{\min}(\bar{\boldsymbol{M}}^{-1}\bar{\boldsymbol{D}})(t-t_0)}, \quad \forall t \geq t_0 \geq 0, \end{aligned} \tag{8.120}$$

where Lemma 2.2 has been used. To investigate stability of the $(\sigma_{\bar{\psi}e}, \bar{\psi}_e, u_e, v_e, r_e)$-subsystem, we take the Lyapunov function

$$V_1 = \sqrt{1+\bar{\psi}_e^2} - 1 + \frac{1}{2}(u_e^2 + v_e^2 + r_e^2) + \frac{1}{2\gamma_3}\sigma_{\bar{\psi}e}^2, \tag{8.121}$$

whose derivative along the solutions of the third, sixth, seventh, eighth, and ninth equations of (8.113) satisfied

$$\begin{aligned} \dot{V}_1 \leq\ & -(k_3 - \varepsilon_{06})\frac{\bar{\psi}_e^2}{1+\bar{\psi}_e^2} - \left(c_1 + \frac{d_{11}}{m_{11}} - \varepsilon_{06}\right)u_e^2 - \\ & \left(\frac{d_{22}}{m_{22}} - \varepsilon_{06}\right)v_e^2 - \left(c_2 + \frac{d_{33}}{m_{33}} - \varepsilon_{06}\right)r_e^2 + \Xi_1 e^{-\delta_1(t-t_0)} \\ \leq\ & \Xi_1 e^{-\delta_1(t-t_0)}, \end{aligned} \tag{8.122}$$

with $\Xi_1 = A_4\|\tilde{\boldsymbol{X}}(t_0)\| + A_5\|\tilde{\boldsymbol{X}}(t_0)\|^2$, where

$$\begin{aligned} A_4 =\ & \max(\underline{b}_1, k_2\underline{u}_{d0}^{-1}), \\ A_5 =\ & 0.25\varepsilon_{06}^{-1}\max(k_1^2 + \underline{b}_1^{-1}\underline{u}_{d0}^{-1}(k_1+k_2)^2(\bar{u}_d+\bar{v}_d)^2 + \\ & (m_{11}m_{33}^{-1})^2(1+\bar{v}_d), 2k_1^2 + (m_{11}m_{33}^{-1})^2(1+\bar{\alpha}_u) + \\ & \underline{b}_1^{-1}\underline{u}_{d0}^{-1}(k_1+k_2)^2(\bar{u}_d+\bar{v}_d)^2, 2k_1^2 + (m_{11}m_{22}^{-1})^2 \times \\ & (1+\bar{\alpha}_u) + \underline{b}_1^{-1}\underline{u}_{d0}^{-1}(k_1+k_2)^2(\bar{u}_d+\bar{v}_d)^2 + k_3^2), \end{aligned} \tag{8.123}$$

and $\delta_1 = \lambda_{\min}(\bar{\boldsymbol{M}}^{-1}\bar{\boldsymbol{D}}) - \phi_0$, ε_{06} is a positive constant that is strictly smaller than $\min(k_3, c_1 + d_{11}/m_{11}, d_{22}/m_{22}, c_2 + d_{33}/m_{33})$. Since we have already proved that ϕ_0 is strictly less than $\lambda_{\min}(\bar{\boldsymbol{M}}^{-1}\bar{\boldsymbol{D}})$, δ_1 is a strictly positive constant. The second inequality of (8.122) implies that $V_1(t)$, i.e., $(\sigma_{\bar{\psi}e}(t), \bar{\psi}_e(t), u_e(t), v_e(t), r_e(t))$, is bounded. By integrating both sides of the first inequality of (8.122) and applying Barbalat's lemma, we have $\lim_{t\to\infty}(\bar{\psi}_e(t), u_e(t), v_e(t), r_e(t)) = 0$.

To show that v_d is bounded, we take the Lyapunov function

$$V_2 = \frac{1}{2}v_d^2,$$

whose derivative along the solutions of the last equation of (8.113) yields

$$\dot{V}_2 = b_2^{-1}\left(-d_{n2}m_{22}^{-1}|v_d|v_d^2 - B_1 v_d^2 + B_2|v_d| + B_3\right), \tag{8.124}$$

where $B_i, 0 \le i \le 3$ are given in (8.142). Since d_{n2}/m_{22} is a positive constant, an application of Theorem 4.18 in [6] shows that v_d is bounded by some constant. It is further noted that if nonlinear damping terms are ignored, i.e., $d_{ni} = 0, i = 1,2,3$, then the path and desired surge velocity must satisfy an additional condition such that B_1 is strictly positive otherwise the filtered sway velocity dynamics will be unstable. This will result in an unstable closed loop system.

To investigate stability of the $(\sigma_{xe}, \sigma_{ye}, x_e, y_e)$-dynamics, we take the Lyapunov function

$$V_3 = \sqrt{1 + x_e^2 + y_e^2} - 1 + \frac{1}{2}\left(\frac{\sigma_{xe}^2}{\gamma_1} + \frac{\sigma_{ye}^2}{\gamma_2}\right),$$

whose derivative along the solutions of the first four equations of (8.113) satisfies

$$\begin{aligned}\dot{V}_3 \le {} & -\frac{(k_1 - 3\varepsilon_{07})x_e^2 + (k_2 - 3\varepsilon_{07})y_e^2}{\Delta^2} + \frac{u_e^2 + v_e^2 + 8u_d^2\bar{\psi}_e^2}{4\varepsilon_{07}} + \\ & \frac{\Xi_2}{\varepsilon_{07}} e^{-\lambda_{\min}(\bar{\mathbf{M}}^{-1}\bar{\mathbf{D}})(t-t_0)} \\ \le {} & \frac{u_e^2 + v_e^2 + 8u_d^2\bar{\psi}_e^2}{4\varepsilon_{07}} + \frac{\Xi_2}{\varepsilon_{07}} e^{-\lambda_{\min}(\bar{\mathbf{M}}^{-1}\bar{\mathbf{D}})(t-t_0)},\end{aligned} \tag{8.125}$$

where ε_{07} is a positive constant, Ξ_2 is a class-K function of $\|\boldsymbol{\chi}_0(t_0)\|$. We have picked this constant such that $\varepsilon_{07} < \min(k_1, k_2)/3$. Since we have proved that $(\sigma_{\bar{\psi}e}(t), \bar{\psi}_e(t), u_e(t), v_e(t), v_d(t))$ is bounded and u_d is given by (8.96), the second inequality of (8.125) implies that $(\sigma_{xe}(t), \sigma_{ye}(t), x_e(t), y_e(t))$ is bounded. Hence integrating both sides of the first inequality of (8.125) and applying Barbalat's lemma yield $\lim_{t\to\infty}(x_e(t), y_e(t)) = 0$.

It now follows from (8.97) and the second equation of (8.92) that $|\psi_e(t)|$ converges to $\left|\arctan((\varphi_2 + \sigma_{ye} + v_d)/\Upsilon_{ud0})\right| < 0.5\pi$, since we have already proven that $\lim_{t\to\infty}(x_e(t), y_e(t)) = 0$, $v_d(t)$ is bounded and $u_{d0}(t)$ is larger than some positive constant by assumption, see Assumption 8.2. Proof that the ship's total linear velocity is tangential to the path follows readily, since $\psi_e(t)$ converges to

$\arctan((\varphi_2 + \sigma_{ye} + v_d)/\Upsilon_{u_{d0}})$, and the proven fact that $\lim_{t\to\infty}(v(t) - v_d(t)) = 0$. Finally, we note that all constants ε_{0i}, $1 \le i \le 7$, A_j, $j = 1,2,3$, K_1, and K_2 are only used in the proof. They are not needed in controller implementation.

8.2.6 Discussion

8.2.6.1 Adding Integrator Approach Versus Adaptive Approach

We have used the projection integral actions instead of an adaptive approach to compensate for the disturbances. The trade-off is that the integral action approach does not give deep insight into the ship dynamics with disturbances. However, it results in a simple controller that is suitable for practical implementation. An adaptive approach would be extremely complicated (even when nonlinear damping terms are ignored) if the controller design follows the methodology proposed in this chapter. In an adaptive approach, the disturbances, $\boldsymbol{\tau}_E$, can be considered directly in the ship dynamics as:

$$\bar{\boldsymbol{M}}\dot{\boldsymbol{v}} = -\bar{\boldsymbol{C}}(\boldsymbol{v})\boldsymbol{v} - \bar{\boldsymbol{D}}\boldsymbol{v} + \boldsymbol{\tau} + \boldsymbol{\tau}_E, \tag{8.126}$$

where the nonlinear damping terms are ignored. Using the same coordinate transformation (8.73) with $\boldsymbol{P}(\eta)$ given in (8.80) gives

$$\dot{\boldsymbol{X}} = -\boldsymbol{P}(\eta)\bar{\boldsymbol{M}}^{-1}\bar{\boldsymbol{D}}\boldsymbol{P}^{-1}(\eta)\boldsymbol{X} + \boldsymbol{P}(\eta)\bar{\boldsymbol{M}}^{-1}(\boldsymbol{\tau} + \boldsymbol{\tau}_E). \tag{8.127}$$

Because of the unknown $\boldsymbol{\tau}_E$, we would design a dynamic observer for $\boldsymbol{X}$ by defining

$$\tilde{\boldsymbol{X}} = \boldsymbol{X} - \hat{\boldsymbol{X}} - \boldsymbol{W}\boldsymbol{\tau}_E, \tag{8.128}$$

where $\boldsymbol{W} \in \mathbb{R}^{3\times 3}$ is a matrix to be determined and $\hat{\boldsymbol{X}}$ is an estimate of $\boldsymbol{X}$. Differentiating both sides of (8.128) and choosing

$$\begin{aligned} \dot{\hat{\boldsymbol{X}}} &= -\boldsymbol{P}(\eta)\bar{\boldsymbol{M}}^{-1}\bar{\boldsymbol{D}}\boldsymbol{P}^{-1}(\eta)\hat{\boldsymbol{X}} + \boldsymbol{P}(\eta)\bar{\boldsymbol{M}}^{-1}\boldsymbol{\tau}, \\ \dot{\boldsymbol{W}} &= -\boldsymbol{P}(\eta)\bar{\boldsymbol{M}}^{-1}\bar{\boldsymbol{D}}\boldsymbol{P}^{-1}(\eta)\boldsymbol{W} + \boldsymbol{P}(\eta)\bar{\boldsymbol{M}}^{-1} \end{aligned} \tag{8.129}$$

results in

$$\dot{\tilde{\boldsymbol{X}}} = -\boldsymbol{Q}(\eta)\bar{\boldsymbol{M}}^{-1}\bar{\boldsymbol{D}}\boldsymbol{Q}^{-1}(\eta)\tilde{\boldsymbol{X}}. \tag{8.130}$$

From (8.129) and the first equation of (8.71), a controller can be designed but this will be very complicated since some elements of $\boldsymbol{P}(\eta)$ depend on (x, y) and $\boldsymbol{W}$ must be generated by the second equation of (8.129).

8.2.6.2 Dealing with Parameter Uncertainties

In the control design, it was assumed that all the system parameters are known. The ship's mass and added mass and linear damping coefficients are determined quite accurately by using semi-empirical methods or hydrodynamic programs (such as MARINTEK). However, it is difficult to obtain nonlinear damping coefficients accurately. The purpose of this section is to discuss how the proposed observer and controller can be modified so that the inaccuracies of the nonlinear damping coefficients can be taken care of. Dealing with inaccuracies of the linear damping coefficients can be carried out similarly if these coefficients are not known accurately. The idea is to replace the real coefficients d_{ni}, $i = 1,2,3$ in the nonlinear damping matrix $\bar{\boldsymbol{D}}_{\boldsymbol{n}}(\boldsymbol{v})$, see (8.72), by their estimates. These estimates will be used in the observer (8.82) instead of the real nonlinear damping coefficients. The estimates of the nonlinear damping coefficients are updated in such a way that the modified observer guarantees that the observer error vector $\tilde{\boldsymbol{X}}(t)$ globally asymptotically converges to zero. Then the proposed control design can be simply modified by replacing all the nonlinear damping coefficients d_{ni}, $i = 1,2,3$ by their estimates to take care of inaccuracies in the nonlinear damping coefficients.

Observer Design Modifications

Letting $\hat{d}_{ni}$, $i = 1,2,3$ be estimates of d_{ni} and defining the estimate errors as $\tilde{d}_{ni} = d_{ni} - \hat{d}_{ni}$, we can write (8.81) as

$$\begin{aligned}\dot{\boldsymbol{X}} = &-\boldsymbol{P}(\eta)\bar{\boldsymbol{M}}^{-1}(\bar{\boldsymbol{D}} + \hat{\bar{\boldsymbol{D}}}_{\boldsymbol{n}}(\boldsymbol{P}^{-1}(\eta)\boldsymbol{X}))\boldsymbol{P}^{-1}(\eta)\boldsymbol{X} + \boldsymbol{P}(\eta)\bar{\boldsymbol{M}}^{-1}\boldsymbol{\tau} - \\ &\boldsymbol{P}(\eta)\bar{\boldsymbol{M}}^{-1}\tilde{\bar{\boldsymbol{D}}}_{\boldsymbol{n}}(\boldsymbol{P}^{-1}(\eta)\boldsymbol{X}))\boldsymbol{P}^{-1}(\eta)\boldsymbol{X},\end{aligned} \tag{8.131}$$

where

$$\hat{\bar{\boldsymbol{D}}}_{\boldsymbol{n}}(\boldsymbol{v}) = \begin{bmatrix} \hat{d}_{n1}|u| & 0 & 0 \\ 0 & \hat{d}_{n2}|v| & 0 \\ 0 & 0 & \hat{d}_{n3}|r| \end{bmatrix}, \quad \tilde{\bar{\boldsymbol{D}}}_{\boldsymbol{n}}(\boldsymbol{v}) = \begin{bmatrix} \tilde{d}_{n1}|u| & 0 & 0 \\ 0 & \tilde{d}_{n2}|v| & 0 \\ 0 & 0 & \tilde{d}_{n3}|r| \end{bmatrix},$$

with (8.73) having been used for a short notation. From (8.131), we design an observer to estimate $\boldsymbol{X}$ as

$$\dot{\hat{\boldsymbol{X}}} = -\boldsymbol{P}(\eta)\bar{\boldsymbol{M}}^{-1}(\bar{\boldsymbol{D}} + \hat{\bar{\boldsymbol{D}}}_{\boldsymbol{n}}(\boldsymbol{P}^{-1}(\eta)\hat{\boldsymbol{X}}))\boldsymbol{P}^{-1}(\eta)\hat{\boldsymbol{X}} + \boldsymbol{P}(\eta)\bar{\boldsymbol{M}}^{-1}\boldsymbol{\tau}. \tag{8.132}$$

From (8.131) and (8.132), we have the observer error dynamics

$$\begin{aligned}\dot{\tilde{\boldsymbol{X}}} =& -\boldsymbol{P}(\eta)\bar{\boldsymbol{M}}^{-1}\bar{\boldsymbol{D}}\boldsymbol{P}^{-1}(\eta)\tilde{\boldsymbol{X}} - \boldsymbol{P}(\eta)\bar{\boldsymbol{M}}^{-1} \times \\ &[\hat{\bar{\boldsymbol{D}}}_{\boldsymbol{n}}(\boldsymbol{P}^{-1}(\eta)\boldsymbol{X})\boldsymbol{P}^{-1}(\eta)\boldsymbol{X} - \hat{\bar{\boldsymbol{D}}}_{\boldsymbol{n}}(\boldsymbol{P}^{-1}(\eta)\hat{\boldsymbol{X}}))\boldsymbol{P}^{-1}(\eta)\hat{\boldsymbol{X}}] - \\ &\boldsymbol{P}(\eta)\bar{\boldsymbol{M}}^{-1}\tilde{\bar{\boldsymbol{D}}}_{\boldsymbol{n}}(\boldsymbol{P}^{-1}(\eta)\boldsymbol{X}))\boldsymbol{P}^{-1}(\eta)\boldsymbol{X} \\ =& -\boldsymbol{P}(\eta)\bar{\boldsymbol{M}}^{-1}\bar{\boldsymbol{D}}\boldsymbol{P}^{-1}(\eta)\tilde{\boldsymbol{X}} - \boldsymbol{Q}(\eta)\bar{\boldsymbol{M}}^{-1} \times \\ &[\bar{\boldsymbol{D}}_{\boldsymbol{n}}(\boldsymbol{P}^{-1}(\eta)\boldsymbol{X})\boldsymbol{P}^{-1}(\eta)\boldsymbol{X} - \bar{\boldsymbol{D}}_{\boldsymbol{n}}(\boldsymbol{P}^{-1}(\eta)\hat{\boldsymbol{X}}))\boldsymbol{P}^{-1}(\eta)\hat{\boldsymbol{X}}] - \boldsymbol{\Phi}\tilde{\boldsymbol{\Theta}},\end{aligned} \tag{8.133}$$

where $\tilde{\boldsymbol{X}} = \boldsymbol{X} - \hat{\boldsymbol{X}}$ and we have defined

$$\boldsymbol{\Phi} = \boldsymbol{P}(\eta)\bar{\boldsymbol{M}}^{-1}\begin{bmatrix} |\hat{u}|\hat{u} & 0 & 0 \\ 0 & |\hat{v}|\hat{v} & 0 \\ 0 & 0 & |\hat{r}|\hat{r} \end{bmatrix}, \ \tilde{\boldsymbol{\Theta}} = \begin{bmatrix} \tilde{d}_{n1} \\ \tilde{d}_{n2} \\ \tilde{d}_{n3} \end{bmatrix}.$$

It is noted that we have used (8.84). To determine an update law for $\hat{d}_{ni}$, $i = 1,2,3$, we take the following Lyapunov function

$$L_0 = \frac{1}{2}\tilde{\boldsymbol{X}}^T\tilde{\boldsymbol{X}} + \frac{1}{2}\tilde{\boldsymbol{\Theta}}^T\boldsymbol{\Gamma}^{-1}\tilde{\boldsymbol{\Theta}}, \tag{8.134}$$

where $\boldsymbol{\Gamma}$ is a positive definite matrix. Using Lemma 2.2, the derivative of L_0 along the solutions of (8.133) satisfies

$$\begin{aligned} \dot{L}_0 = & -\tilde{\boldsymbol{X}}^T\boldsymbol{P}(\eta)\bar{\boldsymbol{M}}^{-1}\bar{\boldsymbol{D}}\boldsymbol{P}^{-1}(\eta)\tilde{\boldsymbol{X}} - \tilde{\boldsymbol{X}}^T\boldsymbol{P}(\eta)\bar{\boldsymbol{M}}^{-1}\times \\ & [\bar{\boldsymbol{D}}_n(\boldsymbol{P}^{-1}(\eta)\boldsymbol{X})\boldsymbol{P}^{-1}(\eta)\boldsymbol{X} - \bar{\boldsymbol{D}}_n(\boldsymbol{P}^{-1}(\eta)\hat{\boldsymbol{X}})\boldsymbol{P}^{-1}(\eta)\hat{\boldsymbol{X}}] - \\ & \tilde{\boldsymbol{X}}^T\boldsymbol{\Phi}\tilde{\boldsymbol{\Theta}}, \\ \leq & -\lambda_{\min}(\bar{\boldsymbol{M}}^{-1}\bar{\boldsymbol{D}})||\tilde{\boldsymbol{X}}||^2 - \tilde{\boldsymbol{X}}^T\boldsymbol{\Phi}\tilde{\boldsymbol{\Theta}} - \dot{\hat{\boldsymbol{\Theta}}}^T\boldsymbol{\Gamma}^{-1}\tilde{\boldsymbol{\Theta}}, \end{aligned} \tag{8.135}$$

where

$$\hat{\boldsymbol{\Theta}} = \left[\hat{d}_{n1}\ \hat{d}_{n2}\ \hat{d}_{n3}\right]^T,$$

and $\lambda_{\min}(\bar{\boldsymbol{M}}^{-1}\bar{\boldsymbol{D}}) > 0$ is the minimum eigenvalue of $\bar{\boldsymbol{M}}^{-1}\bar{\boldsymbol{D}}$. From (8.135), choosing an update law for $\hat{\boldsymbol{\Theta}}$ as

$$\dot{\hat{\boldsymbol{\Theta}}} = -\boldsymbol{\Gamma}\boldsymbol{\Phi}^T(\cdot)\tilde{\boldsymbol{X}} \tag{8.136}$$

results in

$$\dot{L}_0 \leq -\lambda_{\min}(\bar{\boldsymbol{M}}^{-1}\bar{\boldsymbol{D}})||\tilde{\boldsymbol{X}}||^2, \tag{8.137}$$

which means that $\lim_{t\to\infty 0}\tilde{\boldsymbol{X}}(t) = 0$. Indeed, we can use the projection algorithm instead of (8.136). Now it is important to note that the update law (8.136) cannot be used since $\tilde{\boldsymbol{X}}$ contains the unknown vector $\boldsymbol{X}$. We need to get around this problem. Substituting $\tilde{\boldsymbol{X}} = \boldsymbol{X} - \hat{\boldsymbol{X}}$ into the right-hand side of (8.136) and integrating both sides gives

$$\hat{\boldsymbol{\Theta}}(t) = \hat{\boldsymbol{\Theta}}(t_0) - \boldsymbol{\Gamma}\int_{t_0}^{t}\boldsymbol{\Phi}^T(\boldsymbol{X}(\tau) - \hat{\boldsymbol{X}}(\tau))\mathrm{d}\tau. \tag{8.138}$$

On the other hand, from the first equation of (8.71) and (8.73), we have

$$\boldsymbol{X} = (\boldsymbol{J}(\psi)\boldsymbol{P}(\eta))^{-1}\dot{\eta}. \tag{8.139}$$

Substituting (8.139) into (8.138) gives

$$\hat{\boldsymbol{\Theta}}(t) = \hat{\boldsymbol{\Theta}}(t_0) + \boldsymbol{\Gamma}\int_{t_0}^{t}\boldsymbol{\Phi}^T\hat{\boldsymbol{X}}(\tau)\mathrm{d}\tau - \boldsymbol{\Gamma}\int_{\eta(t_0)}^{\eta(t)}\boldsymbol{\Phi}^T(\boldsymbol{J}(\sigma)\boldsymbol{P}(\sigma))^{-1}\mathrm{d}\sigma, \tag{8.140}$$

where we have abused a notation of $\boldsymbol{J}(\psi)$ as $\boldsymbol{J}(\eta)$. It is seen that (8.140) is a linear Volterra equation, which can be numerically solved by using a number of methods available in the literature, see [119]. Noticing that the right-hand side of (8.140) contains only known terms, we can use (8.140) to calculate $\hat{\boldsymbol{\Theta}}(t)$ instead of (8.136). We summarize this section as follows: The observer (8.82) is replaced by the adaptive observer consisting of (8.132) and (8.140) for the case where the nonlinear damping coefficients are not known accurately.

Control Design Modifications

The modified observer, see (8.132) and (8.140), guarantees that $\lim_{t\to\infty 0}\tilde{\boldsymbol{X}}(t) = 0$. The control design is the same as in Section 8.2.4 but the nonlinear damping coefficients $d_{ni},\ i = 1,2,3$ are replaced by their estimates $\hat{d}_{ni},\ i = 1,2,3$. The result with the modified observer and controller as discussed above is almost the same as the one stated in Theorem 8.3. The only difference is that in the case where the nonlinear damping coefficients are known accurately, the observer error $\tilde{\boldsymbol{X}}(t)$ globally exponentially converges to zero but only globally asymptotically converges to zero in the case where the nonlinear damping coefficients are not known accurately due to the adaptation. Proof of the result under the modified observer and controller follows the same lines as that of Theorem 8.3, and is therefore excluded here.

8.2.6.3 Dealing with Actuator Saturation

In Section 8.2.4, the controller was designed without imposing any limitations on the magnitude and rate of the ship actuators. Actuator saturation restricts the reference path, deteriorates the overall path-tracking performance, and can even destabilize the closed loop system. These problems are difficult to deal with but it is important that they are taken care of. In many cases, global stabilization and/or tracking results cannot be achieved when actuators are subject to saturation. For example, consider the scalar system $\dot{x} = x^2 + u$ with x the state and u the control. For this system, no bounded control u can globally exponentially/asymptotically stabilize the system at the origin. In fact, any bounded control u can result in a finite escape time. On the other hand, only restrictive tracking can be achieved for many systems such as a simple system consisting of an integrator chain, see [120]. In this section, we will discuss how to use the controller proposed in Section 8.2.4 when the actuator saturation presents. We will focus on the magnitude saturation. A discussion on the rate saturation can be carried out similarly. The idea is to set $|\tau_u| \le \tau_u^{\max}$ and $|\tau_r| \le \tau_r^{\max}$ where $\tau_u^{\max}$ and $\tau_r^{\max}$ are the maximum values that the actuators can supply, then to find restrictions on the reference path, control gains, and initial conditions such that $|\tau_u| \le \tau_u^{\max}$ and $|\tau_r| \le \tau_r^{\max}$ hold. From the control design in Subsection 8.2.4, we calculate the upper bounds of α_u, $\alpha_{\psi e}$, and α_r as

$$\bar{\alpha}_u = k_1 + \bar{\sigma}_{xe} + \sqrt{\bar{u}_{d0}^2 + \bar{\sigma}_{xe}^2},\ \bar{\alpha}_{\psi e} = \arctan\left(\frac{k_2 + \bar{\sigma}_{ye} + \bar{v}_d}{\sqrt{\bar{u}_{d0}^2 + \bar{\sigma}_{xe}^2}}\right), \quad (8.141)$$
$$\bar{\alpha}_r = \underline{b}_1^{-1}(k_3 + \bar{\sigma}_{\bar{\psi}e} + \bar{r}_d + \underline{u}_{d0}^{-1}((k_2+1)(\bar{v}_d + \bar{u}_d) + \bar{\alpha}_u + \bar{u}_d + \bar{\dot{v}}_d + \bar{\dot{\sigma}}_{ye}) + \underline{u}_{d0}^{-2}(\bar{\dot{u}}_{d0}\bar{u}_{d0} + \bar{\dot{\sigma}}_{xe}\bar{\sigma}_{xe})),$$

where

$$\bar{u}_d = \sqrt{\bar{u}_{d0}^2 + \bar{\sigma}_{xe}^2 + (k_2 + \bar{\sigma}_{ye} + \bar{v}_d)^2},$$

$$\bar{v}_d = \max\left(\sqrt{\frac{B_3}{B_1}}, \sqrt{\frac{B_2 m_{22}}{d_{n2}}}\right),$$
$$\bar{\dot{v}}_d = \frac{(m_{11}\bar{\alpha}_u + \bar{d}_{23})\bar{\alpha}_r + d_{22}\bar{v}_d + d_{n2}\bar{v}_d^2 + k_2\underline{u}_{d0}^{-1}}{m_{22}},$$
$$\bar{r}_d = \frac{|x_d'(s)y_d''(s) - x_d''(s)y_d'(s)|}{\sqrt{(x_d'^2(s) + y_d'^2(s))^3}}\bar{u}_d,$$
$$B_1 = \frac{d_{22}}{m_{22}} - 2B_0,$$
$$B_2 = \frac{m_{11}(k_1 + \bar{u}_{d0}) + |d_{23}|}{\underline{b}_1 m_{22}}\left(k_3 + \bar{\sigma}_{\bar{\psi}e} + \frac{k_2(2 + k_1 + k_2) + \bar{\dot{\sigma}}_{ye}}{\underline{u}_{d0}} + \frac{\bar{u}_{d0}\bar{\dot{u}}_{d0} + \bar{\sigma}_{xe}\bar{\dot{\sigma}}_{xe}}{\underline{u}_{d0}^2}\right) + \frac{k_2}{\underline{u}_{d0}},$$
$$B_3 = \frac{1}{2}B_0(\bar{u}_{d0}^2 + \bar{\sigma}_{xe}^2 + 3k_2^2 + 3\bar{\sigma}_{ye}^2),$$
$$B_0 = \frac{m_{11}(k_1 + \bar{u}_{d0}) + \bar{d}_{23}}{\underline{b}_1 m_{22}}\frac{\left|x_d'(s)y_d''(s) - x_d''(s)y_d'(s)\right|}{\sqrt{(x_d'^2(s) + y_d'^2(s))^3}}, \quad (8.142)$$

and the notations $\bar{\bullet}$ and $\underline{\bullet}$ denote the upper and lower bounds of $|\bullet|$, respectively. By substituting $\hat{u} = u_e + \alpha_u$, $\hat{v} = v_e + v_d$, and $\hat{r} = r_e + \alpha_r$ into (8.112), we can calculate the upper bounds of τ_u and τ_r as

$$|\tau_u| \le \vartheta_1 u_e^2 + \vartheta_2 v_e^2 + \vartheta_3 r_e^2 + \vartheta_0,$$
$$|\tau_r| \le \upsilon_1 u_e^2 + \upsilon_2 v_e^2 + \upsilon_3 r_e^2 + \upsilon_0, \quad (8.143)$$

where

$$\vartheta_1 = 0.5(m_{11}(k_1 + c_1) + d_{n1}),$$
$$\vartheta_2 = m_{22} + 0.5k_2 + m_{11}m_{22}^{-1},$$

$$\begin{aligned}
\vartheta_3 = & \ m_{22}+0.5(k_1+k_2+m_{11}m_{22}^{-1}),\\
\vartheta_0 = & \ 0.5c_1m_{11}+m_{22}(0.5(\bar{\alpha}_r^2+\bar{v}_d^2+\bar{v}_d\bar{\alpha}_r)+d_{11}\bar{\alpha}_u+1.5d_{n1}\bar{\alpha}_u^2+m_{11}\times\\
& ((k_1+k_2)(1+\bar{u}_d+\bar{\alpha}_r)+\bar{\dot{2}}\sigma_{xe}+\bar{\dot{u}}_{d0}+k_2\underline{u}_{d0}^{-1}+0.5m_{11}m_{22}^{-1}\bar{\alpha}_r^2),\\
\upsilon_1 = & \ |m_{11}-m_{22}|+\underline{b}_1^{-1}\underline{u}_{d0}^{-1}(k_1+k_2)(\bar{u}_d+\bar{v}_d)+m_{11}m_{22}^{-1},\\
\upsilon_2 = & \ 0.5\bar{d}_{32}+|m_{11}-m_{22}|+\underline{b}_1^{-1}\underline{u}_{d0}^{-1}m_{33}(k_1+k_2)(\bar{u}_d+\bar{v}_d)+\\
& m_{11}m_{33}m_{22}^{-1}+0.5m_{33}m_{22}^{-1}\bar{d}_{23},\\
\upsilon_3 = & \ 0.5c_2m_{33}+0.5d_{n3}+2_1^{-1}\underline{u}_{d0}^{-1}m_{33}(k_1+k_2)(\bar{u}_d+\bar{v}_d)+0.5\underline{b}_1^{-1}k_3,\\
\upsilon_0 = & \ 0.5c_2m_{33}+1.5d_{n3}\bar{\alpha}_r^2+0.5\bar{d}_{32}+\bar{d}_{23}\bar{v}_d+|m_{11}-m_{22}|(0.5\bar{v}_d^2+\\
& 0.5\bar{\alpha}_u^2+\bar{\alpha}_u\bar{v}_d)+m_{33}(\underline{b}_1^{-1}\underline{u}_{d0}^{-1}(k_1+k_2)(\bar{u}_d+\bar{v}_d)(2+\bar{\alpha}_u+2\bar{u}_d+\\
& 2\bar{\alpha}_r+\bar{v}_d)+\underline{b}_1^{-1}\max(|(x_d^{'}(s)y_d^{''}(s)-x_d^{'''}(s)y_d^{'}(s))(x_d^{'2}(s)+y_d^{'2}(s))-\\
& 2(x_d^{'}(s)y_d^{''}(s)-x_d^{''}(s)y_d^{'}(s))(x_d^{'}(s)x_d^{''}(s)-y_d^{'}(s)y_d^{''}(s)))|\times\\
& (x_d^{'2}(s)+y_d^{'2}(s))^{-2})+\bar{u}_d+_1^{-1}k_3(\bar{r}_d+0.5+\bar{\alpha}_r)+\underline{b}_1^{-1}(\underline{u}_{d0}^{-1}(\bar{\dot{\sigma}}_{xe}^2+\\
& \bar{\dot{\sigma}}_{ye})+\bar{\dot{\sigma}}_{\bar{\psi}e}+\bar{\dot{u}}_{d0}^2+\bar{\dot{v}}_d)+0.5m_{11}m_{22}^{-1}\bar{\alpha}_u^2+0.5m_{22}^{-1}\bar{d}_{23}).
\end{aligned} \tag{8.144}$$

On the other hand, integrating both sides of the second inequality of (8.122) results in

$$u_e^2(t)+v_e^2(t)+r_e^2(t)\le\Phi(t_0),\ \forall t\ge t_0\ge 0, \tag{8.145}$$

where

$$\begin{aligned}
\Phi(t_0) = & \ 2\Big(\sqrt{1+\bar{\psi}_e^2(t_0)}-1+0.5(u_e^2(t_0)+v_e^2(t_0)+r_e^2(t_0))+\\
& \frac{1}{2}\gamma_3\sigma_{\bar{\psi}_e}^2(t_0)+\delta_1^{-1}A_4\|\tilde{X}(t_0)\|+\delta_1^{-1}A_5\|\tilde{X}(t_0)\|^2\Big),
\end{aligned} \tag{8.146}$$

with A_4, A_5 and δ_1 given in (8.123). Now using (8.143) and (8.145), and setting $|\tau_u|\le\tau_u^{\text{Max}}$ and $|\tau_r|\le\tau_r^{\text{Max}}$, we deduce that

$$\Phi(t_0)\le\min\left(\frac{\tau_u^{\max}-\vartheta_0}{\max(\vartheta_1,\vartheta_2,\vartheta_3)},\frac{\tau_r^{\max}-\upsilon_0}{\max(\upsilon_1,\upsilon_2,\upsilon_3)}\right), \tag{8.147}$$

which implies that under the actuator magnitude saturation, restrictions must be placed on the type of reference paths (both desired velocity of the vessel on the path and the path curvature cannot be arbitrarily large), control gains, and initial conditions such that (8.147) holds, i.e., the proposed controller under the actuator magnitude saturation is not only local but also restricts the commanded paths.

8.2.7 Experimental Results

This section describes the experimental set-up and tested results performed on a model ship in the Swan River, Western Australia. The model ship has a length of

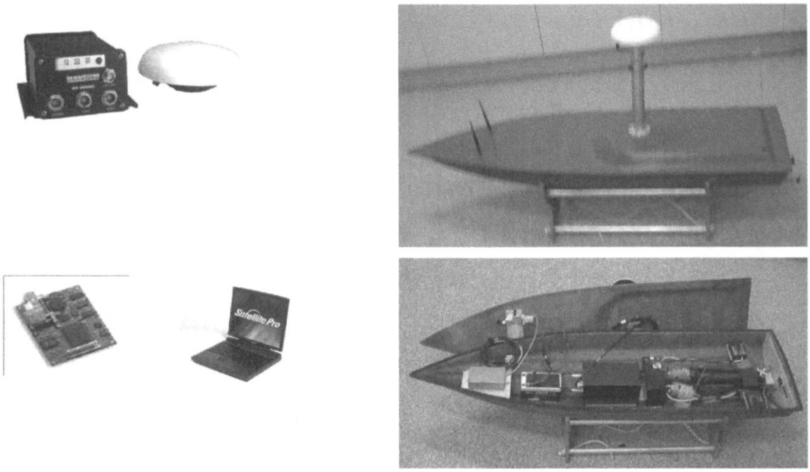

Figure 8.7 Equipment for experimental tests

1.2 m and a mass of 17.5 kg. The model ship is equipped with one differential global positioning system (DGPS) SF-2050G from NavCom, which can provide an accuracy of 25 cm, to obtain longitude, latitude and altitude, one compass TCM2-50 from PNI to measure yaw angle, two direct current motors Torpedo 850 with two remote speed controllers driving two propellers to provide surge force and yaw moment, three batteries to supply power for motors, DGPS and wireless communication equipment, and two sets of wireless transceivers to transmit and receive signals between the model ship and the host computer, see Figure 8.7. The parameters of the ship model are calculated by VERES:

$$\begin{aligned}
&m_{11} = 25.8, m_{22} = 33.8, m_{33} = 2.76, m_{23} = m_{32} = 6.2, \\
&D_{11} = 12 + 2.5|u|, \; D_{22} = 17 + 4.5|v|, \; D_{33} = 0.5 + 0.1|r|, \\
&D_{23} = 0.2, \; D_{32} = 0.5.
\end{aligned}$$

The global coordinates (longitude, latitude, altitude) are transformed to the "river coordinates" based on a three-point algorithm as follows, see Figure 8.8:

1. The DGPS is first used to measure the global coordinates of two fixed points $M_1(h_1, l_1, \mu_1)$ and $M_2(h_2, l_2, \mu_2)$ with h_i, l_i, μ_i, $i = 1, 2$, being the altitude, longitude and latitude of the point $M_i(h_i, l_i, \mu_i)$. These points are chosen along the river bank.
2. We calculate the distance, $\overline{M_1 M_2}$ between these two fixed points by the following formula, see [11]:

$$\overline{M_1M_2} = \sqrt{(x_{e1} - x_{e2})^2 + (y_{e1} - y_{e2})^2 + (z_{e1} - z_{e2})^2}, \tag{8.148}$$

where

$$\begin{aligned} x_{ei} &= (N_i + h_i)\cos(\mu_i)\cos(l_i), \\ y_{ei} &= (N_i + h_i)\cos(\mu_i)\sin(l_i), \\ z_{ei} &= \left(\frac{r_p^2}{r_e^2}N_i + h_i\right)\sin(\mu_i), \\ N_i &= \frac{r_e^2}{\sqrt{r_e^2\cos^2(\mu_i) + r_p^2\sin^2(\mu_i)}}, i = 1,2, \\ r_p &= 6356752, \\ r_e &= 6378137. \end{aligned} \tag{8.149}$$

In the experimental set-up, we adjusted the fixed points $M_1(h_1, l_1, \mu_1)$ and $M_2(h_2, l_2, \mu_2)$ such that $\overline{M_1M_2}$=100 m.

3. The river coordinate system, $O_E X_E Y_E$, is formed such that the $O_E X_E$-axis coincides with the line connecting between $M_1(h_1, l_1, \mu_1)$ and $M_2(h_2, l_2, \mu_2)$, and that the origin O coincides with $M_1(h_1, l_1, \mu_1)$. The OY axis is perpendicular to the OX axis.
4. From this coordinate system, the position coordinates (x, y) of the point, say $M_j(h_j, l_j, \mu_j)$, of interest on the model ship are obtained by calculating the distances, $\overline{M_jM_1}$ and $\overline{M_jM_2}$, from this point to the points $M_1(h_1, l_1, \mu_1)$ and $M_2(h_2, l_2, \mu_2)$.

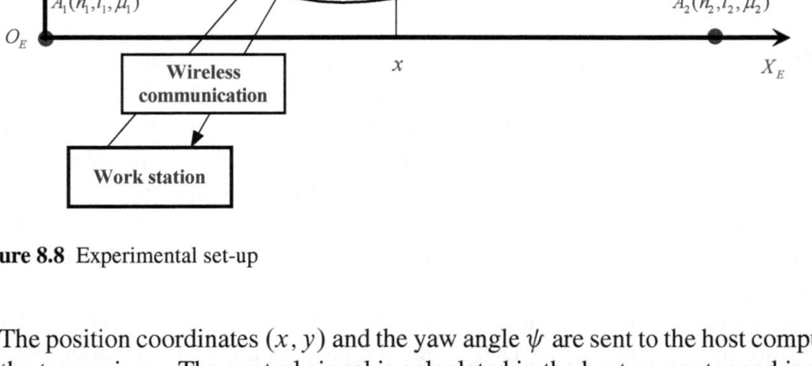

Figure 8.8 Experimental set-up

The position coordinates (x, y) and the yaw angle ψ are sent to the host computer by the transceivers. The control signal is calculated in the host computer and is sent back to the remote speed controllers by a Futaba® transmitter connecting to the host computer via a data acquisition card. A program is written in LabWindows™/CVI, a product of National Instruments™, to implement the control algorithm developed in this chapter. The reference path $\boldsymbol{\Omega}$ is chosen as follows: For the first 90 seconds, $x_d = s$, $y_d = 60$, and $x_d = 20\sin(0.1s) + 60$, $y_d = 20\cos(0.1s) + 40$ for the remainder of testing time. This choice implies that the reference path is a straight line for the first 90 seconds and is a circle centered at (60 m, 40 m) with a radius of 20 m. The desired forward speed is chosen as $u_{d0} = 0.5$ (m/s). Based on conditions specified in Theorem 8.3, the design constants and initial conditions of the observer and filters are chosen as $c_1 = c_2 = 4$, $k_1 = k_2 = k_3 = 1.5$, $\gamma_i = 0.1$, $\xi_i = 0.1$, $i = 1,2,3$, $\upsilon_d(0) = 0$, $\sigma_{xe}(0) = 0$, $\sigma_{ye}(0) = 0$, $\sigma_{\bar{\psi}e}(0) = 0$, $\hat{\boldsymbol{X}}(0) = (0,0,0)^T$, $\bar{\sigma}_{xe} = \bar{\sigma}_{ye} = 10$, and $\bar{\sigma}_{\bar{\psi}e} = 6.5$.

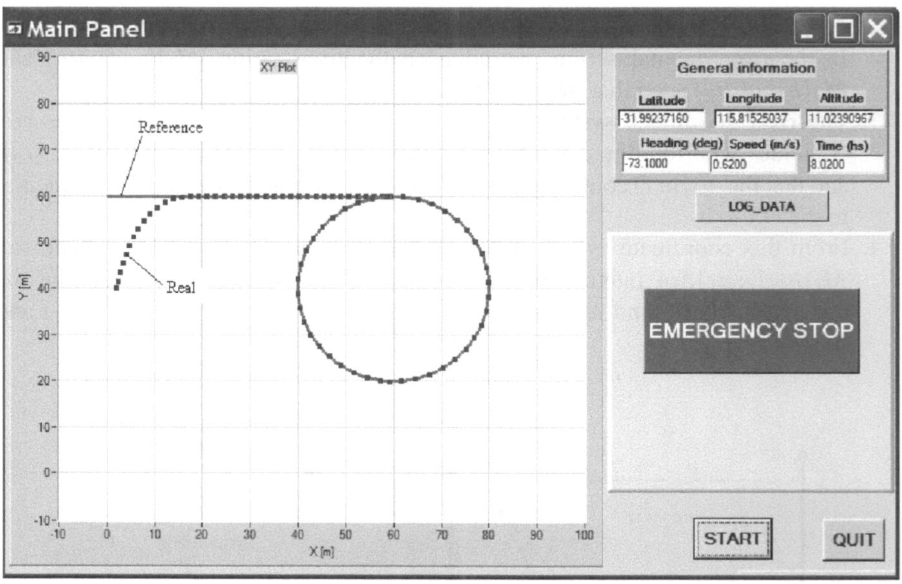

Figure 8.9 A screen shot of the user interface with integral actions

The initial conditions of the model ship are $x = 2$, $y = 40$, $\psi = 0.5$, and $u = \upsilon = r = 0$. Experimental results are given in Figures 8.9 and 8.10. Figure 8.9 is a screen shot of the user interface of the developed program. The user interface displays general information of the ship model such as global coordinates, forward speed, and heading angle. The history of the ship position and the reference path are also displayed in the user interface panel. Figure 8.10 plots the position and orientation tracking errors. It can be seen that these errors are actually not zero as proved in Theorem 8.3. These nonzero errors are due to inaccuracy of the DGPS and compass. A close look at this figure shows that the position and orientation errors are in the range of the DGPS and compass accuracy. To illustrate the role

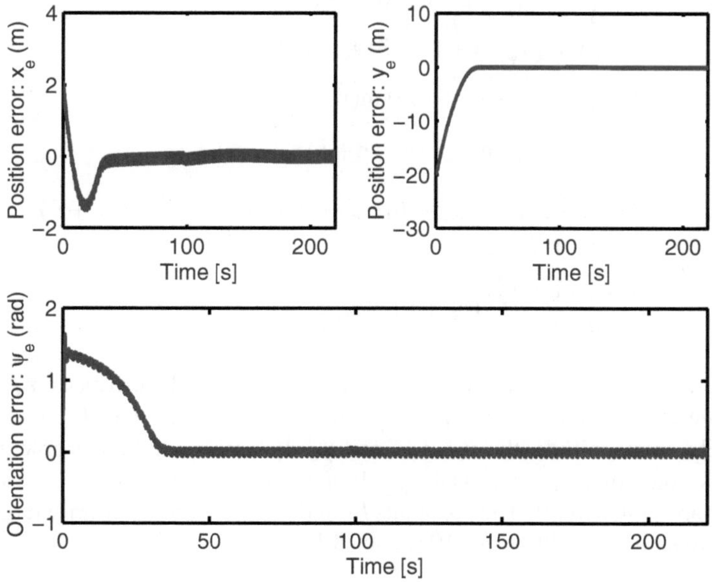

Figure 8.10 Position and orientation errors with integral actions

of integral actions, we also tested the control algorithm without integral actions by setting $\sigma_{xe} = \sigma_{ye} = \sigma_{\bar{\psi}e} = 0$. The tested results without integral actions are given in Figures 8.11 and 8.12. It can be seen that the integral actions play an important role in experimental tests. Poor performance in the test without integral actions can be understood as follows: The disturbances result in velocities, for which the controller does not compensate since the integral actions are switched off. The velocities induced by disturbances result in poor performance and even destroy the stability of the path-tracking error dynamics when these disturbing velocities are large enough (larger than $\min(k_i)$, $i = 1, 2, 3$ in magnitude). Theoretically, we can see this phenomenon by looking at the fourth, fifth, and sixth equations of the closed loop system (8.113):

$$\begin{aligned}
\dot{x}_e = & -k_1\frac{x_e}{\Delta} - \sigma_{xe} + u_e - u_d((\cos(\bar{\psi}_e) - 1)\cos(\alpha_{\psi_e}) - \\
& \sin(\bar{\psi}_e)\sin(\alpha_{\psi_e})) + (\hat{r} + \tilde{r})y_e + \tilde{u} + u_{\tau_E}, \\
\dot{y}_e = & -k_2\frac{y_e}{\Delta} - \sigma_{ye} + v_e + u_d(\sin(\bar{\psi}_e)\cos(\alpha_{\psi_e}) + \\
& (\cos(\bar{\psi}_e) - 1)\sin(\alpha_{\psi_e})) - (\hat{r} + \tilde{r})x_e + \tilde{v} + v_{\tau_E}, \\
\dot{\bar{\psi}}_e = & -k_3\frac{\bar{\psi}_e}{\sqrt{1+\bar{\psi}_e^2}} - \sigma_{\bar{\psi}e} + b_1(r_e + \tilde{r}) - \frac{k_2\Upsilon_{u_{d0}}x_e y_e}{u_d^2\Delta^3}(u_e + \tilde{u}) + \\
& \frac{k_2\Upsilon_{u_{d0}}(1+x_e^2)}{u_d^2\Delta^3}(v_e + \tilde{v}) + r_{\tau_E},
\end{aligned} \tag{8.150}$$

where u_{τ_E}, v_{τ_E}, and r_{τ_E} are velocities induced by the disturbances $\boldsymbol{\tau_E}$. From (8.150), we can see that when u_{τ_E}, v_{τ_E} and r_{τ_E} are larger than k_1, k_2 and k_3, respectively, in magnitude, the x_e-, y_e-, and $\bar{\psi}_e$-dynamics will be unstable, respectively. It is noted that σ_{xe}, σ_{ye}, and $\sigma_{\bar{\psi}e}$ do not compensate for u_{τ_E}, v_{τ_E}, and r_{τ_E} since they are switched off. The constants k_1 and k_2 cannot be arbitrarily large since they have to satisfy conditions (8.103) and (8.111).

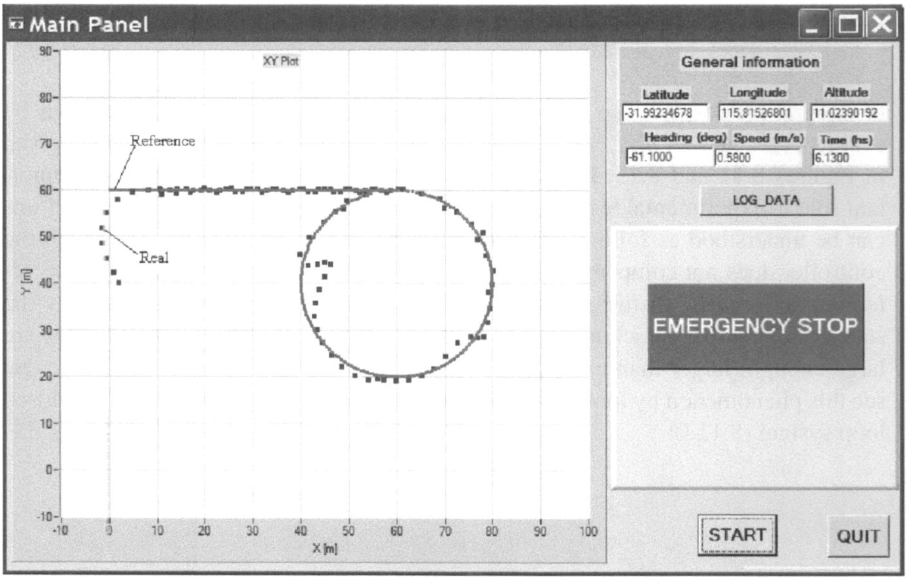

Figure 8.11 A screen shot of the user interface without integral actions

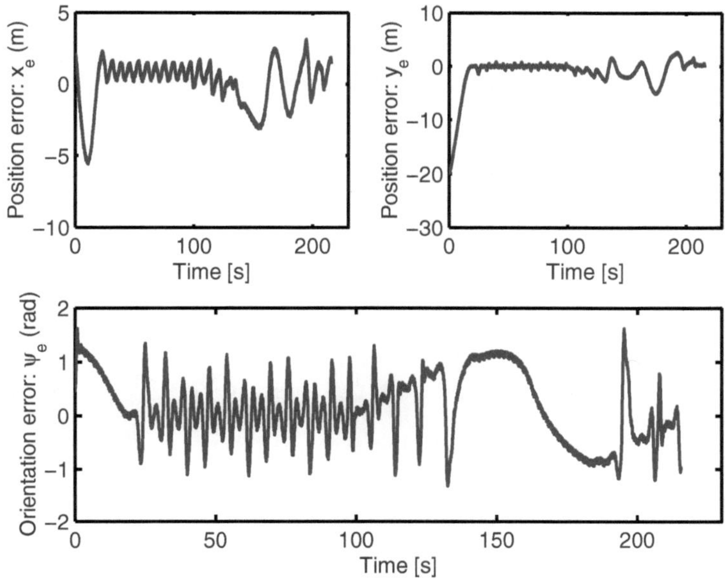

Figure 8.12 Position and orientation errors without integral actions

8.3 Conclusions

This chapter provides global path-tracking controllers for underactuated ships. Both state feedback and output feedback cases have been considered. Of particular note is the use of a "nontraditional" adaptive observer and a simple modification of the proposed controller to deal with inaccuracies of the nonlinear damping coefficients. Moreover, the proposed controllers in this chapter are much simpler than those in the preceding chapters. The simplicity of the proposed controllers in this chapter makes them more suitable for use in practice than those proposed in the previous chapters. The work presented in this chapter is based on [121–125].

Chapter 9
Way-point Tracking Control of Underactuated Ships

This chapter presents state feedback and output feedback controllers that force underactuated ships to globally ultimately track a straight line under environmental disturbances induced by waves, wind, and ocean currents. When there are no environmental disturbances, the controllers are able to drive the heading angle and cross-tracking error to zero asymptotically. Based on the backstepping technique and several technical lemmas introduced for a nonlinear system with nonvanishing disturbances, a full state feedback controller is first designed. An output feedback controller is then developed by using a nonlinear observer, which globally exponentially estimates the unmeasured sway and yaw velocities from the measured sway displacement and the measured yaw angle.

9.1 Control Objective

In addition to the assumptions made in Section 3.4.1.1, we assume that the surge velocity is controlled by the main propulsion control system. As such, the resulting mathematical model of the underactuated ship moving in sway and yaw is rewritten as

$$\begin{aligned}
\dot{y} &= u\sin(\psi) + \cos(\psi)v, \\
\dot{\psi} &= r, \\
\dot{v} &= -\frac{m_{11}u}{m_{22}}r - \frac{d_{22}}{m_{22}}v - \sum_{i\geq 2}\frac{d_{vi}}{m_{22}}|v|^{i-1}v + \frac{1}{m_{22}}\tau_{wv}(t), \\
\dot{r} &= \frac{(m_{11}-m_{22})u}{m_{33}}v - \frac{d_{33}}{m_{33}}r - \sum_{i\geq 2}\frac{d_{ri}}{m_{33}}|r|^{i-1}r + \frac{1}{m_{33}}\tau_r + \frac{1}{m_{33}}\tau_{wr}(t),
\end{aligned} \tag{9.1}$$

where y, v, ψ, r, and u are sway displacement, sway velocity, yaw angle, yaw velocity, and forward speed controlled by the main thruster control system, respectively. Without loss of generality, we assume that the forward speed u is positive and if time-varying, has a bounded derivative $\dot{u}(t)$, i.e., $0 < u_{\min} \leq u(t) \leq u_{\max} < \infty$ and $|\dot{u}(t)| \leq M < \infty$, $\forall t \geq 0$. The positive constant terms m_{jj}, $1 \leq j \leq 3$ denote the ship's inertia including added mass. The positive constant terms d_{22}, d_{33}, d_{vi} and $d_{ri}, i \geq 2$ represent the hydrodynamic damping in sway and yaw. The bounded time-varying terms, $\tau_{wv}(t)$ and $\tau_{wr}(t)$, are the environmental disturbance moments induced by wave, wind, and ocean current with an assumption that $|\tau_{wv}(t)| \leq \tau_{wv\max} < \infty$ and $|\tau_{wr}(t)| \leq \tau_{wr\max} < \infty$. In this chapter, we study two control objectives. The first is full state feedback. In this case, we assume that all states y, v, ψ, and r are available for feedback. In the design of an output feedback controller, only sway and yaw displacements are measurable. For both full state and output feedback cases, we design a control law, τ_r, that forces the ship to track a linear course with ultimate boundedness, i.e. the tracking errors are globally ultimately bounded. When there are no environmental disturbances, the sway displacement and velocity, y and v, yaw angle and velocity, ψ and r, asymptotically converge to zero.

9.2 Full-state Feedback

9.2.1 Control Design

We define the following coordinate transformation

$$z_1 = \psi + \arcsin\left(\frac{ky}{\sqrt{1+(ky)^2}}\right), \tag{9.2}$$

where k is a positive constant to be selected later. Note that the convergence of z_1 and y to zero implies that of ψ. Upon application of the coordinate transformation (9.2), the ship dynamics (9.1) are rewritten as

$$\begin{aligned}
\dot{y} &= -\frac{kuy}{\sqrt{1+(ky)^2}} + \frac{v}{\sqrt{1+(ky)^2}} + \frac{u\left(\sin(z_1) - (\cos(z_1)-1)ky\right)}{\sqrt{1+(ky)^2}} + \\
&\quad \frac{v\left((\cos(z_1)-1) + ky\sin(z_1)\right)}{\sqrt{1+(ky)^2}}, \\
\dot{v} &= -\frac{m_{11}u}{m_{22}}r - \frac{d_{22}}{m_{22}}v - \sum_{i\geq 2}\frac{d_{vi}}{m_{22}}|v|^{i-1}v + \frac{1}{m_{22}}\tau_{wv}(t), \\
\dot{z}_1 &= r - \frac{k^2uy}{(1+(ky)^2)^{3/2}} + \frac{kv}{(1+(ky)^2)^{3/2}} + \frac{ku\left(\sin(z_1) - (\cos(z_1)-1)ky\right)}{(1+(ky)^2)^{3/2}} + \\
&\quad \frac{kv\left((\cos(z_1)-1) + ky\sin(z_1)\right)}{(1+(ky)^2)^{3/2}}, \\
\dot{r} &= \frac{(m_{11}-m_{22})u}{m_{33}}v - \frac{d_{33}}{m_{33}}r - \sum_{i\geq 2}\frac{d_{ri}}{m_{33}}|r|^{i-1}r + \frac{1}{m_{33}}\tau_r + \frac{1}{m_{33}}\tau_{wr}(t).
\end{aligned} \tag{9.3}$$

Therefore the problem of stabilizing (9.1) at the origin becomes that of stabilizing (9.3) at the origin. The structure of the model (9.3) suggests that we design the control τ_r in two stages by applying the popular backstepping technique. At the first step, we design an intermediate control r_d for r and at the second step the actual control τ_r will be designed to eliminate the error between r_d and r.

Step 1

Define

$$z_2 = r - r_d, \tag{9.4}$$

where r_d is an intermediate control designed as

$$\begin{aligned} r_d = {} & -k_1 z_1 + \frac{k^2 uy}{(1+(ky)^2)^{3/2}} - \frac{kv}{(1+(ky)^2)^{3/2}} - \\ & \frac{ku\,(\sin(z_1) - (\cos(z_1)-1)ky)}{(1+(ky)^2)^{3/2}} - \frac{kv\,((\cos(z_1)-1) + ky\sin(z_1))}{(1+(ky)^2)^{3/2}}, \end{aligned} \tag{9.5}$$

where k_1 is a positive constant to be selected later.

Step 2

With (9.5), the time derivative of (9.4) along the solutions of the last equation of (9.3) is

$$\begin{aligned} \dot{z}_2 = {} & \frac{(m_{11}-m_{22})u}{m_{33}} v - \frac{d_{33}}{m_{33}} r - \sum_{i\geq 2} \frac{d_{ri}}{m_{33}} |r|^{i-1} r + \frac{1}{m_{33}} \tau_r + \frac{1}{m_{33}} \tau_{wr}(t) - \\ & \frac{\partial r_d}{\partial u} \dot{u} - \frac{\partial r_d}{\partial z_1}(-k_1 z_1 + z_2) - \frac{\partial r_d}{\partial y}\left(- \frac{kuy}{\sqrt{1+(ky)^2}} + \frac{v}{\sqrt{1+(ky)^2}} + \right. \\ & \left. \frac{u\,(\sin(z_1) - (\cos(z_1)-1)ky)}{\sqrt{1+(ky)^2}} + \frac{v\,((\cos(z_1)-1) + ky\sin(z_1))}{\sqrt{1+(ky)^2}} \right) - \\ & \frac{\partial r_d}{\partial v}\left(- \frac{m_{11}u}{m_{22}} r - \frac{d_{22}}{m_{22}} v - \sum_{i\geq 2} \frac{d_{vi}}{m_{22}} |v|^{i-1} v + \frac{1}{m_{22}} \tau_{wv}(t) \right), \end{aligned} \tag{9.6}$$

where

$$\begin{aligned} \frac{\partial r_d}{\partial u} &= \frac{k^2 y}{(1+(ky)^2)^{3/2}} - \frac{k}{(1+(ky)^2)^{3/2}} (\sin(z_1) - (\cos(z_1)-1)ky), \\ \frac{\partial r_d}{\partial z_1} &= -k_1 - \frac{ku\,(\cos(z_1) + ky\sin(z_1))}{(1+(ky)^2)^{3/2}} - \frac{kv\,(-\sin(z_1) + ky\cos(z_1))}{(1+(ky)^2)^{3/2}}, \end{aligned}$$

$$\frac{\partial r_d}{\partial y} = \frac{-3k^3 y\left(kuy - u\left(\sin(z_1) - y(\cos(z_1)-1)\right)\right)}{\left(1+(ky)^2\right)^{5/2}} - \frac{v\left((\cos(z_1)-1)+y\sin(z_1)\right)}{\left(1+(ky)^2\right)^{5/2}} + \frac{k^2\left(u\cos(z_1) - v\sin(z_1)\right)}{\left(1+(ky)^2\right)^{3/2}},$$

$$\frac{\partial r_d}{\partial v} = -\frac{k}{\left(1+(ky)^2\right)^{3/2}}\left(\cos(z_1)+\sin(z_1)ky\right). \tag{9.7}$$

We now choose the actual control without canceling the useful damping terms as

$$\begin{aligned}\tau_r = m_{33}\Bigg[&-z_1 - k_2 z_2 - \frac{(m_{11}-m_{22})u}{m_{33}}v + \frac{d_{33}}{m_{33}}r_d + \sum_{i\geq 2}\frac{d_{ri}}{m_{33}}|r|^{i-1} r_d + \\ &\frac{\partial r_d}{\partial u}\dot{u} + \frac{\partial r_d}{\partial z_1}(-k_1 z_1 + z_2) + \frac{\partial r_d}{\partial y}\Bigg(-\frac{kuy}{\sqrt{1+(ky)^2}} + \frac{v}{\sqrt{1+(ky)^2}} + \\ &\frac{u\left(\sin(z_1) - (\cos(z_1)-1)ky\right)}{\sqrt{1+(ky)^2}} + \frac{v\left((\cos(z_1)-1)+ky\sin(z_1)\right)}{\sqrt{1+(ky)^2}}\Bigg) + \\ &\frac{\partial r_d}{\partial v}\left(-\frac{m_{11}u}{m_{22}}r - \frac{d_{22}}{m_{22}}v - \sum_{i\geq 2}\frac{d_{vi}}{m_{22}}|v|^{i-1} v\right) - \\ &\frac{1}{m_{33}}\tau_{wr\max}\tanh\left(\frac{z_2}{\rho_1}\right) - \frac{1}{m_{22}}\tau_{wv\max}\frac{\partial r_d}{\partial v}\tanh\left(\frac{\partial r_d}{\partial v}\frac{z_2}{\rho_2}\right)\Bigg],\end{aligned} \tag{9.8}$$

where k_2, ρ_1, and ρ_2 are positive constants to be chosen later. Substituting (9.4), (9.5), and (9.8) into (9.3) results in the following closed loop system:

$$\begin{aligned}\dot{y} &= -\frac{kuy}{\sqrt{1+(ky)^2}} + \frac{v}{\sqrt{1+(ky)^2}} + \frac{u\left(\sin(z_1) - (\cos(z_1)-1)ky\right)}{\sqrt{1+(ky)^2}} + \\ &\quad \frac{v}{\sqrt{1+(ky)^2}}\left((\cos(z_1)-1)+ky\sin(z_1)\right), \\ \dot{v} &= -\frac{d_{22}}{m_{22}}v - \sum_{i\geq 2}\frac{d_{vi}}{m_{22}}|v|^{i-1} v - \frac{m_{11}u}{m_{22}}\frac{k^2uy - kv}{\left(1+(ky)^2\right)^{3/2}} - \\ &\quad \frac{m_{11}u}{m_{22}}\Bigg(-k_1 z_1 + z_2 - \frac{ku}{\left(1+(ky)^2\right)^{3/2}}\left(\sin(z_1) - (\cos(z_1)-1)ky\right) - \\ &\quad \frac{kv}{\left(1+(ky)^2\right)^{3/2}}\left((\cos(z_1)-1)+ky\sin(z_1)\right)\Bigg) + \frac{1}{m_{22}}\tau_{wv}(t), \\ \dot{z}_1 &= -k_1 z_1 + z_2, \\ \dot{z}_2 &= -z_1 - k_2 z_2 - \frac{d_{33}}{m_{33}}z_2 - \sum_{i\geq 2}\frac{d_{ri}}{m_{33}}|r|^{i-1} z_2 + \frac{1}{m_{33}}\Big(\tau_{wr}(t) - \tau_{wr\max}\times \\ &\quad \tanh\left(\frac{z_2}{\rho_1}\right)\Big) + \frac{1}{m_{22}}\left(-\frac{\partial r_d}{\partial v}\tau_{wv}(t) - \frac{\partial r_d}{\partial v}\tau_{wv\max}\tanh\left(\frac{\partial r_d}{\partial v}\frac{z_2}{\rho_2}\right)\right).\end{aligned} \tag{9.9}$$

9.2.2 Stability Analysis

The following two lemmas will be used extensively in stability analysis.

Lemma 9.1. *Consider the following nonlinear system:*

$$\dot{x} = f(t,x) + g(t,x,\xi(t)), \tag{9.10}$$

where $x \in \mathbb{R}^n$, $\xi(t) \in \mathbb{R}^m$, $f(t,x)$ is piecewise continuous in t and locally Lipschitz in x. If there exist positive constants c_i, $1 \le i \le 4$, λ_j, $1 \le j \le 2$, σ_0, ε_0, μ_0, c_0, and a class-K function α_0 such that the following conditions are satisfied:

C1. There exists a proper function $V(t,x)$ satisfying:

$$\begin{aligned}
&c_1 \|x\|^2 \le V(t,x) \le c_2 \|x\|^2, \\
&\left\| \frac{\partial V}{\partial x}(t,x) \right\| \le c_3 \|x\|, \\
&\frac{\partial V}{\partial t} + \frac{\partial V}{\partial x} f(t,x) \le -c_4 \|x\|^2 + c_0.
\end{aligned}$$

C2. The vector function $g(t,x,\xi(t))$ satisfies:

$$\|g(t,x,\xi(t))\| \le (\lambda_1 + \lambda_2 \|x\|) \|\xi(t)\|.$$

C3. $\xi(t)$ globally exponentially converges to a ball centered at the origin:

$$\|\xi(t)\| \le \alpha_0 (\|\xi(t_0)\|) e^{-\sigma_0(t-t_0)} + \varepsilon_0, \ \forall t \ge t_0 \ge 0.$$

C4. The following gain condition is satisfied:

$$c_4 - \lambda_2 c_3 \varepsilon_0 - \frac{\lambda_1 c_3 \varepsilon_0}{4\mu_0} > 0.$$

Then the solution $x(t)$ of (9.10) globally exponentially converges to a ball centered at the origin, i.e.,

$$\|x(t)\| \le \alpha (\|(x(t_0), \xi(t_0))\|) e^{-\sigma(t-t_0)} + \varepsilon, \ \forall t \ge t_0 \ge 0, \tag{9.11}$$

where $\varepsilon = \sqrt{a_4/c_1 a_1}$ and

if $a_1 = \sigma_0$ then

$$\begin{aligned}
\alpha(s) &= \sqrt{\frac{e^{\frac{a_2(s)}{\sigma_0}}}{c_1} \left(c_2 s^2 + \left(a_3(s) + a_1^{-1} a_2(s) a_4\right) \phi\right)} \\
\sigma &= 0.5(a_1 - d);
\end{aligned}$$

$$\begin{aligned}
&\textit{if } a_1 \neq \sigma_0 \textit{ then} \\
&\quad \alpha(s) = \sqrt{\frac{e^{\frac{a_2(s)}{\sigma_0}}}{c_1}\left(c_2 s^2 + \frac{a_1 a_3(s) + a_2(s) a_4}{a_1 |a_1 - \sigma_0|}\right)} \\
&\quad \sigma = 0.5 \min(a_1, |a_1 - \sigma_0|);
\end{aligned}$$

with

$$\begin{aligned}
a_1 &= \frac{1}{c_2}\left(c_4 - \lambda_2 c_3 \varepsilon_0 - \frac{\lambda_1 c_3 \varepsilon_0}{4\mu_0}\right), \\
a_2(s) &= \frac{c_3}{c_1}(\lambda_1 + \lambda_2)\alpha_0(s), \\
a_3(s) &= \frac{\lambda_1 c_3}{4}\alpha_0(s), \\
a_4 &= c_0 + \lambda_1 c_3 \varepsilon_0 \mu_0, \\
0 &< d < a_1, \quad \phi \geq (t - t_0)e^{-d(t-t_0)}, \quad \forall t \geq t_0 \geq 0, s \geq 0.
\end{aligned}$$

When $c_0 = 0$ and $\varepsilon_0 = 0$, we have $\varepsilon = 0$ and the system (9.10) is globally K-exponentially stable. Note that a finite value of the constant ϕ exists for an arbitrarily small positive d.

Proof. From conditions C1, C2, and C3, we have

$$\begin{aligned}
\dot{V} &= \frac{\partial V}{\partial t} + \frac{\partial V}{\partial x} f(t,x) + \frac{\partial V}{\partial x} g(t,x,\xi(t)) \\
&\leq -(c_4 - \lambda_2 c_3 \varepsilon_0 - \lambda_1 c_3 \varepsilon_0 / 4\mu_0) \|x\|^2 + \\
&\quad c_3 \|x\| (\lambda_1 + \lambda_2 \|x\|) \alpha_0 (\|\xi(t_0)\|) e^{-\sigma_0(t-t_0)} + \\
&\quad \lambda_1 c_3 \varepsilon_0 \mu_0 + c_0.
\end{aligned} \tag{9.12}$$

Upon application of the completing square, (9.12) can be rewritten as

$$\dot{V} \leq -\left(a_1 - a_2 e^{-\sigma_0(t-t_0)}\right) V + a_3 e^{-\sigma_0(t-t_0)} + a_4. \tag{9.13}$$

Solving the above differential inequality results in

$$V(t) \leq V(t_0) e^{\frac{a_2}{\sigma_0}} e^{-a_1(t-t_0)} + \left(a_3 + \frac{a_2 a_4}{a_1}\right) e^{\frac{a_2}{\sigma_0}} e^{-a_1 t + \sigma_0 t_0} \int_{t_0}^{t} e^{(a_1 - \sigma_0)\tau} d\tau + \frac{a_4}{a_1}, \tag{9.14}$$

which yields (9.11) readily. □

Lemma 9.2. *Consider the following nonlinear system:*

$$\dot{x} = f(t,x) + g(t,x,\xi(t)), \tag{9.15}$$

where $x_1 \in \mathbb{R}^{n_1}$, $x_2 \in \mathbb{R}^{n_2}$, $x = [x_1\ x_2]^T \in \mathbb{R}^{n_1+n_2}$, $\xi(t) \in \mathbb{R}^m$, $f(t,x)$ *is piecewise continuous in* t *and locally Lipschitz in* x. *If there exist positive constants* c_0, c_1, c_2, c_{31}, c_{32}, λ_i, $0 \le i \le 2$, σ_0, ε_0, c_0, *and a class-K function* α_0 *such that the following conditions are satisfied.*

C1. There exists a proper function $V(t,x)$ *such that:*

$$c_1 \|x\|^2 \le V(t,x) \le c_2 \|x\|^2,$$
$$\frac{\partial V}{\partial t} + \frac{\partial V}{\partial x} f(t,x) \le -c_{31} \|x_1\|^2 - \frac{c_{32} \|x_2\|^2}{\sqrt{1+c_4 \|x_2\|^2}} + c_0,$$

$$\left\| \frac{\partial V}{\partial x} g(t,x,\xi(t)) \right\| \le \left(\lambda_0 + \lambda_1 \|x_1\|^2 + \frac{\lambda_2 \|x_2\|^2}{\sqrt{1+c_4 \|x_2\|^2}} \right) \|\xi(t)\|.$$

C2. $\xi(t)$ *globally exponentially converges to a ball centered at the origin:*

$$\|\xi(t)\| \le \alpha_0 (\|\xi(t_0)\|) e^{-\sigma_0 (t-t_0)} + \varepsilon_0, \ \forall t \ge t_0 \ge 0.$$

C3. The following gain conditions are satisfied:

$$c_{31} - \lambda_1 \varepsilon_0 > 0 \quad \text{and} \quad c_{32} - \lambda_2 \varepsilon_0 > 0.$$

C4. $x_2(t)$ *is bounded:*

$$\|x_2(t)\| \le \varpi,$$

where ϖ *is a nondecreasing function of* $\|(x(t_0), \xi(t_0))\|$,

then the solution $x(t)$ *of (9.15) globally asymptotically converges to a ball centered at the origin, i.e.,*

$$\|x(t)\| \le \alpha (\|(x(t_0), \xi(t_0))\|) e^{-\sigma(\|(x(t_0),\xi(t_0))\|)(t-t_0)} + \varepsilon(s), \ \forall t \ge t_0 \ge 0, \quad (9.16)$$

where $\varepsilon(s) = \sqrt{\frac{a_4}{c_1 a_1(s)}}$ *and*

if $a_1(s) = \sigma_0$ *then*

$$\alpha(s) = \sqrt{c_1^{-1} e^{a_2(s)/\sigma_0} \left(c_2 s^2 + \left(a_3(s) + a_1^{-1}(s) a_2(s) a_4\right) \phi\right)}$$
$$\sigma(s) = 0.5(a_1(s) - d);$$

if $a_1(s) \ne \sigma_0$ *then*

$$\alpha(s) = \sqrt{c_1^{-1} e^{a_2(s)/\sigma_0} \left(c_2 s^2 + \frac{a_1(s) a_3(s) + a_2(s) a_4}{a_1(s) |a_1(s) - \sigma_0|} \right)}$$
$$\sigma(s) = 0.5 \min (a_1(s), |a_1(s) - \sigma_0|);$$

with

$$
\begin{aligned}
&a_1(s) = \frac{1}{c_2}\min\left(c_{31} - \lambda_1\varepsilon_0, \frac{c_{32} - \lambda_2\varepsilon_0}{\sqrt{1 + c_4\varpi^2(s)}}\right),\\
&a_2(s) = \frac{1}{c_1}\max(\lambda_1, \lambda_2)\alpha_0(s),\\
&a_3(s) = \lambda_0\alpha_0(s),\\
&a_4 = c_0 + \lambda_0\varepsilon_0,\\
&0 < d < a_1(s),\ \phi \geq (t - t_0)e^{-d(t - t_0)}, \quad \forall t \geq t_0 \geq 0,\ s \geq 0.
\end{aligned}
$$

When $c_0 = 0$ and $\varepsilon_0 = 0$, we have $\varepsilon = 0$ and the system (9.15) is GAS. Note that a finite value of the constant ϕ exists for an arbitrarily small positive d.

Proof. The proof of this lemma is similar to that of Lemma 9.1. □

Remark 9.1. It is important to note that the rate $\sigma > 0$ in (9.16) and a_1 depend on the initial conditions. In addition, around the origin, both σ and a_1 are bounded below from zero.

We first need to show that the closed loop system (9.9) is forward complete. It is straightforward to show that the derivative of the function $V_0 = z_1^2 + z_2^2 + v^2 + y^2$ along the solutions of the closed loop system (9.9) satisfies $\dot{V}_0 \leq a_0 V_0 + b_0$ where a_0 and b_0 are nonnegative constants. The inequality $\dot{V}_0 \leq a_0 V_0 + b_0$ implies that the closed loop system (9.9) is forward complete. We now apply Lemmas 9.1 and 9.2 to analyze the closed loop system (9.9). We view (z_1, z_2) as $\xi(t)$, v as x in Lemma 9.1, and (v, y) as x in Lemma 9.2. Hence it is necessary to verify all the conditions of Lemmas 9.1 and 9.2.

(z_1, z_2)-subsystem

We take the following quadratic function:

$$
V_1 = \frac{1}{2}(z_1^2 + z_2^2), \tag{9.17}
$$

whose time derivative along the solutions of the last two equations of (9.9) satisfies

$$
\begin{aligned}
\dot{V}_1 = &-k_1 z_1^2 - k_2 z_2^2 - \frac{d_{33}}{m_{33}} z_2^2 - \sum_{i \geq 2} \frac{d_{ri}}{m_{33}} |r|^{i-1} z_2^2 + \frac{z_2}{m_{33}} (\tau_{wr}(t) - \tau_{wr\max} \times \\
&\tanh\left(\frac{z_2}{\rho_1}\right)) + \frac{z_2}{m_{22}}\left(-\frac{\partial r_d}{\partial v}\tau_{wv}(t) - \frac{\partial r_d}{\partial v}\tau_{wv\max}\tanh\left(\frac{\partial r_d}{\partial v}\frac{z_2}{\rho_2}\right)\right)
\end{aligned}
$$

$$
\begin{aligned}
&\leq -k_1 z_1^2 - k_2 z_2^2 + \frac{\tau_{wr\max}}{m_{33}} \times \\
&\left(|z_2| - z_2 \tanh\left(\frac{z_2}{\rho_1}\right)\right) + \frac{\tau_{wv\max}}{m_{22}} \left(\left|\frac{\partial r_d}{\partial v} z_2\right| - \frac{\partial r_d}{\partial v} z_2 \tanh\left(\frac{\partial r_d}{\partial v}\frac{z_2}{\rho_2}\right)\right) \\
&\leq -k_1 z_1^2 - k_2 z_2^2 + 0.2785\left(\frac{1}{m_{33}}\tau_{wr\max}\rho_1 + \frac{1}{m_{22}}\tau_{wv\max}\rho_2\right),
\end{aligned}
\tag{9.18}
$$

where we have used $|x| - x\tanh(x/\lambda) \leq 0.2785\lambda, \;\; \forall x \in \mathbb{R}$ and $\lambda > 0$. From (9.17) and (9.18), it can be shown that

$$
\|z(t)\| \leq \|z(t_0)\| e^{-\sigma_0 (t-t_0)} + \varepsilon_0 \forall t \geq t_0 \geq 0\,, \tag{9.19}
$$

where $z = [z_1 \; z_2]^T$ and

$$
\begin{aligned}
\sigma_0 &= \min(k_1, k_2), \\
\varepsilon_0 &= \sqrt{\frac{0.2785\left(\tau_{wr\max}\rho_1/m_{33} + \tau_{wv\max}\rho_2/m_{22}\right)}{\sigma_0}}.
\end{aligned}
\tag{9.20}
$$

Therefore the (z_1, z_2)-subsystem is globally ultimately stable at the origin. Furthermore, (9.19) implies that $\xi(t) := (z_1, z_2)^T$ globally exponentially converges to a ball centered at the origin. The radius of this ball can be made arbitrarily small by increasing k_1 and k_2 and/or reducing ρ_1 and ρ_2.

Boundedness of v

To prove that v is bounded, we consider the second equation of (9.9). In order to apply Lemma 9.1, define $x = v$, $\xi(t) = [z_1 z_2]^T$ and consider y as a function of time t,

$$
\begin{aligned}
f(\bullet) &= -\frac{d_{22}}{m_{22}} v - \sum_{i \geq 2} \frac{d_{vi}}{m_{22}} |v|^{i-1} v - \\
&\quad \frac{m_{11}u}{m_{22}} \left(\frac{k^2 u y}{(1+(ky)^2)^{3/2}} - \frac{kv}{(1+(ky)^2)^{3/2}}\right) + \frac{1}{m_{22}} \tau_{wv}(t), \\
g(\bullet) &= -\frac{m_{11}u}{m_{22}} (-k_1 z_1 + z_2 - \\
&\quad \left.\frac{k\left(u\left(\sin(z_1) - (\cos(z_1)-1)ky\right) + v\left((\cos(z_1)-1) + ky\sin(z_1)\right)\right)}{(1+(ky)^2)^{3/2}}\right).
\end{aligned}
\tag{9.21}
$$

This abuse of notation is introduced for simplicity and is possible because:

$$
0 \leq \frac{1}{(1+(ky)^2)^{3/2}} \leq 1, \; 0 \leq \left|\frac{ky}{(1+(ky)^2)^{3/2}}\right| < 1, \; \forall \; y \in \mathbb{R},
$$

and we have shown that the closed loop system is forward complete. We now verify all of the conditions of Lemma 9.1.

Verifying Condition C1. We take the function $V_2 = 0.5v^2$ whose time derivative along the solutions of the differential equation $\dot{v} = f(t, v)$, see (9.21), satisfies

$$\begin{aligned}\dot{V}_2 &= -\frac{d_{22}}{m_{22}}v^2 - \sum_{i\geq 2}\frac{d_{vi}}{m_{22}}|v|^{i-1}v^2 - \frac{m_{11}uv}{m_{22}}\frac{k^2uy - kv}{(1+(ky)^2)^{3/2}} + \frac{v}{m_{22}}\tau_{wv}(t)\\ &\leq -\left(\frac{d_{22}}{m_{22}} - \frac{m_{11}ku_{\max}}{m_{22}} - \frac{m_{11}ku^2_{\max}\mu_1}{m_{22}} - \frac{\mu_1}{m_{22}}\right)v^2 + \frac{m_{11}ku^2_{\max} + \tau^2_{wv\max}}{4\mu_1 m_{22}}.\end{aligned} \tag{9.22}$$

Hence, the condition C1 is satisfied with

$$\begin{aligned}c_0 &= \frac{1}{4\mu_1 m_{22}}\left(m_{11}ku^2_{\max} + \tau^2_{wv\max}\right),\ c_1 = c_2 = 0.5,\ c_3 = 1,\\ c_4 &= \frac{d_{22}}{m_{22}} - \frac{m_{11}ku^2_{\max}}{m_{22}}\mu_1 - \frac{m_{11}ku_{\max}}{m_{22}} - \frac{\mu_1}{m_{22}},\end{aligned} \tag{9.23}$$

where $\mu_1 > 0$ and $k > 0$ are chosen such that $c_4 > 0$.

Verifying Condition C2. It is directly shown from (9.21) that

$$|g(t, v, z(t))| \leq (\lambda_1 + \lambda_2|v|)\,\|z(t)\|, \tag{9.24}$$

where

$$\lambda_1 = \frac{m_{11}u_{\max}}{m_{22}}(1 + k_1 + 2ku_{\max}),\ \lambda_2 = \frac{2km_{11}u_{\max}}{m_{22}}. \tag{9.25}$$

Verifying Condition C3. This condition follows directly from (9.19).

Verifying Condition C4. It can be shown from (9.20), (9.23), and (9.25) that we can find positive constant k such that the condition C4 is satisfied, i.e.,

$$c_4 - \lambda_2 c_3\varepsilon_0 - \frac{\lambda_1 c_3 \varepsilon_0}{4\mu_0} > 0. \tag{9.26}$$

All of the conditions of Lemma 9.1 have been verified, hence the sway velocity is bounded and satisfies

$$|v(t)| \leq \alpha_1\left(\|(v(t_0), z(t_0))\|\right)e^{\sigma_1(t-t_0)} + \varepsilon_1, \tag{9.27}$$

where ε_1, σ_1, and α_1 are calculated as in Lemma 9.1, and the constants c_i, $1 \leq i \leq 4$, λ_j, $1 \leq j \leq 2$, σ_0, ε_0, μ_0, and c_0 are given in (9.20), (9.23), and (9.26).

(v, y)-subsystem

In this section, we will apply Lemma 9.2 to prove global ultimate boundedness of the (v, y)-subsystem. It can be seen that the first two equations of (9.9) are in the form of the system in Lemma 9.2 with $x_1 = v$, $x_2 = y$, $\xi(t) = z(t)$, and

$$f(\bullet) = \begin{bmatrix} -\dfrac{d_{22}}{m_{22}}v - \sum\limits_{i\geq 2} \dfrac{d_{vi}}{m_{22}} |v|^{i-1} v - \dfrac{m_{11}u}{m_{22}} \dfrac{k^2uy - kv}{(1+(ky)^2)^{3/2}} + \dfrac{1}{m_{22}}\tau_{wv}(t) \\ -\dfrac{kuy}{\sqrt{1+(ky)^2}} + \dfrac{v}{\sqrt{1+(ky)^2}} \end{bmatrix},$$

$$g(\bullet) = \begin{bmatrix} \dfrac{m_{11}u}{m_{22}} \left(k \dfrac{\sin(z_1)(u+kvy) + (\cos(z_1)-1)(v-kuy)}{(1+(ky)^2)^{3/2}} + k_1 z_1 - z_2 \right) \\ \dfrac{u\left(\sin(z_1) - (\cos(z_1)-1)ky\right)}{\sqrt{1+(ky)^2}} + \dfrac{v\left((\cos(z_1)-1) + \sin(z_1)ky\right)}{\sqrt{1+(ky)^2}} \end{bmatrix}. \tag{9.28}$$

We now need to verify all of the conditions of Lemma 9.2.

Verifying Condition C1. To verify this condition, we take the function $V_3 = 0.5(v^2 + y^2)$. It can be directly shown that this function satisfies condition C1 with $|v(t)| \leq \alpha_1 \left(\|(v(t_0), z(t_0))\|\right) e^{-\sigma_1(t-t_0)} + \varepsilon_1$ and

$$\begin{aligned} c_0 &= \frac{\tau_{wv\max}^2}{4\mu_3 m_{22}}, c_1 = c_2 = 0.5, \\ c_{31} &= \frac{d_{22}}{m_{22}} - \frac{m_{11}ku_{\max}}{m_{22}} - \mu_2 - \frac{m_{11}k^2u_{\max}^2}{m_{22}}\mu_2 - \frac{\mu_3}{m_{22}}, \\ c_{32} &= ku_{\min} - \frac{1}{4\mu_2}\left(1 + \frac{m_{11}k^2u_{\max}^2}{m_{22}}\right), \\ \lambda_0 &= \frac{m_{11}u_{\max}}{4\mu_4 m_{22}}(1 + k_1 + 2ku_{\max}) + \frac{u_{\max}}{4\mu_4}, \\ \lambda_1 &= \frac{m_{11}u_{\max}}{m_{22}}(2ku_{\max} + k_1 + 1)\mu_4 + \frac{2km_{11}u_{\max}}{m_{22}} + \mu_4, \\ \lambda_2 &= \frac{1}{4\mu_4} + u_{\max}\mu_4 + ku_{\max} + k(\alpha_1 + \varepsilon_1), \end{aligned} \tag{9.29}$$

where $k > 0$ and $\mu_2 > 0$ are chosen such that $c_{31} > 0$ and $c_{32} > 0$.

Verifying Condition C2. This condition follows directly from (9.19).

Verifying Condition C3. It can be shown that there exists a positive constant k such that the condition C3 satisfies, i.e.,

$$c_{31} - \lambda_1 \varepsilon_0 > 0, \; c_{32} - \lambda_2 \varepsilon_0 > 0. \tag{9.30}$$

Verifying Condition C4. From the boundedness of the sway velocity $v(t)$ proven in the previous subsection and noting that ε_0 in (9.19) can be made arbitrarily small, it is shown that there exists a nondecreasing function ϖ of $\|((v(t_0), y(t_0)), z(t_0))\|$ such that $|y(t)| \leq \varpi$ by applying Lemma 9.1 to the first equation of (9.9) with the Lyapunov function $V_y = 0.5y^2$.

All of the conditions of Lemma 9.2 have been verified. Therefore we have

$$\|(v(t), y(t))\| \leq \alpha_2 (\|((v(t_0), y(t_0)), z(t_0))\|) e^{-\sigma_2 (t-t_0)} + \varepsilon_2, \; \forall t \geq t_0 \geq 0, \tag{9.31}$$

where ε_2, σ_2, and α_2 are calculated as in Lemma 9.2, and all other constants given in (9.29). It can be seen that when there are no environmental disturbances, since $\varepsilon_2 = 0$, $(v(t), y(t))$ globally asymptotically converges to zero. We have thus proven the first main result of this chapter.

Theorem 9.1. *The full-state feedback control problem stated in Section 9.1 is solved by the control law (9.8) as long as the design constants k, k_1, and k_2 are chosen such that (9.26) and (9.30) hold.*

9.3 Output Feedback

This section is devoted to the development of an output feedback controller to fulfill the output feedback control objective. A nonlinear observer is first designed so that it globally exponentially drives the observer error dynamics to a ball centered at the origin. When there are no environmental disturbances, the observer error dynamics are GES at the origin. A controller is then designed based on the approach in the preceding section and the proposed observer. Before designing an observer and output feedback controller, we impose the following assumption, see [12].

Assumption 9.1. *For the ship model (9.1), the matrix*

$$\boldsymbol{K}_2 = \begin{bmatrix} -\dfrac{d_{22}}{m_{22}} & -\dfrac{m_{11}u}{m_{22}} \\ \dfrac{(m_{11}-m_{22})u}{m_{33}} & -\dfrac{d_{33}}{m_{33}} \end{bmatrix} \tag{9.32}$$

is Hurwitz.

The above assumption implies that the ship (when the nonlinear damping terms $\sum\limits_{i\geq 2} \frac{d_{vi}}{m_{33}} |v|^{i-1} v$ and $\sum\limits_{i\geq 2} \frac{d_{ri}}{m_{33}} |r|^{i-1} r$ are ignored) is dynamic stable in straight-line motion. Straight-line stability physically implies that a new path of the ship will be a straight line after an action in yaw. The direction of the new path will usually be different from that of the initial path, as mentioned in [12]. On the other hand, unstable ships will go into a starboard or port turn without any rudder deflection. We impose Assumption 9.1 to make our observer design possible. Note that this assumption does not hold for several types of surface ships such as large tankers and high-speed crafts with sufficiently small ratios d_{22}/m_{22} and d_{33}/m_{33}, and the added mass in the sway axis sufficiently larger than the added mass in the surge axis. Consequently, for these ships the real part of at least one of the eigenvalues of the matrix $\boldsymbol{K_2}$ is positive.

9.3.1 Observer Design

The ship dynamics (9.1) represent some difficulties for output feedback control design. These difficulties are mainly due to the nonlinear terms $\sum_{i\geq 2} \frac{d_{vi}}{m_{22}} |v|^{i-1} v$ and $\sum_{i\geq 2} \frac{d_{ri}}{m_{33}} |r|^{i-1} r$, the nonlinear kinematic term $\cos(\psi)$, and the underactuated situation. However we first observe that the nonlinear terms are monotonic, i.e., they satisfy

$$
\begin{aligned}
&(v_1 - v_2)\left(\sum_{i\geq 2} \frac{d_{vi}}{m_{22}} |v_1|^{i-1} v_1 - \sum_{i\geq 2} \frac{d_{vi}}{m_{22}} |v_2|^{i-1} v_2\right) \geq 0, \forall\ v_1 \in \mathbb{R}, v_2 \in \mathbb{R},\\
&(r_1 - r_2)\left(\sum_{i\geq 2} \frac{d_{ri}}{m_{33}} |r_1|^{i-1} r_1 - \sum_{i\geq 2} \frac{d_{ri}}{m_{33}} |r_2|^{i-1} r_2\right) \geq 0, \forall r_1 \in \mathbb{R}, r_2 \in \mathbb{R}.
\end{aligned}
\tag{9.33}
$$

Based on the structure of the underatuated ship dynamics (9.1) and property (9.33), we propose the following nonlinear observer:

$$
\begin{aligned}
\dot{\hat{y}} &= u\sin(\psi) + \cos(\psi)\hat{v} + k_{11}(y - \hat{y}) + k_{12}(\psi - \hat{\psi}),\\
\dot{\hat{\psi}} &= \hat{r} + k_{21}(y - \hat{y}) + k_{22}(\psi - \hat{\psi}),\\
\dot{\hat{v}} &= -\frac{m_{11}u}{m_{22}}\hat{r} - \frac{d_{22}}{m_{22}}\hat{v} - \sum_{i\geq 2} \frac{d_{vi}}{m_{22}} |\hat{v}|^{i-1} \hat{v} + k_{31}(y - \hat{y}) +\\
&\quad (k_{13} + \cos(\psi))(y - \hat{y}),\\
\dot{\hat{r}} &= \frac{(m_{11} - m_{22})u}{m_{33}}\hat{v} - \frac{d_{33}}{m_{33}}\hat{r} - \sum_{i\geq 2} \frac{d_{ri}}{m_{33}} |\hat{r}|^{i-1} \hat{r} + \frac{1}{m_{33}}\tau_r +\\
&\quad k_{42}(\psi - \hat{\psi}) + (k_{24} + 1)(\psi - \hat{\psi}),
\end{aligned}
\tag{9.34}
$$

where $\hat{y}, \hat{\psi}, \hat{v}$, and $\hat{r}$ are the estimate of y, ψ, v and r respectively. All the constants $k_{11}, k_{12}, k_{21}, k_{22}, k_{13}, k_{31}$, and k_{42} will be chosen later.

By defining the observer errors as $\tilde{y} = y - \hat{y}$, $\tilde{\psi} = \psi - \hat{\psi}$, $\tilde{\upsilon} = \upsilon - \hat{\upsilon}$, and $\tilde{r} = r - \hat{r}$, the observer error dynamics can be rewritten as

$$\begin{aligned}
\dot{\tilde{y}} &= -k_{11}\tilde{y} - k_{12}\tilde{\psi} - k_{13}\tilde{\upsilon} + (k_{13} + \cos(\psi))\tilde{\upsilon},\\
\dot{\tilde{\psi}} &= -k_{21}\tilde{y} - k_{22}\tilde{\psi} - k_{24}\tilde{r} + (k_{24}+1)\tilde{r},\\
\dot{\tilde{\upsilon}} &= -k_{31}\tilde{y} - \frac{d_{22}}{m_{22}}\tilde{\upsilon} - \sum_{i\geq 2}\frac{d_{\upsilon i}}{m_{22}}\left(|\upsilon|^{i-1}\upsilon - |\hat{\upsilon}|^{i-1}\hat{\upsilon}\right) - \frac{m_{11}u}{m_{22}}\tilde{r} -\\
&\quad (k_{13} + \cos(\psi))\tilde{y} + \frac{1}{m_{22}}\tau_{w\upsilon}(t),\\
\dot{\tilde{r}} &= -k_{42}\tilde{\psi} - (k_{24}+1)\tilde{\psi} + \frac{(m_{11}-m_{22})u}{m_{33}}\tilde{\upsilon} - \frac{d_{33}}{m_{33}}\tilde{r} -\\
&\quad \sum_{i\geq 2}\frac{d_{ri}}{m_{33}}\left(|r|^{i-1}r - |\hat{r}|^{i-1}\hat{r}\right) + \frac{1}{m_{33}}\tau_{wr}(t).
\end{aligned} \tag{9.35}$$

We now show that there exist suitable observer gains k_{11}, k_{12}, k_{13}, k_{21}, k_{22}, k_{24}, k_{31}, and k_{42} such that the observer error dynamics (9.35) is globally ultimately stable. Consider the Lyapunov function

$$V_{\text{obs}} = \frac{1}{2}\tilde{x}^T\tilde{x} \tag{9.36}$$

where $\tilde{x} = \left[\,\tilde{y}\ \tilde{\psi}\ \tilde{\upsilon}\ \tilde{r}\,\right]^T$. The time derivative of (9.36) along the solutions of (9.35) and property (9.33) results in

$$\dot{V}_{\text{obs}} \leq -p_0\|\tilde{x}\|^2 + q_0, \tag{9.37}$$

where

$$\begin{aligned}
p_0 &= -\lambda_{\max}(A) - \max\left(\frac{\tau_{w\upsilon\max}}{m_{22}}, \frac{\tau_{wr\max}}{m_{33}}\right)\frac{1}{4\mu_0},\\
q_0 &= \max\left(\frac{\tau_{w\upsilon\max}}{m_{22}}, \frac{\tau_{wr\max}}{m_{33}}\right)\mu_0,\ \mu_0 > 0,
\end{aligned}$$

$$A = \begin{bmatrix} -k_{11} & -k_{12} & -k_{13} & 0\\ -k_{21} & -k_{22} & 0 & -k_{24}\\ -k_{31} & 0 & -\frac{d_{22}}{m_{22}} & -\frac{m_{11}u}{m_{22}}\\ 0 & -k_{42} & \frac{(m_{11}-m_{22})u}{m_{33}} & -\frac{d_{33}}{m_{33}} \end{bmatrix}.$$

The above matrix A is made negative definite by choosing

$$
\begin{aligned}
&k_{13} = k_{24} = k_{31} = k_{42}, \\
&K_1 := \begin{bmatrix} -k_{11} & -k_{12} \\ -k_{21} & -k_{22} \end{bmatrix} < 0, \\
&K_2 - K_{12} K_1^{-1} K_{12} < 0,
\end{aligned}
\tag{9.38}
$$

where

$$
K_{12} = \begin{bmatrix} -k_{13} & 0 \\ 0 & -k_{24} \end{bmatrix}, \tag{9.39}
$$

and K_2 is defined in (9.32). Here are details of choosing the observer gains such that (9.38) holds. The condition (9.38) is expanded as

$$
\begin{aligned}
&\begin{bmatrix} -k_{11} & -k_{12} \\ -k_{21} & -k_{22} \end{bmatrix} < 0, \\
&\begin{bmatrix} -\dfrac{d_{22}}{m_{22}} + \dfrac{k_{13}^2 k_{22}}{k_{11}k_{22} - k_{12}k_{21}} & -\dfrac{m_{11}u}{m_{22}} - \dfrac{k_{13}k_{12}k_{24}}{k_{11}k_{22} - k_{12}k_{21}} \\ \dfrac{(m_{11} - m_{22})u}{m_{33}} - \dfrac{k_{24}k_{21}k_{13}}{k_{11}k_{22} - k_{12}k_{21}} & -\dfrac{d_{33}}{m_{33}} + \dfrac{k_{24}^2 k_{11}}{k_{11}k_{22} - k_{12}k_{21}} \end{bmatrix} < 0.
\end{aligned}
\tag{9.40}
$$

From (9.40), it suffices that

$$
\begin{aligned}
&\frac{d_{22}}{m_{22}} - \frac{k_{13}^2 k_{22}}{k_{11}k_{22} - k_{12}k_{21}} > 0, \\
&\frac{d_{33}}{m_{33}} - \frac{k_{24}^2 k_{11}}{k_{11}k_{22} - k_{12}k_{21}} > 0, \\
&\frac{m_{11}u}{m_{22}} = -\frac{k_{13}k_{12}k_{24}}{k_{11}k_{22} - k_{12}k_{21}}, \\
&\frac{(m_{11} - m_{22})u}{m_{33}} = \frac{k_{13}k_{21}k_{24}}{k_{11}k_{22} - k_{12}k_{21}}, \\
&k_{11} > 0,\ k_{22} > 0, \\
&k_{11}k_{22} - k_{12}k_{21} > 0.
\end{aligned}
\tag{9.41}
$$

For simplicity, we choose

$$
\begin{aligned}
&k_{13} = k_{24} = \rho\sqrt{u}, \\
&k_{11}k_{22} - k_{12}k_{21} = \rho,
\end{aligned}
\tag{9.42}
$$

where $\rho > 0$ is to be selected later. Substituting (9.42) into (9.41) yields

$$
0 < k_{22} < \frac{d_{22}}{\rho u_{\max} m_{22}},\ 0 < k_{11} < \frac{d_{33}}{\rho u_{\max} m_{33}},
$$

$$k_{12} = -\frac{m_{11}}{\rho m_{22}},\; k_{21} = \frac{(m_{11}-m_{22})}{\rho m_{33}},$$
$$k_{11}k_{22} > -\frac{m_{11}}{\rho^2 m_{22}}\frac{(m_{11}-m_{22})}{m_{33}}. \tag{9.43}$$

Hence, under Assumption 9.1, we can always pick a suitable constant $\rho > 0$ such that (9.43) holds. In summary, the observer gains $k_{11}, k_{12}, k_{13}, k_{21}, k_{22}, k_{24}, k_{31}$, and k_{42} are chosen such that (9.42) and (9.43) hold.

We choose A and μ_0 such that $p_0 > 0$. Hence (9.36) and (9.37) yield

$$\|\tilde{x}(t)\| \leq \|\tilde{x}(t_0)\| e^{-\eta(t-t_0)} + \eta_0, \forall t \geq t_0 \geq 0, \tag{9.44}$$

with $\eta_0 = \sqrt{q_0/p_0}$ and $\eta = p_0$. When there are no environmental disturbances, we have $\eta_0 = 0$. The observer error dynamics (9.35) is thus GES at the origin.

9.3.2 Control Design

We use the coordinate transformation (9.2) to rewrite the ship dynamics (9.3) in conjunction with (9.34) as follows

$$\begin{aligned}
\dot{y} &= -\frac{kuy}{\sqrt{1+(ky)^2}} + \frac{\hat{v}}{\sqrt{1+(ky)^2}} + \frac{u\left(\sin(z_1) - ky(\cos(z_1)-1)\right)}{\sqrt{1+(ky)^2}} + \\
&\quad \frac{\hat{v}\left((\cos(z_1)-1)+\sin(z_1)ky\right)}{\sqrt{1+(ky)^2}} + \frac{\tilde{v}}{\sqrt{1+(ky)^2}}\left(\cos(z_1)+\sin(z_1)ky\right), \\
\dot{z}_1 &= \hat{r} + \tilde{r} - \frac{k^2uy - k\hat{v}}{(1+(ky)^2)^{3/2}} + \frac{ku\left(\sin(z_1) - ky(\cos(z_1)-1)\right)}{(1+(ky)^2)^{3/2}} + \\
&\quad \frac{k\hat{v}\left((\cos(z_1)-1)+\sin(z_1)ky\right)}{(1+(ky)^2)^{3/2}} + \frac{k\tilde{v}\left(\cos(z_1)+\sin(z_1)ky\right)}{(1+(ky)^2)^{3/2}}, \\
\dot{\hat{v}} &= -\frac{m_{11}u}{m_{22}}\hat{r} - \frac{d_{22}}{m_{22}}\hat{v} - \sum_{i\geq 2}\frac{d_{vi}}{m_{22}}|\hat{v}|^{i-1}\hat{v} + (k_{31}+k_{13}+\cos(\psi))\tilde{y}, \\
\dot{\hat{r}} &= \frac{(m_{11}-m_{22})u}{m_{33}}\hat{v} - \frac{d_{33}}{m_{33}}\hat{r} - \sum_{i\geq 2}\frac{d_{ri}}{m_{33}}|\hat{r}|^{i-1}\hat{r} + \frac{1}{m_{33}}\tau_r + (k_{42}+k_{24}+1)\tilde{\psi}.
\end{aligned} \tag{9.45}$$

Similarly to the full state feedback case, we design the control law τ_r in two steps.

Step 1

Define

$$z_2 = \hat{r} - \hat{r}_d, \tag{9.46}$$

where $\hat{r}_d$ is an intermediate control designed as

$$\hat{r}_d = -k_1 z_1 - \frac{ku\left(\sin(z_1) - ky\cos(z_1)\right)}{\left(1+(ky)^2\right)^{3/2}} - \frac{k\hat{v}\left(\cos(z_1) + \sin(z_1)ky\right)}{\left(1+(ky)^2\right)^{3/2}}, \tag{9.47}$$

with k_1 being a positive constant to be selected later.

Step 2

The first time derivative of (9.46) along the solutions of the last equation of (9.45) together with (9.47) is

$$\begin{aligned}
\dot{z}_2 = {} & \frac{(m_{11}-m_{22})u}{m_{33}}\hat{v} - \frac{d_{33}}{m_{33}}\hat{r} - \sum_{i\geq 2}\frac{d_{ri}}{m_{33}}\left|\hat{r}\right|^{i-1}\hat{r} + \frac{1}{m_{33}}\tau_r + \\
& (k_{42}+k_{24}+1)\tilde{\psi} - \frac{\partial \hat{r}_d}{\partial u}\dot{u} - \frac{\partial \hat{r}_d}{\partial z_1}(-k_1 z_1 + z_2) - \\
& \frac{\partial \hat{r}_d}{\partial y}\left(\frac{u\left(\sin(z_1) - ky\cos(z_1)\right)}{\sqrt{1+(ky)^2}} + \frac{\hat{v}\left(\cos(z_1) + \sin(z_1)ky\right)}{\sqrt{1+(ky)^2}}\right) - \\
& \frac{\partial \hat{r}_d}{\partial \hat{v}}\left(-\frac{m_{11}u}{m_{22}}\hat{r} - \frac{d_{22}}{m_{22}}\hat{v} - \sum_{i\geq 2}\frac{d_{vi}}{m_{22}}\left|\hat{v}\right|^{i-1}\hat{v}\right) - \\
& \frac{\partial \hat{r}_d}{\partial z_1}\left(\frac{k}{\left(1+(ky)^2\right)^{3/2}}(\cos(z_1)+\sin(z_1)ky)\tilde{v} + \tilde{r}\right) - \\
& \frac{\partial \hat{r}_d}{\partial y}\frac{1}{\sqrt{1+(ky)^2}}\left(\cos(z_1)+\sin(z_1)ky\right)\tilde{v} - \frac{\partial \hat{r}_d}{\partial \hat{v}}\left(k_{31}+k_{13}+\cos(\psi)\right)\tilde{y},
\end{aligned} \tag{9.48}$$

where

$$\begin{aligned}
\frac{\partial \hat{r}_d}{\partial u} &= -\frac{k\left(-\cos(z_1)ky + \sin(z_1)\right)}{\left(1+(ky)^2\right)^{3/2}}, \quad \frac{\partial \hat{r}_d}{\partial \hat{v}} = -\frac{k\left(\cos(z_1)+\sin(z_1)ky\right)}{\left(1+(ky)^2\right)^{3/2}}, \\
\frac{\partial \hat{r}_d}{\partial z_1} &= -k_1 - \frac{ku\left(\cos(z_1)+ky\sin(z_1)\right)}{\left(1+(ky)^2\right)^{3/2}} - \frac{k\hat{v}\left(-\sin(z_1)+ky\cos(z_1)\right)}{\left(1+(ky)^2\right)^{3/2}}, \\
\frac{\partial \hat{r}_d}{\partial y} &= -\frac{3k^3 y\left(kuy - u\left(\sin(z_1) - (y+\hat{v})(\cos(z_1)-1)\right) - \hat{v}y\sin(z_1)\right)}{\left(1+(ky)^2\right)^{5/2}} + \\
& \quad \frac{k^2\left(u\cos(z_1) - \hat{v}\sin(z_1)\right)}{\left(1+(ky)^2\right)^{3/2}}.
\end{aligned} \tag{9.49}$$

We now choose the actual control without canceling the useful damping terms as

$$\begin{aligned}\tau_r = m_{33}\Bigg(&-z_1 - k_2 z_2 - \frac{(m_{11}-m_{22})u}{m_{33}}\hat{v} + \frac{d_{33}}{m_{33}}\hat{r}_d + \\ &\sum_{i\geq 2}\frac{d_{ri}}{m_{33}}|\hat{r}|^{i-1}\hat{r}_d + \frac{\partial \hat{r}_d}{\partial u}\dot{u} + \frac{\partial \hat{r}_d}{\partial z_1}(-k_1 z_1 + z_2) + \\ &\frac{\partial \hat{r}_d}{\partial y}\left(\frac{u(\sin(z_1) - ky\cos(z_1))}{\sqrt{1+(ky)^2}} + \frac{\hat{v}(\cos(z_1)+\sin(z_1)ky)}{\sqrt{1+(ky)^2}}\right) + \\ &\frac{\partial \hat{r}_d}{\partial \hat{v}}\left(-\frac{m_{11}u}{m_{22}}\hat{r} - \frac{d_{22}}{m_{22}}\hat{v} - \sum_{i\geq 2}\frac{d_{vi}}{m_{22}}|\hat{v}|^{i-1}\hat{v}\right)\Bigg),\end{aligned} \tag{9.50}$$

where k_2 is a positive constant to be chosen later. Substituting (9.46), (9.47), and (9.50) into (9.45) results in the following closed loop system:

$$\begin{aligned}\dot{y} =& -\frac{kuy}{\sqrt{1+(ky)^2}} + \frac{\hat{v}}{\sqrt{1+(ky)^2}} + \frac{u(\sin(z_1) - ky(\cos(z_1)-1))}{\sqrt{1+(ky)^2}} + \\ &\frac{\hat{v}((\cos(z_1)-1)+\sin(z_1)ky)}{\sqrt{1+(ky)^2}} + \frac{\tilde{v}(\cos(z_1)+\sin(z_1)ky)}{\sqrt{1+(ky)^2}}, \\ \dot{\hat{v}} =& -\frac{d_{22}}{m_{22}}\hat{v} - \sum_{i\geq 2}\frac{d_{vi}}{m_{22}}|\hat{v}|^{i-1}\hat{v} - \frac{m_{11}u}{m_{22}}\Bigg(-k_1 z_1 + z_2 - \\ &\frac{ku(\sin(z_1) - ky\cos(z_1))}{(1+(ky)^2)^{3/2}} - \frac{k\hat{v}(\cos(z_1)+\sin(z_1)ky)}{(1+(ky)^2)^{3/2}}\Bigg) + \\ &(k_{31} + k_{13} + \cos(\psi))\tilde{y}, \\ \dot{z}_1 =& -k_1 z_1 + z_2 + \frac{k(\cos(z_1)+\sin(z_1)ky)\tilde{v}}{(1+(ky)^2)^{3/2}} + \tilde{r}, \\ \dot{z}_2 =& -z_1 - k_2 z_2 - \frac{d_{33}}{m_{33}}z_2 - \sum_{i\geq 2}\frac{d_{ri}}{m_{33}}|\hat{r}|^{i-1}z_2 - \\ &\frac{\partial \hat{r}_d}{\partial z_1}\left(\frac{k(\cos(z_1)+\sin(z_1)ky)\tilde{v}}{(1+(ky)^2)^{3/2}} + \tilde{r}\right) + (k_{42} + k_{24} + 1)\tilde{\psi} - \\ &\frac{\partial \hat{r}_d}{\partial \hat{v}}(k_{31} + k_{13} + \cos(\psi))\tilde{y} - \frac{\partial \hat{r}_d}{\partial y}\frac{1(\cos(z_1)+\sin(z_1)ky)\tilde{v}}{\sqrt{1+(ky)^2}}.\end{aligned} \tag{9.51}$$

9.3.3 Stability Analysis

It is not difficult to show that the closed loop system (9.51) is forward complete. We now use Lemmas 9.1 and 9.2 to prove that the closed loop (9.51) is globally ultimately stable. From (9.49), it can be seen that the closed loop (9.51) is different from (9.9) since $\hat{v}$ enters the (z_1, z_2)-subsystem. To remove this obstacle, we first prove that $\hat{\upsilon}$ is bounded. We then prove the convergence of (z_1, z_2) and finally $\hat{\upsilon}$ and y.

Boundedness of $\hat{\upsilon}$

To prove that $\hat{\upsilon}$ is bounded, we view the last three equations of (9.51) as the system studied in Lemma 9.1 with $x_1 = \begin{bmatrix} \hat{\upsilon} & z_1 & z_2 \end{bmatrix}^T$ as x, $\tilde{x}$ as $\xi(t)$ and

$$\dot{x}_1 = f_1(t, x_1) + g_1(t, x_1, \tilde{x}), \tag{9.52}$$

where

$$f_1(t, x_1) = \begin{bmatrix} \Omega_1 \\ -k_1 z_1 + z_2 \\ -z_1 - \left(k_2 + \dfrac{d_{33}}{m_{33}} + \sum_{i \geq 2} \dfrac{d_{ri}}{m_{33}} |\hat{r}|^{i-1} \right) z_2 \end{bmatrix},$$

$$g_1(t, x_1, \tilde{x}) = \begin{bmatrix} (k_{31} + k_{13} + \cos(\psi))\tilde{y} \\ \dfrac{k\tilde{\upsilon}\,(\cos(z_1) + \sin(z_1) k y)}{(1 + (ky)^2)^{3/2}} + \tilde{r} \\ \Omega_2 \end{bmatrix}, \tag{9.53}$$

with

$$\Omega_1 = -\frac{d_{22}}{m_{22}} \hat{\upsilon} - \sum_{i \geq 2} \frac{d_{\upsilon i}}{m_{22}} |\hat{\upsilon}|^{i-1} \hat{\upsilon} - \frac{m_{11} u}{m_{22}} \times$$
$$\left(-k_1 z_1 + z_2 - \frac{k\,(u\,(\sin(z_1) - ky\cos(z_1)) + \hat{\upsilon}\,(\cos(z_1) + \sin(z_1) k y))}{(1 + (ky)^2)^{3/2}} \right),$$
$$\Omega_2 = (k_{42} + k_{24} + 1)\,\tilde{\psi} - \frac{\partial \hat{r}_d}{\partial z_1} \left(\frac{k\,(\cos(z_1) + \sin(z_1) k y)\,\tilde{\upsilon}}{(1 + (ky)^2)^{3/2}} + \tilde{r} \right) -$$
$$\frac{\partial \hat{r}_d}{\partial \hat{\upsilon}} (k_{31} + k_{13} + \cos(\psi))\tilde{y} - \frac{\partial \hat{r}_d}{\partial y} \frac{(\cos(z_1) + \sin(z_1) k y)\,\tilde{\upsilon}}{\sqrt{1 + (ky)^2}},$$

where again with abuse of notation, $\hat{r}$ is considered as a function of time. We now need to verify all of the conditions of Lemma 9.1.

Verifying Condition C1. We take the following Lyapunov function:

$$V_1 = \frac{1}{2}\left(\hat{v}^2 + \delta_1 (z_1^2 + z_2^2)\right), \tag{9.54}$$

where δ_1 is a positive constant. The first time derivative of (9.54) along the solutions of the differential equation $\dot{x}_1 = f_1(t, x_1)$, see (9.52) and (9.53), satisfies

$$\begin{aligned}
\dot{V}_1 &= -\frac{d_{22}}{m_{22}}\hat{v}^2 - \sum_{i\geq 2}\frac{d_{vi}}{m_{22}}|\hat{v}|^{i-1}\hat{v}^2 - \frac{m_{11}u\hat{v}}{m_{22}}(-k_1 z_1 + z_2 - \\
&\quad \left.\frac{k\left(u\left(\sin(z_1) - ky\cos(z_1)\right) + \hat{v}\left(\cos(z_1) + \sin(z_1)ky\right)\right)}{(1+(ky)^2)^{3/2}}\right) - \\
&\quad \delta_1 k_1 z_1^2 - \delta_1 k_2 z_2^2 - \delta_1 \frac{d_{33}}{m_{33}} z_2^2 - \delta_1 \sum_{i\geq 2}\frac{d_{ri}}{m_{33}}|\hat{r}|^{i-1} z_2^2 \\
&\leq -\left(\frac{d_{22}}{m_{22}} - \frac{m_{11}u_{\max}}{m_{22}}(\mu_1 + k_1\mu_1 + 4k + 2k\mu_1 u_{mx})\right)\hat{v}^2 - \\
&\quad \left(\delta_1 k_1 - \frac{m_{11}k_1 u_{\max}}{4\mu_1 m_{22}}\right) z_1^2 - \left(\delta_1 k_2 - \frac{m_{11}u_{\max}}{4\mu_1 m_{22}}\right) z_2^2 + \frac{5m_{11}ku_{\max}^2}{2m_{22}}.
\end{aligned} \tag{9.55}$$

Hence the condition C1 of Lemma 9.1 is verified with

$$\begin{aligned}
c_0 &= \frac{5m_{11}ku_{\max}^2}{2m_{22}}, c_1 = \frac{1}{2}\min(1, \delta_1), \\
c_2 &= \frac{1}{2}\max(1, \delta_1), c_3 = \max(1, \delta_1), \\
c_4 &= \min\left(\left(\frac{d_{22}}{m_{22}} - \frac{m_{11}u_{\max}}{m_{22}}(\mu_1 + k_1\mu_1 + 4k + 2k\mu_1 u_{mx})\right),\right. \\
&\quad \left.\left(\delta_1 k_1 - \frac{m_{11}k_1 u_{\max}}{4\mu_1 m_{22}}\right), \left(\delta_1 k_2 - \frac{m_{11}u_{\max}}{4\mu_1 m_{22}}\right)\right),
\end{aligned} \tag{9.56}$$

where $\mu_1 > 0$ and $k > 0$ are chosen such that $c_4 > 0$.

Verifying Condition C2. To verify this condition of Lemma 9.1, we note from (9.49) that

$$\begin{aligned}
&\left|\frac{\partial \hat{r}_d}{\partial z_1}\right| \leq k_1 + 2k\left(u_{\max} + |\hat{v}|\right), \\
&\left|\frac{\partial \hat{r}_d}{\partial \hat{v}}\right| \leq 2k, \left|\frac{\partial \hat{r}_d}{\partial y}\right| \leq 3k^2\left(u_{\max} + 3ku_{\max} + 3k|\hat{v}|\right).
\end{aligned} \tag{9.57}$$

From (9.55) and (9.57), a simple calculation shows that the condition C2 of Lemma 9.1 is satisfied with

$$\lambda_1 = (2k+1)(k_{31}+k_{13}+1)+k_{42}+k_{24}+1+$$
$$(2k+1)(k_1+2ku_{\max}+1)+6k^2u_{\max}(3k+1),$$
$$\lambda_2 = 2k(2k+1)+18k^3. \tag{9.58}$$

Verifying Condition C3. This condition follows directly from (9.44).

Verifying Condition C4. It can be shown that there exists a positive constant k such that the condition C4 of Lemma 9.1 satisfies

$$c_4 - \lambda_2 c_3 \varepsilon_0 - \frac{\lambda_1 c_3 \varepsilon_0}{4\mu_0} > 0, \tag{9.59}$$

where c_4, c_3, ε_0, λ_1, and λ_2 given in (9.44), (9.56) and (9.58).

All of the conditions of Lemma 9.1 have been verified, therefore we have

$$|\hat{\upsilon}| \leq \|x_1(t)\| \leq \alpha_1\left(\|(x_1(t_0),\tilde{x}(t_0))\|\right)e^{-\sigma_1(t-t_0)}+\varepsilon_1,\ \forall t \geq t_0 \geq 0, \tag{9.60}$$

where α_1, σ_1, and ε_1 are in the form of α, σ, and ε in Lemma 9.1 with all constants given in (9.44), (9.56), and (9.58).

(z_1, z_2)-subsystem

Having proven that $\hat{\upsilon}$ is bounded in the previous section, we now apply Lemma 9.1 to the (z_1, z_2)-subsystem. It is clear that the last two equations of (9.51) are in the form of the system studied in Lemma 9.1 with $z = [z_1\ z_2]^T$ as x, $\tilde{x}$ as $\xi(t)$ and

$$\dot{z} = f_z(t,z)+g_z(t,y,z,\tilde{x}), \tag{9.61}$$

where

$$f_z(t,z) = \begin{bmatrix} -k_1z_1+z_2 \\ -z_1-\left(k_2+\dfrac{d_{33}}{m_{33}}+\sum_{i\geq 2}\dfrac{d_{ri}}{m_{33}}|\hat{r}|^{i-1}\right)z_2 \end{bmatrix},$$
$$g_z(t,y,z,\tilde{x}) = \begin{bmatrix} \dfrac{k\left(\cos(z_1)+\sin(z_1)ky\right)\tilde{\upsilon}}{(1+(ky)^2)^{3/2}}+\tilde{r} \\ \Omega_2 \end{bmatrix}. \tag{9.62}$$

Proceeding with the same steps as in the previous section, it is shown that all the conditions of Lemma 9.1 hold with the Lyapunov function $V_2 = 0.5(z_1^2+z_2^2)$ and

$$c_0 = 0,\ c_1 = c_2 = 0.5,\ c_3 = 1,\ c_4 = \min(k_1,k_2),$$
$$\lambda_1 = 2k\,(k_{13}+k_{31}+1)+(2k+1)(2+2ku_{\max})+k_{42}+$$

$$k_{24}+1+6k^2u_{\max}(1+3k)+(18k^3+2k(2k+1))(\alpha_1+\varepsilon_1),$$
$$\lambda_2=0, \tag{9.63}$$

where α_1 and ε_1 are given in (9.60). The condition C4 of Lemma 9.1 becomes

$$c_4-\frac{\lambda_1 c_3\varepsilon_0}{4\mu_0}>0, \tag{9.64}$$

where c_4, c_3, ε_0, and λ_2 are calculated from in (9.44) and (9.63).

All of the conditions of Lemma 9.1 have been verified, therefore we have

$$\|z(t)\|\leq\alpha_2\left(\|(z(t_0),\tilde{x}(t_0))\|\right)e^{-\sigma_2(t-t_0)}+\varepsilon_2,\ \forall t\geq t_0\geq 0 \tag{9.65}$$

where α_2, σ_2, and ε_2 are in the form of α, σ, and ε in Lemma 9.1 with all constants given in (9.63).

$(y,\hat{v})$-subsystem

It can be seen that the first two equations of (9.51) are in the form of the system studied in Lemma 9.2, i.e., $x_3=\begin{bmatrix}\hat{v} & y\end{bmatrix}^T$, $\tilde{x}_3=\begin{bmatrix}z_1 & z_2 & \tilde{y} & \tilde{v}\end{bmatrix}^T$, and

$$\dot{x}_3=f_3(t,x_3)+g_3(t,x_3,\tilde{x}_3), \tag{9.66}$$

where

$$f_3(t,x_3)=\begin{bmatrix}-\frac{d_{22}}{m_{22}}\hat{v}-\sum_{i\geq 2}\frac{d_{vi}}{m_{22}}|\hat{v}|^{i-1}\hat{v}-\frac{m_{11}u}{m_{22}}\frac{k^2uy-k\hat{v}}{(1+(ky)^2)^{3/2}}\\ -\frac{kuy}{\sqrt{1+(ky)^2}}+\frac{\hat{v}}{\sqrt{1+(ky)^2}}\end{bmatrix},$$
$$g_3(t,x_3,\tilde{x}_3)=\begin{bmatrix}\Omega_{31}\\ \Omega_{32}\end{bmatrix}, \tag{9.67}$$

with

$$\Omega_{31}=-\frac{m_{11}u}{m_{22}}\left(-k_1z_1+z_2-\frac{ku\left(\sin(z_1)-ky(\cos(z_1)-1)\right)}{(1+(ky)^2)^{3/2}}-\frac{k\hat{v}\left((\cos(z_1)-1)+\sin(z_1)ky\right)}{(1+(ky)^2)^{3/2}}\right)+(k_{31}+k_{13}+\cos(\psi))\tilde{y},$$
$$\Omega_{32}=\frac{u\left(\sin(z_1)-ky(\cos(z_1)-1)\right)}{\sqrt{1+(ky)^2}}+\frac{\hat{v}\left((\cos(z_1)-1)+\sin(z_1)ky\right)}{\sqrt{1+(ky)^2}}+\frac{(\cos(z_1)-\sin(z_1)ky)\,\tilde{v}}{\sqrt{1+(ky)^2}}.$$

We now need to verify all conditions of Lemma 9.2 for the system (9.66).

Verifying Condition C1. To verify this condition of Lemma 9.2, we take the following proper function

$$V_3 = \frac{1}{2}\left(\hat{\upsilon}^2 + y^2\right), \tag{9.68}$$

whose time derivative along (9.67) satisfies

$$\dot{V}_3 \leq -c_{31}\hat{\upsilon}^2 - c_{32}\frac{y^2}{\sqrt{1+c_4 y^2}}, \tag{9.69}$$

where

$$\begin{aligned} c_{31} &= \frac{d_{22}}{m_{22}} - \frac{m_{11}ku_{\max}}{m_{22}} - \mu_2 - \frac{m_{11}k^2u_{\max}^2}{m_{22}}\mu_3, \\ c_{32} &= ku_{\min} - \frac{1}{4\mu_2} - \frac{1}{4\mu_3}\frac{m_{11}k^2u_{\max}^2}{m_{22}}, \end{aligned} \tag{9.70}$$

with $\mu_2 > 0$ and $\mu_3 > 0$ chosen such that $c_{31} > 0$ and $c_{32} > 0$.

From (9.67) and (9.68), it is easy to show that

$$\left|\frac{\partial V_3}{\partial x_3} g_3(t, x_3, \tilde{x}_3)\right| \leq \left(\lambda_0 + \lambda_1\hat{\upsilon}^2 + \lambda_2\frac{y^2}{\sqrt{1+c_4y^2}}\right)\|\tilde{x}_3\|, \tag{9.71}$$

with

$$\begin{aligned} \lambda_0 &= \frac{1}{4\mu_4}\left(\frac{m_{11}u_{\max}}{m_{22}}(k_1+1+2ku_{\max}) + (k_{31}+k_{13}+1) + \right. \\ &\quad \left. u_{\max} + 1 + (\alpha_1+\varepsilon_1)^2\right), \\ \lambda_1 &= \frac{m_{11}u_{\max}\mu_4}{m_{22}}(k_1+1+2ku_{\max}) + \frac{2m_{11}ku_{\max}}{m_{22}} + \mu_4(k_{31}+k_{13}+1), \\ \lambda_2 &= (k+\mu_4)u_{\max} + 2\mu_4 + k(\alpha_1+\varepsilon_1+\alpha_2+\varepsilon_2), \end{aligned} \tag{9.72}$$

where $\mu_4 > 0$, α_1, and ε_1 are given in (9.60), and α_2 and ε_2 are given in (9.65).

Verifying Condition C2. To verify this condition, we note that

$$\|\tilde{x}_3(t)\| \leq \left\|\begin{pmatrix} z_1(t) \\ z_2(t) \end{pmatrix}\right\| + \left\|\begin{pmatrix} \tilde{\upsilon}(t) \\ \tilde{y}(t) \end{pmatrix}\right\|. \tag{9.73}$$

Therefore we can write (9.73) from (9.44) and (9.65) as

$$\|\tilde{x}_3(t)\| \leq \alpha_3\left(\|(z(t_0), \tilde{x}(t_0))\|\right) e^{-\sigma_3(t-t_0)} + \varepsilon_3, \tag{9.74}$$

where

$$\alpha_3(\|(z(t_0),\tilde{x}(t_0))\|) = \|(z(t_0),\tilde{x}(t_0))\| + \alpha_2(\|(z(t_0),\tilde{x}(t_0))\|),$$
$$\sigma_3 = \min(\eta,\sigma_2),\ \varepsilon_3 = \eta_0 + \varepsilon_2, \tag{9.75}$$

with α_2 and ε_2 given in (9.65), η_0 and η given in (9.44).

Verifying Condition C3. This condition is satisfied if

$$c_{31} - \lambda_1\varepsilon_3 > 0 \text{ and } c_{32} - \lambda_2\varepsilon_3 > 0, \tag{9.76}$$

where c_{31}, c_{32}, λ_1, λ_2, and ε_3 are given in (9.72) and (9.75). After some lengthy but simple calculation, it can be shown that, under the assumption of small enough environmental disturbances, the condition (9.76) holds for a suitable choice of the observer gains k_{11}, k_{12}, k_{13}, k_{21}, k_{22}, k_{24}, k_{31}, and k_{42}, and the control gains k, k_1, and k_2.

Verifying Condition C4. From the boundedness of the sway velocity estimate, $\hat{v}(t)$, proven in the previous section and noting that ε_2 in (9.65) can be made arbitrarily small, it is directly shown that there exists a nondecreasing function ϖ of $\|((v(t_0),y(t_0)),z(t_0))\|$ such that $|y(t)| \leq \varpi$ by applying Lemma 9.1 to the first equation of (9.51) with the Lyapunov function $V_y = 0.5y^2$.

All of the conditions of Lemma 9.2 have been verified, the closed loop (9.51) is globally ultimately stable, i.e.,

$$\|x_3(t)\| \leq \alpha_4(\|(x(t_0),\xi(t_0))\|)e^{-\sigma_4(t-t_0)} + \varepsilon_4,\ \forall t \geq t_0 \geq 0, \tag{9.77}$$

where α_4, σ_4, and ε_4 are calculated as in Lemma 9.2.

It is noted that when there are no environmental disturbances, $\varepsilon = 0$. Therefore the closed loop (9.51) is GAS. We note that the convergence of z_1 and z_2 implies the convergence of $\hat{r}$ and ψ. The convergence of v and r results from that of $\hat{v}$ and $\hat{r}$ due to the global exponential property of the observer. We have thus proven the second main result of this chapter.

Theorem 9.2. *Under Assumption 9.1, the output feedback control problem stated in Section 9.1 is solved by the observer (9.34) and the control law (9.50) as long as the observer gains k_{11}, k_{12}, k_{13}, k_{21}, k_{22}, k_{24}, k_{13}, k_{31}, and k_{42}, and the control gains k, k_1, and k_2 are chosen such that (9.64), (9.70), (9.76), and (9.43) hold.*

Remark 9.2. Due to underactuaction and nonzero-mean environmental disturbances in the sway dynamic, our controller is only able to force the sway and its velocity to converge to a ball centered at the origin. The radius of this ball cannot be made arbitrarily small. This phenomenon should not be surprising since there is no control force in the sway direction. In addition, the yaw angle cannot be made arbitrarily small due to the effect of the sway. In fact, to guarantee the sway displacement bounded under nonzero-mean environmental disturbances acting on the sway dynamics, our controller forces the heading angle to a small value. This value together with the forward speed will prevent the sway from growing unbounded.

Remark 9.3. The choice of k depends on the ship parameters and forward speed, which coincides with the steering practice of a helmsman. The helmsman uses the ship's course angle to steer the ship toward the straight line rather than use the sway velocity, which will cause the ship to glide sideways. Furthermore, the design constant k is reduced when the ship forward speed is large, see (9.2), (9.23), (9.30), (9.56), and (9.59), otherwise the ship will miss the point on the straight line and slide in the sway direction.

Remark 9.4. By setting the value of k equal to zero, our proposed controller reduces to a course-keeping controller. In this case, the heading angle can be made arbitrarily small. However the sway will grow linearly unbounded under nonzero-mean environmental disturbances, see Figures 9.3 and 9.6.

9.4 Simulations

This section validates the control laws (9.8) and (9.50) for both cases of state and output feedback on a monohull ship with the parameters given in Section 5.4. The ship surge velocity is chosen as $u = 10 + 0.5\sin(3t)\,\text{ms}^{-1}$. The environmental disturbances $\tau_{wv}(t)$ and $\tau_{wr}(t)$ are taken as $\tau_{wv}(t) = 10^5 \times 0.5 \times (1 + \text{rand}(\cdot))$ and $\tau_{wr}(t) = 1.5 \times 10^7 \times \text{rand}(\cdot)$, with rand$(\cdot)$ being zero mean random noise with the uniform distribution on the interval $[-0.5\ 0.5]$. We run simulations for both state feedback and output feedback cases.

9.4.1 State Feedback Simulation Results

The control design parameters are chosen as $k = 0.05$, $k_1 = 0.2$, $k_2 = 0.5$, and $\rho_1 = \rho_2 = 0.05$. It can be directly verified that this choice satisfies all the conditions stated in Theorem 9.1. The initial values are

$$[y(0), v(0), \psi(0), r(0)] = \left[15\,\text{m}, 0.2\ \text{ms}^{-1}, -0.5\,\text{rad}, 0.1\,\text{rads}^{-1}\right].$$

Simulation results are plotted in Figure 9.1 for the case without disturbances. In this case, it can be seen that all sway displacement, sway velocity, and yaw angle converge to zero as desired. The large control effort is due to the fact that we simulate our controllers on a real surface ship but it is within the limit of the maximum yaw moment. For the case with disturbances, simulation results are plotted in Figure 9.2. In this case, all the states converge to a ball centered at the origin as proven in Theorem 9.1. To illustrate Remark 9.4, we simulate our controller with the design constant $k = 0$. The simulation results for this case are given in Figure 9.3. The sway displacement y grows linearly unbounded due to nonvanishing environmental disturbances. It should be noted that all of the course-keeping controllers, see for

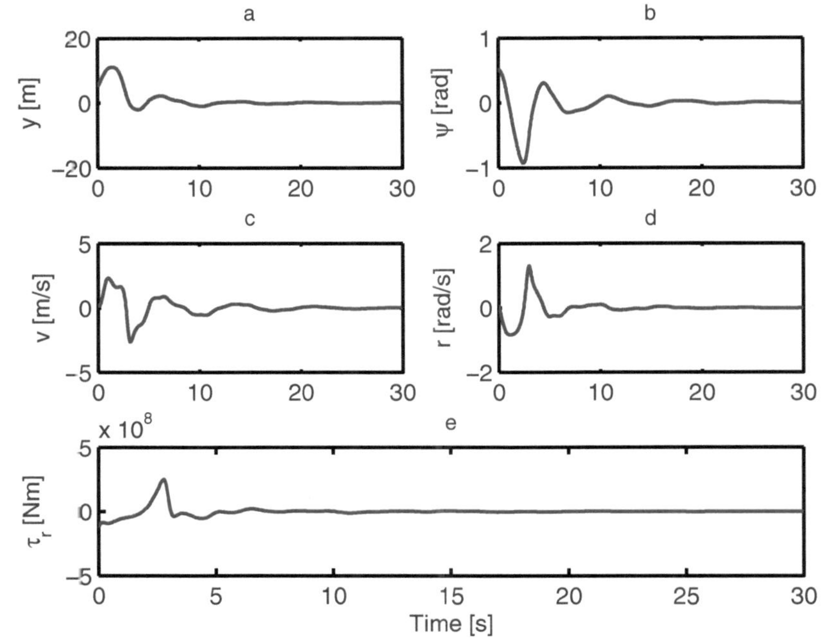

Figure 9.1 State feedback control results without disturbances: **a**. Sway displacement y; **b**. Heading angle ψ; **c**. Sway velocity v; **d**. Yaw velocity r; **e**. Yaw moment τ_r

example [12], which do not take the sway displacement into account, will result in similar unboundedness of the sway that was pointed out in Remark 9.4.

9.4.2 Output Feedback Simulation Results

The control design parameters are chosen as $k = 0.05$, $k_1 = 0.2$, $k_2 = 0.5$, and $\rho_1 = \rho_2 = 0.05$. The observer gains are selected as $k_{11} = k_{22} = 2$, $k_{12} = -\frac{m_{11}}{\rho m_{22}}$, $k_{21} = \frac{m_{11}-m_{22}}{\rho m_{33}}$, $k_{31} = k_{13} = k_{24} = k_{42} = \rho\sqrt{u}$, and $\rho = 0.015$. A calculation shows that this choice satisfies all the conditions stated in Theorem 9.2. The initial values are

$$[y(0), v(0), \psi(0), r(0)] = \left[15\,\text{m}, 0.2\ \text{ms}^{-1}, -0.5\,\text{rad}, 0.1\,\text{rads}^{-1}\right],$$
$$\left[\hat{y}(0), \hat{v}(0), \hat{\psi}(0), \hat{r}(0)\right] = \left[10\,\text{m}, 0\,\text{ms}^{-1}, -0.2\,\text{rad}, 0.2\,\text{rads}^{-1}\right].$$

Simulation results are plotted in Figure 9.4 for the case without disturbances and in Figure 9.5 for the case with disturbances. From Figure 9.4, it is seen that all sway displacement, sway velocity, and yaw angle converge to zero asymptotically. It is also observed that the observer states (dotted lines) exponentially converge to

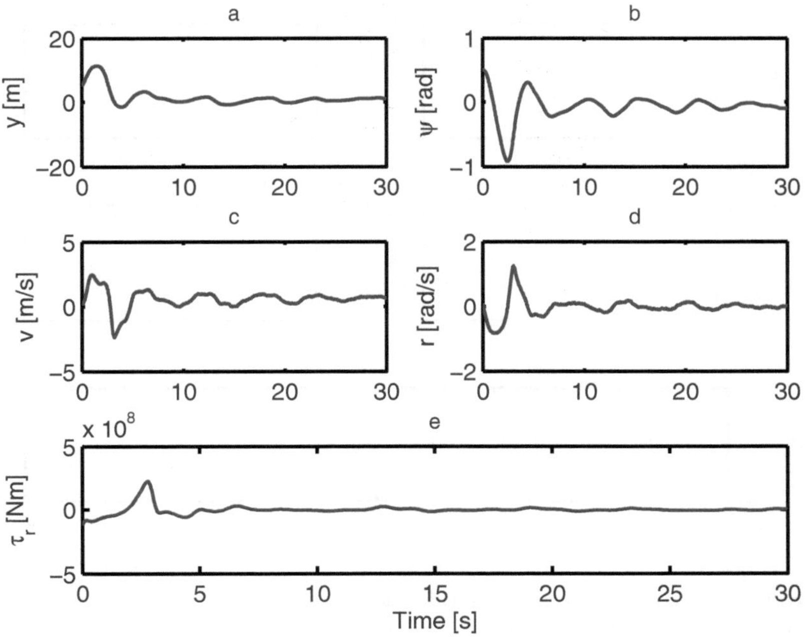

Figure 9.2 State feedback control results with disturbances: **a**. Sway displacement y; **b**. Heading angle ψ; **c**. Sway velocity v; **d**. Yaw velocity r; **e**. Yaw moment τ_r

their unknown estimated ones (solid lines). For the case with disturbances, all the states converge to a ball centered at the origin as proven in Theorem 9.2. The simulation results with the design constant $k = 0$ are plotted in Figure 9.6. Again, the sway displacement y grows linearly unbounded due to nonvanishing environmental disturbances as mentioned in Remark 9.4.

9.5 Conclusions

The control design was based on the idea of an interaction between the ship behavior and the action of a helmsman on a linear course. Although our proposed state feedback controller has been designed by using precise knowledge of the ship parameters, we can easily change them to an adaptive version to take inaccurate knowledge of the system parameters into account, see (9.6). However, for the case of output feedback, an adaptive observer will be required, see (9.34) and (9.45). This chapter is based on [116, 126, 127].

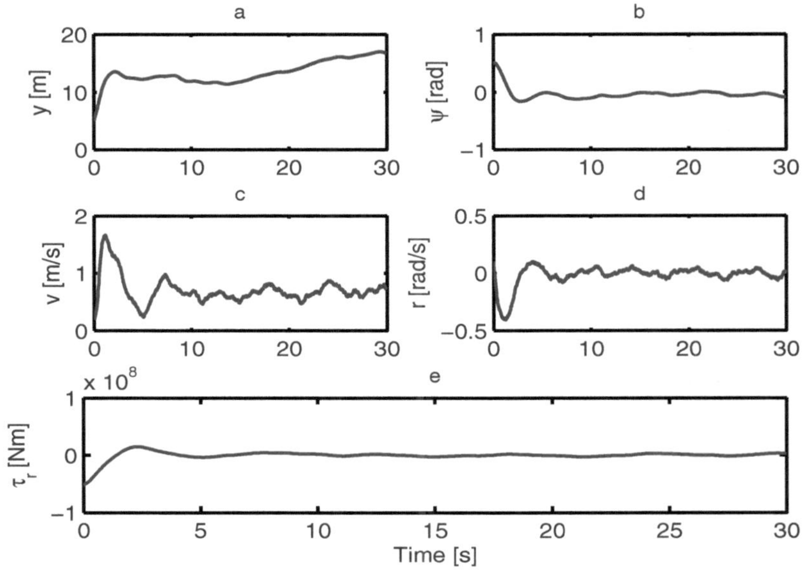

Figure 9.3 State feedback control results with disturbances and $k = 0$: **a**. Sway displacement y; **b**. Heading angle ψ; **c**. Sway velocity v; **d**. Yaw velocity r; **e**. Yaw moment τ_r

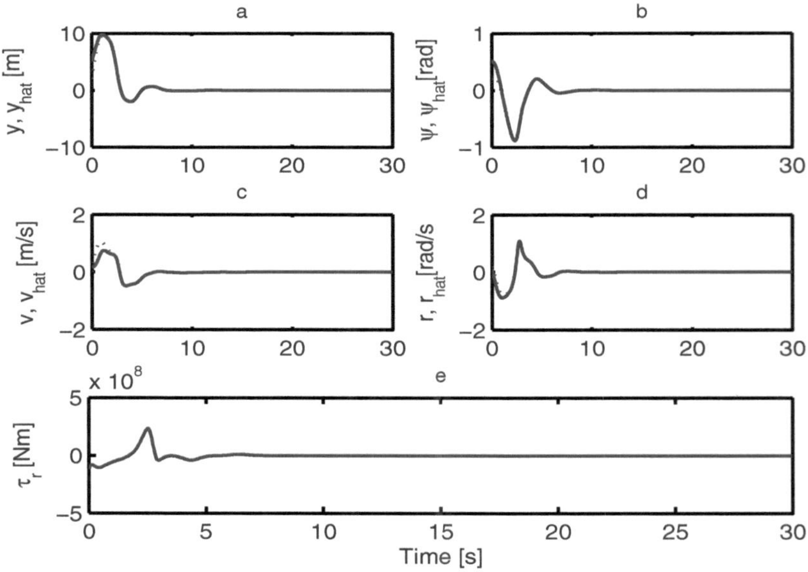

Figure 9.4 Output feedback control results without disturbances: **a**. Sway displacement y; **b**. Heading angle ψ; **c**. Sway velocity v (*solid line*) and its estimate $\hat{v}$ (*dotted line*); **d**. Yaw velocity r (*solid line*) and its estimate $\hat{r}$ (*dotted line*); **e**. Yaw moment τ_r

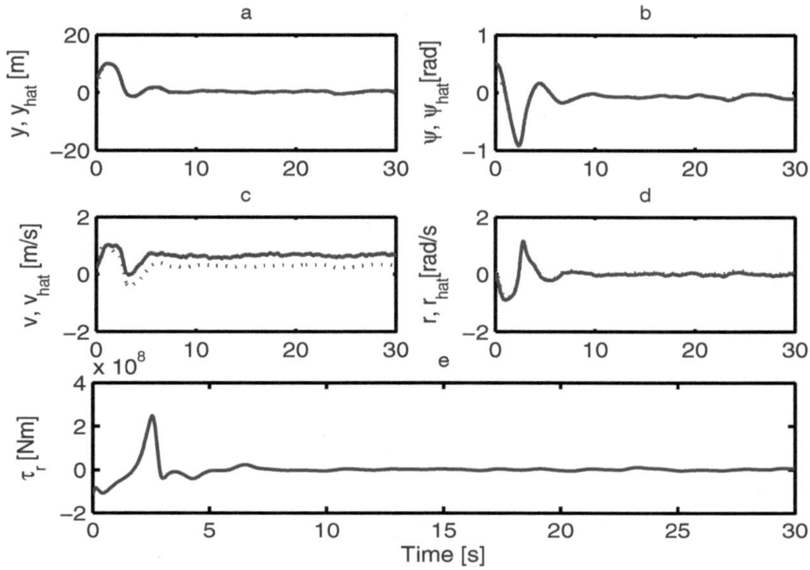

Figure 9.5 Output feedback control results with disturbances: **a**. Sway displacement y; **b**. Heading angle ψ; **c**. Sway velocity v (solid) and its estimate $\hat{v}$ (dot); **d**. Yaw velocity r (solid) and its estimate $\hat{r}$ (dot); **e**. Yaw moment τ_r

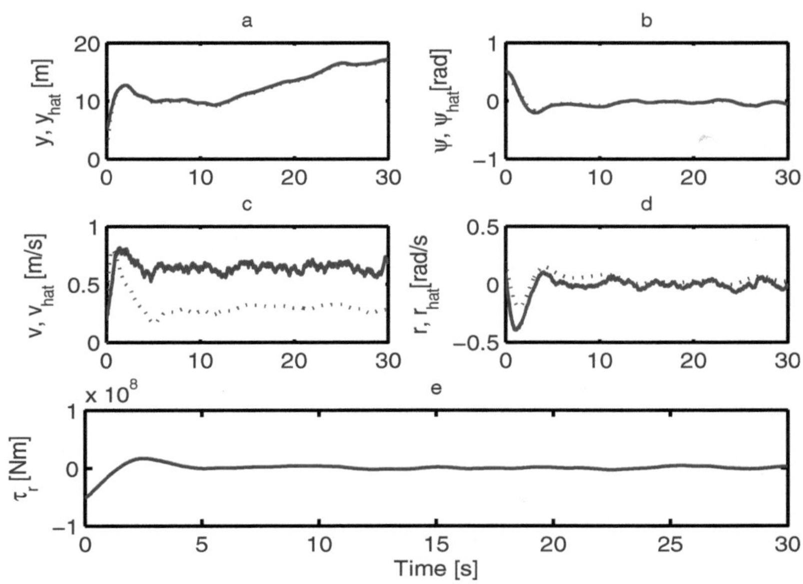

Figure 9.6 Output feedback control results with disturbances and $k = 0$: **a**. Sway displacement y; **b**. Heading angle ψ; **c**. Sway velocity v (*solid line*) and its estimate $\hat{v}$ (*dotted line*); **d**. Yaw velocity r (*solid line*) and its estimate $\hat{r}$ (*dotted line*); **e**. Yaw moment τ_r

Chapter 10
Path-following of Underactuated Ships Using Serret–Frenet Coordinates

This chapter presents state feedback and output feedback controllers that force an underactuated surface ship to follow a predefined path under the presence of environmental disturbances induced by waves, wind, and ocean currents. The control solutions originated an observation that it is reasonable in practice to steer a vessel such that it is on the reference path and its total velocity is tangent to the path, and that the vessel's forward speed is independently controlled by the main propulsion control system. The proposed controllers are designed using Lyapunov's direct method, the popular backstepping technique, and the Serret–Frenet frame. The unmeasured sway and yaw velocities are estimated by introducing a novel nonlinear passive observer.

10.1 Control Objective

In addition to the assumptions made in Section 3.4.1.1, we assume that the surge velocity is a positive constant and is independently controlled by the main propulsion control system. The resulting mathematical model of the underactuated ship moving in surge, sway, and yaw is rewritten as

$$\begin{aligned}
\dot{x} &= u\cos(\psi) - v\sin(\psi),\\
\dot{y} &= u\sin(\psi) + v\cos(\psi),\\
\dot{\psi} &= r,\\
\dot{v} &= -\frac{m_{11}}{m_{22}}ur - \frac{d_{22}}{m_{22}}v - \sum_{i=2}^{3}\frac{d_{vi}}{m_{22}}|v|^{i-1}v + \frac{1}{m_{22}}\tau_{wv}(t),\\
\dot{r} &= \frac{(m_{11}-m_{22})}{m_{33}}uv - \frac{d_{33}}{m_{33}}r - \sum_{i=2}^{3}\frac{d_{ri}}{m_{33}}|r|^{i-1}r + \frac{1}{m_{33}}\tau_r + \frac{1}{m_{33}}\tau_{wr}(t),
\end{aligned} \tag{10.1}$$

where all symbols in (10.1) are defined in Section 3.4.1.2. The positive constant terms d_v, d_r, d_{vi}, and d_{ri}, $i = 2,3$ represent the hydrodynamic damping in surge, sway and yaw. The bounded time-varying terms, $\tau_{wv}(t)$ and $\tau_{wr}(t)$, are the environmental disturbances induced by wave, wind, and ocean current with $|\tau_{wv}(t)| \leq \tau_{wv\max} < \infty$ and $|\tau_{wr}(t)| \leq \tau_{wr\max} < \infty$. The available control is the yaw moment τ_r. Since the sway control force is not available in the sway dynamics, the ship model (10.1) is again underactuated.

The control objective of this chapter is to design the yaw moment τ_r to force the underactuated ship (10.1) to follow a specified path $\boldsymbol{\Omega}$, see Figure 10.1, where M is the orthogonal projection of the ship point P on $\boldsymbol{\Omega}$, x_n and x_t are the normal and the tangent unit vectors to the path at M, respectively. We assume that this point is uniquely defined. This assumption holds if the interior of any circle tangenting $\boldsymbol{\Omega}$ at two or more points does not contain any point of the path and the distance between the ship and $\boldsymbol{\Omega}$ is not too large. Furthermore, it is assumed that the radius of any osculating circle of the path is larger than or equal to $R_{\min}$ which is feasible for the ship to follow. Let the signed distance between M and P be z_e. Also s is the signed distance along the path between some arbitrary fixed point on the path and M, ψ_d is the angle between x_t and X_b, $c(s)$ is the curvature of the path at the point M. We assume that $c(s)$ is uniformly bounded and differentiable. Let the total velocity of the ship be u_t and $\psi_e = \psi - \psi_d$. The variables s, z_e, and ψ_e form a new set of state coordinates for the ship. It can be seen that when the path $\boldsymbol{\Omega}$ coincides with the X_E-axis, the above variables coincide with the ship variables x, y, and ψ.

By using the above parameterization, it is straightforward, see [128, 129], to transform the kinematics of (10.1) to

$$\begin{aligned}
\dot{z}_e &= u\sin(\psi_e) + v\cos(\psi_e), \\
\dot{\psi}_e &= r - \frac{c(s)}{1-c(s)z_e}\left(u\cos(\psi_e) - v\sin(\psi_e)\right), \\
\dot{s} &= \frac{1}{1-c(s)z_e}\left(u\cos(\psi_e) - v\sin(\psi_e)\right).
\end{aligned} \tag{10.2}$$

Note that the above transformation is singular when $z_e c(s) = 1$. We first assume that

$$1 - z_e c(s) \geq \delta^* > 0. \tag{10.3}$$

We then find the initial conditions after the controllers are designed such that this hypothesis holds. Ideally, we want the ship to be on the path and tangent to it, i.e., $z_e = \psi_e = 0$.

However when the curvature of the path is different from zero, the sway velocity v is also different from zero. Hence $z_e = \psi_e = 0$ cannot be the equilibrium point of (10.2) with r as the control input. This feature distinguishes the ship from mobile robots, see [15, 130–133] for some work on controlling mobile robots. If one designs a controller as was proposed in [129], to achieve $\psi_e = 0$, then the error z_e might be very large (depending on v). This phenomenon can be seen by substituting $\psi_e = 0$ into the first equation of (10.2). Therefore, in this chapter, we formulate the

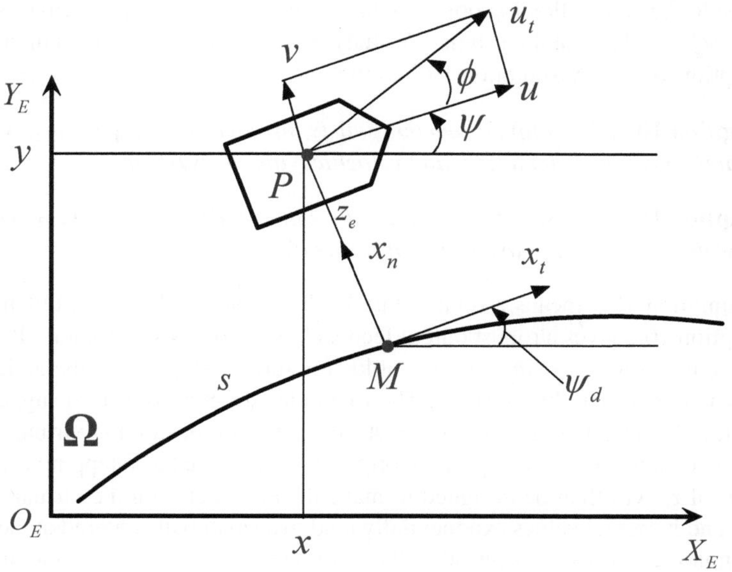

Figure 10.1 General framework of ship path-following

control problem such that $(z_e, \psi_e) = (0, -\phi)$ is the equilibrium point of (10.2), with ϕ being the angle between the surge velocity and the total velocity. This is desirable in practice since it guarantees that the ship is on the path and the ship's total velocity is tangent to the path. Define

$$\psi_e^* = \psi_e + \phi, \tag{10.4}$$

where ψ_e^* is referred to as the modified heading error. With (10.4), the last four equations of the ship model (10.1) with (10.2) are transformed to

$$\begin{aligned}
\dot{z}_e &= u_t \sin(\psi_e^*), \\
\dot{\psi}_e^* &= r\left(1 - \frac{m_{11}}{m_{22}}\frac{u^2}{u_t^2}\right) - \frac{c(s)u_t}{1-c(s)z_e}\cos(\psi_e^*) - \\
&\quad \frac{u}{u_t^2}\left(\frac{d_{22}}{m_{22}}v + \sum_{i=2}^{3}\frac{d_{vi}}{m_{22}}|v|^{i-1}v - \frac{1}{m_{22}}\tau_{wv}(t)\right), \\
\dot{v} &= -\frac{m_{11}}{m_{22}}ur - \frac{d_{22}}{m_{22}}v - \sum_{i=2}^{3}\frac{d_{vi}}{m_{22}}|v|^{i-1}v + \frac{1}{m_{22}}\tau_{wv}(t), \\
\dot{r} &= \frac{(m_{11}-m_{22})}{m_{33}}uv - \frac{d_{33}}{m_{33}}r - \sum_{i=2}^{3}\frac{d_{ri}}{m_{33}}|r|^{i-1}r + \frac{1}{m_{33}}\tau_r + \frac{1}{m_{33}}\tau_{wr}(t),
\end{aligned} \tag{10.5}$$

where

$$u_t = \sqrt{u^2 + v^2}. \tag{10.6}$$

As a result, the path-following objective has been converted to a problem of stabilizing (z_e, ψ_e^*) in (10.5) at the origin. We study this stabilization problem under either Assumption 10.1 or Assumption 10.2 as follows.

Assumption 10.1. *The ship parameters satisfy the condition $m_{11} < m_{22}$. All of the ship states (position, orientation, and velocities) are measurable.*

Assumption 10.2. *The ship parameters satisfy the condition $m_{11} < m_{22}$. The sway and yaw velocities are not available for feedback.*

Assumption 10.1 means a robust state feedback path-following problem, while Assumption 10.2 gives a robust output feedback path-following problem. It is noted that the condition $m_{11} < m_{22}$ always holds for surface ships since the added mass in sway is larger than that in surge. The triangular structure of (10.5) suggests that we design the actual control τ_r in two stages. First, we design the virtual velocity control r to stabilize z_e and ψ_e^* at the origin. Based on the backstepping technique, the control τ_r will then be designed to make the error between the virtual velocity control and its actual values exponentially tend to a small ball centered at the origin. A new nonlinear observer is introduced to estimate the sway and yaw velocities, and an output feedback controller is then designed.

10.2 State Feedback

10.2.1 Control Design

Observing from the first equation of (10.5) that the z_e-dynamics have to be stabilized by using the angle ψ_e^*, we introduce the following coordinate transformation

$$w_e^* = \psi_e^* + \arcsin\left(\frac{kz_e}{\sqrt{1+(kz_e)^2}}\right), \tag{10.7}$$

where k is a positive constant to be selected later.

Remark 10.1. The above coordinate transformation is well defined and convergence of w_e^* and z_e implies that of ψ_e^*. The function arcsin is not unique. It can be replaced by some other smooth bounded functions such as tanh, arctan or an identity function. We here use the arcsin function due to its simplicity. Using (10.7) instead of a linear coordinate change such as $w_e^* = \psi_e^* + kz_e$, we avoid the ship whirling around when z_e is large. In addition the nonlinear change of coordinate (10.7) will result in a "global design".

Upon an application of the coordinate transformation (10.7), the system (10.5) is rewritten as

$$
\begin{aligned}
\dot{z}_e &= -\frac{ku_t z_e}{\sqrt{1+(kz_e)^2}} + \frac{u_t \sin(w_e^*)}{\sqrt{1+(kz_e)^2}} - \frac{ku_t z_e}{\sqrt{1+(kz_e)^2}}\left(\cos(w_e^*)-1\right), \\
\dot{w}_e^* &= r\left(1-\frac{m_{11}}{m_{22}}\frac{u^2}{u_t^2}\right) + \frac{ku_t \sin(\psi_e^*)}{1+(kz_e)^2} - \frac{c(s)u_t\cos(\psi_e^*)}{1-c(s)z_e} - \\
&\quad \frac{u}{u_t^2}\left(\frac{d_{22}}{m_{22}}v + \sum_{i=2}^{3}\frac{d_{vi}}{m_{22}}|v|^{i-1}v - \frac{1}{m_{22}}\tau_{wv}(t)\right), \\
\dot{v} &= -\frac{m_{11}}{m_{22}}ur - \frac{d_{22}}{m_{22}}v - \sum_{i=2}^{3}\frac{d_{vi}}{m_{22}}|v|^{i-1}v + \frac{1}{m_{22}}\tau_{wv}(t), \\
\dot{r} &= \frac{(m_{11}-m_{22})}{m_{33}}uv - \frac{d_{33}}{m_{33}}r - \sum_{i=2}^{3}\frac{d_{ri}}{m_{33}}|r|^{i-1}r + \frac{1}{m_{33}}\tau_r + \frac{1}{m_{33}}\tau_{wr}(t),
\end{aligned}
\tag{10.8}
$$

where we leave the variable ψ_e^* in the second equation of (10.8) for simplicity of presentation. As discussed in Remark 10.1, we will design the control τ_r to stabilize (10.8) at the origin.

Step 1

Define

$$b = 1 - \frac{m_{11}}{m_{22}}\frac{u^2}{u_t^2}. \tag{10.9}$$

The first condition of Assumptions 10.1 implies that

$$b \geq 1 - \frac{m_{11}}{m_{22}} := b^* > 0. \tag{10.10}$$

Introduce the yaw velocity control error $\tilde{r}$ as

$$\tilde{r} = r - r_d, \tag{10.11}$$

where r_d is a virtual control of the yaw velocity r. From the second equation of (10.8), we choose the virtual control r_d as

$$
\begin{aligned}
r_d = &-\frac{k_1}{b}w_e^* - \frac{1}{b}\frac{ku_t\sin(\psi_e^*)}{1+(kz_e)^2} + \frac{1}{b}\frac{c(s)u_t\cos(\psi_e^*)}{1-c(s)z_e} + \\
&\frac{1}{b}\frac{u}{u_t^2}\left(\frac{d_{22}}{m_{22}}v + \sum_{i=2}^{3}\frac{d_{vi}}{m_{22}}|v|^{i-1}v\right) - \\
&\frac{1}{b}\frac{u}{u_t^2}\frac{\tau_{wv\max}}{m_{22}}\tanh\left(\frac{w_e^*}{\varepsilon_1}\frac{\tau_{wv\max}}{m_{22}}\right),
\end{aligned}
\tag{10.12}
$$

where k_1 and ε_1 are positive design constants to be selected later. It is of interest to note that r_d is a smooth function of s, z_e, v, and ψ_e. Substituting (10.12) and

(10.11) into the second equation of (10.8) yields

$$\dot{w}_e^* = -k_1 w_e^* + b\tilde{r} + \frac{u}{u_t^2}\frac{\tau_{wv}(t)}{m_{22}} - \frac{u}{u_t^2}\frac{\tau_{wv\max}}{m_{22}}\tanh\left(\frac{w_e^*}{\varepsilon_1}\frac{\tau_{wv\max}}{m_{22}}\right). \tag{10.13}$$

Step 2

Differentiating both sides of (10.11) along the solutions of the last equation of (10.8) and (10.12) yields

$$\begin{aligned}\dot{\tilde{r}} = & \frac{(m_{11}-m_{22})}{m_{33}}uv - \frac{d_{33}}{m_{33}}r - \sum_{i=2}^{3}\frac{d_{ri}}{m_{33}}|r|^{i-1}r + \frac{1}{m_{33}}\tau_r + \frac{1}{m_{33}}\tau_{wr}(t) - \\ & \frac{\partial r_d}{\partial s}\frac{1}{1-c(s)z_e}(u\cos(\psi_e) - v\sin(\psi_e)) - \frac{\partial r_d}{\partial z_e}(u\sin(\psi_e) + v\cos(\psi_e)) - \\ & \frac{\partial r_d}{\partial v}\left(-\frac{m_{11}}{m_{22}}ur - \frac{d_{22}}{m_{22}}v - \sum_{i=2}^{3}\frac{d_{vi}}{m_{22}}|v|^{i-1}v + \frac{1}{m_{22}}\tau_{wv}(t)\right) - \\ & \frac{\partial r_d}{\partial \psi_e}\left(r - \frac{c(s)}{1-c(s)z_e}(u\cos(\psi_e) - v\sin(\psi_e))\right).\end{aligned} \tag{10.14}$$

From (10.14), we choose the actual control τ_r without canceling the useful damping terms as

$$\begin{aligned}\tau_r = & m_{33}\Bigg(-k_2\tilde{r} - bw_e^* - \frac{(m_{11}-m_{22})}{m_{33}}uv + \frac{d_{33}}{m_{33}}r_d + \sum_{i=2}^{3}\frac{d_{ri}}{m_{33}}|r|^{i-1}r_d + \\ & \frac{\partial r_d}{\partial s}\frac{1}{1-c(s)z_e}(u\cos(\psi_e) - v\sin(\psi_e)) + \frac{\partial r_d}{\partial z_e}(u\sin(\psi_e) + v\cos(\psi_e)) + \\ & \frac{\partial r_d}{\partial v}\left(-\frac{m_{11}}{m_{22}}ur - \frac{d_{22}}{m_{22}}v - \sum_{i=2}^{3}\frac{d_{vi}}{m_{22}}|v|^{i-1}v\right) + \\ & \frac{\partial r_d}{\partial \psi_e}\left(r - \frac{c(s)}{1-c(s)z_e}(u\cos(\psi_e) - v\sin(\psi_e))\right) - \\ & \frac{\tau_{ur\max}}{m_{33}}\tanh\left(\frac{\tilde{r}}{\varepsilon_2}\frac{\tau_{wr\max}}{m_{33}}\right) - \frac{\partial r_d}{\partial v}\frac{\tau_{wv\max}}{m_{22}}\tanh\left(\frac{\partial r_d}{\partial v}\frac{\tilde{r}}{\varepsilon_3}\frac{\tau_{wv\max}}{m_{22}}\right)\Bigg),\end{aligned} \tag{10.15}$$

where k_2, ε_2, and ε_3 are positive constants to be selected later. Substituting (10.15) into (10.14) yields

$$\dot{\tilde{r}} = -\left(k_2 + \frac{d_{33}}{m_{33}}\right)\tilde{r} - bw_e^* - \sum_{i=2}^{3}\frac{d_{ri}}{m_{33}}|r|^{i-1}\tilde{r} + \frac{1}{m_{33}}\tau_{wr}(t) -$$

$$\frac{\tau_{wr\max}}{m_{33}}\tanh\left(\frac{\tilde{r}}{\varepsilon_2}\frac{\tau_{wr\max}}{m_{33}}\right)-\frac{\partial r_d}{\partial v}\frac{1}{m_{22}}\tau_{wv}(t)-$$
$$\frac{\partial r_d}{\partial v}\frac{\tau_{wv\max}}{m_{22}}\tanh\left(\frac{\partial r_d}{\partial v}\frac{\tilde{r}}{\varepsilon_3}\frac{\tau_{wv\max}}{m_{22}}\right). \tag{10.16}$$

We now present the first main result of this chapter, the proof of which is given in the next section.

Theorem 10.1. *Under Assumption 10.1, if the state feedback control law (10.15) is applied to the ship system (10.1) then there exist feasible initial conditions such that the regulation errors* $(z_e(t),\psi_e^*(t))$ *converge to a small ball centered at the origin with an appropriate choice of the design constants* k, k_1, k_2 *and* ε_i, $1\le i\le 3$. *Furthermore if there are no environmental disturbances, the regulation errors converge to zero asymptotically. In addition, the sway velocity* $v(t)$ *is always bounded.*

10.2.2 Stability Analysis

For the reader's convenience we write the closed loop consisting of the first and third equations of (10.8), (10.13), and (10.16) as follows:

$$\begin{aligned}
\dot{z}_e &= -\frac{ku_t z_e}{\sqrt{1+(kz_e)^2}}+\frac{u_t\sin(w_e^*)}{\sqrt{1+(kz_e)^2}}-\frac{ku_t z_e}{\sqrt{1+(kz_e)^2}}\left(\cos(w_e^*)-1\right),\\
\dot{w}_e^* &= -k_1 w_e^*+b\tilde{r}+\frac{u}{u_t^2}\frac{\tau_{wv}(t)}{m_{22}}-\frac{u}{u_t^2}\frac{\tau_{wv\max}}{m_{22}}\tanh\left(\frac{w_e^*}{\varepsilon_1}\frac{\tau_{wv\max}}{m_{22}}\right),\\
\dot{v} &= -\frac{m_{11}}{m_{22}}u(\tilde{r}+r_d)-\frac{d_{22}}{m_{22}}v-\sum_{i=2}^{3}\frac{d_{vi}}{m_{22}}|v|^{i-1}v+\frac{1}{m_{22}}\tau_{wv}(t),\\
\dot{\tilde{r}} &= -\left(k_2+\frac{d_{33}}{m_{33}}\right)\tilde{r}-bw_e^*-\sum_{i=2}^{3}\frac{d_{ri}}{m_{33}}|r|^{i-1}\tilde{r}+\frac{1}{m_{33}}\tau_{wr}(t)-\\
&\quad\frac{\tau_{wr\max}}{m_{33}}\tanh\left(\frac{\tilde{r}}{\varepsilon_2}\frac{\tau_{wr\max}}{m_{33}}\right)-\frac{\partial r_d}{\partial v}\frac{1}{m_{22}}\tau_{wv}(t)-\\
&\quad\frac{\partial r_d}{\partial v}\frac{\tau_{wv\max}}{m_{22}}\tanh\left(\frac{\partial r_d}{\partial v}\frac{\tilde{r}}{\varepsilon_3}\frac{\tau_{wv\max}}{m_{22}}\right),
\end{aligned} \tag{10.17}$$

where r_d is given in (10.12). A direct calculation of the Lyapunov function candidate $V_0=z_e^2+w_e^{*2}+v^2+\tilde{r}^2$ along the solutions of the closed loop system (10.17) shows that there exist nonnegative constants a_0 and b_0 such that $\dot{V}_0\le a_0V_0+b_0$. This means that the closed loop system (10.17) is forward complete. Therefore, to prove Theorem 10.1, we can first consider the $(w_e^*,\tilde{r})$-dynamics, then move to the (z_e,v)-dynamics.

$(w_e^*, \tilde{r})$-dynamics

Consider the following Lyapunov function:

$$V_1 = \frac{1}{2}\left(w_e^{*2} + \tilde{r}^2\right). \tag{10.18}$$

It is not hard to show that the time derivative of (10.18) along the solutions of (10.13) and (10.16) satisfies

$$\dot{V}_1 \leq -k_1 w_e^{*2} - \left(k_2 + \frac{d_{33}}{m_{33}}\right)\tilde{r}^2 + 0.2785\sum_{i=1}^{3}\varepsilon_i, \tag{10.19}$$

which in turn implies that

$$\left\|\left(w_e^*(t), \tilde{r}(t)\right)\right\| \leq \alpha_{wr}(\cdot)e^{-\sigma_{wr}(t-t_0)} + \rho_{wr}, \tag{10.20}$$

where

$$\begin{aligned}
\alpha_{wr}(\cdot) &= \left\|\left(w_e^*(t_0), \tilde{r}(t_0)\right)\right\|, \\
\sigma_{wr} &= \min\left(k_1, \left(k_2 + \frac{d_{33}}{m_{33}}\right)\right), \\
\rho_{wr} &= \sqrt{\frac{0.2785}{\sigma_1}\sum_{i=1}^{3}\varepsilon_i}.
\end{aligned} \tag{10.21}$$

From (10.21), we can see that ρ_{wr} can be made arbitrarily small by adjusting the design constants, k_1, k_2 and ε_i, $1 \leq i \leq 3$. This observation is important in stability analysis of the (z_e, v)-dynamics.

(z_e, v)-dynamics

For the reader's convenience, we rewrite the (z_e, v)-dynamics from the first and third equations of the closed loop system (10.17) with the virtual control r_d substituted in the third equation of the closed loop system (10.17) from (10.12) as

$$\begin{aligned}
\dot{z}_e =& -\frac{k u_t z_e}{\sqrt{1+(kz_e)^2}} + \frac{u_t \sin(w_e^*)}{\sqrt{1+(kz_e)^2}} - \frac{k u_t z_e}{\sqrt{1+(kz_e)^2}}\left(\cos(w_e^*) - 1\right), \\
\dot{v} =& -\frac{d_{22}}{m_{22}}\left(1 + \frac{m_{11}u^2}{b m_{22} u_t^2}\right)v - \sum_{i=2}^{3}\frac{d_{vi}}{m_{22}}\left(1 + \frac{m_{11}u^2}{b m_{22} u_t^2}\right)|v|^{i-1}v + \\
& \frac{1}{m_{22}}\tau_{wv}(t) - \frac{m_{11}u\tilde{r}}{m_{22}} - \frac{m_{11}u}{m_{22}b}\left(-k_1 w_e^* - \frac{k u_t \sin(\psi_e^*)}{1+(kz_e)^2} + \right.
\end{aligned}$$

$$\frac{c(s)u_t\cos(\psi_e^*)}{1-c(s)z_e} - \frac{u}{u_t^2}\frac{\tau_{wv\max}}{m_{22}}\tanh\left(\frac{w_e^*}{\varepsilon_1}\frac{\tau_{wv\max}}{m_{22}}\right)\Bigg). \tag{10.22}$$

If we view that (10.22) and the $(w_e^*, \tilde{r})$-subsystem are in a cascade form, we might think that stability results developed for cascade systems, see Section 2.1.3 can be applied. However, stability results in that section are developed for the cascade systems without nonvanishing disturbances. In fact, nonvanishing disturbances may destroy stability of a cascade system that satisfies all conditions stated in Section 2.1.3. To illustrate this fact, we give the following simple example.

$$\begin{aligned}\dot{x}_1 &= -\frac{x_1}{\sqrt{1+x_1^2}} + \frac{x_1 x_2}{\sqrt{1+x_1^2}},\\ \dot{x}_2 &= -x_2 + d(t).\end{aligned} \tag{10.23}$$

When there is no disturbance $d(t)$, by applying stability results in Section 2.1.3, the cascade system (10.23) is GAS at the origin. However, whenever the magnitude of bounded disturbance $d(t)$ is larger than 1, we have the fact that x_1 grows unbounded although x_2 is bounded. This fact results from the first equation being not globally ISS with respect to x_2 as input, see Section 2.2. Therefore, to analyze stability of (10.22) we first present the following lemma.

Lemma 10.1. *Consider the following nonlinear system:*

$$\dot{x} = f(t,x) + g(t,x,\xi(t)) \tag{10.24}$$

where $x \in \mathbb{R}^n$, $\xi(t) \in \mathbb{R}^m$, $f(t,x)$ is piecewise continuous in t and locally Lipschitz in x. If there exist $\sigma_0 > 0$, positive constants c_i, $1 \le i \le 4$, λ_j, $1 \le j \le 2$, c, ε_0, mu_0, and c_0, and a class-K function α_0 such that the following conditions hold:
C1. There exists a proper function $V(t,x)$ such that:

$$\begin{aligned}&c_1\|x\|^2 \le V(t,x) \le c_2\|x\|^2,\\ &\left\|\frac{\partial V}{\partial x}(t,x)\right\| \le c_3\|x\|,\\ &\frac{\partial V}{\partial t} + \frac{\partial V}{\partial x}f(t,x) \le -\frac{c_4\|x\|^2}{\sqrt{1+c\|x\|^2}} + \frac{c_0}{\sqrt{1+c\|x\|^2}}.\end{aligned}$$

C2. $g(t,x,\xi(t))$ satisfies:

$$\|g(t,x,\xi(t))\| \le \frac{1}{\sqrt{1+c\|x\|^2}}(\lambda_1+\lambda_2\|x\|)\|\xi(t)\|.$$

C3. $\xi(t)$ globally asymptotically converges to a ball centered at the origin:

$$\|\xi(t)\| \le \alpha_0(\|\xi(t_0)\|)e^{-\sigma_0(t-t_0)} + \varepsilon_0,\ \forall t \ge t_0 \ge 0.$$

C4. The following gain condition is satisfied:

$$c_4 - c_3\varepsilon_0\left(\lambda_2 + \frac{\lambda_1}{4\mu_0}\right) > 0.$$

Then the solution $x(t)$ of (10.24) globally asymptotically converges to a ball centered at the origin, i.e.,

$$\|x(t)\| \le \alpha\left(\|(x(t_0),\xi(t_0))\|\right)e^{-\sigma(t-t_0)} + \varepsilon\left(\|(x(t_0),\xi(t_0))\|\right), \ \forall t \ge t_0 \ge 0, \tag{10.25}$$

where $\varepsilon(s) = \sqrt{\frac{a_4}{c_1 a_1}\sqrt{1+cx_m^2(s)}}$, σ and α are given as follows:

$$\text{if } \frac{a_1}{\sqrt{1+cx_m^2}} = \sigma_0 \textit{ then}$$

$$\alpha(s) = \sqrt{\frac{e^{\frac{a_2}{\sigma_0}}}{c_1}\left(c_2 s^2 + \left(a_3 + \frac{a_2 a_4}{a_1}\sqrt{1+cx_m^2}\right)\phi\right)},$$

$$\sigma = \frac{1}{2}(a_1 - d);$$

$$\text{if } \frac{a_1}{\sqrt{1+cx_m^2}} \ne \sigma_0 \textit{ then}$$

$$\alpha(s) = \sqrt{\frac{e^{\frac{a_2}{\sigma_0}}}{c_1}\left(c_2 s^2 + \frac{\left(a_1 a_3 + a_2 a_4\sqrt{1+cx_m^2}\right)}{a_1\left|a_1 - \sigma_0\sqrt{1+cx_m^2}\right|}\sqrt{1+cx_m^2}\right)},$$

$$\sigma(s) = \frac{1}{2}\min\left(\frac{a_1}{\sqrt{1+cx_m^2}}, \left|\frac{a_1}{\sqrt{1+cx_m^2}} - \sigma_0\right|\right);$$

with

$$a_1 = \frac{1}{c_2}\left(c_4 - \lambda_2 c_3\varepsilon_0 - \frac{\lambda_1 c_3\varepsilon_0}{4\mu_0}\right), \ a_2 = \frac{c_3}{c_1}(\lambda_1+\lambda_2)\alpha_0(s),$$

$$a_3 = \frac{\lambda_1 c_3}{4}\alpha_0(s), \ a_4 = c_0 + \lambda_1 c_3\varepsilon_0\mu_0,$$

$$0 < d < a_1, \ \phi \ge (t-t_0)e^{-d(t-t_0)},$$

$$x_m(s) = \sqrt{\frac{1}{c_1}\left(\frac{a_4}{a_1} + \sqrt{W_m(s)}\right)},$$

$$W_m(s) = \left(V(s) - \frac{a_4}{a_1}\right)e^{\left(\frac{2(a_2+a_3+a_2a_4/a_1)}{\sigma_0}\right)} + \frac{a_3 + a_2a_4/a_1}{8(a_2(s)+a_3(s)+a_2(s)a_4/a_1)}\left(e^{\left(\frac{2(a_2+a_3+a_2a_4/a_1)}{\sigma_0}\right)} - 1\right).$$

When $c_0 = 0$ and $\varepsilon_0 = 0$, we have $\varepsilon = 0$ and the system (10.24) is GAS. Note that a finite value of ϕ exists for an arbitrarily small positive d and the convergence rate σ depends on the initial conditions.

Proof. We first prove that x is bounded. From conditions C1, C2, and C3, we have

$$\dot{V} \leq -\left(c_4 - c_3\varepsilon_0\left(\lambda_2 + \frac{\lambda_1}{4\mu_0}\right)\right)\frac{\|x\|^2}{\sqrt{1+c\|x\|^2}} + \frac{c_3\|x\|}{\sqrt{1+c\|x\|^2}} \times$$
$$(\lambda_1 + \lambda_2\|x\|)\alpha_0(\|\xi(t_0)\|)e^{-\sigma_0(t-t_0)} + \frac{\lambda_1 c_3\varepsilon_0\mu_0 + c_0}{\sqrt{1+c\|x\|^2}}. \tag{10.26}$$

Upon application of completing squares, (10.26) can be rewritten as

$$\begin{aligned}
\dot{V} &\leq -\left(a_1 - a_2 e^{-\sigma_0(t-t_0)}\right)\frac{V}{\sqrt{1+c\|x\|^2}} + \frac{a_3 e^{-\sigma_0(t-t_0)}}{\sqrt{1+c\|x\|^2}} + \frac{a_4}{\sqrt{1+c\|x\|^2}} \\
&\leq -\left(a_1 - a_2 e^{-\sigma_0(t-t_0)}\right)\frac{(V - a_4/a_1)}{\sqrt{1+c\|x\|^2}} + \frac{(a_3 + a_2a_4/a_1)e^{-\sigma_0(t-t_0)}}{\sqrt{1+c\|x\|^2}}.
\end{aligned} \tag{10.27}$$

Now consider the differential equation

$$\dot{\kappa} = -\left(a_1 - a_2 e^{-\sigma_0(t-t_0)}\right)\frac{\kappa}{\sqrt{1+c\|x\|^2}} + \frac{(a_3 + a_2a_4/a_1)e^{-\sigma_0(t-t_0)}}{\sqrt{1+c\|x\|^2}}. \tag{10.28}$$

Take the following Lyapunov function:

$$W = \frac{1}{2}\kappa^2, \tag{10.29}$$

whose time derivative along the solutions of (10.28) with the use of condition C4 satisfies

$$\begin{aligned}
\dot{W} &\leq a_2 e^{-\sigma_0(t-t_0)}\frac{\kappa^2}{\sqrt{1+c\|x\|^2}} + \frac{(a_3 + a_2a_4/a_1)\kappa e^{-\sigma_0(t-t_0)}}{\sqrt{1+c\|x\|^2}} \\
&\leq 2(a_2 + a_3 + \frac{a_2a_4}{a_1})We^{-\sigma_0(t-t_0)} + \frac{(a_3 + \frac{a_2a_4}{a_1})}{4}e^{-\sigma_0(t-t_0)}.
\end{aligned} \tag{10.30}$$

Hence $W(t) \leq W_m$ which implies from (10.30) and the comparison principle found in [6] that $V(t) \leq a_4/a_1 + \sqrt{2W_m}$ or $\|x(t)\| \leq x_m$. Substituting this inequality into (10.27) yields

$$\dot{V} \leq -\left(\frac{a_1}{\sqrt{1+cx_m^2}} - a_2 e^{-\sigma_0(t-t_0)}\right) V + a_3 e^{-\sigma_0(t-t_0)} + a_4. \tag{10.31}$$

Solving the inequality (10.31) readily yields (10.25). □

To investigate stability of the (z_e, v)-dynamics, we apply Lemma 10.1.

z_e-dynamics

We will apply Lemma 10.1 to estimate convergence of the path-following error, z_e, dynamics. Indeed, it is seen that z_e- dynamics, the first equation of (10.22), is in the form of the system studied in Lemma 10.1, i.e.,

$$\dot{z}_e = f_{ze}(\cdot) + g_{ze}(\cdot), \tag{10.32}$$

where

$$\begin{aligned} f_{ze}(\cdot) &= -\frac{k u_t z_e}{\sqrt{1+(kz_e)^2}}, \\ g_{ze}(\cdot) &= \frac{u_t \sin(w_e^*)}{\sqrt{1+(kz_e)^2}} - \frac{k u_t z_e}{\sqrt{1+(kz_e)^2}} \left(\cos(w_e^*) - 1\right). \end{aligned} \tag{10.33}$$

We now verify all conditions of Lemma 10.1.

Verifying Condition C1. By taking the proper function $V_2 = 0.5 z_e^2$, it is not hard to show that this condition holds with

$$c = k^2,\ c_1 = c_2 = 0.5,\ c_3 = 1,\ c_4 = k u_t,\ c_0 = 0. \tag{10.34}$$

Verifying Condition C2. By noting that

$$\begin{aligned} &\left| \frac{u_t \sin(w_e^*)}{\sqrt{1+(kz_e)^2}} - \frac{k u_t z_e}{\sqrt{1+(kz_e)^2}} \left(\cos(w_e^*) - 1\right) \right| \leq \\ &\qquad \frac{1}{\sqrt{1+(kz_e)^2}} u_t (1 + k|z_e|) |w_e^*|, \end{aligned} \tag{10.35}$$

condition C2 holds with

$$\begin{aligned} \lambda_1 &= u_t, \\ \lambda_2 &= k u_t. \end{aligned} \tag{10.36}$$

Verifying Condition C3. This condition holds directly from (10.20).

Verifying Condition C4. From (10.34), (10.36), and (10.20), this condition becomes

$$ku_t - \rho_{wr}\left(u_t + \frac{ku_t}{4\mu_0}\right) > 0 \Leftrightarrow k - \rho_{wr}\left(1 + \frac{k}{4\mu_0}\right) > 0. \tag{10.37}$$

Since μ_0 is an arbitrarily positive constant, see Lemma 10.1, and ρ_{wr} can be made arbitrarily small, we can always pick a positive constant k such that (10.37) holds. Therefore, we have

$$|z_e(t)| \leq \alpha_z(\cdot)e^{-\sigma_z(t-t_0)} + \rho_z, \tag{10.38}$$

where $\alpha_z(\cdot)$, σ_z and ρ_z are calculated as in Lemma 10.1. From (10.38), the requirement of initial conditions such that $1 - c(s)z_e \geq \delta^* > 0$ becomes

$$\alpha_z(\cdot) \leq \frac{1-\delta^*}{|c(s)|} - \rho_z. \tag{10.39}$$

Boundedness of v

To show that v is bounded, we take the following Lyapunov function

$$V_3 = \frac{1}{2}v^2, \tag{10.40}$$

whose time derivative along the solutions of the second equation of (10.22), after some simple calculation, satisfies

$$\begin{aligned}\dot{V}_3 \leq\ & -\frac{d_{22}}{m_{22}}\left(1 + \frac{m_{11}u^2}{bm_{22}u_t^2}\right)v^2 + v\left(\frac{1}{m_{22}}\tau_{wv}(t) - \frac{m_{11}u\tilde{r}}{m_{22}} - \right.\\ & \frac{m_{11}u}{m_{22}b}\left(-k_1 w_e^* - \frac{ku_t\sin(\psi_e^*)}{1+(kz_e)^2} + \frac{c(s)u_t\cos(\psi_e^*)}{1-c(s)z_e} - \right.\\ & \left.\left.\frac{u}{u_t^2}\frac{\tau_{wv\max}}{m_{22}}\tanh\left(\frac{w_e^*}{\varepsilon_1}\frac{\tau_{wv\max}}{m_{22}}\right)\right)\right).\end{aligned} \tag{10.41}$$

From (10.41), (10.20), (10.38), and (10.7), it can be shown that

$$|v(t)| \leq \alpha_v(\cdot)e^{-\sigma_v(t-t_0)} + \rho_v, \tag{10.42}$$

where $\alpha_v(\cdot)$ is a nondecreasing function of $\left\|\left(w_e^*(t_0), \tilde{r}(t_0), z_e(t_0)\right)\right\|$, and σ_v and ρ_v are positive constants. The constant ρ_v depends on the environmental disturbances and the path curvature. Indeed, this constant is equal to zero when there are no environmental disturbances and $c(s) = 0$.

Remark 10.2. From (10.39), it is seen that the initial conditions are mainly limited by the path curvature $c(s)$. If the initial conditions do not satisfy (10.39), one can generate a smooth curve segment, $\boldsymbol{\Omega}_0$, such that it is tangent to the path $\boldsymbol{\Omega}$ and that (10.39) holds, see Figure 10.2.

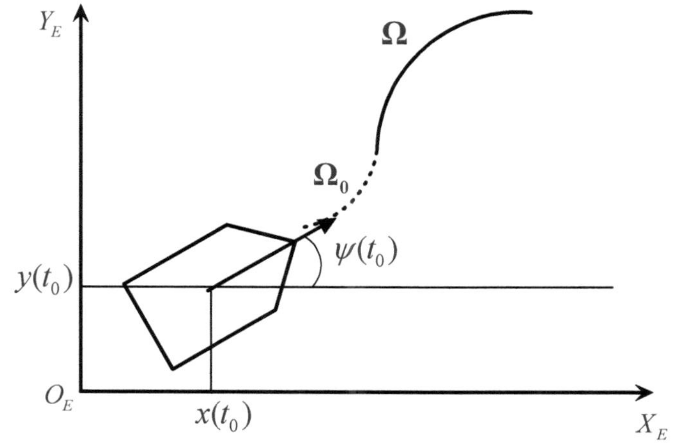

Figure 10.2 Overcoming "bad" initial conditions

10.3 Output Feedback

10.3.1 Observer Design

To design an observer that estimates the sway and yaw velocities from the ship position and orientation measurements, we rewrite (10.1) as follows

$$\begin{aligned}
\dot{x} &= u\cos(\psi) - v\sin(\psi),\\
\dot{\boldsymbol{x}}_1 &= \boldsymbol{F}_1(\boldsymbol{x}_1) + \boldsymbol{J}_1(\boldsymbol{x}_1)\boldsymbol{x}_2,\\
\dot{\boldsymbol{x}}_2 &= \boldsymbol{K}_2\boldsymbol{x}_2 - \boldsymbol{D}_2(\boldsymbol{x}_2)\boldsymbol{x}_2 + \boldsymbol{\tau}_2 + \boldsymbol{\tau}_{w2}(t),
\end{aligned} \tag{10.43}$$

where

$$\boldsymbol{x}_1 = \begin{bmatrix} y \\ \psi \end{bmatrix},\ \boldsymbol{x}_2 = \begin{bmatrix} y \\ \psi \end{bmatrix},\ \boldsymbol{F}_1(\boldsymbol{x}_1) = \begin{bmatrix} u\sin(\psi) \\ 0 \end{bmatrix},$$

$$\boldsymbol{\tau}_2 = \begin{bmatrix} 0 \\ \dfrac{1}{m_{33}}\tau_r \end{bmatrix},\ \boldsymbol{\tau}_{w2}(t) = \begin{bmatrix} \dfrac{1}{m_{22}}\tau_{wv}(t) \\ \dfrac{1}{m_{33}}\tau_{wr}(t) \end{bmatrix},\ \boldsymbol{J}_1(\boldsymbol{x}_1) = \begin{bmatrix} \cos(\psi) & 0 \\ 0 & 1 \end{bmatrix},$$

$$\boldsymbol{D}_2(\boldsymbol{x}_2) = \begin{bmatrix} \sum_{i=2}^{3} \dfrac{d_{vi}}{m_{22}}|v|^{i-1} & 0 \\ 0 & \sum_{i=2}^{3} \dfrac{d_{ri}}{m_{33}}|r|^{i-1} \end{bmatrix},\ \boldsymbol{K}_2 = \begin{bmatrix} -\dfrac{d_{22}}{m_{22}} & -\dfrac{m_{11}u}{m_{22}} \\ \dfrac{(m_{11}-m_{22})u}{m_{33}} & -\dfrac{d_{33}}{m_{33}} \end{bmatrix}. \tag{10.44}$$

If we use the observer designed in [116, 126] and Chapter 9, then the matrix $\boldsymbol{K}_2$ has to be Hurwitz, i.e., the ship has to possess straight-line stability. This limitation motivates us to seek a new observer in this section. The idea is to find a coordinate transformation to transform the system (10.1) to a form that does not require $\boldsymbol{K}_2$ to be a Hurwitz matrix to design an observer. To this end, we define

$$\boldsymbol{z}_2 = \boldsymbol{x}_2 - \boldsymbol{\chi}_2(x, \boldsymbol{x}_1), \tag{10.45}$$

where $\boldsymbol{\chi}_2(x, \boldsymbol{x}_1)$ is a locally Lipschitz vector function, which will be determined later. Using (10.45), we write the last two equations of (10.43) as follows:

$$\begin{aligned}
\dot{\boldsymbol{x}}_1 &= \boldsymbol{F}_1(\boldsymbol{x}_1) + \boldsymbol{J}_1(\boldsymbol{x}_1)\boldsymbol{\chi}_2(x, \boldsymbol{x}_1) + \boldsymbol{J}_1(\boldsymbol{x}_1)\boldsymbol{z}_2, \\
\dot{\boldsymbol{z}}_2 &= \left(\boldsymbol{K}_2 + \frac{\partial \boldsymbol{\chi}_2(x, \boldsymbol{x}_1)}{\partial x}\boldsymbol{g}_2(\boldsymbol{x}_1) - \frac{\partial \boldsymbol{\chi}_2(x, \boldsymbol{x}_1)}{\partial \boldsymbol{x}_1}\boldsymbol{J}_1(\boldsymbol{x}_1)\right)\boldsymbol{z}_2 - \boldsymbol{D}_2(\boldsymbol{z}_2 + \\
&\quad \boldsymbol{\chi}_2(x, \boldsymbol{x}_1))(\boldsymbol{z}_2 + \boldsymbol{\chi}_2(x, \boldsymbol{x}_1)) + \boldsymbol{\Phi}(x, \boldsymbol{x}_1) + \boldsymbol{\tau}_2 + \boldsymbol{\tau}_{w2}(t),
\end{aligned} \tag{10.46}$$

where

$$\begin{aligned}
\boldsymbol{\Phi}(x, \boldsymbol{x}_1) &= \boldsymbol{K}_2\boldsymbol{\chi}_2(x, \boldsymbol{x}_1) - \frac{\partial \boldsymbol{\chi}_2(x, \boldsymbol{x}_1)}{\partial x}(u\cos(\psi) - \boldsymbol{g}_2(\boldsymbol{x}_1)\boldsymbol{\chi}_2(x, \boldsymbol{x}_1)) - \\
&\quad \frac{\partial \boldsymbol{\chi}_2(x, \boldsymbol{x}_1)}{\partial \boldsymbol{x}_1}(\boldsymbol{F}_1(\boldsymbol{x}_1) + \boldsymbol{J}_1(\boldsymbol{x}_1)\boldsymbol{\chi}_2(x, \boldsymbol{x}_1)), \\
\boldsymbol{g}_2(\boldsymbol{x}_1) &= \left[\sin(\psi)\ 0\right]^T, \\
\boldsymbol{D}_2(\boldsymbol{z}_2 + \boldsymbol{\chi}_2(x, \boldsymbol{x}_1)) &= \\
&\begin{bmatrix} \sum_{i=2}^{3} \frac{d_{vi}}{m_{22}}|z_{21} + \chi_{21}(x, \boldsymbol{x}_1)|^{i-1} & 0 \\ 0 & \sum_{i=2}^{3} \frac{d_{ri}}{m_{33}}|z_{22} + \chi_{21}(x, \boldsymbol{x}_1)|^{i-1} \end{bmatrix},
\end{aligned} \tag{10.47}$$

where z_{2i} and $\chi_{2i}(x, \boldsymbol{x}_1)$, $i = 1, 2$, are the first and second elements of $\boldsymbol{z}_2$ and $\boldsymbol{\chi}_2(x, \boldsymbol{x}_1)$, respectively. From (10.46), one can design a reduced order observer. However it is often noise-sensitive. We here propose the following observer:

$$\begin{aligned}
\dot{\hat{\boldsymbol{x}}}_1 &= \boldsymbol{F}_1(\boldsymbol{x}_1) + \boldsymbol{J}_1(\boldsymbol{x}_1)\boldsymbol{\chi}_2(x, \boldsymbol{x}_1) + \boldsymbol{J}_1(\boldsymbol{x}_1)\hat{\boldsymbol{z}}_2 + \boldsymbol{K}_{01}(\boldsymbol{x}_1 - \hat{\boldsymbol{x}}_1), \\
\dot{\hat{\boldsymbol{z}}}_2 &= \left(\boldsymbol{K}_2 + \frac{\partial \boldsymbol{\chi}_2(x, \boldsymbol{x}_1)}{\partial x}\boldsymbol{g}_2(\boldsymbol{x}_1) - \frac{\partial \boldsymbol{\chi}_2(x, \boldsymbol{x}_1)}{\partial \boldsymbol{x}_1}\boldsymbol{J}_1(\boldsymbol{x}_1)\right)\hat{\boldsymbol{z}}_2 - \boldsymbol{D}_2(\hat{\boldsymbol{z}}_2 + \\
&\quad \boldsymbol{\chi}_2(x, \boldsymbol{x}_1))(\hat{\boldsymbol{z}}_2 + \boldsymbol{\chi}_2(x, \boldsymbol{x}_1)) + \boldsymbol{\Phi}(x, \boldsymbol{x}_1) + \boldsymbol{\tau}_2 + \boldsymbol{K}_{02}(\boldsymbol{x}_1 - \hat{\boldsymbol{x}}_1)
\end{aligned} \tag{10.48}$$

where $\hat{\boldsymbol{x}}_1 := [\hat{y}\ \hat{\psi}]^T$ and $\hat{\boldsymbol{z}}_2$ are estimates of $\boldsymbol{x}_1$ and $\boldsymbol{z}_2$, respectively, and $\boldsymbol{K}_{01}$ and $\boldsymbol{K}_{02}$ are the observer gains to be selected later. From (10.46) and (10.48), we have the observer error dynamics:

$$\begin{aligned}
\dot{\tilde{\boldsymbol{x}}}_1 &= -\boldsymbol{K}_{01}\tilde{\boldsymbol{x}}_1 + \boldsymbol{J}_1(\boldsymbol{x}_1)\tilde{\boldsymbol{z}}_2,\\
\dot{\tilde{\boldsymbol{z}}}_2 &= -\boldsymbol{K}_{02}\tilde{\boldsymbol{x}}_1 + \left(\boldsymbol{K}_2 + \frac{\partial \boldsymbol{\chi}_2(x,\boldsymbol{x}_1)}{\partial x}\boldsymbol{g}_2(\boldsymbol{x}_1) - \frac{\partial \boldsymbol{\chi}_2(x,\boldsymbol{x}_1)}{\partial \boldsymbol{x}_1}\boldsymbol{J}_1(\boldsymbol{x}_1)\right)\tilde{\boldsymbol{z}}_2 -\\
&\quad \Big(\boldsymbol{D}_2(\boldsymbol{z}_2 + \boldsymbol{\chi}_2(x,\boldsymbol{x}_1))(\boldsymbol{z}_2 + \boldsymbol{\chi}_2(x,\boldsymbol{x}_1)) - \boldsymbol{D}_2(\hat{\boldsymbol{z}}_2 +\\
&\quad \boldsymbol{\chi}_2(x,\boldsymbol{x}_1))(\hat{\boldsymbol{z}}_2 + \boldsymbol{\chi}_2(x,\boldsymbol{x}_1))\Big) + \boldsymbol{\tau}_{w2}(t),
\end{aligned} \tag{10.49}$$

where $\tilde{\boldsymbol{x}}_1 := [\tilde{y}\ \tilde{\psi}]^T = \boldsymbol{x}_1 - \hat{\boldsymbol{x}}_1$, and $\tilde{\boldsymbol{z}}_2 = \boldsymbol{z}_2 - \hat{\boldsymbol{z}}_2$. Now, considering the Lyapunov function

$$V_0 = \tilde{\boldsymbol{x}}_1^T \boldsymbol{P}_{01}\tilde{\boldsymbol{x}}_1 + \tilde{\boldsymbol{z}}_2^T \boldsymbol{P}_{02}\tilde{\boldsymbol{z}}_2, \tag{10.50}$$

where $\boldsymbol{P}_{01}$ and $\boldsymbol{P}_{02}$ are positive definite matrices, whose time derivative along the solutions of (10.49) satisfies:

$$\dot{V}_0 \leq -\tilde{\boldsymbol{x}}_1^T \boldsymbol{Q}_{01}\tilde{\boldsymbol{x}}_1 - \tilde{\boldsymbol{z}}_2^T \boldsymbol{Q}_{02}\tilde{\boldsymbol{z}}_2 + \boldsymbol{\tau}_{w2}^T(t)\boldsymbol{P}_{02}\tilde{\boldsymbol{z}}_2 + \tilde{\boldsymbol{z}}_2^T \boldsymbol{P}_{02}\boldsymbol{\tau}_{w2}(t), \tag{10.51}$$

where

$$\begin{aligned}
\boldsymbol{Q}_{01} &= \boldsymbol{K}_{01}^T\boldsymbol{P}_{01} + \boldsymbol{P}_{01}\boldsymbol{K}_{01},\\
\boldsymbol{Q}_{02} &= -\left(\boldsymbol{K}_2 + \frac{\partial \boldsymbol{\chi}_2(x,\boldsymbol{x}_1)}{\partial x}\boldsymbol{g}_2(\boldsymbol{x}_1) - \frac{\partial \boldsymbol{\chi}_2(x,\boldsymbol{x}_1)}{\partial \boldsymbol{x}_1}\boldsymbol{J}_1(\boldsymbol{x}_1)\right)^T \boldsymbol{P}_{02} -\\
&\quad \boldsymbol{P}_{02}\left(\boldsymbol{K}_2 + \frac{\partial \boldsymbol{\chi}_2(x,\boldsymbol{x}_1)}{\partial x}\boldsymbol{g}_2(\boldsymbol{x}_1) - \frac{\partial \boldsymbol{\chi}_2(x,\boldsymbol{x}_1)}{\partial \boldsymbol{x}_1}\boldsymbol{J}_1(\boldsymbol{x}_1)\right),
\end{aligned} \tag{10.52}$$

and we have chosen

$$\boldsymbol{P}_{02}\boldsymbol{K}_{02} - \boldsymbol{J}_1^T(\boldsymbol{x}_1)\boldsymbol{P}_{01} = 0 \tag{10.53}$$

and used the following inequality:

$$\begin{aligned}
&(\boldsymbol{D}_2(\boldsymbol{z}_2 + \boldsymbol{\chi}_2(x,\boldsymbol{x}_1))(\boldsymbol{z}_2 + \boldsymbol{\chi}_2(x,\boldsymbol{x}_1)) - \boldsymbol{D}_2(\hat{\boldsymbol{z}}_2 + \boldsymbol{\chi}_2(x,\boldsymbol{x}_1))\\
&(\hat{\boldsymbol{z}}_2 + \boldsymbol{\chi}_2(x,\boldsymbol{x}_1)))\,\boldsymbol{P}_{02}\tilde{\boldsymbol{z}}_2 + \tilde{\boldsymbol{z}}_2^T\boldsymbol{P}_{02}\,(\boldsymbol{D}_2(\boldsymbol{z}_2 + \boldsymbol{\chi}_2(x,\boldsymbol{x}_1))(\boldsymbol{z}_2 +\\
&\boldsymbol{\chi}_2(x,\boldsymbol{x}_1)) - \boldsymbol{D}_2(\hat{\boldsymbol{z}}_2 + \boldsymbol{\chi}_2(x,\boldsymbol{x}_1))(\hat{\boldsymbol{z}}_2 + \boldsymbol{\chi}_2(x,\boldsymbol{x}_1))) \geq 0.
\end{aligned} \tag{10.54}$$

It is not hard to show that there always exist $\boldsymbol{K}_{01}$, $\boldsymbol{K}_{02}$ and $\boldsymbol{\chi}_2(x,\boldsymbol{x}_1)$ such that $\boldsymbol{Q}_{01}$ and $\boldsymbol{Q}_{02}$ are positive definite and (10.53) holds. For example, one can take

$$\begin{aligned}
\boldsymbol{K}_{01} &= \boldsymbol{K}_{01}^T > 0,\\
\boldsymbol{K}_{02} &= \boldsymbol{P}_{02}^{-1}\boldsymbol{J}_1^T(\boldsymbol{x}_1)\boldsymbol{P}_{01},\\
\boldsymbol{\chi}_2(x,\boldsymbol{x}_1) &= \left[-\frac{m_{11}u}{m_{22}}\psi\ \ 0\right]^T.
\end{aligned} \tag{10.55}$$

Therefore (10.51) implies that

$$\|\tilde{\mathbf{x}}_0(t)\| \leq \varphi_0\,\|\tilde{\mathbf{x}}_0(t_0)\|\,e^{-\eta_0(t-t_0)} + \xi_0, \tag{10.56}$$

where φ_0, η_0, and ξ_0 are some positive constants, and $\tilde{\boldsymbol{x}}_{\boldsymbol{0}} = [\tilde{\boldsymbol{x}}_{\boldsymbol{1}}\ \tilde{\boldsymbol{z}}_{\boldsymbol{2}}]^{\boldsymbol{T}}$. It is noted that ξ_0 cannot be made arbitrarily small since the ratio $\frac{\lambda_{\min}(\boldsymbol{Q}_{\boldsymbol{02}})}{\lambda_{\max}(\boldsymbol{P}_{\boldsymbol{02}})}$ is maximized with the choice of $\boldsymbol{Q}_{\boldsymbol{02}} = \mathrm{diag}(1,1)$. By defining an estimate of the velocity vector as

$$\hat{\boldsymbol{x}}_{\boldsymbol{2}} = \hat{\boldsymbol{z}}_{\boldsymbol{2}} + \boldsymbol{\chi}_{\boldsymbol{2}}(x, \boldsymbol{x}_{\boldsymbol{1}}), \tag{10.57}$$

where $\hat{\mathbf{x}}_2 = [\hat{v}\ \hat{r}]^T$, then the velocity observer errors $\tilde{\mathbf{x}}_2 = [\tilde{v}\ \tilde{r}]^T$ satisfy:

$$\|(\tilde{v}(t), \tilde{r}(t))\| \leq \varphi_0 \|\tilde{\boldsymbol{x}}_{\boldsymbol{0}}(t_0)\| e^{-\eta_0(t-t_0)} + \xi_0. \tag{10.58}$$

It is seen that when there are no environmental disturbances, the observer errors globally exponentially tend to zero.

10.3.2 Control Design

The control design in this section is very similar to the one of the state feedback controller. We therefore present it briefly. Define

$$\begin{aligned}
\hat{\phi} &= \arctan\left(\frac{\hat{v}}{u}\right), \\
\hat{u}_t &= \sqrt{u^2 + \hat{v}^2}, \\
\hat{\psi}_e^* &= \psi_e + \hat{\phi}.
\end{aligned} \tag{10.59}$$

By denoting $\tilde{\phi} = \phi - \hat{\phi}$, using the Taylor series expansion, it can be shown that

$$|\tilde{\phi}(t)| = \left| \frac{\arctan(v/u) - \arctan(\hat{v}/u)}{(v - \hat{v})/u} \frac{v - \hat{v}}{u} \right| \leq \frac{1}{u} |\tilde{v}(t)|. \tag{10.60}$$

Similar to the state feedback control design, we apply the following coordinate transformation

$$\hat{w}_e^* = \hat{\psi}_e^* + \arcsin\left(\frac{kz_e}{\sqrt{1 + (kz_e)^2}}\right) \tag{10.61}$$

to (10.48) in conjunction with (10.59). After some simple calculation, we have

$$\begin{aligned}
\dot{z}_e &= -\frac{k\hat{u}_t z_e}{\sqrt{1+(kz_e)^2}} + \frac{\hat{u}_t \sin(\hat{w}_e^*)}{\sqrt{1+(kz_e)^2}} - \frac{(\cos(\hat{w}_e^*) - 1)k\hat{u}_t z_e}{\sqrt{1+(kz_e)^2}} + \\
&\quad \frac{\tilde{v}\cos(\hat{w}_e^* - \hat{\phi})}{\sqrt{1+(kz_e)^2}} + \frac{\tilde{v}\sin(\hat{w}_e^* - \hat{\phi})kz_e}{\sqrt{1+(kz_e)^2}}, \\
\dot{\hat{w}}_e^* &= \hat{r}\left(1 - \frac{m_{11}}{m_{22}} \frac{u^2}{\hat{u}_t^2}\right) - \frac{c(s)\hat{u}_t \cos(\hat{\psi}_e^*)}{1 - c(s)z_e} -
\end{aligned}$$

$$
\begin{aligned}
&\frac{u}{\hat{u}_t^2}\left(\frac{d_{22}}{m_{22}}\hat{v}+\sum_{i=2}^{3}\frac{d_{vi}}{m_{22}}|\hat{v}|^{i-1}\hat{v}\right)+\frac{k\hat{u}_t\sin(\hat{\psi}_e^*)}{1+(kz_e)^2}+\\
&\frac{u}{\hat{u}_t^2}\cos(\psi)\tilde{y}+\tilde{r}\left(1-\frac{m_{11}}{m_{22}}\frac{u^2}{\hat{u}_t^2}\right)+\frac{k\tilde{v}\cos(\psi_e)}{1+(kz_e)^2}+\frac{c(s)\tilde{v}\sin(\psi_e)}{1-c(s)z_e},\\
\dot{\hat{v}} &= -\frac{m_{11}u}{m_{22}}\hat{r}-\frac{d_{22}}{m_{22}}\hat{v}-\sum_{i=2}^{3}\frac{d_{vi}}{m_{22}}|\hat{v}|^{i-1}\hat{v}+\cos(\psi)\tilde{y}-\frac{m_{11}}{m_{22}}u\tilde{r},\\
\dot{\hat{r}} &= \frac{(m_{11}-m_{22})u}{m_{33}}\hat{v}-\frac{d_{33}}{m_{33}}\hat{r}-\sum_{i=2}^{3}\frac{d_{ri}}{m_{33}}|\hat{r}|^{i-1}\hat{r}+\frac{1}{m_{33}}\tau_r+\tilde{\psi}
\end{aligned}
\tag{10.62}
$$

where for simplicity, we have taken

$$
\begin{aligned}
\boldsymbol{K}_{02} &= \boldsymbol{J}_1^T(\boldsymbol{x}_1),\\
\boldsymbol{\chi}_2(x,\boldsymbol{x}_1) &= \left[-\frac{m_{11}u}{m_{22}}\psi \;\; 0\right]^T.
\end{aligned}
\tag{10.63}
$$

The control design consists of two steps as follows:

Step 1

Introduce the virtual yaw velocity error as

$$
r_e = \hat{r}-\hat{r}_d, \tag{10.64}
$$

where $\hat{r}_d$ is the virtual yaw velocity control which is chosen as

$$
\begin{aligned}
\hat{r}_d = &\frac{1}{1-\frac{m_{11}}{m_{22}}\frac{u^2}{\hat{u}_t^2}}\left(-k_1\hat{w}_e^*+\frac{c(s)\hat{u}_t}{1-c(s)z_e}\cos(\hat{\psi}_e^*)+\right.\\
&\left.\frac{u}{\hat{u}_t^2}\left(\frac{d_{22}}{m_{22}}\hat{v}+\sum_{i=2}^{3}\frac{d_{vi}}{m_{22}}|\hat{v}|^{i-1}\hat{v}\right)-\frac{k\hat{u}_t\sin(\hat{\psi}_e^*)}{1+(kz_e)^2}-\frac{d_1\hat{w}_e^*}{(1-c(s)z_e)^2}\right),
\end{aligned}
\tag{10.65}
$$

where k_1 and d_1 are positive constants. The term multiplied by d_1 is the nonlinear damping term to overcome the effect of the observer errors. It is seen that $\hat{r}_d$ is a smooth function of s, z_e, ψ_e, and $\hat{v}$. Substituting (10.64) and (10.65) into the second equation of (10.62) results in

$$
\dot{\hat{w}}_e^* = -k_1\hat{w}_e^*+r_e\left(1-\frac{m_{11}}{m_{22}}\frac{u^2}{\hat{u}_t^2}\right)+\hat{\boldsymbol{f}}_1^T(\cdot)\tilde{\boldsymbol{x}}-\frac{d_1\hat{w}_e^*}{(1-c(s)z_e)^2}, \tag{10.66}
$$

where

$$\hat{\boldsymbol{f}}_{\mathbf{1}}(\cdot)=\begin{bmatrix}\dfrac{u\cos(\psi)}{\hat{u}_t^2}\\0\\\dfrac{c(s)\sin(\psi_e)}{1-c(s)z_e}+\dfrac{k\cos(\psi_e)}{1+(kz_e)^2}\\1-\dfrac{m_{11}}{m_{22}}\dfrac{u^2}{\hat{u}_t^2}\end{bmatrix},\ \tilde{\boldsymbol{x}}=\begin{bmatrix}\tilde{y}\\\tilde{\psi}\\\tilde{v}\\\tilde{r}\end{bmatrix}. \tag{10.67}$$

Step 2

Differentiating both sides of (10.64) yields

$$\begin{aligned}\dot{r}_e=&\ \frac{(m_{11}-m_{22})u}{m_{33}}\hat{v}-\frac{d_{33}}{m_{33}}\hat{r}-\sum_{i=2}^{3}\frac{d_{ri}}{m_{33}}|\hat{r}|^{i-1}\hat{r}+\frac{1}{m_{33}}\tau_r-\\&\frac{\partial\hat{r}_d}{\partial s}\frac{\cos(\hat{\psi}_e^*)}{1-c(s)z_e}-\frac{\partial\hat{r}_d}{\partial z_e}\sin(\hat{\psi}_e^*)-\frac{\partial\hat{r}_d}{\partial\psi_e}\left(\hat{r}-\frac{c(s)\cos(\hat{\psi}_e^*)}{1-c(s)z_e}\right)-\\&\frac{\partial\hat{r}_d}{\partial\hat{v}}\left(-\frac{m_{11}u}{m_{22}}\hat{r}-\frac{d_{22}}{m_{22}}\hat{v}-\sum_{i=2}^{3}\frac{d_{vi}}{m_{22}}|\hat{v}|^{i-1}\hat{v}\right)+\hat{\boldsymbol{f}}_{\mathbf{2}}^{\boldsymbol{T}}(\cdot)\tilde{\boldsymbol{x}},\end{aligned} \tag{10.68}$$

where

$$\hat{\boldsymbol{f}}_{\mathbf{2}}(\cdot)=\begin{bmatrix}-\dfrac{\partial\hat{r}_d}{\partial\hat{v}}\cos(\psi)\\1\\\dfrac{\partial\hat{r}_d}{\partial s}\dfrac{\sin(\psi_e)}{1-c(s)z_e}-\dfrac{\partial\hat{r}_d}{\partial z_e}\cos(\psi_e)-\dfrac{\partial\hat{r}_d}{\partial\psi_e}\dfrac{c(s)\sin(\psi_e)}{1-c(s)z_e}\\-\dfrac{\partial\hat{r}_d}{\partial\psi_e}-\dfrac{\partial\hat{r}_d}{\partial\hat{v}}\dfrac{m_{11}}{m_{22}}u\end{bmatrix}. \tag{10.69}$$

From (10.68), the control τ_r is designed without canceling the useful nonlinear damping terms as follows:

$$\begin{aligned}\tau_r=&\ m_{33}\left(-k_2r_e-\frac{(m_{11}-m_{22})u}{m_{33}}\hat{v}+\frac{d_{33}}{m_{33}}\hat{r}_d+\sum_{i=2}^{3}\frac{d_{ri}}{m_{33}}|\hat{r}|^{i-1}\hat{r}_d+\right.\\&\frac{\partial\hat{r}_d}{\partial s}\frac{\cos(\hat{\psi}_e^*)}{1-c(s)z_e}+\frac{\partial\hat{r}_d}{\partial z_e}\hat{u}_t\sin(\hat{\psi}_e^*)+\frac{\partial\hat{r}_d}{\partial\psi_e}\left(\hat{r}-\frac{c(s)\cos(\hat{\psi}_e^*)}{1-c(s)z_e}\right)-\frac{\partial\hat{r}_d}{\partial\hat{v}}\times\\&\left.\left(\frac{m_{11}u}{m_{22}}\hat{r}+\frac{d_{22}}{m_{22}}\hat{v}+\sum_{i=2}^{3}\frac{d_{vi}}{m_{22}}|\hat{v}|^{i-1}\hat{v}\right)-\sum_{i=1}^{4}d_2\hat{f}_{2i}^2(\cdot)r_e-\hat{w}_e^*\left(1-\frac{m_{11}}{m_{22}}\frac{u^2}{\hat{u}_t^2}\right)\right),\end{aligned} \tag{10.70}$$

where k_2 and d_2 are positive design constants, $\hat{f}_{2i}(\cdot)$ is the ith component of $\hat{\boldsymbol{f}}_2(\cdot)$. Substituting (10.70) into (10.68) yields

$$\begin{aligned}\dot{r}_e = & -k_2 r_e - \frac{d_{33}}{m_{33}} r_e - \sum_{i=2}^{3} \frac{d_{ri}}{m_{33}} |\hat{r}|^{i-1} r_e - \left(1 - \frac{m_{11}}{m_{22}} \frac{u^2}{\hat{u}_t^2}\right) \hat{w}_e^* + \\ & \hat{\boldsymbol{f}}_2^T(\cdot)\tilde{\boldsymbol{x}} - \sum_{i=1}^{4} d_2 \hat{f}_{2i}^2(\cdot) r_e. \qquad (10.71)\end{aligned}$$

We now present the second main result of this chapter, the proof of which is given in the next section.

Theorem 10.2. *Under Assumption 10.2, if the output feedback control law (10.70) and the observer (10.48) are applied to the ship system (10.1) then there exist feasible initial conditions such that the regulation errors $(z_e(t), \psi_e^*(t))$ converge to a ball centered at the origin with an appropriate choice of the design constants k, k_i and d_i with $i = 1, 2$. Furthermore when there are no environmental disturbances, the regulation errors converge to zero asymptotically. In addition, the velocity $v(t)$ is always bounded.*

10.3.3 Stability Analysis

To prove Theorem 10.2, we first consider the $(\hat{w}_e^*, r_e)$-dynamics then move to the $(\hat{v}, z_e)$ dynamics.

$(\hat{w}_e^*, r_e)$-dynamics

Consider the following Lyapunov function

$$V_1 = \frac{1}{2}\hat{w}_e^{*2} + \frac{1}{2} r_e^2, \qquad (10.72)$$

whose time derivative along the solutions of (10.66) and (10.71) satisfies

$$\begin{aligned}\dot{V}_1 = & -k_1 \hat{w}_e^{*2} + \hat{w}_e^* \hat{\boldsymbol{f}}_1^T(\cdot)\tilde{\boldsymbol{x}} - \frac{d_1 \hat{w}_e^{*2}}{(1 - c(s) z_e)^2} - k_2 r_e^2 - \frac{d_{33}}{m_{33}} r_e^2 - \\ & \sum_{i=2}^{3} \frac{d_{ri}}{m_{33}} |\hat{r}|^{i-1} r_e^2 + r_e \hat{\boldsymbol{f}}_2^T(\cdot)\tilde{\boldsymbol{x}} - d_2 r_e^2 \sum_{i=1}^{4} \hat{f}_{2i}^2(\cdot) \\ \leq & -(k_1 - \mu_{11}) \hat{w}_e^{*2} - k_2 r_e^2 + \mu_{12} \|\tilde{\boldsymbol{x}}\|^2, \qquad (10.73)\end{aligned}$$

where μ_{11} and μ_{12} are some positive constants. From (10.73) and (10.56), it is direct to show that

$$\left\|\left(\hat{w}_e^*(t), r_e(t)\right)\right\| \leq \left\|\left(\hat{w}_e^*(t_0), r_e(t_0)\right)\right\| e^{-\sigma_1(t-t_0)} + \rho_1, \tag{10.74}$$

for some positive constants ρ_1 and σ_1. It is noted that ρ_1 can be made arbitrarily small by adjusting the design constants k_1, k_2, d_1, and d_2.

$(\hat{v}, z_e)$-dynamics

To obtain convergence of z_e, we apply Lemma 10.1 to the first equation of (10.62) by verifying all conditions of this lemma. Indeed, the first equation of (10.62) is in the form of the system studied in Lemma 10.1 by writing this equation as

$$\dot{z}_e = f_{ze}(\cdot) + g_{ze}(\cdot), \tag{10.75}$$

where

$$\begin{aligned} f_{ze}(\cdot) &= -\frac{k\hat{u}_t z_e}{\sqrt{1+(kz_e)^2}}, \\ g_{ze}(\cdot) &= \frac{\hat{u}_t \sin(\hat{w}_e^*)}{\sqrt{1+(kz_e)^2}} - (\cos(\hat{w}_e^*)-1)\frac{k\hat{u}_t z_e}{\sqrt{1+(kz_e)^2}} + \\ &\quad \frac{\tilde{v}\cos(\hat{w}_e^* - \hat{\phi})}{\sqrt{1+(kz_e)^2}} + \frac{\tilde{v}\sin(\hat{w}_e^* - \hat{\phi})kz_e}{\sqrt{1+(kz_e)^2}}. \end{aligned} \tag{10.76}$$

Verifying Condition C1. By taking the Lyapunov function $V_2 = 0.5z_e^2$, it can be shown that this condition holds with

$$c_0 = 0,\ c_1 = c_2 = 0.5,\ c_3 = 1,\ c = k^2,\ c_4 = k\hat{u}_t. \tag{10.77}$$

Verifying Condition C2. From (10.76), we have

$$|g_{ze}(\cdot)| \leq \frac{1}{\sqrt{1+(kz_e)^2}}\left(\hat{u}_t + 1 + (\hat{u}_t+1)k\,|z_e|\right)\left\|\left(\tilde{v}, \hat{w}_e^*\right)\right\|, \tag{10.78}$$

which implies that condition C2 holds with

$$\lambda_1 = \hat{u}_t + 1, \lambda_2 = k\,(\hat{u}_t + 1). \tag{10.79}$$

Verifying Condition C3. This condition directly holds from (10.74) and (10.56).

Verifying Condition C4. From (10.77), (10.74), and (10.56), this condition becomes

$$k\hat{u}_t - (\rho_1 + \xi_0)\left(\frac{\hat{u}_t + 1}{4\mu_0} + k(\hat{u}_t + 1)\right) > 0. \tag{10.80}$$

It is noted that ρ_1 can be made arbitrarily small but ξ_0 cannot, see (10.56). Therefore Condition C4 is satisfied if the environmental disturbances are not too large and the ship forward speed u is high enough such that (10.80) holds for some positive constant μ_0. This can be explained as follows: When there are large environmental disturbances, the observer cannot estimate sufficiently accurate velocities. In addition, the low ship speed cannot compensate for the environmental disturbances. Consequently, the ship under large environmental disturbances in question will diverge from the path. All the conditions of Lemma 10.1 have been verified, we have

$$|z_e(t)| \leq \alpha_z(\cdot)e^{-\sigma_z(t-t_0)} + \rho_z, \tag{10.81}$$

where α_z, σ_z and ρ_z are calculated as in Lemma 10.1. The feasibility of initial conditions such that $1 - c(s)z_e \geq \delta^* > 0$ is similar to the case of the state feedback design, see (10.39).

Similar to the state feedback control design, it is not hard to show that

$$|v(t)| \leq \alpha_v(\cdot)e^{-\sigma_v(t-t_0)} + \rho_v \tag{10.82}$$

where α_v, σ_v and ρ_v are calculated similarly.

10.4 Simulations

This section validates the control laws (10.15) and (10.70) by simulating them on a monohull ship with the parameters given in Section 5.4. In the simulation, we assume that the disturbances are $\tau_{wv} = 26 \times 10^4(1 + \text{rand}(\cdot))$, and $\tau_{wr} = 950 \times 10^5(1 + \text{rand}(\cdot))$, where rand$(\cdot)$ is random zero-mean noise with the uniform distribution on the interval $[-0.5\ 0.5]$. This choice results in nonzero-mean disturbances. It should be noted that only boundaries of the environmental disturbances are needed in our proposed controllers. We simulate both state feedback and output feedback cases. For both cases, the ship forward speed is 4 m/s.

10.4.1 State Feedback Simulation Results

The initial conditions are chosen as

$$[x(0), y(0), \psi(0), v(0), r(0)] = [-250,\ 50,\ -0.5,\ 0,\ 0].$$

The reference path is a circle centered at the origin with a radius of 200 m. The control parameters are taken as $k = 0.5$, $k_1 = k_2 = 10$, and $\varepsilon_i = 0.2$. Figure 10.3 plots the ship positions and orientation in the (x, y)-plane. The path-following errors are plotted in Figure 10.4. It can be seen from this figure that the orientation error ψ_e^* converges close to zero but not ψ_e due to nonzero sway velocity, and that the position path-following error, z_e, also converges closely to zero. The sway and yaw velocities and control torque τ_r are drawn in Figure 10.5. The high control effort is due to the fact that we simulate on data of a real ship and large environmental disturbances. However, the control magnitude is within the limit of the maximum yaw moment. It is clearly seen from Figure 10.3 that the controller forces the vessel to move from its initial position to the path in the direction perpendicular to the circle. When the vessel closes to the path, the controller is able to drive it along the path in the desired direction. These observations coincide with our control objective. All the simulation results plotted in Figures 10.3, 10.4, and 10.5 illustrate that our control goal is achieved as stated in Theorem 10.1. It is noted that for clarity, we only plot the path-following errors, sway, and yaw velocities, and control input for the first 20 seconds.

10.4.2 Output Feedback Simulation Results

The initial conditions are

$$[x(0), y(0), \psi(0), v(0), r(0)] = [-250,\ 50,\ -0.5,\ -0.2,\ -0.2],$$
$$\left[\hat{x}(0), \hat{y}(0), \hat{\psi}(0), \hat{v}(0), \hat{r}(0)\right] = [-220,\ 40,\ -0.2,\ -0.2,\ 0.1,\ 0.1].$$

The observer gain matrix is matrix chosen as $\boldsymbol{K_{01}} = \text{diag}(1.5, 1.5)$. The control gains are $k = 0.5$, $k_1 = k_2 = 15$, $d_1 = d_2 = 0.5$, and $\varepsilon_i = 0.2$. The environmental disturbances are the same as full state feedback case. Simulation results are plotted in Figures 10.6– 10.8. All the comments on the results are similar to the state feedback simulation results. However, due to the effect of the observer, the performance of the output feedback case is slightly worse than the performance of the state feedback case.

10.5 Conclusions

Although the sway velocity v is controlled in some conventional course-keeping control systems, see for example [12, 134], using Nomoto's second-order model, the sway displacement y is not controlled and can be unbounded. This phenomenon can be seen from the second equation of (10.1), i.e., $\dot{y} = u\sin(\psi) + v\cos(\psi)$, that boundedness of the sway velocity v does not imply that of the sway displacement y. The controllers proposed in this chapter did control the sway displacement y and kept the sway velocity v bounded, and cover the ones in [126] where the linear course stabilization of the underactuated surface ships is addressed. This can be seen from (10.2) by setting $z_e = y$ and the curvature of the path $c(s)$ equal zero. The work in this chapter is based on [135]. A combination of the results in this chapter with the results in [127] is straightforward to cover the case where the roll and pitch modes are not ignored. By setting the value of k equal zero, the proposed controllers in this chapter reduce to the conventional course-keeping controllers,

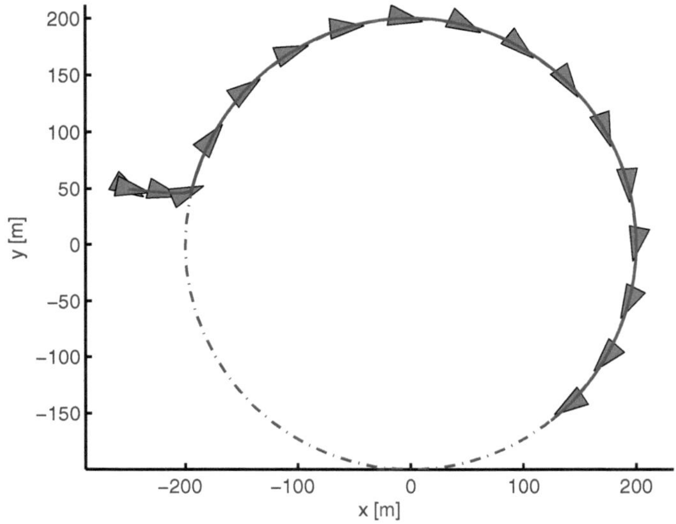

Figure 10.3 State feedback, ship positions and orientation in the (x, y)-plane: Real path (*solid line*) and reference path (*dash-dotted line*)

see for example [11, 12, 134]. However, the sway displacement will grow linearly unbounded under nonzero-mean environmental disturbances.

The main limitation of both state feedback and output feedback controllers designed in this chapter is that the ship must not be too far away from the reference path at the initial time, i.e., the condition (10.3) holds at the initial time t_0. This limitation will be removed in the next chapter where a different approach for solving the path-following problem is proposed.

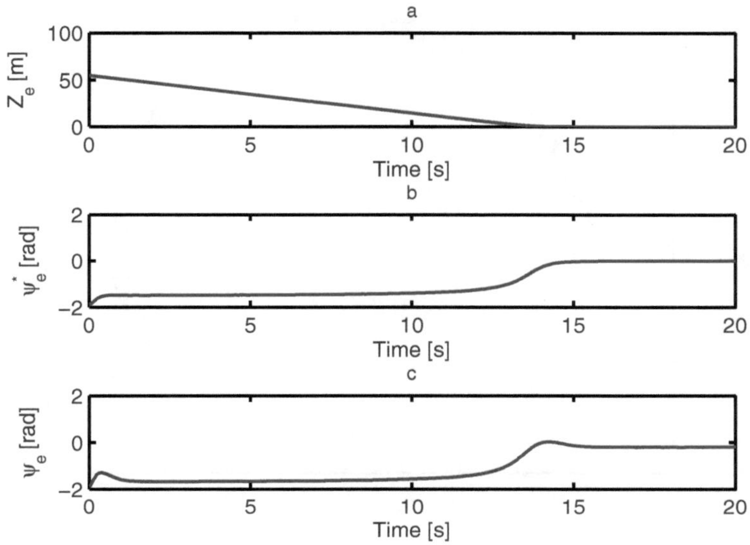

Figure 10.4 State feedback, path-following errors: **a**. Position error z_e; **b**. Modified heading error ψ_e^*; **c**. True heading error ψ_e

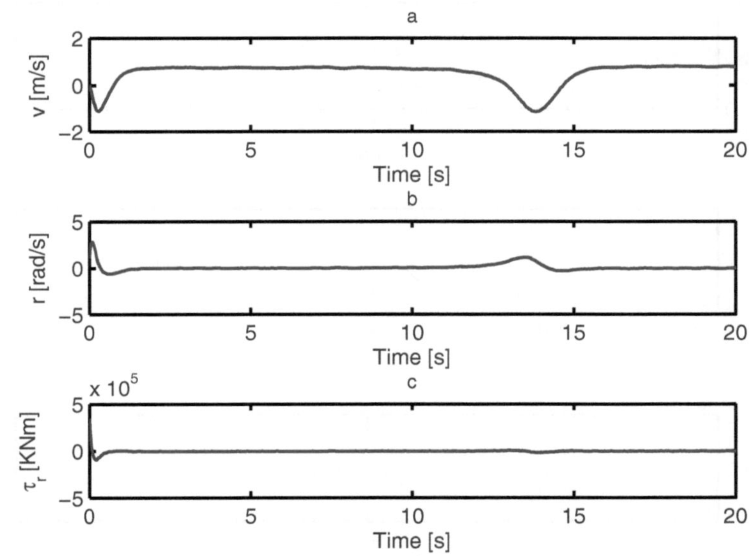

Figure 10.5 State feedback: **a**. Sway velocity v; **b**. Yaw velocity r; **c**. Control torque τ_r

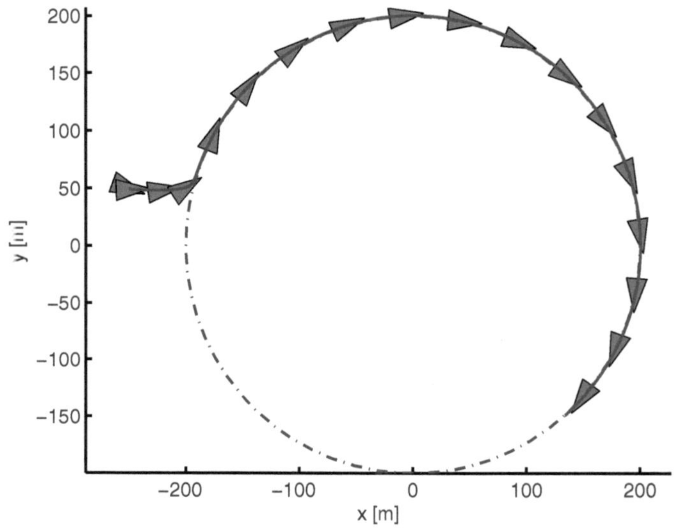

Figure 10.6 Output feedback, ship positions and orientation in the (x, y)-plane: Real path (*solid line*) and reference path (*dash-dotted line*)

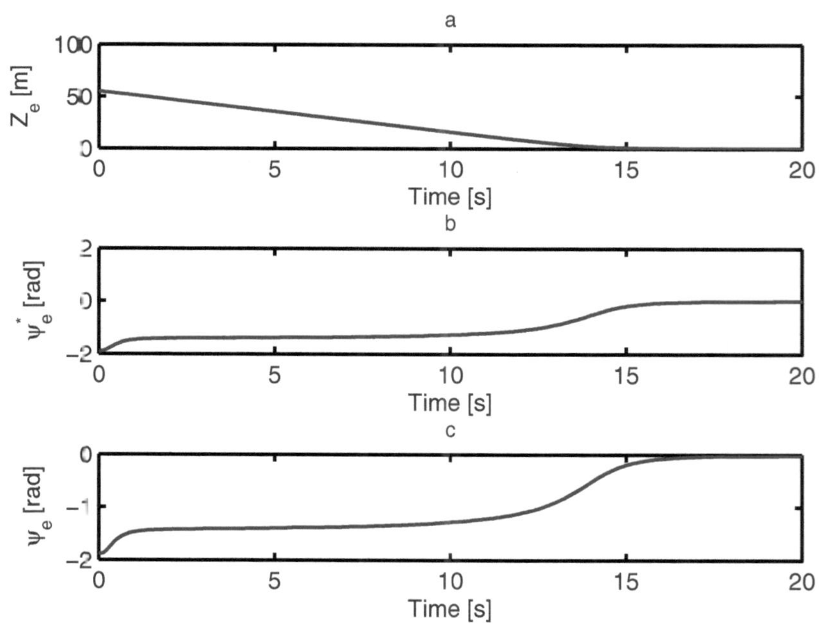

Figure 10.7 Output feedback, path-following errors: **a**. Position error z_e; **b**. Modified heading error ψ_e^*; **c**. True heading error ψ_e

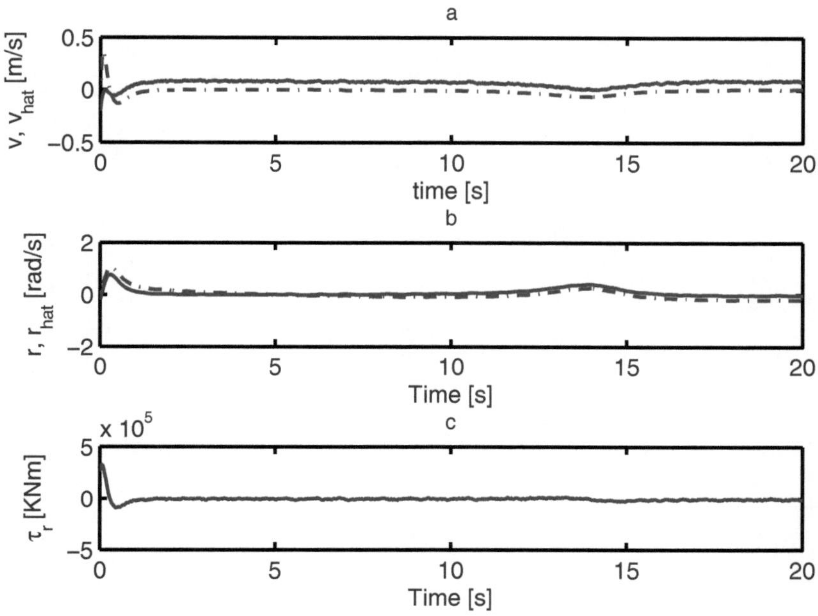

Figure 10.8 Output feedback results: **a**. Sway velocity v (*solid line*) and its estimate $\hat{v}$ (*dash-dotted line*); **b**. Yaw velocity r (*solid line*) and its estimate $\hat{r}$ (*dash-dotted line*); **c**. Control torque τ_r

Chapter 11
Path-following of Underactuated Ships Using Polar Coordinates

This chapter is devoted to a different path-following approach from the preceding chapter. The approach is motivated by the practical experience of steering a ship in the sense that when traveling in open sea the helmsman first looks at the weather map, then generates way-points to avoid the vessel moving into bad weather areas. A continuous reference path curve is then generated so that it goes via (almost) all of the way-points. It is then practical to steer a vessel so that it is in a tube of nonzero adjustable diameter centered on the reference path, and moves along the path with the desired speed. In comparison with the approach in the preceding chapter, the control system developed in this chapter allows the ship to be far away from the desired path at the initial time, and covers situations of practical importance such as parking and point-to-point navigation.

11.1 Control Objective

For the reader's convenience, the mathematical model of the underactuated ship moving in surge, sway, and yaw, see Sections 3.4.1.1 and 3.4.1.2, is rewritten as

$$\begin{aligned}
\dot{x} &= u\cos(\psi) - v\sin(\psi),\\
\dot{y} &= u\sin(\psi) + v\cos(\psi),\\
\dot{\psi} &= r,\\
\dot{u} &= \frac{m_{22}}{m_{11}}vr - \frac{d_{11}}{m_{11}}u - \sum_{i=2}^{3}\frac{d_{ui}}{m_{11}}|u|^{i-1}u + \frac{1}{m_{11}}\tau_u + \frac{1}{m_{11}}\tau_{wu}(t),\\
\dot{v} &= -\frac{m_{11}}{m_{22}}ur - \frac{d_{22}}{m_{22}}v - \sum_{i=2}^{3}\frac{d_{vi}}{m_{22}}|v|^{i-1}v + \frac{1}{m_{22}}\tau_{wv}(t),\\
\dot{r} &= \frac{(m_{11}-m_{22})}{m_{33}}uv - \frac{d_{33}}{m_{33}}r - \sum_{i=2}^{3}\frac{d_{ri}}{m_{11}}|r|^{i-1}r + \frac{1}{m_{33}}\tau_r + \frac{1}{m_{33}}\tau_{wr}(t),
\end{aligned} \tag{11.1}$$

where all of symbols have been defined in Section 3.3. The bounded time-varying terms, $\tau_{wu}(t)$, $\tau_{wv}(t)$, and $\tau_{wr}(t)$, are the environmental disturbances induced by waves, wind, and ocean currents with $|\tau_{wu}(t)| \leq \tau_{wu\max} < \infty$, $|\tau_{wv}(t)| \leq \tau_{wv\max} < \infty$, and $\tau_{wr}(t)| \leq \tau_{wr\max} < \infty$.

In this chapter, we consider a control objective of designing the surge force τ_u and the yaw moment τ_r to force the underactuated ship (11.1) to follow a specified path $\boldsymbol{\Omega}$, see Figure 11.1. In this figure, P is the ship's center of mass and P_d is a point attached to the virtual ship, which moves along the path with speed of u_0. If we are able to steer the ship to closely follow a virtual ship that moves along the path with a desired speed u_0, then the control objective is fulfilled, i.e., the ship is in a tube of nonzero diameter centered on the reference path and moves along the specified path at the speed u_0. Roughly speaking, the approach is to steer the ship so that it heads toward the virtual ship and eliminates the distance between itself and the virtual ship.

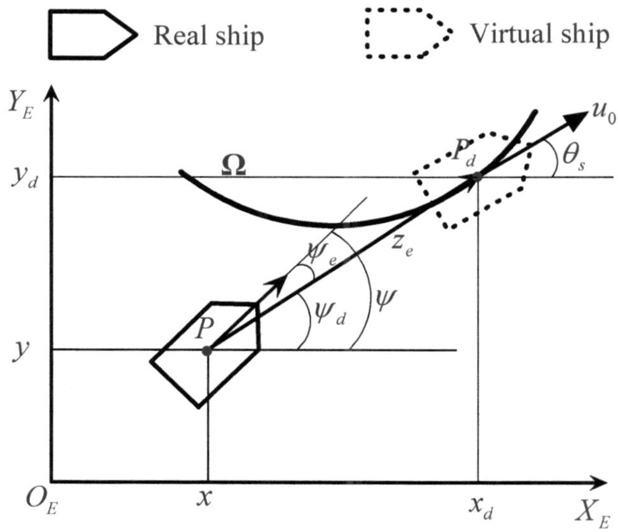

Figure 11.1 General framework of ship path-following

Define

$$\begin{aligned} x_e &= x_d - x, \\ y_e &= y_d - x, \\ \psi_e &= \psi - \psi_d, \\ z_e &= \sqrt{x_e^2 + y_e^2}, \end{aligned} \tag{11.2}$$

where

$$\psi_d = \arcsin\left(\frac{y_e}{z_e}\right). \tag{11.3}$$

With (11.2), the above control objective is mathematically stated as follows.

Control Objective. Under Assumption 11.1, design the surge force τ_u and the yaw moment τ_r to make the underactuated ship (11.1) follow the path $\boldsymbol{\Omega}$ given by

$$x_d = x_d(s),\ y_d = y_d(s), \tag{11.4}$$

where s is the path parameter variable, such that

$$\begin{aligned} &\lim_{t\to\infty} z_e(t) \leq \overline{z_e}, \\ &\lim_{t\to\infty} |\psi_e(t)| \leq \overline{\psi_e}, \end{aligned} \tag{11.5}$$

with $\overline{z_e}$ and $\overline{\psi_e}$ being arbitrarily small positive constants.

Assumption 11.1.

1. *The reference path is regular, i.e.,* $0 < R_{\min} \leq \left(\frac{\partial x_d}{\partial s}\right)^2 + \left(\frac{\partial y_d}{\partial s}\right)^2 \leq R_{\max} < \infty$.
2. *The minimum radius of the osculating circle of the path is larger than or equal to the minimum possible turning radius of the ship.*

Remark 11.1.

1. Assumption 11.1 ensures that the path is feasible for the ship to follow.
2. The angle ψ_d is not defined at $z_e = 0$ but $\lim_{z_e\to 0}\psi_d = \theta_s$ with θ_s being the orientation angle of the virtual ship on the path, see Figure 11.1. Therefore, the fulfillment of the control objective guarantees that the ship closely follows the path in terms of both position and orientation. If the reference path is not regular, then we can often split it into regular pieces and consider each of them separately. This is a case of point-to-point navigation.
3. The path parameter, s, is not the arclength of the path in general. For example, a circle with radius of R centered at the origin can be described as $x_d = R\cos(s)$ and $y_d = R\sin(s)$, see [136] for more details.

If one differentiates both sides of $\psi_e = \psi - \psi_d$ to get the $\dot{\psi}_e$-dynamics, there will be discontinuity in the $\dot{\psi}_e$-dynamics on the y_e-axis. This discontinuity will cause difficulties in applying the backstepping technique. To get around this problem, we compute the $\dot{\psi}_e$-dynamics based on

$$\begin{aligned} \sin(\psi_e) &= \frac{x_e\sin(\psi) - y_e\cos(\psi)}{z_e}, \\ \cos(\psi_e) &= \frac{x_e\cos(\psi) + y_e\sin(\psi)}{z_e}. \end{aligned} \tag{11.6}$$

We now use (11.2) and (11.6) to transform (11.1) to

$$\dot{z}_e = \ -\cos(\psi_e)u + \sin(\psi_e)v + \left(\frac{x_e}{z_e}\frac{\partial x_d}{\partial s} + \frac{y_e}{z_e}\frac{\partial y_d}{\partial s}\right)\dot{s},$$

$$\begin{aligned}
\dot{\psi}_e &= r + \frac{\dot{s}}{\cos(\psi_e)}\left(\frac{\partial x_d}{\partial s}\left(\frac{\sin(\psi)}{z_e} - \frac{x_e\sin(\psi_e)}{z_e^2}\right) - \right.\\
&\quad \left.\frac{\partial y_d}{\partial s}\left(\frac{\cos(\psi)}{z_e} + \frac{y_e\sin(\psi_e)}{z_e^2}\right)\right) + \frac{\sin(\psi_e)u + \cos(\psi_e)v}{z_e},\\
\dot{u} &= \frac{m_{22}}{m_{11}}vr - \frac{d_{11}}{m_{11}}u - \sum_{i=2}^{3}\frac{d_{ui}}{m_{11}}|u|^{i-1}u + \frac{1}{m_{11}}\tau_u + \frac{1}{m_{11}}\tau_{wu}(t),\\
\dot{v} &= -\frac{m_{11}}{m_{22}}ur - \frac{d_{22}}{m_{22}}v - \sum_{i=2}^{3}\frac{d_{vi}}{m_{22}}|v|^{i-1}v + \frac{1}{m_{22}}\tau_{wv}(t),\\
\dot{r} &= \frac{(m_{11}-m_{22})}{m_{33}}uv - \frac{d_{33}}{m_{33}}r - \sum_{i=2}^{3}\frac{d_{ri}}{m_{11}}|r|^{i-1}r + \frac{1}{m_{33}}\tau_r + \frac{1}{m_{33}}\tau_{wr}(t).
\end{aligned} \tag{11.7}$$

It is noted that the $\dot{\psi}_e$-dynamic is not defined at $\psi_e = \pm 0.5\pi$. However, our controller will guarantee $|\psi_e(t)| < 0.5\pi$, $\forall\, 0 \le t < \infty$ for feasible initial conditions. Therefore, we will design the surge force τ_u and the yaw moment τ_r for (11.7) to yield the control objective. A procedure to design a stabilizer for the path-following error system (11.7) will be presented in detail. The triangular structure of (11.7) suggests that we design the actual controls τ_u and τ_r in two stages. First, we design the virtual velocity controls for u, r and choose $\dot{s}$ to ultimately stabilize z_e and ψ_e at the origin. Based on the backstepping technique, controls τ_u and τ_r will then be designed to make the errors between the virtual velocity controls, and their actual values exponentially tend to a small ball centered at the origin. Since the ship parameters are unknown, an adaptation scheme is also introduced in this step to estimate their values used in the control laws. The nonzero lower bound of z_e to guarantee the existence of ψ_d and the boundedness of the sway velocity, v, are analyzed.

11.2 Control Design

11.2.1 Step 1

The z_e and ψ_e dynamics have three inputs that can be chosen to stabilize z_e and ψ_e, namely $\dot{s}$, u, and r. The input r should be designed to stabilize the ψ_e-dynamics at the origin. Therefore, two inputs, $\dot{s}$, u, can be used to ultimately stabilize z_e at the origin. We can either choose the input u or $\dot{s}$ and then design the remaining input. If we fix $\dot{s}$, then the virtual ship is allowed to move at a desired speed. The real ship will follow the virtual one on the path by the controller, and vice versa. In this chapter we choose to fix $\dot{s}$. This allows us to adjust the initial conditions in most cases without moving the ships, see Section 11.3. Since the transformed system (11.7) is not defined at $z_e = 0$ and $\psi_e = \pm 0.5\pi$, we first assume in the following that $z_e(t) \geq z_e^* > 0$ and $|\psi_e(t)| < 0.5\pi$, $\forall\ 0 \leq t < \infty$. In Section 11.3, we will then show that there exist initial conditions such that this hypothesis holds. Define

$$\begin{aligned} \tilde{u} &= u - u_d, \\ \tilde{r} &= r - r_d, \end{aligned} \tag{11.8}$$

where u_d and r_d are the virtual controls of u and r, respectively. As discussed above, we choose the virtual controls u_d and r_d, and $\dot{s}$ as follows:

$$u_d = k_1(z_e - \delta_e) + \left(\frac{x_e}{z_e}\frac{\partial x_d}{\partial s} + \frac{y_e}{z_e}\frac{\partial y_d}{\partial s}\right)\frac{u_0(t,z_e)}{\sqrt{\left(\frac{\partial x_d}{\partial s}\right)^2 + \left(\frac{\partial y_d}{\partial s}\right)^2}} + \tan(\psi_e)\,v, \tag{11.9}$$

where

$$\begin{aligned} r_d = & -\left(\frac{\partial x_d}{\partial s}\left(\frac{\sin(\psi)}{z_e} - \frac{x_e \sin(\psi_e)}{z_e^2}\right) - \frac{\partial y_d}{\partial s}\left(\frac{\cos(\psi)}{z_e} + \frac{y_e \sin(\psi_e)}{z_e^2}\right)\right) \times \\ & \frac{u_0(t,z_e)}{\sqrt{\left(\frac{\partial x_d}{\partial s}\right)^2 + \left(\frac{\partial y_d}{\partial s}\right)^2}} - \frac{\sin(\psi_e)u_d + \cos(\psi_e)v}{z_e} - k_2\psi_e, \end{aligned} \tag{11.10}$$

and

$$\dot{s} = \frac{u_0(t,z_e)\cos(\psi_e)}{\sqrt{\left(\frac{\partial x_d}{\partial s}\right)^2 + \left(\frac{\partial y_d}{\partial s}\right)^2}}, \tag{11.11}$$

where k_1, k_2, and δ_e are positive constants to be selected later, $u_0(t,z_e) \neq 0$, $\forall t \geq t_0 \geq 0$, and $z_e \in \mathbb{R}$, is the speed of the virtual ship on the path. Indeed, one can choose this speed to be a constant. However, the time-varying speed and position path-following dependence of the virtual ship on the path is more desirable, especially when the ship starts to follow the path. For example, one might choose

$$u_0(t,z_e) = u_0^*(1 - \chi_1 e^{-\chi_2(t-t_0)})e^{-\chi_3 z_e}, \tag{11.12}$$

where $u_0^* \neq 0$, $\chi_i > 0$, $i = 1,2,3$, $\chi_1 < 1$. The choice of $u_0(t,z_e)$ in (11.12) has the following desired feature: When the path-following error, z_e, is large, the virtual ship will wait for the real one; when z_e is small, the virtual ship will move along the path at the speed closed to u_0^* and the real one follows it within the specified look-ahead distance. This feature is suitable in practice because it avoids using a high gain control for large signal z_e.

Remark 11.2. If the sway velocity is assumed to be bounded by the surge velocity as in [129], the term $\sin(\psi_e)v$ is not required to be canceled. This controller can be designed similarly to the one in this chapter. Lemma 10.1 can be directly applied to the stability analysis. It is noted that the sway velocity is not needed to be bounded by the surge velocity with a relatively small constant.

Substituting (11.8)–(11.11) into the first two equations of (11.7) results in

$$\begin{aligned}\dot{z}_e &= -k_1\cos(\psi_e)(z_e-\delta_e)-\cos(\psi_e)\tilde{u},\\ \dot{\psi}_e &= -k_2\psi_e+\frac{\sin(\psi_e)}{z_e}\tilde{u}+\tilde{r}.\end{aligned} \tag{11.13}$$

11.2.2 Step 2

By noting that with $z_e(t)\geq z_e^*>0$ and $|\psi_e(t)|<0.5\pi,\ \forall\, 0\leq t<\infty$, the virtual controls u_d and r_d are smooth functions of x_e, y_e, s, u_0, ψ and v, differentiating both sides of (11.8) with (11.9) and (11.10) yields

$$\begin{aligned}\dot{\tilde{u}} =\ & \frac{m_{22}}{m_{11}}vr-\frac{d_{11}}{m_{11}}u-\sum_{i=2}^{3}\frac{d_{ui}}{m_{11}}|u|^{i-1}u+\frac{1}{m_{11}}\tau_u+\frac{1}{m_{11}}\tau_{wu}(t)-\\ & \frac{\partial u_d}{\partial x_e}\dot{x}_e-\frac{\partial u_d}{\partial y_e}\dot{y}_e-\frac{\partial u_d}{\partial s}\dot{s}-\frac{\partial u_d}{\partial\psi}\dot{\psi}-\frac{\partial u_d}{\partial u_0}\dot{u}_0-\frac{\partial u_d}{\partial v}\times\\ & \left(-\frac{m_{11}}{m_{22}}ur-\frac{d_{22}}{m_{22}}v-\sum_{i=2}^{3}\frac{d_{vi}}{m_{22}}|v|^{i-1}v\right)-\frac{\partial u_d}{\partial v}\frac{1}{m_{22}}\tau_{wv}(t),\\ \dot{\tilde{r}} =\ & \frac{(m_{11}-m_{22})}{m_{33}}uv-\frac{d_{33}}{m_{33}}r-\sum_{i=2}^{3}\frac{d_{ri}}{m_{11}}|r|^{i-1}r+\frac{1}{m_{33}}\tau_r+\\ & \frac{1}{m_{33}}\tau_{wr}(t)-\frac{\partial r_d}{\partial x_e}\dot{x}_e-\frac{\partial r_d}{\partial y_e}\dot{y}_e-\frac{\partial r_d}{\partial s}\dot{s}-\frac{\partial r_d}{\partial\psi}\dot{\psi}-\frac{\partial r_d}{\partial u_0}\dot{u}_0-\\ & \frac{\partial r_d}{\partial v}\left(-\frac{m_{11}}{m_{22}}ur-\frac{d_{22}}{m_{22}}v-\sum_{i=2}^{3}\frac{d_{vi}}{m_{22}}|v|^{i-1}v\right)-\frac{\partial r_d}{\partial v}\frac{1}{m_{22}}\tau_{wv}(t),\end{aligned} \tag{11.14}$$

where for convenience of choosing u_0, the terms $\frac{\partial u_d}{\partial x_e}$, $\frac{\partial u_d}{\partial y_e}$, $\frac{\partial r_d}{\partial x_e}$ and $\frac{\partial r_d}{\partial y_e}$ do not include $\frac{\partial u_0}{\partial x_e}$ and $\frac{\partial u_0}{\partial y_e}$, which are lumped into $\dot{u}_0$. From (11.14), we choose the actual controls τ_u and τ_r without canceling useful nonlinear damping terms as

$$\begin{aligned}\tau_u =\ & -k_3\tilde{u}-\hat{\theta}_1^T f_1(\cdot)-\frac{\sin(\psi_e)}{z_e}\psi_e-\hat{\theta}_{31}\tanh\left(\frac{\tilde{u}\hat{\theta}_{31}}{\varepsilon_1}\right)-\\ & \hat{\theta}_{32}\frac{\partial u_d}{\partial v}\tanh\left(\frac{\partial u_d}{\partial v}\frac{\tilde{u}\hat{\theta}_{32}}{\varepsilon_2}\right),\\ \tau_r =\ & -k_4\tilde{r}-\hat{\theta}_2^T f_2(\cdot)-\psi_e-\hat{\theta}_{33}\tanh\left(\frac{\tilde{r}\hat{\theta}_{33}}{\varepsilon_3}\right)-\hat{\theta}_{34}\frac{\partial r_d}{\partial v}\tanh\left(\frac{\partial r_d}{\partial v}\frac{\tilde{r}\hat{\theta}_{34}}{\varepsilon_4}\right),\end{aligned} \tag{11.15}$$

and the update laws as

$$\begin{aligned}
\dot{\hat{\theta}}_{1j} &= \gamma_{1j}\text{proj}\left(\tilde{u} f_{1j}(\cdot), \hat{\theta}_{1j}\right), \ 1 \leq j \leq 9, \\
\dot{\hat{\theta}}_{2j} &= \gamma_{2j}\text{proj}\left(\tilde{r} f_{2j}(\cdot), \hat{\theta}_{2j}\right), \ 1 \leq j \leq 9, \\
\dot{\hat{\theta}}_{31} &= \gamma_{31}\text{proj}\left(|\tilde{u}|, \hat{\theta}_{31}\right), \\
\dot{\hat{\theta}}_{32} &= \gamma_{32}\text{proj}\left(\left|\tilde{u}\tfrac{\partial u_d}{\partial v}\right|, \hat{\theta}_{32}\right), \\
\dot{\hat{\theta}}_{33} &= \gamma_{33}\text{proj}\left(|\tilde{r}|, \hat{\theta}_{33}\right), \\
\dot{\hat{\theta}}_{34} &= \gamma_{34}\text{proj}\left(\left|\tilde{r}\tfrac{\partial r_d}{\partial v}\right|, \hat{\theta}_{34}\right)
\end{aligned} \tag{11.16}$$

where k_3, k_4, ε_i, γ_{1j}, γ_{2j}, γ_{3i}, $1 \leq i \leq 4$, and $1 \leq j \leq 9$ are positive constants to be selected later, and $f_{1j}(\cdot)$ and $f_{2j}(\cdot)$ are the jth elements of $f_1(\cdot)$ and $f_2(\cdot)$, respectively, with

$$\begin{aligned}
f_1(\cdot) = \Bigg[& vr \ -u_d \ -|u|u_d \ -u^2u_d \ -\left(\frac{\partial u_d}{\partial x_e}\dot{x}_e + \frac{\partial u_d}{\partial y_e}\dot{y}_e + \frac{\partial u_d}{\partial s}\dot{s} + \right. \\
& \left. \frac{\partial u_d}{\partial \psi}\dot{\psi} + \frac{\partial u_d}{\partial u_0}\dot{u}_0\right) \ \frac{\partial u_d}{\partial v}ur \ \frac{\partial u_d}{\partial v}v \ \frac{\partial u_d}{\partial v}|v|v \ \frac{\partial u_d}{\partial v}|v|^2 v \Bigg]^T, \\
f_2(\cdot) = \Bigg[& uv \ -r_d \ -|r|r_d \ -r^2r_d \ -\left(\frac{\partial r_d}{\partial x_e}\dot{x}_e + \frac{\partial r_d}{\partial y_e}\dot{y}_e + \frac{\partial r_d}{\partial s}\dot{s} + \right. \\
& \left. \frac{\partial r_d}{\partial \psi}\dot{\psi} + \frac{\partial r_d}{\partial u_0}\dot{u}_0\right) \ \frac{\partial r_d}{\partial v}ur \ \frac{\partial r_d}{\partial v}v \ \frac{\partial r_d}{\partial v}|v|v \ \frac{\partial r_d}{\partial v}|v|^2 v \Bigg]^T,
\end{aligned} \tag{11.17}$$

$\hat{\theta}_{ij}$, $1 \leq i \leq 3$ is the jth element of $\hat{\theta}_i$, which is an estimate of θ_i with

$$\begin{aligned}
\theta_1 &= \left[m_{22} \ d_{11} \ d_{u2} \ d_{u3} \ m_{11} \ \frac{m_{11}^2}{m_{22}} \ \frac{d_{22}m_{11}}{m_{22}} \ \frac{d_{v2}m_{11}}{m_{22}} \ \frac{d_{v3}m_{11}}{m_{22}} \right]^T, \\
\theta_2 &= \left[(m_{11} - m_{22}) \ d_{33} \ d_{r2} \ d_{r3} \ m_{33} \ \frac{m_{11}m_{33}}{m_{22}} \ \frac{d_{22}m_{33}}{m_{22}} \ \frac{d_{v2}m_{33}}{m_{22}} \ \frac{d_{v3}m_{33}}{m_{22}} \right]^T, \\
\theta_3 &= \left[\tau_{wu\max} \ \frac{m_{11}}{m_{22}}\tau_{wv\max} \ \tau_{wr\max} \ \frac{m_{33}}{m_{22}}\tau_{wv\max} \right]^T.
\end{aligned} \tag{11.18}$$

The operator, proj, is the Lipschitz continuous projection algorithm (repeated here for the reader's convenience) as follows:

$$\begin{aligned}
&\text{proj}(\varpi, \hat{\omega}) = \varpi && \text{if } \ \Xi(\hat{\omega}) \leq 0, \\
&\text{proj}(\varpi, \hat{\omega}) = \varpi && \text{if } \ \Xi(\hat{\omega}) \geq 0 \text{ and } \Xi_{\hat{\omega}}(\hat{\omega})\varpi \leq 0, \\
&\text{proj}(\varpi, \hat{\omega}) = (1 - \Xi(\hat{\omega}))\varpi && \text{if } \Xi(\hat{\omega}) > 0 \text{ and } \Xi_{\hat{\omega}}(\hat{\omega})\varpi > 0,
\end{aligned} \tag{11.19}$$

where $\Xi(\hat{\omega}) = \frac{\hat{\omega}^2 - \omega_M^2}{\mu^2 + 2\mu\omega_M}$, $\Xi_{\hat{\omega}}(\hat{\omega}) = \frac{\partial \Xi(\hat{\omega})}{\partial \hat{\omega}}$, μ is an arbitrarily small positive constant, $\hat{\omega}$ is an estimate of ω and $|\omega| \leq \omega_M$.

The projection algorithm is such that if $\dot{\hat{\omega}} = \text{proj}(\varpi, \hat{\omega})$ and $\hat{\omega}(t_0) \leq \omega_M$ then

1. $\hat{\omega}(t) \leq \omega_M + \xi$, $\forall\, 0 \leq t_0 \leq t < \infty$,
2. $\text{proj}(\varpi, \hat{\omega})$ is Lipschitz continuous,
3. $|\text{proj}(\varpi, \hat{\omega})| \leq |\varpi|$,
4. $\tilde{\omega}\text{proj}(\varpi, \hat{\omega}) \geq \tilde{\omega}\varpi$ with $\tilde{\omega} = \omega - \hat{\omega}$.

Substituting (11.15) into (11.14) yields

$$\begin{aligned}
\dot{\tilde{u}} = &-\left(\frac{k_3}{m_{11}} + \frac{d_{11}}{m_{11}} + \sum_{i=2}^{3} \frac{d_{ui}}{m_{11}} |u|^{i-1}\right)\tilde{u} - \frac{1}{m_{11}} \frac{\sin(\psi_e)}{z_e}\psi_e - \\
&\frac{1}{m_{11}}\hat{\theta}_1^T f_1(\cdot) + \frac{1}{m_{11}}\theta_1^T f_1(\cdot) - \frac{1}{m_{11}}\hat{\theta}_{31}\tanh\left(\frac{\tilde{u}\hat{\theta}_{31}}{\varepsilon_1}\right) + \\
&\frac{1}{m_{11}}\tau_{wu}(t) - \hat{\theta}_{32}\frac{\partial u_d}{\partial v}\tanh\left(\frac{\partial u_d}{\partial v}\frac{\tilde{u}\hat{\theta}_{32}}{\varepsilon_2}\right) - \frac{\partial u_d}{\partial v}\frac{1}{m_{22}}\tau_{wv}(t), \\
\dot{\tilde{r}} = &-\left(\frac{k_4}{m_{33}} + \frac{d_{33}}{m_{33}} - \sum_{i=2}^{3} \frac{d_{ri}}{m_{11}} |r|^{i-1}\right)\tilde{r} - \frac{1}{m_{33}}\psi_e - \frac{1}{m_{33}}\hat{\theta}_2^T f_2(\cdot) + \\
&\frac{1}{m_{33}}\theta_2^T f_2(\cdot) - \frac{1}{m_{33}}\hat{\theta}_{33}\tanh\left(\frac{\tilde{r}\hat{\theta}_{33}}{\varepsilon_3}\right) + \frac{1}{m_{33}}\tau_{wr}(t) - \\
&\frac{1}{m_{33}}\hat{\theta}_{34}\frac{\partial r_d}{\partial v}\tanh\left(\frac{\partial r_d}{\partial v}\frac{\tilde{r}\hat{\theta}_{34}}{\varepsilon_4}\right) - \frac{\partial r_d}{\partial v}\frac{1}{m_{22}}\tau_{wv}(t).
\end{aligned} \tag{11.20}$$

We now present the following result, the proof of which is given in the next section.

Theorem 11.1. *Assume that (a) the ship inertia, added mass and damping matrices are diagonal and (b) Assumption 11.1 is satisfied. If the adaptive state feedback control law (11.15) and adaptation law (11.16) are applied to the ship system (11.1) then there exist feasible initial conditions such that the path-following errors* $(z_e(t), \psi_e(t))$ *converge to a small ball centered at the origin with an appropriate choice of the design constants* k_i, ε_i, γ_{1j}, γ_{2j}, *and* γ_{3i} *for all* $1 \leq i \leq 4$, $1 \leq j \leq 9$.

11.3 Stability Analysis

To prove Theorem 11.1, we first consider the $(\psi_e, \tilde{u}, \tilde{r})$-dynamics, then move to the z_e- and v-dynamics.

$(\psi_e, \tilde{u}, \tilde{r})$-dynamics

Consider the following Lyapunov function

$$V_1 = \frac{1}{2}\psi_e^2 + \frac{m_{11}}{2}\tilde{u}^2 + \frac{m_{33}}{2}\tilde{r}^2 + \frac{1}{2}\sum_{i=1}^{3}\tilde{\theta}_i^T \Gamma_i^{-1}\tilde{\theta}_i, \tag{11.21}$$

where $\tilde{\theta}_i = \theta_i - \hat{\theta}_i$ and $\Gamma_i = \mathrm{diag}\left(\gamma_{ij}\right)$. Differentiating both sides of (11.21) along (11.14), (11.15), and (11.16) yields

$$\dot{V}_1 \leq -k_2\psi_e^2 - (k_3 + d_{11})\tilde{u}^2 - (k_4 + d_{33})\tilde{r}^2 + 0.2785\sum_{i=1}^{4}\varepsilon_i. \tag{11.22}$$

By subtracting and adding $\frac{1}{2}\sum\limits_{i=1}^{3}\tilde{\theta}_i^T \Gamma_i^{-1}\tilde{\theta}_i$ to the right-hand side of (11.22), we arrive at

$$\dot{V}_1 \leq -\delta V_1 + \rho, \tag{11.23}$$

where

$$\delta = \min\left(1, 2k_2, \frac{2(k_3 + d_{11})}{m_{11}}, \frac{2(k_4 + d_{33})}{m_{33}}\right),$$
$$\rho = \frac{1}{2}\sum_{i=1}^{3}\tilde{\theta}_i^T \Gamma_i^{-1}\tilde{\theta}_i + 0.2785\sum_{i=1}^{4}\varepsilon_i. \tag{11.24}$$

From (11.23), it is direct to show that

$$V_1(t) \leq V_1(t_0)e^{-\delta(t-t_0)} + \frac{\rho}{\delta}, \ \forall t \geq t_0 \geq 0, \tag{11.25}$$

which further yields

$$\min\left(\frac{1}{2}, \frac{m_{11}}{2}, \frac{m_{33}}{2}\right)\|X_1(t)\|^2 \leq$$
$$\|X_1^*(t_0)\|^2 \max\left(\frac{1}{2}, \frac{m_{11}}{2}, \frac{m_{33}}{2}, \frac{1}{2\gamma_{1j}}, \frac{1}{2\gamma_{2j}}, \frac{1}{2\gamma_{3j}}\right)e^{-\delta(t-t_0)} + \frac{\rho}{\delta}, \tag{11.26}$$

where

$$X_1(t) = \left[\,\psi_e(t)\ \tilde{u}(t)\ \tilde{r}(t)\,\right]^T,$$
$$X_1^*(t_0) = \left[\,\psi_e(t_0)\ \tilde{u}(t_0)\ \tilde{r}(t_0)\ \tilde{\theta}_1^T(t_0)\ \tilde{\theta}_2^T(t_0)\ \tilde{\theta}_3^T(t_0)\,\right]^T.$$

Therefore we have

$$\|X_1(t)\| \leq \alpha_1 e^{-\sigma_1(t-t_0)} + \rho_1, \ \forall t \geq t_0 \geq 0, \tag{11.27}$$

where

$$
\begin{aligned}
p_1 &= \sqrt{\frac{\max\left(\frac{1}{2},\frac{m_{11}}{2},\frac{m_{33}}{2},\frac{1}{2\gamma_{1j}},\frac{1}{2\gamma_{2j}},\frac{1}{2\gamma_{3j}}\right)}{0.5\min(1,m_{11},m_{33})}},\\
\alpha_1 &= p_1\left\|X_1^*(t_0)\right\|,\\
\sigma_1 &= \frac{\delta}{2},\\
\rho_1 &= \sqrt{\frac{\rho}{0.5\delta\min(1,m_{11},m_{33})}}.
\end{aligned}
\tag{11.28}
$$

Remark 11.3. It is important to note that, due to the use of the projection algorithm, by adjusting $k_3, k_4, \varepsilon_i, \gamma_{1j}, \gamma_{2j}, \gamma_{3i}, 1 \le i \le 4, 1 \le j \le 9$, we can make ρ_1 arbitrarily small. This observation plays a crucial role in the stability analysis of the z_e-dynamics.

z_e-dynamics

Lower-bound of z_e. It can be seen from (11.13) and (11.27) that

$$
\dot{\tilde{z}}_e \ge -k_1\tilde{z}_e - \left(\alpha_1 e^{-\sigma_1(t-t_0)} + \rho_1\right), \tag{11.29}
$$

where $\tilde{z}_e = z_e - \delta_e$, which , with $\sigma_1 \neq k_1$, further yields

$$
\begin{aligned}
\tilde{z}_e(t) \ge{}& \tilde{z}_e(t_0)e^{-k_1(t-t_0)} + \frac{p_1\left\|X_1^*(t_0)\right\|}{\sigma_1-k_1}\left(e^{-\sigma_1(t-t_0)} - e^{-k_1(t-t_0)}\right) -\\
&\frac{\rho_1}{k_1}\left(1-e^{-k_1(t-t_0)}\right).
\end{aligned}
\tag{11.30}
$$

Hence

$$
\begin{aligned}
z_e(t) \ge{}& (z_e(t_0)+\delta_e)e^{-k_1(t-t_0)} + \frac{p_1\left\|X_1^*(t_0)\right\|}{\sigma_1-k_1}\left(e^{-\sigma_1(t-t_0)} - e^{-k_1(t-t_0)}\right) -\\
&\frac{\rho_1}{k_1}\left(1-e^{-k_1(t-t_0)}\right) + \delta_e.
\end{aligned}
\tag{11.31}
$$

Therefore, the condition $z_e(t) \ge z_e^* > 0$ holds when

$$
\begin{aligned}
&\sigma_1 > k_1,\ \delta_e \ge z_e^* + \frac{\rho_1}{k_1},\\
&z_e(t_0) \ge \frac{p_1\left\|X_1^*(t_0)\right\|}{\sigma_1-k_1} - \delta_e.
\end{aligned}
\tag{11.32}
$$

Upper Bound of z_e. We rewrite the first equation of (11.13) as

$$\dot{\tilde{z}}_e = -k_1\tilde{z}_e - (\cos(\psi_e) - 1)k_1\tilde{z}_e - \cos(\psi_e)\tilde{u}. \tag{11.33}$$

It can be seen that (11.33) is of the form of the system studied in Lemma 10.1. Therefore, we will apply Lemma 10.1 to investigate stability of (11.33). We need to verify all conditions C1–C4 of Lemma 10.1 to (11.33).

Verifying Condition C1. Take the following Lyapunov function:

$$V_2 = \frac{1}{2}z_e^2. \tag{11.34}$$

It is direct to show that C1 holds with

$$c_0 = 0,\, c_1 = c_2 = 0.5,\, c_3 = 1,\, c_4 = k_1. \tag{11.35}$$

Verifying Condition C2. By noting that

$$|(\cos(\psi_e) - 1)k_1\tilde{z}_e - \cos(\psi_e)\tilde{u}| \leq (k_1\,|\tilde{z}_e| + 1)\,\|X_1(t)\|\,,$$

we have

$$\lambda_1 = 1,\, \lambda_2 = k_1. \tag{11.36}$$

Verifying Condition C3. This condition directly holds from (11.27).

Verifying Condition C4. From (11.35), (11.36), and (11.27), condition C4 becomes

$$k_1 - k_1\rho_1 - 0.25\rho_1/\mu_0 > 0. \tag{11.37}$$

From Remark 11.3 and noting that μ_0 is an arbitrarily positive constant, we can see that there always exists k_1 such that (11.37) holds. All the conditions of Lemma 10.1 have been verified, and we therefore have

$$|z_e(t)| \leq \alpha_2\left(\left\|\left(X_1^*(t_0), z_e(t_0)\right)\right\|\right)e^{-\sigma_2(t-t_0)} + \rho_2, \tag{11.38}$$

where α_2, σ_2, and ρ_2 are calculated as in Lemma 10.1.

v-dynamics

From (11.9) and (11.10) the sway velocity dynamics can be rewritten as

$$\dot{v} = \phi_1(\cdot)v + \phi_2(\cdot)v^2 - \frac{d_{v2}}{m_{22}}\,|v|\,v - \frac{d_{v3}}{m_{22}}v^3 + \phi_3(\cdot), \tag{11.39}$$

where

$$
\begin{aligned}
\phi_1(\cdot) = & -\frac{d_{22}}{m_{22}} - \frac{m_{11}}{m_{22}}\left(\tan(\psi_e)\tilde{r} - \frac{1}{z_e\cos(\psi_e)}\tilde{u} + \right. \\
& \left. \left(a_2 - \frac{\sin(\psi_e)}{z_e}a_1\right)\tan(\psi_e) - \frac{a_1}{z_e\cos(\psi_e)}\right), \\
\phi_2(\cdot) = & -\frac{m_{11}}{m_{22}}\frac{\tan(\psi_e)}{z_e\cos(\psi_e)}, \\
\phi_3(\cdot) = & -\frac{m_{11}}{m_{22}}\left(\tilde{u}\tilde{r} + a_1\tilde{r} + \left(a_2 + \frac{\sin(\psi_e)}{z_e}a_1\right)\tilde{u} + \right. \\
& \left. \left(a_2 + \frac{\sin(\psi_e)}{z_e}a_1\right)a_1\right) + \frac{1}{m_{22}}\tau_{wv}(t),
\end{aligned} \tag{11.40}
$$

with

$$
\begin{aligned}
a_1 = & \; k_1(z_e - \delta_e) + \left(\frac{x_e}{z_e}\frac{\partial x_d}{\partial s} + \frac{y_e}{z_e}\frac{\partial y_d}{\partial s}\right)\frac{u_0(t,z_e)}{\sqrt{\left(\frac{\partial x_d}{\partial s}\right)^2 + \left(\frac{\partial y_d}{\partial s}\right)^2}}, \\
a_2 = & -\left(\frac{\partial x_d}{\partial s}\left(\frac{\sin(\psi)}{z_e} - \frac{x_e\sin(\psi_e)}{z_e^2}\right) - \frac{\partial y_d}{\partial s}\left(\frac{\cos(\psi)}{z_e} + \frac{y_e\sin(\psi_e)}{z_e^2}\right)\right) \times \\
& \frac{u_0(t,z_e)}{\sqrt{\left(\frac{\partial x_d}{\partial s}\right)^2 + \left(\frac{\partial y_d}{\partial s}\right)^2}} - k_2\psi_e.
\end{aligned}
$$

To show that v is bounded, we take the following Lyapunov function:

$$
V_3 = \frac{1}{2}v^2, \tag{11.41}
$$

whose derivative along the solutions of (11.39), for any $\eta > 0$, satisfies

$$
\dot{V}_3 \le -\left(\frac{d_{v3}}{m_{22}} - \phi_2^M\eta\right)v^4 + \left(\phi_1^M + \frac{\phi_2^M + \phi_3^M}{4\eta}\right)v^2 + \phi_3^M\eta, \tag{11.42}
$$

where ϕ_i^M, $1 \le i \le 3$, are the upper bounds of $\phi_i(\cdot)$. It can be seen from (11.40) that ϕ_i^M exist and are finite since their arguments are bounded as shown above. Therefore, we can pick $\eta > 0$ such that $d_{v3}/m_{22} - \phi_2^M\eta > \eta^* > 0$; then, see [6], the inequality (11.42) guarantees a finite upper-bound of the sway velocity v.

11.4 Discussion of the Initial Condition

We now discuss how to obtain the initial conditions such that $|\psi(t)| < 0.5\pi$ and the last inequality of (11.32) holds. Since $0.5\psi^2(t) \le V_1(t)$, from (11.25), the condition $|\psi(t)| < 0.5\pi$ can be rewritten as

$$\sqrt{\psi_e^2(t_0)+m_{11}\tilde{u}^2(t_0)+m_{33}\tilde{r}^2(t_0)+\sum_{i=1}^{3}\tilde{\theta}_i^T(t_0)\Gamma_i^{-1}\tilde{\theta}_i(t_0)+\rho/\delta}<\frac{1}{2}\pi. \quad (11.43)$$

On the other hand, the term $\sum_{i=1}^{3}\tilde{\theta}_i^T(t_0)\Gamma_i^{-1}\tilde{\theta}_i(t_0)+\frac{\rho}{\delta}$ can be made arbitrarily small. Noticing that $x_e(t_0)=x(t_0)-x_d(s(t_0))$ and $y_e(t_0)=y(t_0)-y_d(s(t_0))$, it can be seen that the initial value, $s(t_0)$, can be adjusted such that (11.43) holds if:

1. the ship heads toward "almost" of the half-plane containing the initial part of the path to be followed,
2. the path satisfies Assumption 11.1, see Figure 11.2,
3. the initial velocities $u(t_0)$, $v(t_0)$ and $r(t_0)$ are not too large, and
4. the design constants are chosen such that k_1 is small and k_2 is large.

The angle δ_0 (see Figure 11.2) should be increased if the initial velocities $u(t_0)$, $v(t_0)$, and $r(t_0)$ are large. Otherwise the ship might cross the edge line of the plane in question, which might result in $\psi_e=\pm 0.5\pi$. Similarly the last inequality of (11.32) is rewritten as

$$\sqrt{(x(t_0)-x_d(s(t_0)))^2+(y(t_0)-y_d(s(t_0)))^2}\geq\frac{p_1\left\|X_1^*(t_0)\right\|}{\sigma_1-k_1}-\delta_e. \quad (11.44)$$

Again $s(t_0)$ can be adjusted such that (11.44) holds. It is noted that if the above-mentioned conditions are not satisfied, one might generate an additional path segment such that it makes the above conditions hold and "smoothly" connects to the path to be followed.

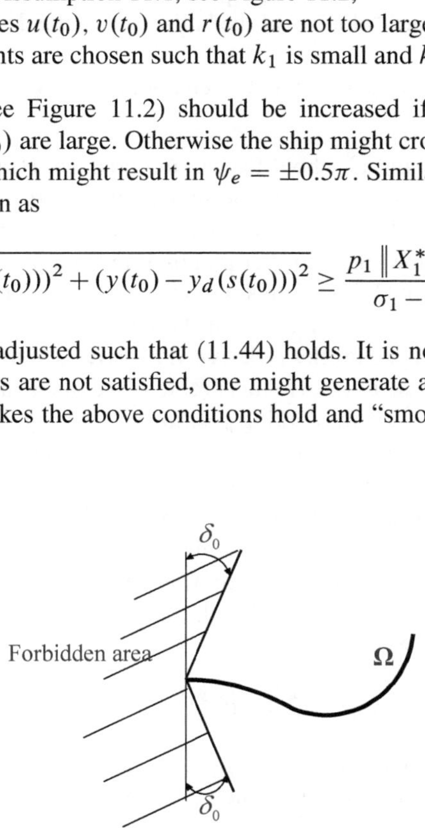

Figure 11.2 Feasible initial conditions

11.5 Parking and Point-to-point Navigation

11.5.1 Parking

Parking Objective. Design the surge force τ_u and the yaw moment τ_r to park the underactuated ship (11.1) from the initial position and orientation, $(x(t_0),\ y(t_0),\ \psi(t_0))$, to the desired parking position and orientation of (x_p, y_p, ψ_p) under the following conditions:

1. There exists a large enough positive constant ϖ_p such that $\sqrt{(x(t_0)-x_p)^2+(y(t_0)-y_p)^2} \geq \varpi_p$.
2. The ship heads toward "almost" of the half-plane of the desired parking orientation.
3. At the desired parking position and orientation, the environmental disturbances are negligible.

The above conditions normally hold for parking practice. However, if the first two conditions do not hold, one can apply the strategy in the preceding section to move the ship until they hold. Having formulated the parking problem as above, one might claim that the path-following controller proposed in Section 11.2 can be applied by setting u_0 equal to zero. However, this will result in a yaw angle that may be very different from the desired parking one, at the desired parking position, since our proposed path-following controller is designed to drive z_e to a small ball, not to zero for reasons of robustness. To resolve this problem, we first generate a regular curve, $\boldsymbol{\Omega}_p(x_d, y_d)$, which goes via the parking position and its tangent angle at the parking position is equal to the desired parking yaw angle, see Figure 11.3. For simplicity of calculation, the curve can be taken as a straight line in almost all cases of the vessel's initial conditions. Then the proposed path-following controller can be used to make the vessel follow $\boldsymbol{\Omega}_p(x_d, y_d)$. In this case, the velocity u_0 should be chosen such that it is equal to zero or tends to zero when the virtual ship tends to the desired parking position, i.e., $\lim_{z_{ep}\to 0} u_0 = 0$ with $z_{ep} = \sqrt{(x_d-x_p)^2+(y_d-y_p)^2}$.

A simple choice can be taken as

$$u_0 = u_0^*(1-e^{-\chi_1 z_{ep}})e^{-\chi_2 z_e}, \tag{11.45}$$

where $\chi_i > 0$, $i = 1,2$. Special care should be taken when choosing the initial values of $(x_d(t_0), y_d(t_0))$ and the sign of u_0 such that they result in a short parking time.

Remark 11.4. At the desired parking position and orientation, if there are large environmental disturbances, there will be an oscillatory behavior in the yaw dynamics and the ship might diverge from its desired position. This phenomenon is well known in dynamic positioning systems, [11]. However, for parking practice, we here assume that the environmental disturbances are negligible since the parking place is usually in a harbor.

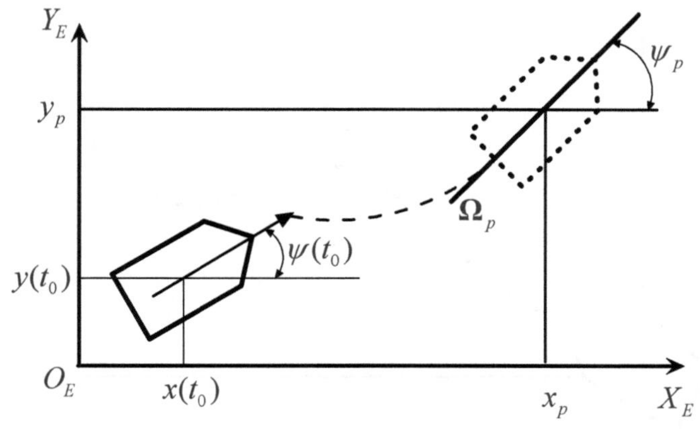

Figure 11.3 Ship parking problem

11.5.2 Point-to-point Navigation

As seen in Section 11.1, the requirement of the reference path to be a regular curve might be too cumbersome in practice, since this curve has to go via desired points generated by the helmsman, and its derivatives are needed in the path-following controller. These restrictions motivate us to consider the point-to-point navigation problem as follows:

Point-to-point Navigation Objective. Design the surge force τ_u and the yaw moment τ_r to make the underactuated ship (11.1) go from the initial position and orientation, $(x(t_0), y(t_0), \psi(t_0))$, via desired points generated by a path planner.

To solve this control objective, we first assume that the path planner generates desired points, which are feasible for the ship to be navigated through. We then apply the proposed path-following controller in Section 11.2 to each regular curve segment connecting desired points in sequence, see Figure 11.4. The regular curve segments can be straight line, arc, or known regular curves. A fundamental difference between point-to-point navigation and the proposed smooth path-following is that there are a finite number of "spikes", equal to the number of points, in the errors z_e and ψ_e. This phenomenon is due to the path being nonsmooth in the orientation at the points.

11.6 Numerical Simulations

The same 32 m long monohull ship used in Section 5.4 is also used in this section. The ship parameters are again listed below for the convenience of the reader:

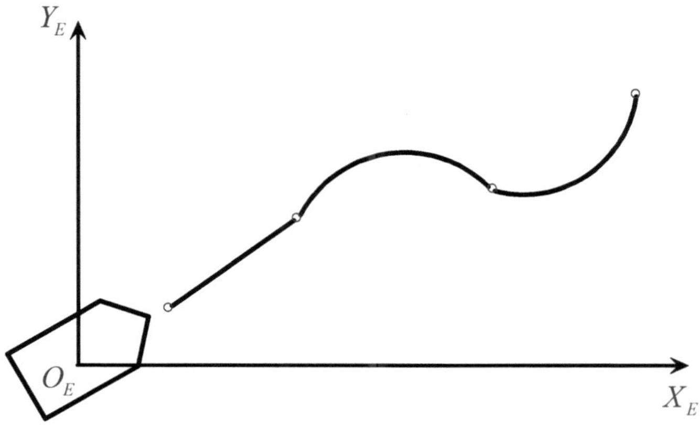

Figure 11.4 Point-to-point navigation

$$
\begin{aligned}
&m_{11} = 120\times 10^3\,\text{kg},\ m_{22} = 177.9\times 10^3\,\text{kg},\ m_{33} = 636\times 10^5\,\text{kgm}^2,\\
&d_u = 215\times 10^2\,\text{kgs}^{-1},\ d_{u2} = 43\times 10^2\,\text{kgm}^{-1},\ d_{u3} = 21.5\times 10^2\,\text{kgsm}^{-2},\\
&d_v = 117\times 10^3\,\text{kgs}^{-1},\ d_{v2} = 23.4\times 10^3\,\text{kgm}^{-1},\ d_{v3} = 11.7\times 10^3\,\text{kgsm}^{-2},\\
&d_r = 802\times 10^4\,\text{kgm}^2 s^{-1},\ d_{r2} = 160.4\times 10^4\,\text{kgm}^2,\ d_{r3} = 80.2\times 10^4\,\text{kgm}^2 s,\\
&d_{ui} = 0,\ d_{vi} = 0,\ d_{ri} = 0,\ \forall i > 3.
\end{aligned}
$$

This ship has a minimum turning circle with a radius of 150 m, a maximum surge force of 5.2×10^9 N, and the maximum yaw moment of 8.5×10^8 Nm. The above ship parameters are assumed to be those of real ships and are estimated on-line by adaptation laws (11.16). We assume that these parameters fluctuate around the above values $\pm 15\%$ to calculate the maximum and minimum values used in the choice of the design constants in (11.37). We assume that the environmental disturbances are $\tau_{wu} = 11\times 10^4(1+\text{rand}(\cdot))$, $\tau_{wv} = 26\times 10^4(1+\text{rand}(\cdot))$ and $\tau_{wr} = 950\times 10^5(1+\text{rand}(\cdot))$, where $\text{rand}(\cdot)$ is random noise with the uniform distribution on the interval $[-0.5\ 0.5]$. This choice results in nonzero-mean disturbances. In practice, the environmental disturbances may be different. We take the above disturbances for simplicity of generation. It should be noted that only the boundaries of the environmental disturbances are needed in our proposed controller. In simulations, the control parameters are taken as $k_1 = 1.5$, $k_2 = 7.5$, $k_3 = 6\times 10^5$, $k_4 = 3\times 10^8$, $\gamma_{1j} = \gamma_{2j} = \gamma_{3j} = 2\times 10^7$, $\varepsilon_i = 0.2$ and $\delta_e = 0.95$. The initial conditions are $[x(0),\ y(0),\ \psi(0),\ u(0),\ v(0),\ r(0),\ s(0)] =[-200, -20, -0.4, 0, 0, 0, 0]$, and all initial values of parameter estimates are taken to be 70% of their assumed true ones.

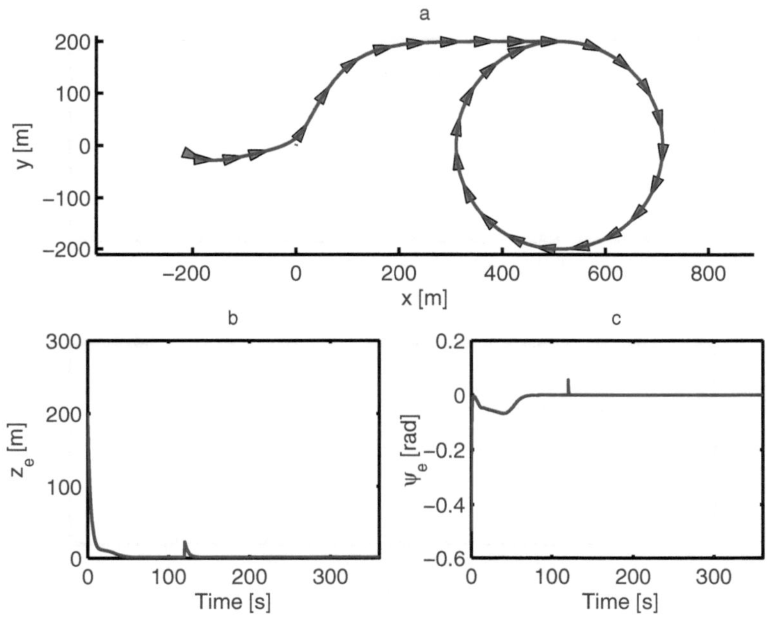

Figure 11.5 Path-following: **a**. Ship position and orientation in the (x, y)-plane; **b**. Ship position error z_e; **c**. Ship orientation error ψ_e

11.6.1 Path-following Simulation Results

The reference is $(x_d = s,\ y_d = 200\tanh(0.02s))$ for the first 120 seconds, then followed by a circle with a radius of 200 m. The virtual ship velocity on the path is taken as $u_0(t, z_e) = 5(1 - 0.8e^{-2t})e^{-0.5z_e}$. The simulation results of the ship's position and orientation, path-following error z_e, and path-following orientation error ψ_e are plotted in Figure 11.5. The control inputs τ_u and τ_r are plotted in Figure 11.6. For simplicity of presentation, we plot some samples of parameter estimates in Figure 11.7. It can be seen that the projection algorithm clearly prevents instability in adaptation due to non-vanishing disturbances.

11.6.2 Point-to-point Simulation Results

For the case of point-to-point navigation, we want the ship to go via points: $((0\,\text{m}, 0\,\text{m}), (400\,\text{m}, 200\,\text{m}), (1000\,\text{m}, 200\,\text{m}), (1400\,\text{m}, 0\,\text{m}))$, then follow a horizontal straight line. For simplicity, we use straight-line segments to connect the above desired points. The virtual ship velocity is taken to be the same as for the case of path-following. The simulation results of the ship position and orientation, path-

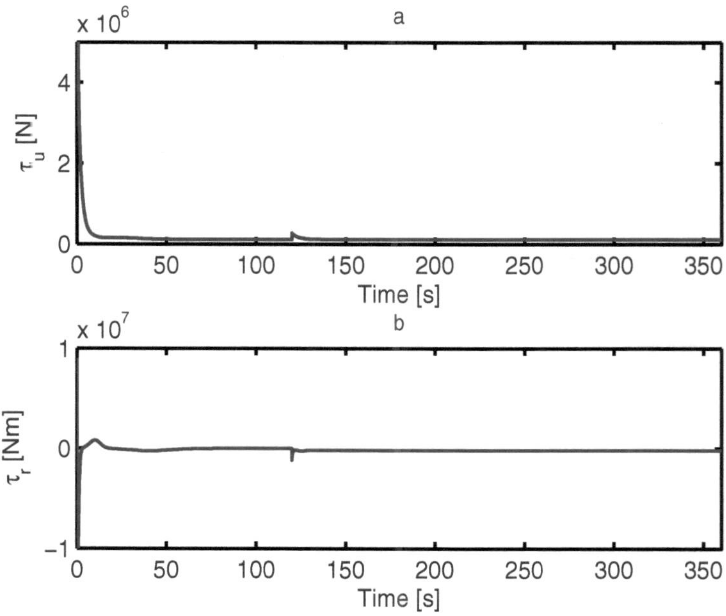

Figure 11.6 Path-following: **a**. Surge force τ_u [N]; **b**.Yaw moment τ_r [Nm]

following error z_e, and path-following orientation error ψ_e are plotted in Figures 11.8 and 11.9. The "spikes" in z_e, ψ_e, τ_u and τ_r are due to the nonsmooth reference path. However, these spikes are moderate in magnitude.

11.6.3 Parking Simulation Results

We assume that the parking position is at the origin and the parking orientation is $\psi_p = \pi/4$ without environmental disturbances at the desired parking place. We choose $u_0 = 5(1 - e^{-0.2z_{ep}})e^{-0.2z_e}$ and a straight-line segment starting at $(-20, -20)$ with a slope of 1. The simulation results are plotted in Figures 11.10 and 11.11.

In summary, it can be seen from Figures 11.5–11.11 that our proposed controller is able to force the underactuated ship in question to follow the predefined paths. The path-following position error z_e converges to a small nonzero value specified by δ_e and does not cross zero as expected in the control design. It can be seen from Figures 11.5 and 11.8 that, under the nonvanishing environmental disturbances, the proposed controller forces the yaw angle to a small value. This value together with the ship forward speed prevents the ship from drifting away from the path. The high magnitude of controls results from the fact that we simulate the proposed con-

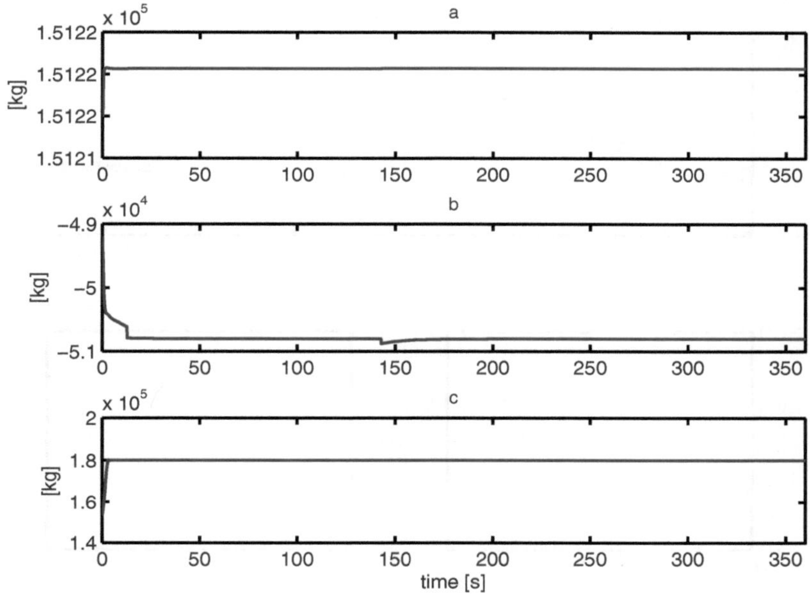

Figure 11.7 Path-following: **a**. Estimate $\hat{\theta}_{11}$; **b**. Estimate $\hat{\theta}_{21}$; **c**. Estimate $\hat{\theta}_{31}$

troller on a real ship and environmental disturbances with large magnitude. Indeed, the magnitude of the control inputs, can be reduced by adjusting the control and adaptation gains. However, this will result in a slow transient response. Finally, the transient response of ψ_e is much shorter than that of z_e due to the control gains chosen such that $k_2 >> k_1$ to guarantee a nonzero lower bound of z_e.

11.7 Conclusions

The proposed results in this chapter can be readily combined with the observer designs presented in Chapter 7 to design an output feedback path-following system for an underactuated ship. It is seen from Section 11.2 that the ship should not be too close to the desired path at the initial time. However, the strategy in Section 11.4 can be used to overcome the mentioned limitation of the control system proposed in this chapter. The work presented in this chapter is based on [137, 138]. The approach in this chapter will be extended to the case of the underactuated underwater vehicle in six degrees of freedom in Chapter 13.

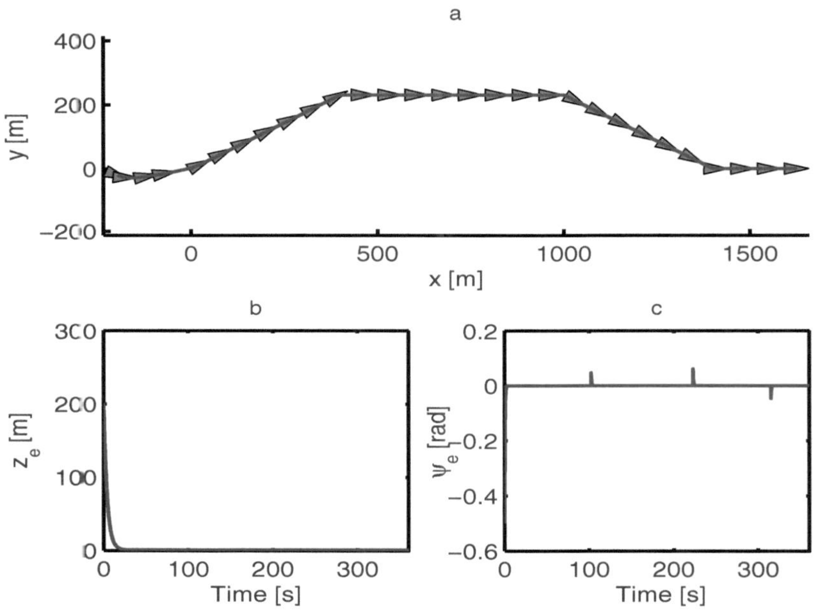

Figure 11.8 Point-to-point navigation: **a**. Ship position and orientation in the (x, y) plane; **b**. Ship position error z_e; **c**. Ship orientation error ψ_e

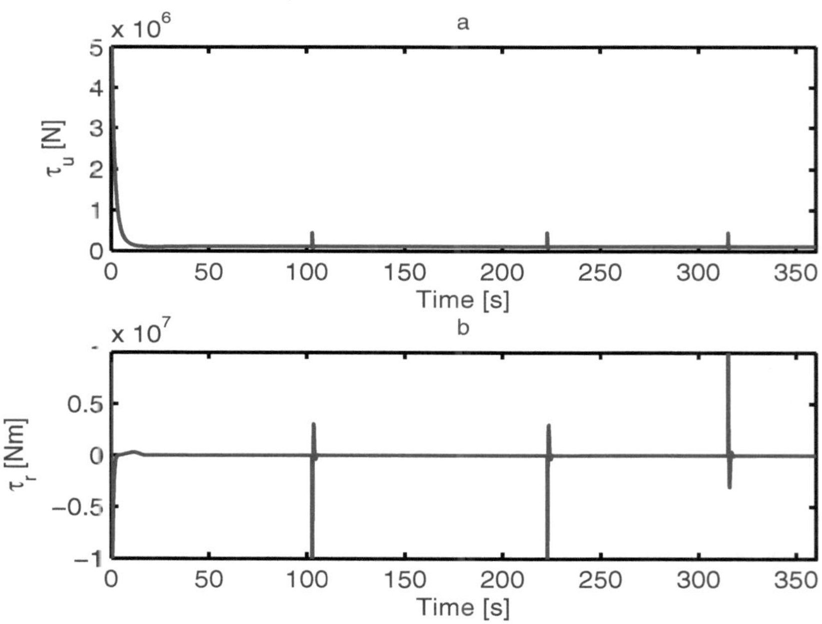

Figure 11.9 Point-to-point navigation: **a**. Surge force τ_u [N]; **b**.Yaw moment τ_r [Nm]

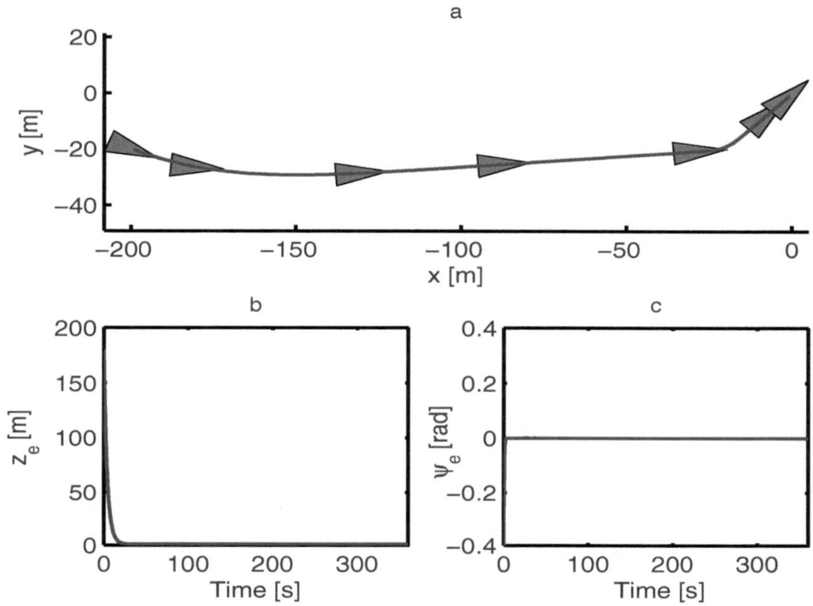

Figure 11.10 Parking: **a**. Ship position and orientation in the (x, y) plane; **b**. Ship position error; **c**. Ship orientation error in the (x, y) plane

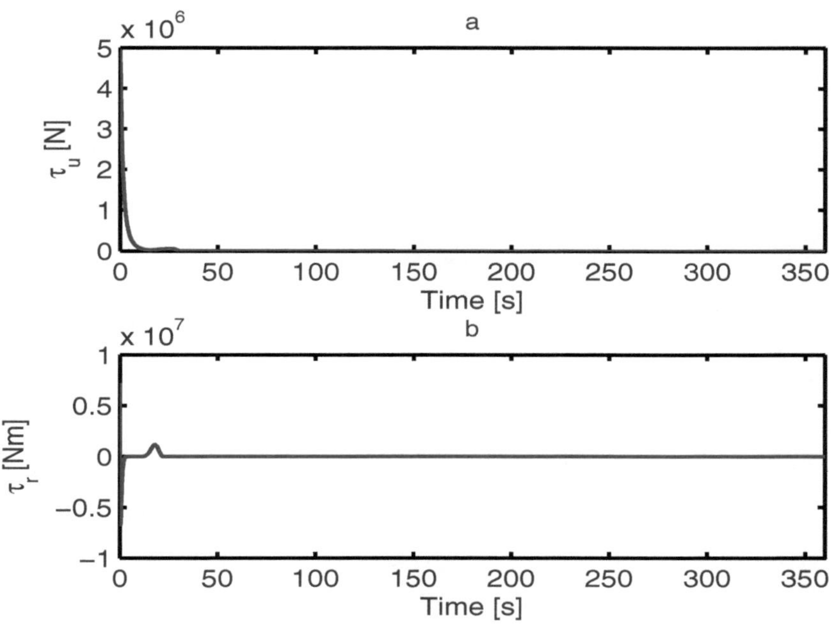

Figure 11.11 Parking: **a**. Surge force τ_u [N]; **b**.Yaw moment τ_r [Nm]

Part IV
Control of Underactuated Underwater Vehicles

Chapter 12
Trajectory-tracking Control of Underactuated Underwater Vehicles

This chapter addresses the problem of trajectory-tracking control of underactuated underwater vehicles. These vehicles do not have independent actuators in the sway and heave axes. Based on the techniques developed for underactuated surface ships in Chapters 5, 6, and 7, we present a method to design a controller for an underactuated underwater vehicle to globally asymptotically track a reference trajectory generated by a suitable virtual underwater vehicle. The yaw and pitch reference velocities do not have to satisfy a persistently exciting condition as was often required in previous literature. Due to the complex dynamics of the underwater vehicles in comparison with that of the ships, the control design and stability analysis require more complicated coordinate transformations and techniques than those developed for underactuated ships in Chapters 5, 6, and 7.

12.1 Control Objective

In this chapter, we consider the following mathematical model of an underactuated underwater vehicle when nonlinear hydrodynamic damping terms and roll motion are ignored, see Section 3.4.2.2:

1. Kinematics

$$\begin{aligned}
\dot{x} &= \cos(\psi)\cos(\theta)u - \sin(\psi)v + \sin(\theta)\cos(\psi)w, \\
\dot{y} &= \sin(\psi)\cos(\theta)u + \cos(\psi)v + \sin(\theta)\sin(\psi)w, \\
\dot{z} &= -\sin(\theta)u + \cos(\theta)w, \\
\dot{\theta} &= q, \\
\dot{\psi} &= \frac{r}{\cos(\theta)}.
\end{aligned} \tag{12.1}$$

2. Kinetics

$$\begin{aligned}
\dot{u} &= \frac{m_{22}}{m_{11}} vr - \frac{m_{33}}{m_{11}} wq - \frac{d_{11}}{m_{11}} u + \frac{1}{m_{11}} \tau_u, \\
\dot{v} &= -\frac{m_{11}}{m_{22}} ur - \frac{d_{22}}{m_{22}} v, \\
\dot{w} &= \frac{m_{11}}{m_{33}} uq - \frac{d_{33}}{m_{33}} w, \\
\dot{q} &= \frac{m_{33} - m_{11}}{m_{55}} uw - \frac{d_{55}}{m_{55}} q - \frac{\rho g \nabla \overline{GM_L} \sin(\theta)}{m_{55}} + \frac{1}{m_{55}} \tau_q, \\
\dot{r} &= \frac{m_{11} - m_{22}}{m_{66}} uv - \frac{d_{66}}{m_{66}} r + \frac{1}{m_{66}} \tau_r.
\end{aligned} \tag{12.2}$$

In (12.1) and (12.2) the symbols (see Figure 12.1) θ, ψ, q, and r denote the roll, pitch, and yaw angles and velocities while x, y, z, u, v, and w are the surge, sway, and heave displacements and velocities, respectively. The available control inputs are τ_u, τ_q, and τ_r. Since the sway and heave control forces are not available in the sway and heave dynamics, the underwater vehicle in question is underactuated. Notice that (12.1) is not defined when the pitch angle is equal to $\pm 90^0$. However, during practical operations with underwater vehicles, this problem is unlikely to happen due to the metacentric restoring forces. One way to avoid the singularities is to use a four-parameter description known as the quaternion. Here, we use the above Euler parameters because of their physical representation and computational efficiency. We first assume that $|\theta(t_0)| < 0.5\pi$. Then we find feasible initial conditions such

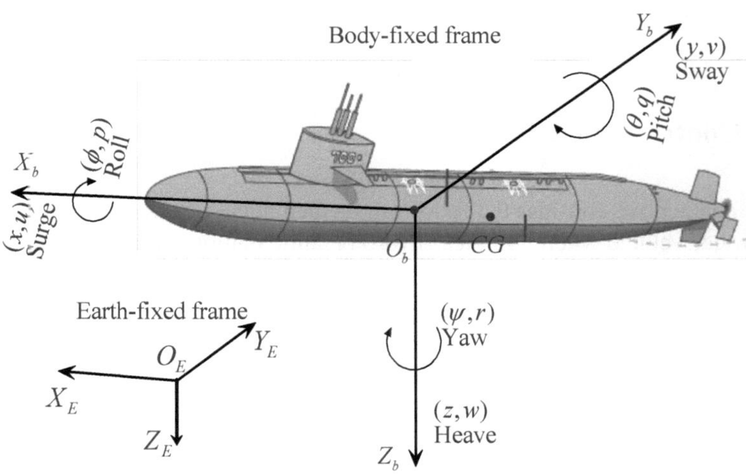

Figure 12.1 Motion variables of an underwater vehicle

that our proposed controller guarantees $|\theta(t)| < 0.5\pi$, $\forall t \geq t_0 \geq 0$.

We consider a control objective of designing the control inputs τ_u, τ_q, and τ_r to force the underactuated vehicle given in (12.1) and (12.2) to asymptotically track a reference trajectory generated by the following virtual vehicle

$$
\begin{aligned}
\dot{x}_d &= \cos(\psi_d)\cos(\theta_d)u_d - \sin(\psi_d)v_d + \sin(\theta_d)\cos(\psi_d)w_d,\\
\dot{y}_d &= \sin(\psi_d)\cos(\theta_d)u_d + \cos(\psi_d)v_d + \sin(\theta_d)\sin(\psi_d)w_d,\\
\dot{z}_d &= -\sin(\theta_d)u_d + \cos(\theta_d)w_d,\\
\dot{\theta}_d &= q_d,\\
\dot{\psi}_d &= \frac{r_d}{\cos(\theta_d)},\\
\dot{u}_d &= \frac{m_{22}}{m_{11}}v_d r_d - \frac{m_{33}}{m_{11}}w_d q_d - \frac{d_{11}}{m_{11}}u_d + \frac{1}{m_{11}}\tau_{ud},\\
\dot{v}_d &= -\frac{m_{11}}{m_{22}}u_d r_d - \frac{d_{22}}{m_{22}}v_d,\\
\dot{w}_d &= \frac{m_{11}}{m_{33}}u_d q_d - \frac{d_{33}}{m_{33}}w_d,\\
\dot{q}_d &= \frac{m_{33}-m_{11}}{m_{55}}u_d w_d - \frac{d_{55}}{m_{55}}q_d - \frac{\rho g \nabla \overline{GM_L}\sin(\theta_d)}{m_{55}} + \frac{1}{m_{55}}\tau_{qd},\\
\dot{r}_d &= \frac{m_{11}-m_{22}}{m_{66}}u_d v_d - \frac{d_{66}}{m_{66}}r_d + \frac{1}{m_{66}}\tau_{rd},
\end{aligned}
\tag{12.3}
$$

where all of the symbols in (12.3) have the same meaning as the real vehicle. In this section, we impose the following assumption on the reference model (12.3):

Assumption 12.1.

1. *The reference signals u_d, q_d, r_d, $\dot{u}_d$, $\dot{r}_d$, and $\dot{q}_d$ are bounded. There exists a strictly positive constant $u_{d\min}$, such that $|u_d(t)| \geq u_{d\min}$, $\forall t \geq 0$. The reference sway and heave velocities satisfy: $|v_d(t)| < |u_d(t)|$, $|w_d(t)| < |u_d(t)|$, $\forall t \geq 0$.*
2. *The reference pitch angle satisfies $|\theta_d(t)| < \frac{1}{2}\pi$, $\forall t \geq 0$.*

Remark 12.1. The condition $|u_d(t)| \geq u_{d\min}$, $\forall t \geq 0$, covers both forward and backward tracking since the reference surge velocity is always nonzero but can be either positive or negative. Indeed, the condition $|u_d(t)| \geq u_{d\min}$, $\forall t \geq 0$, is much less restrictive than a persistently exciting condition on the yaw reference velocity. The condition $|v_d(t)| < |u_d(t)|$ and $|w_d(t)| < |u_d(t)|$ implies that the underactuated underwater vehicle cannot track a helix with an arbitrarily large curvature and twist due to the vehicle's high inertia and underactuation in the sway and heave directions.

12.2 Coordinate Transformations

Since designing the control inputs τ_u, τ_q, and τ_r to achieve the control directly from (12.1) and (12.2) is difficult, we interpret the tracking errors in a frame attached to the vehicle body as follows:

$$\begin{bmatrix} x_e \\ y_e \\ z_e \\ \theta_e \\ \psi_e \end{bmatrix} = \begin{bmatrix} \cos(\psi)\cos(\theta) & \sin(\psi)\cos(\theta) & -\sin(\theta) & 0 & 0 \\ -\sin(\psi) & \cos(\psi) & 0 & 0 & 0 \\ \sin(\theta)\cos(\psi) & \sin(\theta)\sin(\psi) & \cos(\theta) & 0 & 0 \\ 0 & 0 & 0 & 1 & 0 \\ 0 & 0 & 0 & 0 & 1 \end{bmatrix} \begin{bmatrix} x - x_d \\ y - y_d \\ z - z_d \\ \theta - \theta_d \\ \psi - \psi_d \end{bmatrix}. \tag{12.4}$$

Indeed, convergence of $(x_e, y_e, z_e, \theta_e, \psi_e)$ to the origin implies that of $(x - x_d, y - y_d, z - z_d, \theta - \theta_d, \psi - \psi_d)$ since the matrix

$$\begin{bmatrix} \cos(\psi)\cos(\theta) & \sin(\psi)\cos(\theta) & -\sin(\theta) & 0 & 0 \\ -\sin(\psi) & \cos(\psi) & 0 & 0 & 0 \\ \sin(\theta)\cos(\psi) & \sin(\theta)\sin(\psi) & \cos(\theta) & 0 & 0 \\ 0 & 0 & 0 & 1 & 0 \\ 0 & 0 & 0 & 0 & 1 \end{bmatrix}$$

is nonsingular for all $(\psi, \theta) \in \mathbb{R}^2$. We also define the velocity tracking errors as

$$\begin{aligned} u_e &= u - u_d, \\ v_e &= v - v_d, \\ w_e &= w - w_d, \\ q_e &= q - q_d, \\ r_e &= r - r_d. \end{aligned} \tag{12.5}$$

Differentiating (12.4) along the solutions of (12.1) and (12.3) yields:

$$\begin{aligned} \dot{x}_e =\; & u_e - (\cos(\theta_e) - 1 + \cos(\theta)\cos(\theta_d)(\cos(\psi_e) - 1))\, u_d - \\ & \cos(\theta)\sin(\psi_e) v_d + (\sin(\theta_e) - \cos(\theta)\sin(\theta_d)(\cos(\psi_e) - 1))\, w_d + \\ & (r_d + r_e) y_e - (q_d + q_e) z_e, \\ \dot{y}_e =\; & v_e + \cos(\theta_d)\sin(\psi_e) u_d - (\cos(\psi_e) - 1) v_d - \sin(\theta_d)\sin(\psi_e) w_d - \\ & (x_e + \tan(\theta) z_e)(r_d + r_e), \\ \dot{z}_e =\; & w_e - (\sin(\theta_e) + \sin(\theta)\cos(\theta_d)(\cos(\psi_e) - 1))\, u_d - \sin(\theta)\sin(\psi_e) v_d - \\ & (\cos(\theta_e) - 1 + \sin(\theta)\sin(\theta_d)(\cos(\psi_e) - 1))\, w_d + \tan(\theta)(r_d + r_e) y_e + \\ & (q_d + q_e) x_e, \\ \dot{\theta}_e =\; & q_e, \\ \dot{\psi}_e =\; & \frac{r_e}{\cos(\theta)} + \frac{r_d}{\cos(\theta)\cos(\theta_d)} (\cos(\theta_d)(1 - \cos(\theta_e)) + \sin(\theta_d)\sin(\theta_e)). \end{aligned} \tag{12.6}$$

From (12.6), we can directly see that x_e, θ_e, and ψ_e can be stabilized by u_e, q_e, and r_e. There are several options to stabilize y_e and z_e. We can either use q_e, r_e, v_e, w_e, x_e, θ_e, or ψ_e. If q_e and r_e are used, the control design will be extremely complicated since q_e and r_e enter all of the first three equations of (12.6). On the other hand, the use of v_e and w_e to stabilize y_e and z_e will result in an undesired feature of marine vehicle control practice, namely the vessel will slide in the sway and heave directions. If we use x_e and z_e, the reference yaw and heave velocities must then satisfy persistently exciting conditions. Hence we will choose θ_e and ψ_e to stabilize the sway error y_e and the heave error z_e, respectively. This choice also coincides with the ship control practice. Toward this end, we define the following coordinates

$$
\begin{aligned}
z_1 &= \psi_e + \arcsin\left(\frac{k_1 y_e}{\sqrt{1+x_e^2+y_e^2+z_e^2}}\right),\\
z_2 &= \theta_e - \arcsin\left(\frac{k_2 z_e}{\sqrt{1+x_e^2+y_e^2+z_e^2}}\right),
\end{aligned}
\tag{12.7}
$$

where the constants k_i, $i = 1,2$, are such that $|k_i| < 1$. These constants will be specified later. It can be seen that (12.7) is well defined and that convergence of (z_1, y_e) and (z_2, z_e) implies that of θ_e and ψ_e.

Remark 12.2. Using the nonlinear coordinate transformations (12.7) instead of $z_1 = \psi_e + k_1 y_e$ and $z_2 = \theta_e - k_2 z_e$, we avoid the vessel whirling around when y_e and z_e are large. If one uses the transformations $z_1 = \psi_e + \arcsin\left(k_1 y_e/\sqrt{1+y_e^2}\right)$ and $z_2 = \theta_e - \arcsin\left(k_2 z_e/\sqrt{1+z_e^2}\right)$, the problem of the vessel whirling around is avoided. Indeed using these coordinate transformations will result in a much simpler tracking error system than using (12.7). However, these coordinate transformations make the design of the control inputs τ_u, τ_q, and τ_r difficult because the terms $(x_e + \tan(\theta) z_e) r_e$ and $(\tan(\theta) r_e y_e + x_e q_e)$ appear in the y_e- and z_e-dynamics.

Using the coordinate changes (12.7), the tracking error system in the $(x_e, y_e, z_e, z_1, z_2, u_e, v_e, w_e, q_e, r_e)$ coordinates can be written as:

$$
\begin{aligned}
\dot{x}_e =\ & u_e - (\varpi_2 - \varpi + \cos(\theta)\cos(\theta_d)(\varpi_1 - \varpi))\varpi^{-1} u_d + \\
& \cos(\theta) k_1 v_d \varpi^{-1} y_e + (k_2 z_e - \cos(\theta)\sin(\theta_d)(\varpi_1 - \varpi))\varpi^{-1} w_d + \\
& (r_d + r_e) y_e - (q_d + q_e) z_e + p_x,\\
\dot{y}_e =\ & v_e - \cos(\theta_d) k_1 u_d \varpi^{-1} y_e - (\varpi_1 - \varpi)\varpi^{-1} v_d + \sin(\theta_d) k_1 w_d \varpi^{-1} y_e - \\
& (x_e + \tan(\theta) z_e)(r_d + r_e) + p_y,\\
\dot{z}_e =\ & w_e - (k_2 z_e + \sin(\theta)\cos(\theta_d)(\varpi_1 - \varpi))\varpi^{-1} u_d + \sin(\theta) k_1 v_d \varpi^{-1} y_e - \\
& (\varpi_2 - \varpi + \sin(\theta)\sin(\theta_d)(\varpi_1 - \varpi))\varpi^{-1} w_d + \tan(\theta)(r_d + r_e) y_e + \\
& (q_d + q_e) x_e + p_z,\\
\dot{z}_1 =\ & \left(1 - k_1 \varpi_1^{-1}(\cos(\theta) x_e + \sin(\theta) z_e)\right)\cos(\theta)^{-1} r_e - k_1 \varpi_1^{-1} \varpi^{-2} x_e y_e u_e + \\
& f_{z1} + p_{z1},\\
\dot{z}_2 =\ & \left(1 - k_2 \varpi_2^{-1} x_e\right) q_e - k_2 \tan(\theta) \varpi_2^{-1} y_e r_e + k_2 \varpi_2^{-1} \varpi^{-2} x_e z_e u_e + f_{z2} + p_{z2},
\end{aligned}
$$

$$
\begin{aligned}
\dot{u}_e &= \frac{m_{22}}{m_{11}}vr - \frac{m_{33}}{m_{11}}wq - \frac{d_{11}}{m_{11}}u + \frac{1}{m_{11}}\tau_u - \dot{u}_d, \\
\dot{v}_e &= -\frac{m_{11}}{m_{22}}(u_e r_e + u_e r_d + u_d r_e) - \frac{d_{22}}{m_{22}}v_e, \\
\dot{w}_e &= \frac{m_{11}}{m_{33}}(u_e q_e + u_e q_d + u_d q_e) - \frac{d_{33}}{m_{33}}w_e, \\
\dot{q}_e &= \frac{m_{33}-m_{11}}{m_{55}}uw - \frac{d_{55}}{m_{55}}q - \frac{\rho g \nabla \overline{GM_L}\sin(\theta)}{m_{55}} + \frac{1}{m_{55}}\tau_q - \dot{q}_d, \\
\dot{r}_e &= \frac{m_{11}-m_{22}}{m_{66}}uv - \frac{d_{66}}{m_{66}}r + \frac{1}{m_{66}}\tau_r - \dot{r}_d,
\end{aligned}
\tag{12.8}
$$

where, for simplicity, we have defined the following notations:

1. The terms ϖ, ϖ_1 and ϖ_2:

$$
\begin{aligned}
\varpi &= \sqrt{1+x_e^2+y_e^2+z_e^2}, \\
\varpi_1 &= \sqrt{1+x_e^2+(1-k_1^2)y_e^2+z_e^2}, \\
\varpi_2 &= \sqrt{1+x_e^2+y_e^2+(1-k_2^2)z_e^2}.
\end{aligned}
\tag{12.9}
$$

2. The terms exponentially tend to zero when z_1 and z_2 do:

$$
\begin{aligned}
p_x = & -((\cos(z_2)-1)\varpi_2 - \sin(z_2)k_2 z_e + \cos(\theta)\cos(\theta_d)((\cos(z_1)- \\
& 1)\varpi_1 + \sin(z_1)k_1 y_e))\,\varpi^{-1}u_d + (\sin(z_2)\varpi_2 + (\cos(z_2)-1)\times \\
& k_2 z_e - \cos(\theta)\sin(\theta_d)((\cos(z_1)-1)\varpi_1 + \sin(z_1)k_1 y_e))\,\varpi^{-1}\times \\
& w_d - \cos(\theta)(\sin(z_1)\varpi_1 - (\cos(z_1)-1)k_1 y_e)\varpi^{-1}v_d, \\
p_y = & \cos(\theta_d)\,(\sin(z_1)\varpi_1 - (\cos(z_1)-1)k_1 y_e)\,\varpi^{-1}u_d - ((\cos(z_1)- \\
& 1)\varpi_1 + \sin(z_1)k_1 y_e)\,\varpi^{-1}v_d - \sin(\theta_d)(\sin(z_1)\varpi_1 - (\cos(z_1)- \\
& 1)k_1 y_e)\varpi^{-1}w_d, \\
p_z = & -(\sin(z_2)\varpi_2 + (\cos(z_2)-1)k_2 z_e + \sin(\theta)\cos(\theta_d)((\cos(z_1)- \\
& 1)\varpi_1 + \sin(z_1)k_1 y_e))\,\varpi^{-1}u_d - ((\cos(z_2)-1)\varpi_2 - \sin(z_2)\times \\
& k_2 z_e + \sin(\theta)\sin(\theta_d)\,((\cos(z_1)-1)\varpi_1 + \sin(z_1)k_1 y_e))\,\varpi^{-1}\times \\
& w_d - \sin(\theta)(\sin(z_1)\varpi_1 - (\cos(z_1)-1)k_1 y_e)\varpi^{-1}v_d, \\
p_{z1} = & k_1\varpi_1^{-1}\left(p_y - y_e\varpi^{-2}(x_e p_x + y_e p_y + z_e p_z)\right) + \cos(\theta)^{-1}\varpi^{-1}\times \\
& ((1-\cos(z_2))\varpi_2 + \sin(z_2)k_2 z_e + \tan(\theta_d)(\sin(z_2) + \\
& (\cos(z_2)-1)k_2 z_e, \\
p_{z2} = & -k_2\varpi_2^{-1}\left(p_z - z_e\varpi^{-2}(x_e p_x + y_e p_y + z_e p_z)\right).
\end{aligned}
\tag{12.10}
$$

3. The terms f_{z1} and f_{z2}:

$$
\begin{aligned}
f_{z1} &= (\varpi - \varpi_2 + \tan(\theta_d)k_2 z_e)\cos(\theta)^{-1}\varpi^{-1} + k_1\varpi_1^{-1}(v_e - \cos(\theta_d)\times \\
&\quad k_1 u_d \varpi^{-1} y_e - \varpi_1 - \varpi)\varpi^{-1} v_d + \sin(\theta_d)k_1 w_d \varpi^{-1} y_e - (x_e + \\
&\quad \tan(\theta) z_e) r_d - y_e \varpi^{-2}(x_e(-(\varpi_2 - \varpi + \cos(\theta)\cos(\theta_d)(\varpi_1 - \varpi))\times \\
&\quad \varpi^{-1} u_d + \cos(\theta)k_1 v_d \varpi^{-1} y_e + (k_2 z_e - \cos(\theta)\sin(\theta_d)(\varpi_1 - \varpi))\times \\
&\quad \varpi^{-1} w_d) + y_e(v_e - \cos(\theta_d)k_1 u_d \varpi^{-1} y_e - (\varpi_1 - \varpi)\varpi^{-1} v_d + \\
&\quad \sin(\theta_d)k_1 w_d \varpi^{-1} y_e) + z_e(w_e - (k_2 z_e + \sin(\theta)\cos(\theta_d)(\varpi_1 - \varpi))\times \\
&\quad \varpi^{-1} u_d + \sin(\theta)k_1 v_d \varpi^{-1} y_e - (\varpi_2 - \varpi + \sin(\theta)\sin(\theta_d)(\varpi_1 - \varpi))\times \\
&\quad \varpi^{-1} w_d)))), \\
f_{z2} &= -k_2\varpi_2^{-1}(w_e - (k_2 z_e + \sin(\theta)\cos(\theta_d)(\varpi_1 - \varpi))\varpi^{-1} u_d + \sin(\theta)\times \\
&\quad k_1 v_d \varpi^{-1} y_e - (\varpi_2 - \varpi + \sin(\theta)\ \sin(\theta_d)(\varpi_1 - \varpi))\varpi^{-1} w_d + x_e q_d + \\
&\quad \tan(\theta) y_e r_d - z_e \varpi^{-2}(x_e(-(\varpi_2 - \varpi + \cos(\theta)\cos(\theta_d)(\varpi_1 - \varpi))\times \\
&\quad \varpi^{-1} u_d + \cos(\theta)k_1 v_d \varpi^{-1} y_e + (k_2 z_e - \cos(\theta)\sin(\theta_d)(\varpi_1 - \varpi))\times \\
&\quad \varpi^{-1} w_d) + y_e(v_e - \cos(\theta_d)\,k_1 u_d \varpi^{-1} y_e - (\varpi_1 - \varpi)\varpi^{-1} v_d + \\
&\quad \sin(\theta_d)k_1 w_d \varpi^{-1} y_e) + z_e(w_e - (k_2 z_e + \sin(\theta)\cos(\theta_d)\times \\
&\quad (\varpi_1 - \varpi)\varpi^{-1} u_d + \sin(\theta)k_1 v_d \varpi^{-1} y_e - (\varpi_2 - \varpi + \\
&\quad \sin(\theta)\sin(\theta_d)(\varpi_1 - \varpi))\varpi^{-1} w_d)))).
\end{aligned} \tag{12.11}
$$

It is now clear that the problem of forcing the underactuated underwater vehicle given in (12.1) and (12.2) to track the virtual ship (12.3) becomes one of stabilizing the system (12.8). The efforts we have made so far are to put the tracking error dynamics in the triangular form of (12.8), and to have the terms $-\cos(\theta_d)k_1 u_d \varpi^{-1} y_e$ and $-k_2 u_d \varpi^{-1} z_e$ in the y_e and z_e-dynamics, respectively. These terms play an important role in stabilizing the y_e- and z_e-dynamics. In the next section we will design the control inputs τ_u, τ_q, and τ_r to asymptotically stabilize (12.8) at the origin.

12.3 Control Design

The triangular structure of (12.8) suggests that we design the actual controls τ_u, τ_q, and τ_r in two stages. First, we design the virtual velocity controls of u_e, q_e, and r_e to asymptotically stabilize x_e, y_e, z_e, z_1, z_2, w_e, and v_e at the origin. Based on the backstepping technique, the controls τ_u, τ_q, and τ_r will then be designed to force the errors between the virtual velocity controls and their actual values exponentially to zero. Since u_e enters the v_e- and w_e-dynamics, we will design a bounded virtual control of u_e to simplify the stability analysis. The virtual controls of q_e and r_e are chosen to stabilize the z_1- and z_2-dynamics.

12.3.1 Step 1

Define the virtual control errors as

$$\tilde{u}_e = u_e - u_e^d,\ \tilde{q}_e = q_e - q_e^d,\ \tilde{r}_e = r_e - r_e^d, \tag{12.12}$$

where u_e^d, q_e^d, and r_e^d are the virtual velocity controls of u_e, q_e, and r_e, respectively. The virtual controls u_e^d, q_e^d, and r_e^d are chosen as follows:

$$\begin{aligned}
u_e^d &= -k_0\varpi^{-1}x_e + (\varpi_2 - \varpi + \cos(\theta)\cos(\theta_d)(\varpi_1 - \varpi))\varpi^{-1}u_d - \\
&\quad \cos(\theta)k_1 v_d \varpi^{-1} y_e - (k_2 z_e - \cos(\theta)\sin(\theta_d)(\varpi_1 - \varpi))\varpi^{-1}w_d, \\
r_e^d &= r_{1e}^d + r_{2e}^d, \\
q_e^d &= q_{1e}^d + q_{2e}^d,
\end{aligned} \tag{12.13}$$

where

$$\begin{aligned}
r_{1e}^d &= \frac{\cos(\theta)}{1 - k_1\varpi_1^{-1}(\cos(\theta)x_e + \sin(\theta)z_e)}\left(k_1\varpi_1^{-1}\varpi^{-2}x_e y_e u_e^d - f_{z1}\right), \\
r_{2e}^d &= \frac{\cos(\theta)}{1 - k_1\varpi_1^{-1}(\cos(\theta)x_e + \sin(\theta)z_e)}(-c_1 z_1 - p_{z1}), \\
q_{1e}^d &= \frac{1}{1 - k_2\varpi_2^{-1}x_e}\left(k_2\tan(\theta)\varpi_2^{-1}y_e r_{1e}^d - k_2\varpi_2^{-1}\varpi^{-2}x_e z_e u_e^d - f_{z2}\right), \\
q_{2e}^d &= \frac{1}{1 - k_2\varpi_2^{-1}x_e}\left(-c_2 z_2 + k_2\tan(\theta)\varpi_2^{-1}y_e r_{1e}^d - p_{z2}\right),
\end{aligned} \tag{12.14}$$

with k_0 being a positive design constant to be specified later, and c_1 and c_2 being positive constants. It is not hard to show that the virtual control u_e^d is bounded as

$$\left|u_e^d\right| \le k_0 + (1 + k_1^2 + k_2^2)\,|u_d| + |k_1 v_d| + \left(|k_2| + k_1^2\right)|w_d| := u_{em}. \tag{12.15}$$

Remark 12.3. Unlike the standard application of backstepping, in order to reduce complexity of the controller expressions, we have chosen a simple virtual control law u_e^d without canceling some known terms. We have written $r_e^d = r_{1e}^d + r_{2e}^d$ and $q_e^d = q_{1e}^d + q_{2e}^d$ to simplify stability analysis, because q_{2e}^d and r_{2e}^d exponentially vanish when z_1 and z_2 do. From (12.14), we observe that r_e^d and q_e^d are Lipschitz in $(x_e, y_e, z_e, z_1, z_2, v_e, w_e)$. This observation plays a crucial role in the stability analysis of the closed loop system.

12.3.2 Step 2

By differentiating (12.12) along the solutions of (12.14) and (12.6), the actual controls τ_u, τ_q, and τ_r without canceling the useful damping terms are chosen as

$$
\begin{aligned}
\tau_u = {} & m_{11}\Bigg(-\rho_1\tilde{u}_e - \frac{m_{22}}{m_{11}}vr + \frac{m_{33}}{m_{11}}wq + \frac{d_{11}}{m_{11}}(u_e^d + u_d) + \dot{u}_d + \\
& \dot{u}_e^d + k_1\varpi_1^{-1}\varpi^{-2}x_e y_e z_1 - k_2\varpi_2^{-1}\varpi^{-2}x_e z_e z_2\Bigg), \\
\tau_r = {} & m_{66}\Bigg(-\rho_2\tilde{r}_e - \frac{m_{11}-m_{22}}{m_{66}}uv + \frac{d_{66}}{m_{66}}(r_e^d + r_d) + \dot{r}_d + \dot{r}_e^d - \\
& \left(1 - k_1\varpi_1^{-1}(\cos(\theta)x_e + \sin(\theta)z_e)\right)\frac{z_1}{\cos(\theta)} + k_2\tan(\theta)\varpi_2^{-1}y_e z_2\Bigg), \\
\tau_q = {} & m_{55}\Bigg(-\rho_3\tilde{q}_e - \frac{m_{33}-m_{11}}{m_{55}}uw + \frac{d_{55}}{m_{55}}(q_e^d + q_d) + \\
& \frac{\rho g\nabla\overline{GM_L}\sin(\theta)}{m_{55}} + \dot{q}_d + \dot{q}_e^d - \left(1 - k_2\varpi_2^{-1}x_e\right)z_2\Bigg),
\end{aligned} \tag{12.16}
$$

where $\rho_i, i = 1,2,3$, are positive constants. Substituting (12.16), (12.13), and (12.14) into (12.8) yields the closed loop system

$$
\begin{aligned}
\dot{x}_e = {} & -k_0\varpi^{-1}x_e - (q_d + q_e)z_e + (r_d + r_e)y_e + p_x + \tilde{u}_e, \\
\dot{y}_e = {} & v_e - \cos(\theta_d)k_1 u_d\varpi^{-1}y_e - (\varpi_1 - \varpi)\varpi^{-1}v_d + \\
& \sin(\theta_d)k_1 w_d\varpi^{-1}y_e - (x_e + \tan(\theta)z_e)(r_d + r_e) + p_y, \\
\dot{z}_e = {} & w_e - (k_2 z_e + \sin(\theta)\cos(\theta_d)(\varpi_1 - \varpi))\varpi^{-1}u_d + \\
& \sin(\theta)k_1 v_d\varpi^{-1}y_e - (\varpi_2 - \varpi + \sin(\theta)\sin(\theta_d)\times \\
& (\varpi_1 - \varpi))\varpi^{-1}w_d + \tan(\theta)(r_d + r_e)y_e + \\
& (q_d + q_e)x_e + p_z, \\
\dot{v}_e = {} & -m_{11}m_{22}^{-1}(u_e^d r_{1e}^d + u_e^d r_d + u_d r_{1e}^d) - \frac{d_{22}}{m_{22}}v_e - m_{11}m_{22}^{-1}\times \\
& ((u_e^d + u_d)(r_{2e}^d + \tilde{r}_e) + \tilde{u}_e(r_e^d + r_d + \tilde{r}_e)), \\
\dot{w}_e = {} & m_{11}m_{33}^{-1}(u_e^d q_{1e}^d + u_e^d r_d + u_d q_{1e}^d) - \frac{d_{33}}{m_{33}}w_e + m_{11}m_{33}^{-1}\times \\
& ((u_e^d + u_d)(q_{2e}^d + \tilde{q}_e) + \tilde{u}_e(q_e^d + q_d + \tilde{q}_e)), \\
\dot{z}_1 = {} & -c_1 z_1 + \left(1 - k_1\varpi_1^{-1}(\cos(\theta)x_e + \sin(\theta)z_e)\right)\times \\
& \cos(\theta)^{-1}\tilde{r}_e - k_1\varpi_1^{-1}\varpi^{-2}x_e y_e\tilde{u}_e, \\
\dot{z}_2 = {} & -c_2 z_2 + \left(1 - k_2\varpi_2^{-1}x_e\right)\tilde{q}_e - k_2\tan(\theta)\varpi_2^{-1}y_e\tilde{r}_e + k_2\varpi_2^{-1}\varpi^{-2}x_e z_e\tilde{u}_e, \\
\dot{\tilde{u}}_e = {} & -\left(\rho_1 + d_{11}m_{11}^{-1}\right)\tilde{u}_e + k_1\varpi_1^{-1}\varpi^{-2}x_e y_e z_1 - k_2\varpi_2^{-1}\varpi^{-2}x_e z_e z_2, \\
\dot{\tilde{q}}_e = {} & -\left(\rho_2 + d_{55}m_{55}^{-1}\right)\tilde{q}_e - \left(1 - k_2\varpi_2^{-1}x_e\right)z_2 \\
\dot{\tilde{r}}_e = {} & -\left(\rho_3 + d_{66}m_{66}^{-1}\right)\tilde{r}_e + k_2\tan(\theta)\varpi_2^{-1}y_e z_2 - \\
& \left(1 - k_1\varpi_1^{-1}(\cos(\theta)x_e + \sin(\theta)z_e)\right)\cos(\theta)^{-1}z_1.
\end{aligned} \tag{12.17}
$$

We now state the main result of this section, the proof of which is given in the next section.

Theorem 12.1. *Assume that the reference signals $(x_d, y_d, z_d, \theta_d, \psi_d, v_d, w_d)$ generated by the virtual vessel model (12.3), and Assumption 12.1 holds. If the state feedback control law (12.16) is applied to the vessel system (12.1) and (12.2), then the tracking errors $(x(t)-x_d(t),\ y(t)-y_d(t),\ z(t)-z_d(t),\ \theta(t)-\theta_d(t),\ \psi(t)-\psi_d(t),\ v(t)-v_d(t)$, and $w(t)-w_d(t))$ asymptotically converge to zero with an appropriate choice of the design constants k_0, k_1, and k_2, i.e., the closed loop system (12.17) is locally asymptotically stable at the origin.*

12.4 Stability Analysis

To prove Theorem 12.1, we just need to show that the closed loop system (12.17) is asymptotically stable at the origin. To simplify stability analysis of this closed loop system, we observe that (12.17) consists of two subsystems $(x_e, y_e, z_e, v_e, w_e)$ and $(z_1, z_2, \tilde{u}_e, \tilde{q}_e, \tilde{r}_e)$ in an interconnected structure. Therefore we first consider the $(z_1, z_2, \tilde{u}_e, \tilde{q}_e, \tilde{r}_e)$-subsystem then move to $(x_e, y_e, z_e, v_e, w_e)$-subsystem.

$(z_1, z_2, \tilde{u}_e, \tilde{q}_e, \tilde{r}_e)$-subsystem

From the last five equations of (12.17), it is direct to show that this subsystem is exponentially stable at the origin by taking the following Lyapunov function

$$V_1 = \frac{1}{2}\left(z_1^2 + z_2^2 + \tilde{u}_e^2 + \tilde{q}_e^2 + \tilde{r}_e^2\right), \tag{12.18}$$

whose time derivative along the solutions of (12.17) satisfies

$$\dot{V}_1 = -c_1 z_1^2 - c_2 z_2^2 - \left(\rho_1 + d_{11} m_{11}^{-1}\right)\tilde{u}_e^2 - \left(\rho_2 + d_{55} m_{55}^{-1}\right)\tilde{q}_e^2 - \left(\rho_3 + d_{66} m_{66}^{-1}\right)\tilde{r}_e^2, \tag{12.19}$$

which in turn implies that

$$\begin{aligned} &\|(z_1(t), z_2(t), \tilde{u}_e(t), \tilde{q}_e(t), \tilde{r}_e(t))\| \le \\ &\qquad \|(z_1(t_0), z_2(t_0), \tilde{u}_e(t_0), \tilde{q}_e(t_0), \tilde{r}_e(t_0))\|\, e^{-\sigma_1(t-t_0)}, \end{aligned} \tag{12.20}$$

where

$$\sigma_1 = \min\left(c_1, c_2, \left(\rho_1 + d_{11} m_{11}^{-1}\right), \left(\rho_2 + d_{55} m_{55}^{-1}\right), \left(\rho_3 + d_{66} m_{66}^{-1}\right)\right).$$

$(x_e, y_e, z_e, v_e, w_e)$-subsystem

To analyze stability of this subsystem, we consider the following Lyapunov function:

$$V_2 = \sqrt{1 + x_e^2 + y_e^2 + y_e^2} - 1 + \frac{1}{2} k_3 \left(v_e^2 + w_e^2 \right), \tag{12.21}$$

where k_3 is a positive constant to be specified later. The time derivative of (12.21) along the solutions of the first five equations of (12.17), after a lengthy but simple calculation using completed squares, satisfies

$$\begin{aligned} \dot{V}_2 \leq\ & -\mu_x(t)\varpi^{-2} x_e^2 - \mu_y(t)\varpi^{-2} y_e^2 - \mu_z(t)\varpi^{-2} z_e^2 - \mu_v(t) v_e^2 - \mu_w(t) w_e^2 + \\ & (\chi_1(\cdot) V_2 + \chi_2(\cdot)) e^{-\sigma_1(t-t_0)}, \end{aligned} \tag{12.22}$$

where $\chi_i(\cdot)$, $i = 1,2$ are some nondecreasing functions of $\|(z_1(t_0), z_2(t_0), \tilde{u}_e(t_0), \tilde{q}_e(t_0), \tilde{r}_e(t_0))\|$,

$$\begin{aligned} \mu_x(t) =\ & k_0 - \frac{k_3 m_{11} |r_d| \varepsilon_1}{m_{22}} - \frac{k_3 m_{11} u_{em} \varepsilon_1}{m_{22}(1-2|k_1|)} \left(\frac{|k_1 r_d|}{1-k_1^2} + |k_1^3 u_d| + k_1^2 |v_d| + \right. \\ & \left. (k_1^2 + k_2^2) |k_1 w_d| \right) - \frac{k_3 m_{11} |q_d| \varepsilon_1}{m_{33}} - \frac{k_3 m_{11} u_{em} |k_2 \tan(\theta)| \varepsilon_1}{m_{33}(1-|k_2|)} \times \\ & \left(|k_1 r_d| + |k_1^3 u_d| + k_1^2 |v_d| + (k_1^2 + k_2^2) |k_1 w_d| + u_{em} \right) - \\ & \frac{k_3 m_{11} u_{em} \varepsilon_1}{m_{33}(1-|k_2|)} \left(\frac{|k_2 q_d|}{1-k_1^2} + |k_1^3 u_d| + k_1^2 |k_2 u_d| \right), \end{aligned} \tag{12.23}$$

$$\begin{aligned} \mu_y(t) =\ & k_1 u_d \cos(\theta_d) - |k_1 w_d| - \varepsilon_3 - k_1^2 (|v_d| + |u_d| + |w_d|) - \\ & \frac{k_3 m_{11} u_{em} \varepsilon_1}{m_{22}(1-|k_2|)} \left(k_1^2 (|u_d| + |w_d|) + |k_1 v_d| \right) - \frac{k_3 m_{11} u_{em} \varepsilon_1}{m_{22}(1-2|k_1|)} \times \\ & \left(u_{em} |k_1| + k_1^2 (|v_d| + 2|u_d| + 2|w_d|) + |k_1^3 w_d| + 2|k_1^3 v_d| + \right. \\ & \left. (k_1^2 + k_2^2) |k_1 u_d| \right) - \frac{k_3 m_{11} |q_d| \varepsilon_1}{m_{33}} \left(k_1^2 (|u_d| + |w_d|) + |k_1 v_d| \right) - \\ & \frac{k_3 m_{11} u_{em} |k_2 \tan(\theta)| \varepsilon_1}{m_{33}(1-|k_2|)} \left(u_{em} |k_1| + k_1^2 (|v_d| + 2|u_d| + 2|w_d|) + \right. \\ & \left. |k_1^3| (2|v_d| + |w_d|) + (k_1^2 + k_2^2) |k_1 u_d| \right) - \frac{k_3 m_{11} u_{em} \varepsilon_1}{m_{33}(1-|k_2|)} \left(k_1^2 |k_2| \times \right. \\ & \left. (|u_d| + |w_d|) + |k_1 k_2| (|v_d| + |u_d|) + \frac{|k_2 r_d|}{1-k_2^2} \right) - \frac{|k_1 v_d|}{4\varepsilon_2}, \end{aligned} \tag{12.24}$$

$$\begin{aligned} \mu_z(t) =\ & k_2 u_d - \varepsilon_3 - |k_1 v_d| - k_2^2 |w_d| - \frac{k_3 m_{11} |r_d| \varepsilon_1}{m_{22}} \left(k_2^2 |u_d| + |k_2 w_d| \right) - \\ & \left(\frac{k_3 m_{11} u_{em} |k_2 \tan(\theta)| \varepsilon_1}{m_{33}(1-|k_2|)} + \frac{k_3 m_{11} u_{em} \varepsilon_1}{m_{22}(1-2|k_1|)} \right) \left(k_2^2 + \frac{|k_1 r_d|}{1-k_1^2} + \right. \\ & \left. |k_2| + |k_1 k_2| (|u_d| + |w_d|) \right) - \frac{k_3 m_{11} |q_d| \varepsilon_1}{m_{33}} \left(|k_2 w_d| + k_2^2 |u_d| \right) - \end{aligned}$$

$$\frac{k_3 m_{11} u_{em} \varepsilon_1}{m_{33}(1-|k_2|)}\left(2\left|k_2^3 w_d\right|+\left(k_2^2+k_1^2|k_2|\right)(|u_d|+|v_d|+|w_d|)+\right.$$
$$\left.|k_1 k_2|(|u_d|+|v_d|+2|w_d|),\right), \tag{12.25}$$

$$\begin{aligned}
\mu_v(t) = & \frac{k_3 d_{22}}{m_{22}}-\frac{1}{4\varepsilon_3}-\frac{k_3 m_{11}|r_d|}{m_{22}4\varepsilon_1}\left(k_0^2+(k_1^2+k_2^2)|u_d|+|k_1 v_d|+\right.\\
& \left.|k_2 w_d|+k_1^2|w_d|\right)-\frac{k_3 m_{11} u_{em}}{m_{22}(1-2|k_1|)4\varepsilon_1}\left(|k_1|(u_{em}+2.5+\right.\\
& |r_d|(1+\tan^2(\theta)))+k_2^2+k_1^2(|u_d|+3|v_d|+2|w_d|)+(k_1^2+\\
& \left.k_2^2)|k_1|(|u_d|+|w_d|)+|k_1 k_2|(|u_d|+|w_d|)\right)+\left|k_1^3\right|\times(|u_d|+2\times\\
& |v_d|+|w_d|)+|k_2|)-\frac{k_3 k_2^2 m_{11} u_{em}|\tan(\theta)|}{m_{33}(1-|k_2|)8\varepsilon_1}-\frac{k_3 k_2^2 m_{11} u_{em}}{m_{33}(1-|k_2|)8\varepsilon_1},
\end{aligned} \tag{12.26}$$

$$\begin{aligned}
\mu_w(t) = & \frac{k_3 d_{33}}{m_{33}}-\frac{1}{4\varepsilon_3}-\frac{k_3 m_{11}|q_d|}{m_{33}4\varepsilon_1}\left(k_0^2+(k_1^2+k_2^2)|u_d|+|k_1 v_d|+\right.\\
& \left.|k_2 w_d|+k_1^2|w_d|\right)-\frac{k_3 m_{11} u_{em}|k_2\tan(\theta)|}{m_{33}(1-|k_2|)8\varepsilon_1}\left(|k_1|(u_{em}+2.5\varepsilon_1+\right.\\
& |r_d|(1+\tan^2(\theta)))+k_2^2+k_1^2(|u_d|+3|v_d|+2|w_d|)+(k_1^2+\\
& k_2^2)|k_1|(|u_d|+|w_d|)+|k_1 k_2|(|u_d|+|w_d|)+\left|k_1^3\right|(|u_d|+|v_d|+\\
& |w_d|)+|k_2|)-\frac{k_3|k_2|m_{11}u_{em}}{m_{33}(1-|k_2|)4\varepsilon_1}\left(k_0+(k_1^2+k_2^2)|u_d|+|k_1 v_d|+\right.\\
& \left.(|k_2|+k_2^2)|w_d|\right)-\frac{k_3 m_{11} u_{em}}{m_{33}(1-|k_2|)4\varepsilon_1}\left(|k_2|(2.5\varepsilon_1+\tan^2(\theta)+\right.\\
& |q_d|)+k_1^2|k_2|(3|u_d|+2|v_d|+2|w_d|)+k_2^2(|u_d|+2|w_d|)+\\
& \left.\left|k_2^3\right|(|u_d|+2|w_d|)+2|k_1 k_2|(|u_d|+|v_d|+|w_d|)\right),
\end{aligned} \tag{12.27}$$

with $\varepsilon_i, i=1,2,3$ being some positive constants, and

$$\theta(t)=\theta_d(t)+z_2(t)+\arcsin\left(\frac{k_2 z_e(t)}{\sqrt{1+x_e^2(t)+y_e^2(t)+z_e^2(t)}}\right). \tag{12.28}$$

We now choose the design constants k_i, $0\le i\le 3$ such that

$$\begin{aligned}
&\mu_x \ge \mu_x^*,\\
&\mu_y(t) \ge \mu_y^*,\\
&\mu_z(t) \ge \mu_z^*,\\
&\mu_v(t) \ge \mu_v^*,\\
&\mu_w(t) \ge \mu_w^*,
\end{aligned} \tag{12.29}$$

for all $t \geq t_0 \geq 0$ for some positive constants μ_x^*, μ_y^*, μ_z^*, μ_v^*, μ_w^*, and θ being replaced by

$$\theta(t_0) = \theta_d(t_0) + z_2(t_0) + \arcsin\left(\frac{k_2 z_e(t_0)}{\sqrt{1 + x_e^2(t_0) + z_e^2(t_0) + z_e^2(t_0)}}\right). \tag{12.30}$$

Substituting (12.29) into (12.26) yields

$$\begin{aligned} \dot{V}_2 \leq\ & -\mu_x^* \varpi^{-2} x_e^2 - \mu_y^* \varpi^{-2} y_e^2 - \mu_z^* \varpi^{-2} z_e^2 - \mu_v^* v_e^2 - \mu_w^* w_e^2 + \\ & (\chi_1(\cdot) V_2 + \chi_2(\cdot)) e^{-\sigma_1 (t - t_0)}. \end{aligned} \tag{12.31}$$

From (12.31) we have $\dot{V}_2 \leq (\chi_1(\cdot) V_2 + \chi_2(\cdot)) e^{-\sigma_1 (t-t_0)}$, which implies that $V_2(t) \leq \chi_3(\cdot)$ with $\chi_3(\cdot)$ being a class-K function of $\|(x_e(t_0), y_e(t_0), z_e(t_0), v_e(t_0), w_e(t_0), z_1(t_0), z_2(t_0), \tilde{u}_e(t_0), \tilde{q}_e(t_0), \tilde{r}_e(t_0))\|$. Substituting $V_2(t) \leq \chi_3(\cdot)$ into (12.31) results in

$$\begin{aligned} \dot{V}_2 \leq\ & -\mu_x^* \varpi^{-2} x_e^2 - \mu_y^* \varpi^{-2} y_e^2 - \mu_z^* \varpi^{-2} z_e^2 - \mu_v^* v_e^2 - \mu_w^* w_e^2 + \\ & (\chi_1(\cdot) \chi_3(\cdot) + \chi_2(\cdot)) e^{-\sigma_1 (t - t_0)}. \end{aligned} \tag{12.32}$$

From (12.21) and (12.32) it is not hard to show that there exists a nonnegative constant σ_2 and a class-K function $\gamma_2(\cdot)$ depending on the initial conditions such that

$$\|(x_e(t), y_e(t), z_e(t), v_e(t), w_e(t))\| \leq \gamma_2(\cdot) e^{-\sigma_2 (t-t_0)}. \tag{12.33}$$

The dependence of $\sigma_2 > 0$ on the initial conditions implies that the closed loop system (12.17) is asymptotically stable at the origin. However one can straightforwardly show that (12.17) is also locally exponentially stable at the origin. To complete the proof of Theorem 12.1, we need to show that there exist the design constants k_i, $0 \leq i \leq 3$ such that $|k_1| < 1$, $|k_2| < 1$, condition (12.29) holds, and that $|\theta(t)| < 0.5\pi$. Before discussing these conditions, we note the following observations.

1. Under Assumption 12.1, the reference surge velocity u_d is always nonzero, and the magnitudes of the reference sway and heave velocities are always less than that of the reference surge velocity.
2. The mass, including added masses in the sway and heave dynamics, m_{22} and m_{33}, is often larger than that in the surge dynamics, m_{11}, for underwater vehicles, i.e., $m_{11} m_{22}^{-1} < 1$ and $m_{11} m_{33}^{-1} < 1$.

The condition of $|k_1| < 1$ and $|k_2| < 1$ can be satisfied easily by picking small enough k_1 and k_2. A close look at (12.29) shows that it always holds if we pick k_1 and k_2 such that they are small and have the same sign with the surge reference velocity u_d, and pick large enough k_3 for some small positive constants ε_i, $i = 1,2,3$. For example, one can choose $\varepsilon_i = |k_1| = |k_2|$. We can see from the above choice that the value of $|k_1|$ and $|k_2|$ should be decreased if $|u_d|$ is large. This physically means that the distance from the vessel to the point it aims to track should be increased if the surge velocity is large, otherwise the vessel will miss that point. Furthermore when $d_{22}m_{22}^{-1}$ and $d_{33}m_{33}^{-1}$ are small, $|k_1|$ and $|k_2|$ should also be decreased. Small $|k_1|$ and $|k_2|$ also imply a small value of k_0. This can be physically interpreted as follows: If the damping in the sway and heave dynamics is small, the control gain in the surge dynamics should also be small otherwise the vessel will slide in the sway and heave directions.

The condition $|\theta(t)| < 0.5\pi$ can be written as

$$|\theta(t)| \le |\theta_d(t)| + \gamma_2(\cdot) + |\arcsin(k_2\gamma_2(\cdot))| < 0.5\pi.$$

Hence there always exist initial conditions such that this condition holds under Assumption 12.1. Due to complicated expressions of $\mu_x(t)$, $\mu_y(t)$, $\mu_z(t)$, $\mu_v(t)$ and $\mu_w(t)$, we provide some general guidelines to choose the design constants rather than present their extremely complex explicit expressions.

Select small values for $|k_1|$ and $|k_2|$, set $\varepsilon_i = |k_1|$, $i = 1,2,3$, large enough value for k_3. Then increase k_3 and/or decrease $|k_1|$ and $|k_2|$ until (12.29) holds.

Finally, we note that for the ease of choosing the design constants, one can replace the absolute values of u_d, v_d, w_d and r_d in all of the negative terms in $\mu_x(t)$, $\mu_y(t)$, $\mu_z(t)$, $\mu_v(t)$ and $\mu_w(t)$ by their maximum values. The trade-off is that the control gains k_i, $0 \le i \le 2$ may be very small, which results in slow convergence of the tracking errors, if the surge reference velocity $u_d(t)$ varies largely, i.e. $u_{d\max} >> u_{d\min}$ with $u_{d\max}$ and $u_{d\min}$ being the maximum and minimum values of $u_d(t)$, respectively.

12.5 Simulations

This section illustrates the effectiveness of the control law (12.16) by simulating it on an underwater vehicle with a length of 5.56 m, a mass of 1089.8 kg, and other parameters taken from [139] as follows:

$$\begin{aligned}
&m_{11} = 1116\,\text{kg},\ m_{22} = 2133\,\text{kg},\ m_{33} = 2133\,\text{kg},\ m_{44} = 36.7\,\text{kgm}^2,\\
&m_{55} = 4061\,\text{kgm}^2,\ m_{66} = 4061\,\text{kgm}^2,\ d_{11} = 25.5\,\text{kgs}^{-1},\ d_{22} = 138\,\text{kgs}^{-1},\\
&d_{33} = 138\,\text{kgs}^{-1},\ d_{44} = 10\,\text{kgm}^2\text{s}^{-1},\ d_{55} = 490\,\text{kgm}^2\text{s}^{-1},\ d_{66} = 490\,\text{kgm}^2\text{s}^{-1},\\
&d_{u2} = 0,\ d_{u3} = 0,\ d_{p2} = 0,\ d_{p3} = 0,\ d_{q2} = 0,\ d_{q3} = 0,\ d_{r2} = 0,\ d_{r3} = 0,\\
&d_{v2} = 920.1\,\text{kgm}^{-2}\text{s},\ d_{v3} = 750\,\text{kgm}^{-3}\text{s}^2,\ d_{w2} = 920.1\,\text{kgm}^{-2}\text{s},\\
&d_{w3} = 750\,\text{kgm}^{-3}\text{s}^2.
\end{aligned}$$

This vehicle has a minimum turning circle with a radius of 75 m, a maximum surge force of 2×10^4 N, a maximum yaw moment of 1.5×10^4 Nm, a maximum pitch moment of 1.5×10^4 Nm, and the maximum roll moment of 120 Nm.

The reference trajectory is generated by (12.3) with

$$
\begin{aligned}
\tau_{ud} &= 5d_{11} - (m_{22}v_d r_d - m_{33}w_d q_d),\\
\tau_{qd} &= (-\theta_d + 0.2 - \dot{\theta}_d)m_{55} - (m_{33} - m_{11})u_d w_d - d_{55}q_d - \rho g \nabla \overline{GM_L}\sin(\theta_d),\\
\tau_{rd} &= -(m_{11} - m_{22})u_d v_d
\end{aligned} \tag{12.34}
$$

for the first 200 seconds, and

$$
\begin{aligned}
\tau_{ud} &= 5d_{11} - (m_{22}v_d r_d - m_{33}w_d q_d),\\
\tau_{qd} &= (-\theta_d + 0.2 - \dot{\theta}_d)m_{55} - (m_{33} - m_{11})u_d w_d - d_{55}q_d - \rho g \nabla \overline{GM_L}\sin(\theta_d),\\
\tau_{rd} &= -(m_{11} - m_{22})u_d v_d + 0.02d_{66}
\end{aligned} \tag{12.35}
$$

for the rest of simulation time. This choice means that the reference trajectory is a straight line for the first 200 seconds followed by a helix with constant curvature and torsion. The initial conditions are picked as follows:

$$
\begin{aligned}
&(x(t_0), y(t_0), z(t_0), \theta(t_0), \psi(t_0), u(t_0), v(t_0), w(t_0), q(t_0), r(t_0)) =\\
&\qquad (-50, -50, 0, 0, 0, 0, 0, 0, 0, 0),\\
&(x_d(t_0), y_d(t_0), z_d(t_0), \theta_d(t_0), \psi_d(t_0), u_d(t_0), v_d(t_0), w_d(t_0), q_d(t_0), r_d(t_0)) =\\
&\qquad (0, 0, 20, 0, 0, 0, 0, 10, 0, 0, 0).
\end{aligned} \tag{12.36}
$$

Based on the proof of Theorem 12.1, the design constants are chosen as $k_0 = 0.8$, $k_1 = k_2 = 0.4$, $c_i = 2$, and $\rho_i = 5$ with $i = 1, 2, 3$. The reference and real trajectories in three dimensions are plotted in Figure 12.2. The vessel position and orientation are plotted in Figure 12.3a while the tracking errors are plotted in Figures 12.3b-c. The control inputs plotted in Figures 12.4a-c. As proven in Theorem 12.1, the tracking errors asymptotically converge to the origin. Moreover, the control inputs have not reached their limits. This means that we still can further shorten the transient time by increasing the control gains.

12.6 Conclusions

The key to the control development is the coordinate transformations (12.7) to transform the tracking error system, in which the tracking errors are interpreted in the frame attached to the vehicle body, to a triangular form, to which the backstepping technique can be applied. The control development in this chapter is based on the approach developed for underactuated surface ships in Chapter 5.

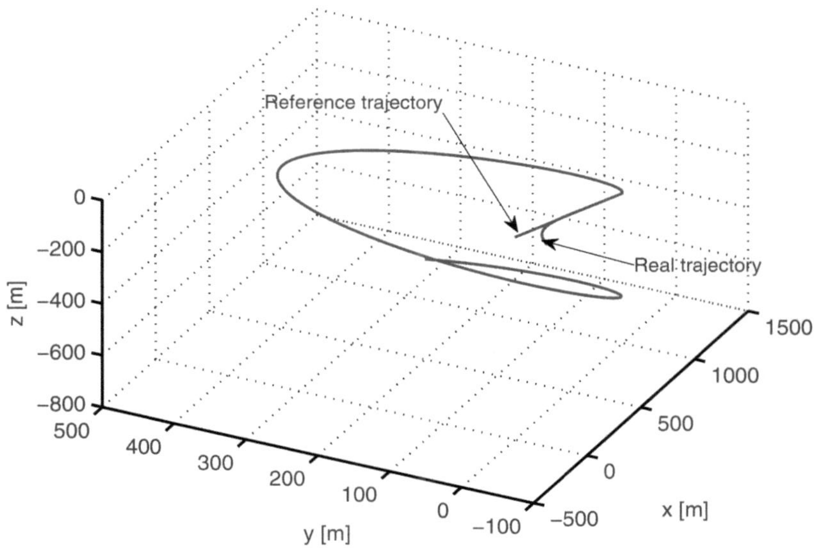

Figure 12.2 Reference and real trajectories in three dimensions

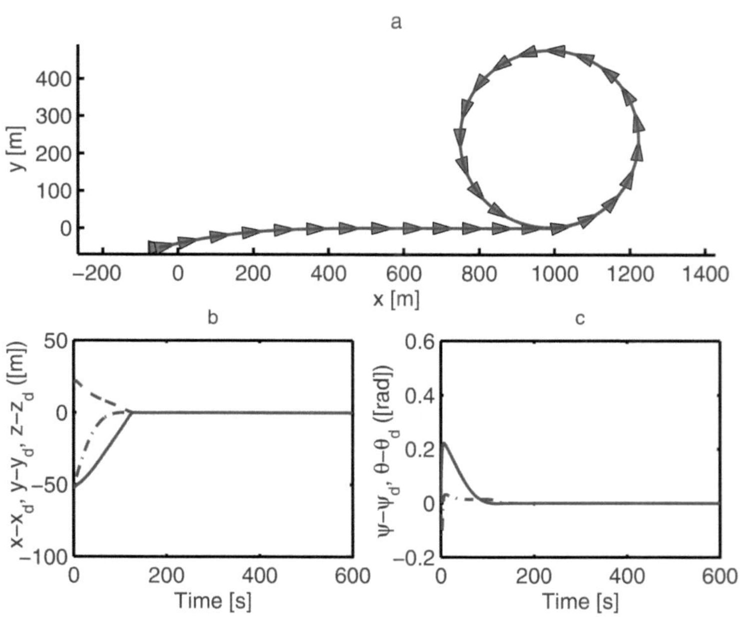

Figure 12.3 **a**. Position and orientation in the (x,y)-plane; **b**. Tracking errors: $x - x_d$ (*solid line*), $y - y_d$ (*dash-dotted line*), $z - z_d$ (dash); **c**. Tracking errors: $\psi - \psi_d$ (*solid line*), $\theta - \theta_d$ (*dash-dotted line*)

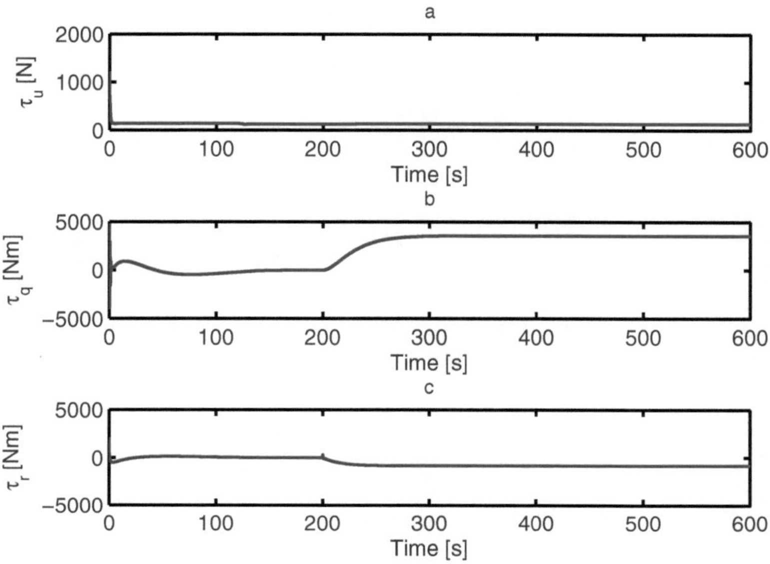

Figure 12.4 **a**. Control input: τ_u; **b**. Control input τ_q; **c**. Control input τ_r

Chapter 13
Path-following of Underactuated Underwater Vehicles

This chapter extends the approach proposed for underactuated surface ships in Chapter 11 to design a path-following system for six degrees of freedom underactuated underwater vehicles. Although the control design is much more involved in comparison with that for underactuated surface ships in Chapter 11, it still guarantees that path-following errors asymptotically converge to a ball, with an adjustable radius, centered on a desired path, and covers both parking and point-to-point navigation problems.

13.1 Control Objective

For the reader's convenience, we rewrite the mathematical model of an underactuated underwater vehicle, which is described in detail in Section 3.4.2.1, moving in six degrees of freedom as follows:

$$\begin{aligned}
\dot{\boldsymbol{\eta}}_1 &= \boldsymbol{J}_1(\boldsymbol{\eta}_2)\boldsymbol{v}_1,\\
\boldsymbol{M}_1\dot{\boldsymbol{v}}_1 &= -\boldsymbol{C}_1(\boldsymbol{v}_1)\boldsymbol{v}_2 - \boldsymbol{D}_1\boldsymbol{v}_1 - \boldsymbol{D}_{n1}(\boldsymbol{v}_1)\boldsymbol{v}_1 + \boldsymbol{\tau}_1 + \boldsymbol{\tau}_{1E},\\
\dot{\boldsymbol{\eta}}_2 &= \boldsymbol{J}_2(\boldsymbol{\eta}_2)\boldsymbol{v}_2,\\
\boldsymbol{M}_2\dot{\boldsymbol{v}}_2 &= -\boldsymbol{C}_1(\boldsymbol{v}_1)\boldsymbol{v}_1 - \boldsymbol{C}_2(\boldsymbol{v}_2)\boldsymbol{v}_2 - \boldsymbol{D}_2\boldsymbol{v}_2 - \boldsymbol{D}_{n2}(\boldsymbol{v}_2)\boldsymbol{v}_2 -\\
&\qquad \boldsymbol{g}_2(\boldsymbol{\eta}_2) + \boldsymbol{\tau}_2 + \boldsymbol{\tau}_{2E},
\end{aligned} \tag{13.1}$$

where $\boldsymbol{J}_1(\boldsymbol{\eta}_2)$ and $\boldsymbol{J}_2(\boldsymbol{\eta}_2)$ are given by

$$\boldsymbol{J}_1(\boldsymbol{\eta}_2) = \begin{bmatrix}
\cos(\psi)\cos(\theta) & -\sin(\psi)\cos(\phi)+\sin(\phi)\sin(\theta)\cos(\psi) & \sin(\psi)\sin(\phi)+\sin(\theta)\cos(\psi)\cos(\phi)\\
\sin(\psi)\cos(\theta) & \cos(\psi)\cos(\phi)+\sin(\phi)\sin(\theta)\sin(\psi) & -\cos(\psi)\sin(\phi)+\sin(\theta)\sin(\psi)\cos(\phi)\\
-\sin(\theta) & \sin(\phi)\cos(\theta) & \cos(\phi)\cos(\theta)
\end{bmatrix},$$

$$J_2(\eta_2) = \begin{bmatrix} 1 & \sin(\phi)\tan(\theta) & \cos(\phi)\tan(\theta) \\ 0 & \cos(\phi) & -\sin(\phi) \\ 0 & \sin(\phi)/\cos(\theta) & \cos(\phi)/\cos(\theta) \end{bmatrix}. \tag{13.2}$$

The matrices M_1 and M_2 are

$$M_1 = \begin{bmatrix} m_{11} & 0 & 0 \\ 0 & m_{22} & 0 \\ 0 & 0 & m_{33} \end{bmatrix}, \; M_2 = \begin{bmatrix} m_{44} & 0 & 0 \\ 0 & m_{55} & 0 \\ 0 & 0 & m_{66} \end{bmatrix}. \tag{13.3}$$

The matrices $C_1(v_1)$ and $C_2(v_2)$ are

$$\begin{aligned} C_1(v_1) &= \begin{bmatrix} 0 & m_{33}w & -m_{22}v \\ -m_{33}w & 0 & m_{11}u \\ m_{22}v & -m_{11}u & 0 \end{bmatrix}, \\ C_2(v_2) &= \begin{bmatrix} 0 & m_{66}r & -m_{55}q \\ -m_{66}r & 0 & m_{44}p \\ m_{55}q & -m_{44}p & 0 \end{bmatrix}. \end{aligned} \tag{13.4}$$

The linear and nonlinear damping matrices D_1, D_2, $D_{n1}(v_1)$, and $D_{n2}(v_2)$ are

$$\begin{aligned} D_1 &= \begin{bmatrix} d_{11} & 0 & 0 \\ 0 & d_{22} & 0 \\ 0 & 0 & d_{33} \end{bmatrix}, \; D_2 = \begin{bmatrix} d_{44} & 0 & 0 \\ 0 & d_{55} & 0 \\ 0 & 0 & d_{66} \end{bmatrix}, \\ D_{n1}(v_1) &= \begin{bmatrix} \sum_{i=2}^{3} d_{ui}|u|^{i-1} & 0 & 0 \\ 0 & \sum_{i=2}^{3} d_{vi}|v|^{i-1} & 0 \\ 0 & 0 & \sum_{i=2}^{3} d_{wi}|w|^{i-1} \end{bmatrix}, \\ D_{n2}(v_2) &= \begin{bmatrix} \sum_{i=2}^{3} d_{pi}|p|^{i-1} & 0 & 0 \\ 0 & \sum_{i=2}^{3} d_{qi}|q|^{i-1} & 0 \\ 0 & 0 & \sum_{i=2}^{3} d_{ri}|r|^{i-1} \end{bmatrix}. \end{aligned} \tag{13.5}$$

The restoring force and moment vector $g_2(\eta_2)$ is given by

$$g_2(\eta_2) = \begin{bmatrix} \rho g \nabla \overline{GM_T} \sin(\phi)\cos(\theta) \\ \rho g \nabla \overline{GM_L} \sin(\theta) \\ 0 \end{bmatrix}. \tag{13.6}$$

The propulsion force and moment vectors τ_1 and τ_2 are

$$\tau_1 = \begin{bmatrix} \tau_u \\ 0 \\ 0 \end{bmatrix}, \quad \tau_2 = \begin{bmatrix} \tau_p \\ \tau_q \\ \tau_r \end{bmatrix}, \tag{13.7}$$

which imply that the vehicle under consideration does not have independent actuators in the sway and heave. The environmental disturbance vectors τ_{1E} and τ_{2E}

are given by

$$\boldsymbol{\tau_{1E}} = \begin{bmatrix} \tau_{Eu}(t) \\ \tau_{Ev}(t) \\ \tau_{Ew}(t) \end{bmatrix}, \boldsymbol{\tau_{2E}} = \begin{bmatrix} \tau_{Ep}(t) \\ \tau_{Eq}(t) \\ \tau_{Er}(t) \end{bmatrix}, \tag{13.8}$$

where $\tau_{Eu}(t)$, $\tau_{Ev}(t)$, $\tau_{Ew}(t)$, $\tau_{Ep}(t)$, $\tau_{Eq}(t)$, and $\tau_{Er}(t)$ are the environmental disturbance forces or moments acting on the surge, sway, heave, roll, pitch, and yaw axes, respectively.

We assume that these disturbances are bounded as follows:

$$\begin{aligned} &|\tau_{Eu}(t)| \le \tau_{Eu}^{\max} < \infty, |\tau_{Ev}(t)| \le \tau_{Ev}^{\max} < \infty, |\tau_{Ew}(t)| \le \tau_{Ew}^{\max} < \infty, \\ &|\tau_{Ep}(t)| \le \tau_{Ep}^{\max} < \infty, |\tau_{Eq}(t)| \le \tau_{Eq}^{\max} < \infty, |\tau_{Er}(t)| \le \tau_{Er}^{\max} < \infty. \end{aligned} \tag{13.9}$$

Since the sway and heave control forces are not available in the sway and heave dynamics, the vehicle model (13.1) is underactuated.

In this chapter, we consider a control objective of designing the control inputs $\boldsymbol{\tau_1}$ and $\boldsymbol{\tau_2}$ to force the underactuated vehicle (13.1) to follow a specified path $\boldsymbol{\Omega}$, see Figure 13.1. If we are able to drive the vehicle to follow closely a virtual vessel that moves along the path with a desired speed u_0, then the control objective is fulfilled, i.e., the vessel is in a tube of nonzero diameter centered on the reference path and moves along the specified path at the speed u_0. Roughly speaking, the approach is to steer the vessel such that it heads toward the virtual one and diminishes the distance between the real and the virtual vessels.

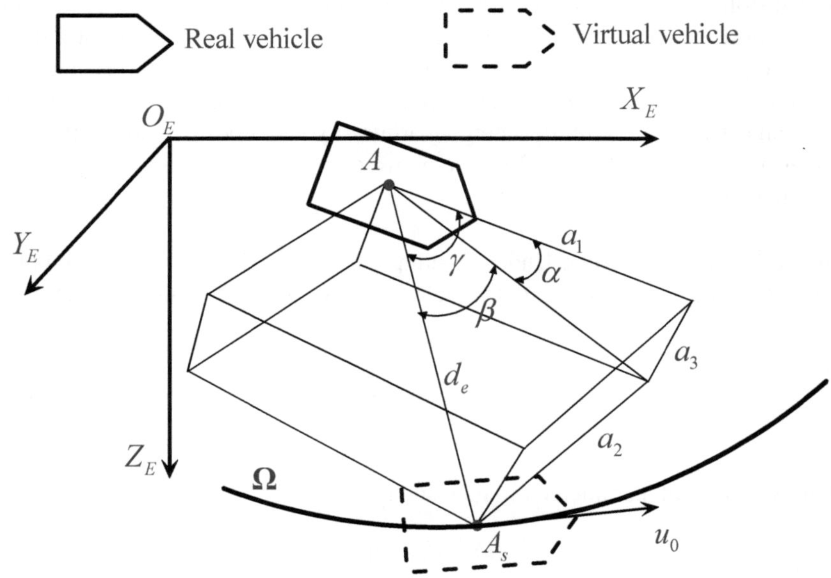

Figure 13.1 General framework of underwater vehicle path-following

In Figure 13.1, A is the center of the real vehicle and A_s is a point on the reference path attached to the virtual vehicle. Define the following path-following errors

$$\begin{aligned} x_e &= x_d - x, \\ y_e &= y_d - y, \\ z_e &= z_d - z, \\ d_e &= \sqrt{x_e^2 + y_e^2 + z_e^2}, \end{aligned} \tag{13.10}$$

where x_d, y_d and z_d are the coordinates of A_s. Then the terms $a_i, 1 \leq i \leq 3$ in Figure 13.1 are obtained from x_e, y_e and z_e by rotating the body frame around the earth-fixed frame $O_E X_E Y_E Z_E$ the roll, pitch, and yaw angles, i.e.,

$$\begin{bmatrix} a_1 \\ a_2 \\ a_3 \end{bmatrix} = \boldsymbol{J_1}(\boldsymbol{\eta_2}) \begin{bmatrix} x_e \\ y_e \\ z_e \end{bmatrix}. \tag{13.11}$$

Expanding (13.11) yields

$$\begin{aligned} a_1 &= x_e J_1^{11}(\boldsymbol{\eta_2}) + y_e J_1^{21}(\boldsymbol{\eta_2}) + z_e J_1^{31}(\boldsymbol{\eta_2}), \\ a_2 &= x_e J_1^{12}(\boldsymbol{\eta_2}) + y_e J_1^{22}(\boldsymbol{\eta_2}) + z_e J_1^{32}(\boldsymbol{\eta_2}), \\ a_3 &= x_e J_1^{13}(\boldsymbol{\eta_2}) + y_e J_1^{23}(\boldsymbol{\eta_2}) + z_e J_1^{33}(\boldsymbol{\eta_2}), \end{aligned} \tag{13.12}$$

where $J_1^{ij}(\eta_2)$ is the element of $\boldsymbol{J_1}(\boldsymbol{\eta_2})$ at the ith row and jth column. Therefore the path-following orientation errors are defined by the angles α and β. It is noted that the angles γ, α, and β are not defined at $d_e = 0$ but with the aid of a desired controller, $\lim_{\|(d_e,\alpha,\beta)\| \to 0}(\theta, \psi) = (\theta_s, \psi_s)$ with θ_s and ψ_s being the orientation angles of the virtual vessel. Hence in this chapter, we will design a controller such that it guarantees $d_e \geq d_e^*$ with d_e^* being an arbitrarily small positive constant to avoid chattering caused by $d_e = 0$. With the above definitions, our control objective can be mathematically stated as follows:

Path-following Objective. Under Assumption 13.1, design the control inputs $\boldsymbol{\tau_1}$ and $\boldsymbol{\tau_2}$ to force the underactuated vehicle (13.1) to follow the path $\boldsymbol{\Omega}$ given by

$$\begin{aligned} x_d &= x_d(s), \\ y_d &= y_d(s), \\ z_d &= z_d(s), \end{aligned} \tag{13.13}$$

where s is the path parameter variable, such that

$$\begin{aligned} &\lim_{t\to\infty} d_e(t) \leq \overline{d_e}, \ \lim_{t\to\infty} |\alpha(t)| \leq \overline{\alpha}, \\ &\lim_{t\to\infty} |\beta(t)| \leq \overline{\beta}, \ \lim_{t\to\infty} |\phi(t)| \leq \overline{\phi}, \end{aligned}$$

with $\overline{d_e}$, $\overline{\phi}$, $\overline{\alpha}$, and $\overline{\beta}$ being arbitrarily small positive constants.

Assumption 13.1. *The reference path is regular, i.e. there exist strictly positive constants* $a_{3\min}$, $a_{3\max}$, $a_{2\min}$ *and* $a_{2\max}$ *such that*

$$a_{3\min} \leq \sqrt{\left(\frac{\partial x_d}{\partial s}\right)^2 + \left(\frac{\partial y_d}{\partial s}\right)^2 + \left(\frac{\partial z_d}{\partial s}\right)^2} \leq a_{3\max},$$
$$a_{2\min} \leq \sqrt{\left(\frac{\partial x_d}{\partial s}\right)^2 + \left(\frac{\partial y_d}{\partial s}\right)^2} \leq a_{2\max}. \tag{13.14}$$

Remark 13.1.

1. We might refer to the above objective as a path-tracking one. However, we use the term "path-following" since our approach is to make the real vehicle follow the virtual one, see Figure 13.1.
2. Assumption 13.1 ensures that the path is feasible for the vessel to follow, see Section 13.3. The condition (13.14) implies that the reference trajectory cannot contain a vertical straight line to avoid singularity of $\boldsymbol{J_2}(\boldsymbol{\eta_2})$ at the pitch angle $\theta = \pm 0.5\pi$.
3. If the reference path is not regular, then we can often split it into regular pieces and consider each of them separately. This is the case of point-to-point navigation, which will be addressed in Section 13.6.
4. The path parameter, s, is not the arc length of the path in general. For example, a circle with radius R centered at the origin can be described as $x_d = R\cos(s)$ and $y_d = R\sin(s)$.

13.2 Coordinate Transformations

From (13.10) and (13.12), we have the position kinematic error dynamics as follows:

$$\dot{d}_e = \frac{1}{d_e}\left(x_e\frac{\partial x_d}{\partial s} + y_e\frac{\partial y_d}{\partial s} + z_e\frac{\partial z_d}{\partial s}\right)\dot{s} - \frac{a_1}{d_e}u - \frac{a_2}{d_e}v - \frac{a_3}{d_e}w. \tag{13.15}$$

For the path-following orientation errors, referring to Figure 13.1 and the control objective stated in the previous section, one can see that the following holds

$$\frac{a_1}{d_e} = \cos(\gamma) = \cos(\alpha)\cos(\beta), \tag{13.16}$$

which in turn implies that

$$\begin{cases} \lim_{t\to\infty} \alpha(t) = 0 \\ \lim_{t\to\infty} \beta(t) = 0 \end{cases} \Leftrightarrow \lim_{t\to\infty} \gamma(t) = 0 \Rightarrow \lim_{t\to\infty}\left(\frac{a_1(t)}{d_e(t)}\right) = 1. \tag{13.17}$$

Hence we can either choose the angles α and β, or the angle γ, or the term a_1/d_e as the orientation coordinates for the control design. We now discuss the above options and then choose one that results in a simple control design and enhance feasible initial conditions.

Using Angles α and β. In this case, the path-following orientation errors are defined as follows, see Figure 13.1:

$$\begin{aligned} \alpha &= e_\alpha - 2\pi n_\alpha(e_\alpha), \\ \beta &= e_\beta - 2\pi n_\beta(e_\beta), \end{aligned} \tag{13.18}$$

where

$$\begin{aligned} e_\alpha &= \begin{cases} 2\arctan\left(\dfrac{a_3}{a_1}\right); \ (a_3,a_1) \neq (0,0), \\ 0; \ (a_3,a_1) = (0,0), \end{cases} \\ e_\beta &= \begin{cases} 2\arctan\left(\dfrac{a_2}{\sqrt{a_1^2+a_3^2}}\right); \ \left(a_2, \sqrt{a_1^2+a_3^2}\right) \neq (0,0), \\ 0 \ ; \ \left(a_2, \sqrt{a_1^2+a_3^2}\right) = (0,0). \end{cases} \end{aligned} \tag{13.19}$$

The functions $n_\alpha(e_\alpha)$ and $n_\beta(e_\beta)$ take values in $(0,\pm1,\pm2,...)$ such that α and β belong to $(-\pi,\pi]$. Hence α and β are periodic and piecewise continuous functions with respect to e_α and e_β. The reason for introducing (13.18) is to convert all equilibrium points of α and β to the origin. It is seen from (13.19) that e_α and e_β are discontinuous on the following surfaces:

$$\begin{aligned} D_\alpha &= \{(a_1,a_3) : a_3 \neq 0, \, a_1 = 0\}, \\ D_\beta &= \left\{(a_1,a_2,a_3) : a_2 \neq 0, \, \sqrt{a_1^2+a_3^2} = 0\right\}. \end{aligned} \tag{13.20}$$

It is also seen from (13.18) that α and β are discontinuous on the surfaces:

$$\begin{aligned} C_\alpha &= \{(a_1,a_3) : \alpha = \pi\}, \\ C_\beta &= \{(a_1,a_2,a_3) : \beta = \pi\}. \end{aligned} \tag{13.21}$$

We now use (13.18) to transform the kinematic part of (13.1) to

$$\dot{\boldsymbol{\eta}}_{\alpha\beta} = \boldsymbol{f}^{s}_{\alpha\beta}(\cdot)\dot{s} + \boldsymbol{f}^{u}_{\alpha\beta}(\cdot)u + \boldsymbol{f}^{v}_{\alpha\beta}(\cdot)v + \boldsymbol{f}^{w}_{\alpha\beta}(\cdot)w + \boldsymbol{A}_{\alpha\beta}(\cdot)\boldsymbol{J}_2(\boldsymbol{\eta}_2)\boldsymbol{v}_2, \tag{13.22}$$

where

$$\begin{aligned} \boldsymbol{\eta}_{\alpha\beta} &= \left[\phi \ \alpha \ \beta\right]^T, \\ \boldsymbol{f}^{s}_{\alpha\beta}(\cdot) &= \boldsymbol{F}_{\alpha\beta}(\cdot)\left[\frac{\partial x_d}{\partial s} \ \frac{\partial y_d}{\partial s} \ \frac{\partial z_d}{\partial s}\right]^T, \end{aligned}$$

$$\begin{aligned}
\boldsymbol{f}^{\boldsymbol{u}}_{\alpha\beta}(\cdot) &= -\boldsymbol{F}_{\alpha\beta}(\cdot)\left[J_1^{11}(\boldsymbol{\eta_2})\ J_1^{21}(\boldsymbol{\eta_2})\ J_1^{31}(\boldsymbol{\eta_2})\right]^T,\\
\boldsymbol{f}^{\boldsymbol{v}}_{\alpha\beta}(\cdot) &= -\boldsymbol{F}_{\alpha\beta}(\cdot)\left[J_1^{12}(\boldsymbol{\eta_2})\ J_1^{22}(\boldsymbol{\eta_2})\ J_1^{32}(\boldsymbol{\eta_2})\right]^T,\\
\boldsymbol{f}^{\boldsymbol{w}}_{\alpha\beta}(\cdot) &= -\boldsymbol{F}_{\alpha\beta}(\cdot)\left[J_1^{13}(\boldsymbol{\eta_2})\ J_1^{23}(\boldsymbol{\eta_2})\ J_1^{33}(\boldsymbol{\eta_2})\right]^T,\\
\boldsymbol{F}_{\alpha\beta}(\cdot) &= \left[0\ \boldsymbol{f}^{\boldsymbol{1}}_{\alpha\beta}(\cdot)\ \boldsymbol{f}^{\boldsymbol{2}}_{\alpha\beta}(\cdot)\right]^T, \qquad (13.23)\\
\boldsymbol{A}_{\alpha\beta}(\cdot) &= \left[\begin{matrix} 1 & 0\\ -\cos(\alpha)\tan(\beta) & \cos(\phi)+\sin(\alpha)\tan(\beta)\sin(\phi)\\ \sin(\alpha) & \sin(\phi)\cos(\alpha)\end{matrix}\right.\\
&\qquad \left.\begin{matrix} 0\\ \cos(\theta)\sin(\phi)+\tan(\beta)(\sin(\theta)\cos(\alpha)-\cos(\theta)\sin(\alpha)\cos(\phi)\\ -(\cos(\alpha)\cos(\theta)\cos(\phi)+\sin(\alpha)\sin(\theta)\end{matrix}\right],
\end{aligned}$$

with

$$\begin{aligned}
\boldsymbol{f}^{\boldsymbol{1}}_{\alpha\beta}(\cdot) = {} & \frac{1}{d_e\cos(\beta)}\left[\cos(\alpha)J_1^{13}(\boldsymbol{\eta_2})-\sin(\alpha)J_1^{11}(\boldsymbol{\eta_2}),\right.\\
& \cos(\alpha)J_1^{23}(\boldsymbol{\eta_2})-\sin(\alpha)J_1^{21}(\boldsymbol{\eta_2}),\\
& \left.\cos(\alpha)J_1^{33}(\boldsymbol{\eta_2})-\sin(\alpha)J_1^{31}(\boldsymbol{\eta_2})\right],\\
\boldsymbol{f}^{\boldsymbol{2}}_{\alpha\beta}(\cdot) = {} & \left[-\frac{\sin(\beta)\cos(\alpha)}{d_e}J_1^{11}(\boldsymbol{\eta_2})+\frac{\cos(\beta)}{d_e}J_1^{12}(\boldsymbol{\eta_2})-\frac{\sin(\beta)\sin(\alpha)}{d_e}J_1^{13}(\boldsymbol{\eta_2}),\right.\\
& -\frac{\sin(\beta)\cos(\alpha)}{d_e}J_1^{21}(\boldsymbol{\eta_2})+\frac{\cos(\beta)}{d_e}J_1^{22}(\boldsymbol{\eta_2})-\frac{\sin(\beta)\sin(\alpha)}{d_e}J_1^{23}(\boldsymbol{\eta_2}),\\
& \left.-\frac{\sin(\beta)\cos(\alpha)}{d_e}J_1^{31}(\boldsymbol{\eta_2})+\frac{\cos(\beta)}{d_e}J_1^{32}(\boldsymbol{\eta_2})-\frac{\sin(\beta)\sin(\alpha)}{d_e}J_1^{33}(\boldsymbol{\eta_2})\right].
\end{aligned} \qquad (13.24)$$

From (13.23), we calculate the determination of the matrix $\boldsymbol{A}_{\alpha\beta}(\cdot)$ as follows:

$$\begin{aligned}
\det(\boldsymbol{A}_{\alpha\beta}(\cdot)) = {} & -\cos(\alpha)\cos(\theta)-\sin(\alpha)\sin(\theta)\cos(\phi)-\\
& \sin(\theta)\sin(\phi)\tan(\beta).
\end{aligned} \qquad (13.25)$$

It can be seen from (13.25) that the matrix $\boldsymbol{A}_{\alpha\beta}(\cdot)$ is not globally invertible even when $\phi = 0$ and $\theta \neq \pm 0.5\pi$. It is also observed that if we choose the angles α and β as the orientation coordinates for control design, there are a number of discontinuous surfaces, see (13.19), (13.20), and (13.21), which make the stability analysis difficult.

Using Angle γ. In this case, the path-following orientation error is defined as follows, see Figure 13.1:

$$\gamma = e_\gamma - 2\pi n_\gamma(e_\gamma), \qquad (13.26)$$

where

$$e_\gamma = \begin{cases} 2\arctan\left(\dfrac{\sqrt{a_2^2+a_3^2}}{a_1}\right); \left(a_1, \sqrt{a_2^2+a_3^2}\right) \neq (0,0), \\ 0; \left(a_1, \sqrt{a_2^2+a_3^2}\right) = (0,0). \end{cases} \tag{13.27}$$

The interpretation of the above expressions is similar to that of (13.18) and (13.19). It can be seen from (13.27) that e_γ is discontinuous on the following surface:

$$D_\gamma = \left\{(a_1,a_2,a_3) : \sqrt{a_2^2+a_3^2} \neq 0, a_1 = 0\right\}. \tag{13.28}$$

It can also be seen from (13.26) that γ is discontinuous on the surface:

$$C_\gamma = \{(a_1,a_2,a_3) : \gamma = \pi\}. \tag{13.29}$$

Differentiating both sides of (13.26) results in

$$\begin{aligned} \dot{\gamma} = & \frac{1}{e^2\sqrt{a_2^2+a_3^2}}\left(a_1a_2\left(\frac{\partial a_2}{\partial x_e}\dot{x}_e + \frac{\partial a_2}{\partial y_e}\dot{y}_e + \frac{\partial a_2}{\partial z_e}\dot{z}_e\frac{\partial a_2}{\partial \phi}\dot{\phi}\right) + \right. \\ & a_1a_3\left(\frac{\partial a_3}{\partial x_e}\dot{x}_e + \frac{\partial a_3}{\partial y_e}\dot{y}_e + \frac{\partial a_3}{\partial z_e}\dot{z}_e\frac{\partial a_3}{\partial \phi}\dot{\phi}\right) - \\ & (a_2^2+a_3^2)\left(\frac{\partial a_1}{\partial x_e}\dot{x}_e + \frac{\partial a_1}{\partial y_e}\dot{y}_e + \frac{\partial a_1}{\partial z_e}\dot{z}_e + \frac{\partial a_1}{\partial \phi}\dot{\phi}\right) + \\ & \left(a_1\left(a_2\frac{\partial a_2}{\partial \theta} + a_3\frac{\partial a_3}{\partial \theta}\right) - (a_2^2+a_3^2)\frac{\partial a_1}{\partial \theta}\right)\dot{\theta} + \\ & \left.\left(a_1\left(a_2\frac{\partial a_2}{\partial \psi} + a_3\frac{\partial a_3}{\partial \psi}\right) - (a_2^2+a_3^2)\frac{\partial a_1}{\partial \psi}\right)\dot{\psi}\right). \end{aligned} \tag{13.30}$$

Since

$$\begin{aligned} & \left[\frac{\left(a_1\left(a_2\frac{\partial a_2}{\partial \theta} + a_3\frac{\partial a_3}{\partial \theta}\right) - (a_2^2+a_3^2)\frac{\partial a_1}{\partial \theta}\right)}{d_e^2\sqrt{a_2^2+a_3^2}}\right]^2 + \\ & \left[\frac{\left(a_1\left(a_2\frac{\partial a_2}{\partial \psi} + a_3\frac{\partial a_3}{\partial \psi}\right) - (a_2^2+a_3^2)\frac{\partial a_1}{\partial}\right)}{d_e^2\sqrt{a_2^2+a_3^2}}\right]^2 \neq 0, \end{aligned} \tag{13.31}$$

for all $(x_e, y_e, z_e, \phi, \psi) \in \mathbb{R}^5$ and $\theta \in \mathbb{R}\backslash \pm 0.5\pi$, the angle γ can be chosen as the orientation coordinate for the control design. However (13.30) will result in a very complicated control law.

Using a_1/d_e. By defining

$$\begin{aligned} a_e &= \sqrt{x_e^2+y_e^2}, \\ \cos(\gamma_1) &= \frac{a_e}{d_e}\cos(\theta)-\frac{z_e}{d_e}\sin(\theta), \\ \cos(\gamma_2) &= \frac{x_e}{a_e}\cos(\psi)+\frac{y_e}{a_e}\sin(\psi), \end{aligned} \tag{13.32}$$

we can write a_1/d_e as follows

$$\frac{a_1}{d_e} = \cos(\gamma_1)+\frac{a_e}{d_e}\cos(\theta)(\cos(\gamma_2)-1). \tag{13.33}$$

Since $\left|\frac{a_e}{d_e}\cos(\theta)\right| \leq 1,\ \forall\,(x_e,y_e,z_e,\theta)\in\mathbb{R}^4$, the conditions

$$\begin{aligned} &\lim_{t\to\infty}\gamma_1 = 0, \\ &\lim_{t\to\infty}\gamma_2 = 0 \end{aligned} \tag{13.34}$$

imply that $\lim_{t\to\infty}(a_1/d_e) = 1$. Furthermore, from (13.32) we can write

$$\begin{aligned} \gamma_1 &= \theta+\theta_d, \\ \gamma_2 &= \psi-\psi_d, \end{aligned} \tag{13.35}$$

where $\theta_d = \arccos(a_e/d_e)$ and $\psi_d = \arccos(x_e/a_e)$ are the desired orientation angles of the vessel in the horizontal and vertical planes, respectively. Hence (13.34) also implies the orientation control objective. If one differentiates both sides of (13.35) to obtain $\dot\gamma_1$ and $\dot\gamma_2$, there will be discontinuity in the γ_1- and γ_2- dynamics at $z_e=0$ and/or $y_e=0$, i.e., on the a_e and/or x_e axes. This discontinuity will cause difficulties in applying the backstepping technique. To get around this problem, we compute $\dot\gamma_1$ and $\dot\gamma_2$ based on (13.32) as follows:

$$\begin{aligned} \dot\gamma_1 &= \dot\theta+\frac{\dot a_e\sin(\theta)+\dot z_e\cos(\theta)}{d_e\cos(\gamma_1)}-\frac{\dot d_e\sin(\gamma_1)}{d_e\cos(\gamma_1)}, \\ \dot\gamma_2 &= \dot\psi+\frac{\dot x_e\sin(\psi)-\dot y_e\cos(\psi)}{a_e\cos(\gamma_2)}-\frac{\dot a_e\sin(\gamma_2)}{a_e\cos(\gamma_2)}. \end{aligned} \tag{13.36}$$

It can be seen that (13.36) is not defined at $\gamma_i=\pm 0.5\pi$, $i=1,2$, and $a_e=0$, $d_e=0$. However, our controller will guarantee that $|\gamma_i(t)|<0.5\pi$ and $d_e(t)\geq d_e^*$, $a_e(t)\geq a_e^*$, $\forall\,t\geq t_0\geq 0$ with arbitrarily small positive constants d_e^* and a_e^* and for feasible initial conditions. From (13.36), we can see that $\dot\theta$ and $\dot\psi$ are decoupled. Hence, designing a controller to achieve the control objective posed in the previous section by using the orientation coordinate a_1/d_e would be much simpler than using the angles γ, α, and β. For convenience of control design, we rewrite the transformed system cynamics (13.15) and (13.36) as follows:

$$\begin{aligned}
\dot{d}_e &= \frac{1}{d_e}\left(x_e\frac{\partial x_d}{\partial s}+y_e\frac{\partial y_d}{\partial s}+z_e\frac{\partial z_d}{\partial s}\right)\dot{s}-\frac{a_1}{d_e}u-\frac{a_2}{d_e}v-\frac{a_3}{d_e}w,\\
\dot{\boldsymbol{\eta}}_{2\gamma} &= \boldsymbol{f}_{2\gamma}(\cdot)+\boldsymbol{J}_2(\boldsymbol{\eta}_2)\boldsymbol{v}_2,\\
\boldsymbol{M}_1\dot{\boldsymbol{v}}_1 &= -\boldsymbol{C}_1(\boldsymbol{v}_1)\boldsymbol{v}_2-\boldsymbol{D}_1(\boldsymbol{v}_1)\boldsymbol{v}_1+\boldsymbol{\tau}_1+\boldsymbol{\tau}_{1E}(t),\\
\boldsymbol{M}_2\dot{\boldsymbol{v}}_2 &= -\boldsymbol{C}_1(\boldsymbol{v}_1)\boldsymbol{v}_1-\boldsymbol{C}_2(\boldsymbol{v}_2)\boldsymbol{v}_2-\boldsymbol{D}_2(\boldsymbol{v}_2)\boldsymbol{v}_2-\boldsymbol{g}_2(\boldsymbol{\eta}_2)+\boldsymbol{\tau}_2+\boldsymbol{\tau}_{2E}(t),
\end{aligned} \tag{13.37}$$

where

$$\begin{aligned}
\boldsymbol{\eta}_{2\gamma} &= [\phi\ \gamma_1\ \gamma_2]^T,\\
\boldsymbol{f}_{2\gamma}(\cdot) &= \begin{bmatrix} 0 \\ f_1^s\dot{s}+f_1^u u+f_1^v v+f_1^w w \\ f_2^s\dot{s}+f_2^u u+f_2^v v+f_2^w w \end{bmatrix},\\
f_1^s &= \varpi_{11}\frac{\partial x_d}{\partial s}+\varpi_{12}\frac{\partial y_d}{\partial s}+\varpi_{13}\frac{\partial z_d}{\partial s},\\
f_1^u &= -\left[\varpi_{11}J_1^{11}(\boldsymbol{\eta}_2)+\varpi_{12}J_1^{21}(\boldsymbol{\eta}_2)+\varpi_{13}J_1^{31}(\boldsymbol{\eta}_2)\right],\\
f_1^v &= -\left[\varpi_{11}J_1^{12}(\boldsymbol{\eta}_2)+\varpi_{12}J_1^{22}(\boldsymbol{\eta}_2)+\varpi_{13}J_1^{32}(\boldsymbol{\eta}_2)\right],\\
f_1^w &= -\left[\varpi_{11}J_1^{13}(\boldsymbol{\eta}_2)+\varpi_{12}J_1^{23}(\boldsymbol{\eta}_2)+\varpi_{13}J_1^{33}(\boldsymbol{\eta}_2)\right],\\
f_2^s &= \varpi_{21}\frac{\partial x_d}{\partial s}+\varpi_{22}\frac{\partial y_d}{\partial s},\\
f_2^u &= -\left[\varpi_{21}J_1^{11}(\boldsymbol{\eta}_2)+\varpi_{22}J_1^{21}(\boldsymbol{\eta}_2)\right],\\
f_2^v &= -\left[\varpi_{21}J_1^{12}(\boldsymbol{\eta}_2)+\varpi_{22}J_1^{22}(\boldsymbol{\eta}_2)\right],\\
f_2^w &= -\left[\varpi_{21}J_1^{13}(\boldsymbol{\eta}_2)+\varpi_{22}J_1^{23}(\boldsymbol{\eta}_2)\right],
\end{aligned} \tag{13.38}$$

with

$$\begin{aligned}
\varpi_{11} &= \left(\frac{x_e\sin(\theta)}{a_e d_e\cos(\gamma_1)}-\frac{x_e\sin(\gamma_1)}{d_e^2\cos(\gamma_1)}\right),\\
\varpi_{12} &= \left(\frac{y_e\sin(\theta)}{a_e d_e\cos(\gamma_1)}-\frac{y_e\sin(\gamma_1)}{d_e^2\cos(\gamma_1)}\right),\\
\varpi_{13} &= \left(\frac{\cos(\theta)}{d_e\cos(\gamma_1)}-\frac{z_e\sin(\gamma_1)}{d_e^2\cos(\gamma_1)}\right),\\
\varpi_{21} &= \left(\frac{\sin(\psi)}{a_e\cos(\gamma_2)}-\frac{x_e\sin(\gamma_2)}{a_e^2\cos(\gamma_2)}\right),\\
\varpi_{22} &= \left(-\frac{\cos(\psi)}{a_e\cos(\gamma_2)}-\frac{y_e\sin(\gamma_2)}{a_e^2\cos(\gamma_2)}\right).
\end{aligned} \tag{13.39}$$

Therefore, we will design the control inputs $\boldsymbol{\tau}_1$ and $\boldsymbol{\tau}_2$ for (13.37) to yield the control objective. In Section 13.3, a procedure to design a stabilizer for the path-following error system (13.37) is presented in detail. The structure of (13.37) suggests that we design the actual controls $\boldsymbol{\tau}_1$ and $\boldsymbol{\tau}_2$ in two stages. First, we design the virtual velocity controls for u and $\boldsymbol{v}_2$ and choose $\dot{s}$ to ultimately stabilize d_e and $\gamma_i, i = 1,2$ at the origin. Based on the backstepping technique, the controls $\boldsymbol{\tau}_1$ and $\boldsymbol{\tau}_2$ will then be designed to make the errors between the virtual velocity controls and their actual values asymptotically tend to a small ball centered at the origin. Since the vessel parameters are unknown, an adaptation scheme is also introduced in this step to estimate their values used in the control laws. We split the control design procedure into two steps. The first step is to design $\boldsymbol{\tau}_1$ while the second step takes care of $\boldsymbol{\tau}_2$. This allows us to simplify the choice of feasible initial conditions.

Since the transformed system (13.37) is not defined at $d_e(t) = 0$, $a_e(t) = 0$, $\gamma_i(t) = \pm 0.5\pi, i = 1,2$, we first assume that

$$d_e(t) \geq d_e^*, a_e(t) \geq a_e^*, |\gamma_i(t)| < 0.5\pi, \ i = 1,2, \ \forall t \geq t_0 \geq 0, \tag{13.40}$$

for some positive constants d_e^* and a_e^*. Our controller design will guarantee (13.40) for feasible initial conditions.

13.3 Control Design

The d_e-dynamics have two inputs that can be chosen to stabilize d_e, namely $\dot{s}$ and u. We can either choose the input u or $\dot{s}$ and then design the remaining input. If we fix $\dot{s}$, then the virtual vessel is allowed to move at a desired speed. The real vessel will follow the virtual one on the path by the controller, and vice versa. In this chapter, we choose to fix $\dot{s}$. This allows us to adjust the initial conditions in most cases without moving the real vessel, see Section 13.4.

Define

$$\tilde{u} = u - u_d, \tag{13.41}$$

where u_d is the intermediate control of u. As discussed above, we choose the intermediate control u_d and $\dot{s}$ as follows:

$$u_d = k_1(d_e - \delta_e) - \frac{1}{a_1}(a_2 v + a_3 w) + \frac{1}{d_e} \frac{u_0(t, d_e)\left(x_e \dfrac{\partial x_d}{\partial s} + y_e \dfrac{\partial y_d}{\partial s} + z_e \dfrac{\partial z_d}{\partial s}\right)}{\sqrt{\left(\dfrac{\partial x_d}{\partial s}\right)^2 + \left(\dfrac{\partial y_d}{\partial s}\right)^2 + \left(\dfrac{\partial z_d}{\partial s}\right)^2}}, \tag{13.42}$$

$$\dot{s} = \frac{a_1}{d_e} \frac{u_0(t, d_e)}{\sqrt{\left(\dfrac{\partial x_d}{\partial s}\right)^2 + \left(\dfrac{\partial y_d}{\partial s}\right)^2 + \left(\dfrac{\partial z_d}{\partial s}\right)^2}}, \tag{13.43}$$

where k_1 and δ_e are positive constants to be selected later, and $u_0(t, d_e) \neq 0, \forall t \geq t_0 \geq 0, d_e(t) \in \mathbb{R}$, is the speed of the virtual vessel on the path. Indeed, one can choose this speed to be a constant. However, the time-varying speed and position path-following error dependence of the virtual vessel on the path is more desirable, especially when the underwater vehicle starts to follow the path. For example, one might choose

$$u_0(t,d_e) = u_0^*(1-\chi_1 e^{-\chi_2(t-t_0)})e^{-\chi_3 d_e}, \tag{13.44}$$

where $u_0^* \neq 0$, $\chi_i > 0$, $i = 1,2,3$, $\chi_1 < 1$. The choice of $u_0(t,d_e)$ in (13.44) has the following desired feature: When the path-following error, d_e, is large, the virtual vessel will wait for the real one; when d_e is small the virtual vessel will move along the path at the speed closed to u_0^* and the real one follows it within the specified look ahead distance. This feature is suitable in practice because it avoids using a high-gain control for large signal d_e. It is noted that u_d is not defined at $a_1 = 0$. Since the terms a_2/a_1 and a_3/a_1 can be written as $(a_2/d_e)/(a_1/d_e)$ and $(a_3/d_e)/(a_1/d_e)$, and recalling that $\cos(\gamma) = a_1/d_e$, the intermediate control u_d is well defined if (13.40) holds and

$$|\gamma(t)| < 0.5\pi, \forall t \geq t_0 \geq 0. \tag{13.45}$$

We will come back to this issue in Section 13.4.

Remark 13.2.

1. If we design the virtual control u_d without canceling the terms $a_2 v$ and $a_3 w$ in the d_e-dynamics, then the condition (13.45) is not required for u_d being well defined. However, an assumption of the sway and heave velocities being bounded is needed in advance in the stability analysis, i.e. assume stability to prove stability.
2. If the sway and heave velocities are assumed to be bounded by the surge velocity, the terms $a_2 v$ and $a_3 w$ are not required to be canceled either. This controller can be designed similarly to the one in this chapter. It is noted that the sway velocity does not require to be bounded by the surge velocity with a relatively small constant as in [129] for the case of path-following in the horizontal plane.

Substituting (13.42) and (13.43) into the first equation of (13.37) results in

$$\dot{d}_e = -k_1 \frac{a_1}{d_e}(d_e - \delta_e) - \frac{a_1}{d_e}\tilde{u}. \tag{13.46}$$

By noticing that under Assumption 13.1, see (13.40) and (13.45), the intermediate control u_d is a smooth function of $x_e, y_e, z_e, s, u_0, \eta_2, v$, and w, differentiating both sides of (13.41) with (13.42) and (13.43) yields

$$\begin{aligned}\dot{\tilde{u}} = {} & \frac{m_{22}}{m_{11}} vr - \frac{m_{33}}{m_{11}} wq - \frac{d_{11}}{m_{11}} u - \sum_{i=2}^{3} \frac{d_{ui}}{m_{11}} |u|^{i-1} u + \frac{1}{m_{11}} \tau_u + \\ & \frac{1}{m_{11}} \tau_{wu}(t) - \frac{\partial u_d}{\partial x_e} \dot{x}_e - \frac{\partial u_d}{\partial y_e} \dot{y}_e - \frac{\partial u_d}{\partial z_e} \dot{z}_e - \frac{\partial u_d}{\partial s} \dot{s} - \frac{\partial u_d}{\partial u_0} \dot{u}_0 - \\ & \frac{\partial u_d}{\partial \boldsymbol{\eta}_2} \dot{\boldsymbol{\eta}}_2 - \frac{\partial u_d}{\partial v} \left(\frac{m_{33}}{m_{22}} wp - \frac{m_{11}}{m_{22}} ur - \frac{d_{22}}{m_{22}} v - \sum_{i=2}^{3} \frac{d_{vi}}{m_{22}} |v|^{i-1} v + \right.\end{aligned}$$

$$\frac{1}{m_{22}}\tau_{wv}(t)\Bigg) - \frac{\partial u_d}{\partial w}\left(\frac{m_{11}}{m_{33}}uq - \frac{m_{22}}{m_{33}}vp - \frac{d_{33}}{m_{33}}w - \right.$$

$$\left.\sum_{i=2}^{3}\frac{d_{wi}}{m_{33}}|w|^{i-1}w + \frac{1}{m_{33}}\tau_{ww}(t)\right), \tag{13.47}$$

where for convenience of choosing u_0, the terms $\frac{\partial u_d}{\partial x_e}$, $\frac{\partial u_d}{\partial y_e}$, and $\frac{\partial u_d}{\partial z_e}$ do not include $\frac{\partial u_0}{\partial x_e}$, $\frac{\partial u_0}{\partial y_e}$ and $\frac{\partial u_0}{\partial z_e}$, which are lumped into $\dot{u}_0$. From (13.47), we choose the actual control $\boldsymbol{\tau}_1$ or τ_u without canceling useful nonlinear damping terms as

$$\tau_u = -c_1\tilde{u} - \hat{\theta}_1^T f_1(\cdot) - \hat{\theta}_{21}\tanh\left(\frac{\tilde{u}\hat{\theta}_{21}}{\varepsilon_{21}}\right) - \hat{\theta}_{22}\frac{\partial u_d}{\partial v}\tanh\left(\frac{\partial u_d}{\partial v}\frac{\tilde{u}\hat{\theta}_{22}}{\varepsilon_{22}}\right) -$$

$$\hat{\theta}_{23}\frac{\partial u_d}{\partial w}\tanh\left(\frac{\partial u_d}{\partial w}\frac{\tilde{u}\hat{\theta}_{23}}{\varepsilon_{23}}\right), \tag{13.48}$$

and the update law as

$$\begin{aligned}
\dot{\hat{\theta}}_{1j} &= \gamma_{1j}\mathrm{proj}\left(\tilde{u}f_{1j}(\cdot), \hat{\theta}_{1j}\right), \ 1 \le j \le 16, \\
\dot{\hat{\theta}}_{21} &= \gamma_{21}\mathrm{proj}\left(|\tilde{u}|, \hat{\theta}_{21}\right), \\
\dot{\hat{\theta}}_{22} &= \gamma_{22}\mathrm{proj}\left(\left|\tilde{u}\frac{\partial u_d}{\partial v}\right|, \hat{\theta}_{22}\right), \\
\dot{\hat{\theta}}_{23} &= \gamma_{23}\mathrm{proj}\left(\left|\tilde{u}\frac{\partial u_d}{\partial w}\right|, \hat{\theta}_{23}\right),
\end{aligned} \tag{13.49}$$

where $c_1, \varepsilon_{2i}, \gamma_{1j}, \gamma_{2i}, 1 \le i \le 3$ and $1 \le j \le 16$, are positive constants to be selected later, and $f_{1j}(\cdot)$ are the jth elements of $f_1(\cdot)$, respectively, with

$$f_1(\cdot) = \left[vr, -wq, -u_d, -|u|u_d, -u^2u_d, -\left(\frac{\partial u_d}{\partial x_e}\dot{x}_e + \frac{\partial u_d}{\partial y_e}\dot{y}_e + \frac{\partial u_d}{\partial z_e}\dot{z}_e + \right.\right.$$

$$\left.\frac{\partial u_d}{\partial s}\dot{s} + \frac{\partial u_d}{\partial u_0}\dot{u}_0 + \frac{\partial u_d}{\partial \boldsymbol{\eta}_2}\dot{\boldsymbol{\eta}}_2\right), -\frac{\partial u_d}{\partial v}wp, \frac{\partial u_d}{\partial v}ur, \frac{\partial u_d}{\partial v}v, \frac{\partial u_d}{\partial v}|v|v,$$

$$\left.\frac{\partial u_d}{\partial v}v^3, -\frac{\partial u_d}{\partial w}uq, \frac{\partial u_d}{\partial w}vp, \frac{\partial u_d}{\partial w}w, \frac{\partial u_d}{\partial w}|w|w, \frac{\partial u_d}{\partial w}w^3\right]^T, \tag{13.50}$$

$\hat{\theta}_{ij}$, $1 \le i \le 2$ is the jth element of $\hat{\theta}_i$, which is an estimate of θ_i with

$$\theta_1 = \left[m_{22}, m_{33}, d_{11}, d_{u2}, d_{u3}, m_{11}, \frac{m_{11}m_{33}}{m_{22}}, \frac{m_{11}^2}{m_{22}}, \frac{m_{11}d_{22}}{m_{22}}, \frac{m_{11}d_{v2}}{m_{22}},\right.$$

$$\left.\frac{m_{11}d_{v3}}{m_{22}}, \frac{m_{11}^2}{m_{33}}, \frac{m_{11}m_{22}}{m_{33}}, \frac{m_{11}d_{33}}{m_{33}}, \frac{m_{11}d_{w2}}{m_{33}}, \frac{m_{11}d_{w3}}{m_{33}}\right]^T,$$

$$\theta_2 = \left[\tau_{wu}^{\max}, \frac{m_{11}}{m_{22}} \tau_{wv}^{\max}, \frac{m_{11}}{m_{33}} \tau_{ww}^{\max} \right]^T . \tag{13.51}$$

The operator, proj, is the Lipschitz continuous projection algorithm repeated here for the convenience of the reader as follows:

$$\begin{array}{ll} \mathrm{proj}(\varpi, \hat{\omega}) = \varpi & \text{if } \Xi(\hat{\omega}) \leq 0, \\ \mathrm{proj}(\varpi, \hat{\omega}) = \varpi & \text{if } \Xi(\hat{\omega}) \geq 0 \text{ and } \Xi_{\hat{\omega}}(\hat{\omega})\varpi \leq 0, \\ \mathrm{proj}(\varpi, \hat{\omega}) = (1 - \Xi(\hat{\omega}))\varpi & \text{if } \Xi(\hat{\omega}) > 0 \text{ and } \Xi_{\hat{\omega}}(\hat{\omega})\varpi > 0, \end{array} \tag{13.52}$$

where $\Xi(\hat{\omega}) = \frac{\hat{\omega}^2 - \omega_M^2}{\mu^2 + 2\mu\omega_M}$, $\Xi_{\hat{\omega}}(\hat{\omega}) = \frac{\partial \Xi(\hat{\omega})}{\partial \hat{\omega}}$, μ is an arbitrarily small positive constant, $\hat{\omega}$ is an estimate of ω and $|\omega| \leq \omega_M$.

The projection algorithm is such that if $\dot{\hat{\omega}} = \mathrm{proj}(\varpi, \hat{\omega})$ and $\hat{\omega}(t_0) \leq \omega_M$ and then

1. $\hat{\omega}(t) \leq \omega_M + \xi$, $\forall\, 0 \leq t_0 \leq t < \infty$,
2. $\mathrm{proj}(\varpi, \hat{\omega})$ is Lipschitz continuous,
3. $|\mathrm{proj}(\varpi, \hat{\omega})| \leq |\varpi|$,
4. $\tilde{\omega}\mathrm{proj}(\varpi, \hat{\omega}) \geq \tilde{\omega}\varpi$ with $\tilde{\omega} = \omega - \hat{\omega}$.

Substituting (13.48) into (13.47) yields the error dynamics

$$\begin{aligned} \dot{\tilde{u}} =& -\frac{1}{m_{11}} \left(c_1 + d_{11} + \sum_{i=2}^{3} d_{ui} |u|^{i-1} \right) \tilde{u} + \frac{1}{m_{11}} \theta_1^T f_1(\cdot) - \frac{1}{m_{11}} \hat{\theta}_1^T f_1(\cdot) + \\ & \frac{1}{m_{11}} \tau_{wu}(t) - \frac{1}{m_{11}} \hat{\theta}_{21} \tanh\left(\frac{\tilde{u}\hat{\theta}_{21}}{\varepsilon_{21}} \right) - \frac{\partial u_d}{\partial v} \frac{1}{m_{22}} \tau_{wv}(t) - \frac{1}{m_{11}} \hat{\theta}_{22} \frac{\partial u_d}{\partial v} \times \\ & \tanh\left(\frac{\partial u_d}{\partial v} \frac{\tilde{u}\hat{\theta}_{22}}{\varepsilon_{22}} \right) - \frac{\partial u_d}{\partial w} \frac{1}{m_{33}} \tau_{ww}(t) - \frac{1}{m_{11}} \hat{\theta}_{23} \frac{\partial u_d}{\partial w} \tanh\left(\frac{\partial u_d}{\partial w} \frac{\tilde{u}\hat{\theta}_{23}}{\varepsilon_{23}} \right). \end{aligned} \tag{13.53}$$

Define

$$\tilde{\boldsymbol{v}}_2 = \boldsymbol{v}_2 - \boldsymbol{v}_{2d}, \tag{13.54}$$

where $\boldsymbol{v}_{2d} = [p_d, q_d, r_d]^T$ is the intermediate control of $\boldsymbol{v}_2$. Recalling that our goal is to ultimately stabilize $\boldsymbol{\eta}_{2\gamma} = [\phi, \gamma_1, \gamma_2]^T$ at the origin, the second equation of (13.37) suggests that we choose this intermediate control as follows:

$$\boldsymbol{v}_{2d} = \boldsymbol{J}_2^{-1}(\eta_2)\left(-f_{2\gamma}(\cdot) - \boldsymbol{K}_2 \eta_{2\gamma}\right), \tag{13.55}$$

where $\boldsymbol{K}_2 = \mathrm{diag}(k_{21}, k_{22}, k_{23})$ is a positive definite diagonal matrix. Substituting (13.54) and (13.55) into the second equation of (13.37) yields

$$\dot{\eta}_{2\gamma} = -\boldsymbol{K}_2 \eta_{2\gamma} + \boldsymbol{J}_2(\eta_2)\tilde{\boldsymbol{v}}_2. \tag{13.56}$$

To design the actual control $\boldsymbol{\tau_2}$, we first note that under Assumption 13.1, (13.40) and (13.45), the intermediate control $\boldsymbol{v_{2d}}$ is a smooth function of $x_e, y_e, z_e, s, u_0, \boldsymbol{\eta_2}$, and $\boldsymbol{v_1}$. Differentiating both sides of (13.54) and multiplying by $\boldsymbol{M_2}$, along the solutions of the last two equations of (13.37) results in

$$\boldsymbol{M_2}\dot{\tilde{\boldsymbol{v}}}_{\boldsymbol{2}} = -\boldsymbol{C_2}(\boldsymbol{v_2})\tilde{\boldsymbol{v}}_{\boldsymbol{2}} - \boldsymbol{D_2}(\boldsymbol{v_2})\tilde{\boldsymbol{v}}_{\boldsymbol{2}} + \boldsymbol{F}(\cdot)\boldsymbol{\theta_3} + \boldsymbol{G}(\cdot)\boldsymbol{\theta_4}(t) + \boldsymbol{\tau_2}, \tag{13.57}$$

where

$$\begin{aligned}
\boldsymbol{F}(\cdot)\boldsymbol{\theta_3} = & -\boldsymbol{C_1}(\boldsymbol{v_1})\boldsymbol{v_1} - \boldsymbol{C_2}(\boldsymbol{v_2})\boldsymbol{v_{2d}} - \boldsymbol{D_2}(\boldsymbol{v_2})\boldsymbol{v_{2d}} - \boldsymbol{g_2}(\boldsymbol{\eta_2}) - \\
& \boldsymbol{M_2}\left(\frac{\partial \boldsymbol{v_{2d}}}{\partial x_e}\dot{x}_e + \frac{\partial \boldsymbol{v_{2d}}}{\partial y_e}\dot{y}_e + \frac{\partial \boldsymbol{v_{2d}}}{\partial z_e}\dot{z}_e + \frac{\partial \boldsymbol{v_{2d}}}{\partial s}\dot{s} + \frac{\partial \boldsymbol{v_{2d}}}{\partial u_0}\dot{u}_0 + \right. \\
& \left.\frac{\partial \boldsymbol{v_{2d}}}{\partial \boldsymbol{\eta_2}}\dot{\boldsymbol{\eta}}_{\boldsymbol{2}}\right) - \boldsymbol{M_2}\frac{\partial \boldsymbol{v_{2d}}}{\partial \boldsymbol{v_1}}\boldsymbol{M_1}^{-1}\left(-\boldsymbol{C_1}(\boldsymbol{v_1})\boldsymbol{v_2} - \boldsymbol{D_1}(\boldsymbol{v_1})\boldsymbol{v_1} + \boldsymbol{\tau_1}\right), \\
\boldsymbol{G}(\cdot)\boldsymbol{\theta_4}(t) = & \, \boldsymbol{\tau_{w2}}(t) - \boldsymbol{M_2}\frac{\partial \boldsymbol{v_{2d}}}{\partial \boldsymbol{v_1}}\boldsymbol{M_1}^{-1}\boldsymbol{\tau_{w1}}(t),
\end{aligned} \tag{13.58}$$

with $\boldsymbol{F}(\cdot) \in \mathbb{R}^{3\times m_3}$ and $\boldsymbol{G}(\cdot) \in \mathbb{R}^{3\times m_4}$ being the regression matrices, $\boldsymbol{\theta_3} \in \mathbb{R}^{3\times m_3}$ and $\boldsymbol{\theta_4}(t) \in \mathbb{R}^{m_4}$ being the vectors of unknown vessel and environmental disturbance parameters. For the sake of simplicity, the regression matrices $\boldsymbol{F}(\cdot)$ and $\boldsymbol{G}(\cdot)$, and the vectors $\boldsymbol{\theta_3}$ and $\boldsymbol{\theta_4}(t)$ are not written down explicitly. From (13.56) and (13.57), we choose the actual control $\boldsymbol{\tau_2}$ and update laws as follows:

$$\boldsymbol{\tau_2} = -\boldsymbol{K_3}\tilde{\boldsymbol{v}}_{\boldsymbol{2}} - \left(\boldsymbol{\eta}_{\boldsymbol{2\gamma}}^{\boldsymbol{T}}\boldsymbol{J_2}(\boldsymbol{\eta_2})\right)^{\boldsymbol{T}} - \boldsymbol{F}(\cdot)\hat{\boldsymbol{\theta}}_{\boldsymbol{3}} - \boldsymbol{G}(\cdot)\hat{\boldsymbol{\theta}}_{\boldsymbol{4}}, \tag{13.59}$$

$$\begin{aligned}
\dot{\hat{\theta}}_{3i} &= \gamma_{3i}\mathrm{proj}\left(\sum_{j=1}^{3}\tilde{v}_{2j}f_{ji}, \hat{\theta}_{3i}\right), \ 1 \le i \le m_3, \\
\dot{\hat{\theta}}_{4i} &= \gamma_{4i}\mathrm{proj}\left(\sum_{j=1}^{3}\left|\tilde{v}_{2j}g_{ji}\right|, \hat{\theta}_{4i}\right), \ 1 \le i \le m_4,
\end{aligned} \tag{13.60}$$

where $\varepsilon_{4i} > 0$, $1 \le i \le 3$, $\gamma_{3j} > 0$, $1 \le j \le m_3$, $\gamma_{4l} > 0$, $1 \le l \le m_4$, $\boldsymbol{K_3} = \mathrm{diag}(k_{31}, k_{32}, k_{33})$ is a positive definite diagonal matrix, $\hat{\theta}_{3i}$ is an estimate of the ith element of $\boldsymbol{\theta_3}$, and $\hat{\theta}_{4i}$ is an estimate of the maximum value of the ith element of $\boldsymbol{\theta_4}(t)$. For simplicity of notation, we have defined

$$\boldsymbol{F}(\cdot)\hat{\boldsymbol{\theta}}_{\boldsymbol{3}} := \begin{bmatrix} \sum_{i=1}^{m_3} f_{1i}\hat{\theta}_{3i} \\ \sum_{i=1}^{m_3} f_{2i}\hat{\theta}_{3i} \\ \sum_{i=1}^{m_3} f_{3i}\hat{\theta}_{3i} \end{bmatrix}, \ \boldsymbol{G}(\cdot)\hat{\boldsymbol{\theta}}_{\boldsymbol{4}} := \begin{bmatrix} \sum_{i=1}^{m_4} g_{1i}\hat{\theta}_{4i}\tanh\left(\varepsilon_{41}^{-1}\tilde{v}_{21}\sum_{i=1}^{m_2} g_{1i}\hat{\theta}_{4i}\right) \\ \sum_{i=1}^{m_4} g_{2i}\hat{\theta}_{4i}\tanh\left(\varepsilon_{42}^{-1}\tilde{v}_{22}\sum_{i=1}^{m_2} g_{2i}\hat{\theta}_{4i}\right) \\ \sum_{i=1}^{m_4} g_{3i}\hat{\theta}_{4i}\tanh\left(\varepsilon_{43}^{-1}\tilde{v}_{23}\sum_{i=1}^{m_2} g_{3i}\hat{\theta}_{4i}\right) \end{bmatrix}, \tag{13.61}$$

with f_{ji}, $1 \le j \le 3$, $1 \le i \le m_3$, being the element in jth column and ith row of the regression matrix $\boldsymbol{F}(\cdot)$. Similarly g_{ji}, $1 \le j \le 3$, $1 \le i \le m_4$, is the element in jth column and ith row of the regression matrix $\boldsymbol{G}(\cdot)$.

Substituting (13.59) into (13.57) yields the error dynamics

$$\begin{aligned}\boldsymbol{M_2}\dot{\tilde{\boldsymbol{v}}}_2 = & -\boldsymbol{C}_2(\boldsymbol{v}_2)\tilde{\boldsymbol{v}}_2 - (\boldsymbol{K}_3 + \boldsymbol{D}_2(\boldsymbol{v}_2))\tilde{\boldsymbol{v}}_2 - \left(\boldsymbol{\eta}_{2\gamma}^T \boldsymbol{J}_2(\boldsymbol{\eta}_2)\right)^T + \boldsymbol{F}(\cdot)\boldsymbol{\theta}_3 + \\ & \boldsymbol{G}(\cdot)\boldsymbol{\theta}_4(t) - \boldsymbol{F}(\cdot)\hat{\boldsymbol{\theta}}_3 - \boldsymbol{G}(\cdot)\hat{\boldsymbol{\theta}}_4. \end{aligned} \tag{13.62}$$

We now present the main result of this chapter, the proof of which is given in the next section.

Theorem 13.1. *Assume that*

1. *the vessel inertia, added mass and damping matrices are diagonal;*
2. *the environmental disturbances are bounded;*
3. *the vessel parameters are unknown but constant;*
4. *the reference path satisfies Assumption 13.1.*

If the state feedback control laws (13.48) and (13.59), and the update laws (13.49) and (13.60) are applied to the vessel system (13.1) then there exist feasible initial conditions such that the path-following errors $(x(t)-x_d(t), y(t)-y_d(t), z(t)-z_d(t), \gamma_1(t), \gamma_2(t))$ *converge to a ball centered on the desired path* $\boldsymbol{\Omega}$ *asymptotically. Furthermore, the radius of this ball can be made arbitrarily small by adjusting the control gains.*

13.4 Stability Analysis

To prove Theorem 13.1, we first consider the $(\boldsymbol{\eta}_{2\gamma}, \tilde{\boldsymbol{v}}_2)$-subsystem and then the $(\boldsymbol{d_e}, \tilde{\boldsymbol{u}})$-subsystem.

$(\boldsymbol{\eta}_{2\gamma}, \tilde{\boldsymbol{v}}_2)$-subsystem

To investigate stability of this subsystem, we consider the following Lyapunov function

$$V_1 = \frac{1}{2}\boldsymbol{\eta}_{2\gamma}^T\boldsymbol{\eta}_{2\gamma} + \frac{1}{2}\tilde{\boldsymbol{v}}_2^T\boldsymbol{M}_2\tilde{\boldsymbol{v}}_2 + \frac{1}{2}\sum_{i=3}^{4}\tilde{\boldsymbol{\theta}}_i^T\boldsymbol{\Gamma}_i^{-1}\tilde{\boldsymbol{\theta}}_i, \tag{13.63}$$

where $\tilde{\boldsymbol{\theta}}_i = \boldsymbol{\theta}_i - \hat{\boldsymbol{\theta}}_i$ and $\boldsymbol{\Gamma}_i = \mathrm{diag}\left(\gamma_{ij}\right)$, $1 \le j \le m_3$ for $i=3$, $1 \le j \le m_4$ for $i=4$. Differentiating both sides of (13.63) along (13.56), (13.60), and (13.62) yields

$$\dot{V}_1 \le -\boldsymbol{\eta}_{2\gamma}^T\boldsymbol{K}_2\boldsymbol{\eta}_{2\gamma} - \tilde{\boldsymbol{v}}_2^T\left(\boldsymbol{D}_{20} + \boldsymbol{K}_3\right)\tilde{\boldsymbol{v}}_2 + 0.2785\sum_{i=1}^{3}\varepsilon_{4i}, \tag{13.64}$$

where $\boldsymbol{D_{20}} = \mathrm{diag}(d_{44}, d_{55}, d_{66})$ and we have used $|x| - x\tanh(x/\lambda) \leq 0.2785\lambda$ for all $x \in \mathbb{R}$ and $\lambda > 0$. From (13.64), we conclude that $\boldsymbol{\eta_{2\gamma}}$ and $\tilde{\boldsymbol{v}}_{\boldsymbol{2}}$ are ultimately asymptotically stable at the origin. To estimate the upper bound of $\boldsymbol{\eta_{2\gamma}}$ and $\tilde{\boldsymbol{v}}_{\boldsymbol{2}}$, we subtract and add $\frac{1}{2}\sum_{i=3}^{4} \tilde{\boldsymbol{\theta}}_{\boldsymbol{i}}^{T} \boldsymbol{\Gamma}_{\boldsymbol{i}}^{-1} \tilde{\boldsymbol{\theta}}_{\boldsymbol{i}}$ to the right-hand side of (13.64) to obtain

$$\dot{V}_1 \leq -\sigma_1 V_1 + \rho_1, \tag{13.65}$$

where

$$\begin{aligned}
\sigma_1 &= \min\left(1, 2\lambda_{\min}(\boldsymbol{K_2}), \frac{2\lambda_{\min}(\boldsymbol{D_{20}} + \boldsymbol{K_3})}{\lambda_{\max}(\boldsymbol{M_2})}\right), \\
\rho_1 &= \frac{1}{2}\sum_{i=3}^{4} \tilde{\boldsymbol{\theta}}_{\boldsymbol{i}}^{T} \boldsymbol{\Gamma}_{\boldsymbol{i}}^{-1} \tilde{\boldsymbol{\theta}}_{\boldsymbol{i}} + 0.2785 \sum_{i=1}^{3} \varepsilon_{4i}.
\end{aligned} \tag{13.66}$$

From (13.65), it is direct to show that

$$V_1(t) \leq V_1(t_0)e^{-\sigma_1(t-t_0)} + \frac{\rho_1}{\sigma_1}, \tag{13.67}$$

which further yields

$$\begin{aligned}
\|\boldsymbol{\eta_{2\gamma}}(\boldsymbol{t})\| &\leq \sqrt{2V_1(t_0)}e^{-\frac{\sigma_1}{2}(t-t_0)} + \sqrt{\frac{2\rho_1}{\sigma_1}} := \alpha_\eta(\cdot)e^{-\frac{\sigma_1}{2}(t-t_0)} + \rho_\eta, \\
\|\tilde{\boldsymbol{v}}_{\boldsymbol{2}}(\boldsymbol{t})\| &\leq \sqrt{\frac{2V_1(t_0)}{\lambda_{\min}(\boldsymbol{M_2})}}e^{-\frac{\sigma_1}{2}(t-t_0)} + \sqrt{\frac{2\rho_1}{\sigma_1\lambda_{\min}(\boldsymbol{M_2})}} := \alpha_v(\cdot)e^{-\frac{\sigma_1}{2}(t-t_0)} + \rho_v.
\end{aligned} \tag{13.68}$$

$(d_e, \tilde{u})$-subsystem

To analyze the stability of this subsystem more easily, we first consider the $\tilde{u}$-dynamics and then the d_e-dynamics.

$\tilde{u}$-dynamics. Consider the following Lyapunov function

$$V_2 = \frac{m_{11}}{2}\tilde{u}^2 + \frac{1}{2}\sum_{i=1}^{2} \tilde{\theta}_i^T \Gamma_i^{-1} \tilde{\theta}_i, \tag{13.69}$$

where $\tilde{\theta}_i = \theta_i - \hat{\theta}_i$ and $\Gamma_i = \mathrm{diag}(\gamma_{ij})$, $1 \leq j \leq 16$ for $i = 1$, $1 \leq j \leq 3$ for $i = 2$. Differentiating both sides of (13.69) along (13.47), (13.48), and (13.49) yields

$$\dot{V}_2 \leq -(c_1 + d_{11})\tilde{u}^2 + 0.2785 \sum_{i=1}^{3} \varepsilon_{2i}. \tag{13.70}$$

Subtracting and adding $\frac{1}{2}\sum_{i=1}^{2}\tilde{\boldsymbol{\theta}}_i^T\boldsymbol{\Gamma}_i^{-1}\tilde{\boldsymbol{\theta}}_i$ to the right-hand side of (13.70) yields

$$\dot{V}_2 \leq -\sigma_2 V_2 + \rho_2, \tag{13.71}$$

where

$$\begin{aligned}\sigma_2 &= \min\left(1, \frac{2(c_1+d_{11})}{m_{11}^{\min}}\right),\\ \rho_2 &= \frac{1}{2}\sum_{i=1}^{2}\tilde{\boldsymbol{\theta}}_i^T\boldsymbol{\Gamma}_i^{-1}\tilde{\boldsymbol{\theta}}_i + 0.2785\sum_{i=1}^{3}\varepsilon_{2i},\end{aligned} \tag{13.72}$$

with $m_{11}^{\min}$ being the minimum value of m_{11}. From (13.71), it is direct to show that

$$V_2(t) \leq V_2(t_0)e^{-\sigma_2(t-t_0)} + \frac{\rho_2}{\sigma_2}, \tag{13.73}$$

which further yields

$$|\tilde{u}(t)| \leq \sqrt{2V_2(t_0)}e^{-\sigma_2/2(t-t_0)} + \sqrt{\frac{2\rho_2}{\sigma_2}} := \alpha_u(\cdot)e^{-\sigma_2/2(t-t_0)} + \rho_u. \tag{13.74}$$

Remark 13.3. It is noted that, due to the use of the projection algorithm, by adjusting $\boldsymbol{K}_2$, $\boldsymbol{K}_3$, c_1, ε_{2i}, ε_{4i}, γ_{1j}, γ_{2i}, γ_{3n}, γ_{4l}, $1 \leq i \leq 3$, $1 \leq j \leq 16$, $1 \leq n \leq m_3$, and $1 \leq l \leq m_4$, we can make ρ_u, ρ_η and ρ_v arbitrarily small. This observation plays an important role in the stability analysis of the d_e-dynamics.

d_e-dynamics. We first calculate the lower-bound of d_e. We now show that there exist initial conditions such that $d_e(t) \geq d_e^* > 0$. From (13.46) and (13.16), we have

$$\dot{\tilde{d}}_e \geq -k_1\tilde{d}_e - \left(\alpha_u e^{-\sigma_2/2(t-t_0)} + \rho_u\right), \tag{13.75}$$

where $\tilde{d}_e = d_e - \delta_e$, which with $\sigma_2 > 2k_1$ further yields

$$\begin{aligned}\tilde{d}_e(t) \geq\ & \tilde{d}_e(t_0)e^{-k_1(t-t_0)} + \frac{\alpha_u(\cdot)e^{-k_1(t-t_0)}}{\sigma_2/2-k_1}\left(-1+e^{-(\sigma_2/2-k_1)(t-t_0)}\right) - \\ & \frac{\rho_u}{k_1}\left(1-e^{-k_1(t-t_0)}\right).\end{aligned} \tag{13.76}$$

Therefore, the condition $d_e(t) \geq d_e^* > 0$ holds when

$$\sigma_2 > 2k_1,\ \delta_e \geq d_e^* + \frac{\rho_u}{k_1}, d_e(t_0) \geq \frac{\alpha_u(\cdot)}{\sigma_2/2-k_1} + \delta_e - \rho_u. \tag{13.77}$$

We will come back to this issue in the next section. We now calculate the upper-bound of d_e. We rewrite (13.46) as

$$\dot{\tilde{d}}_e = -k_1\tilde{d}_e - \cos(\gamma)\tilde{u} - k_1(\cos(\gamma_1) - 1) + \cos(\gamma)(\cos(\gamma_2) - 1))\tilde{d}_e. \tag{13.78}$$

Since (13.78) and the $(\tilde{u}, \gamma_1, \gamma_2)$-subsystem are in the cascade form, one might think that the stability results developed for cascade systems in [17] and [69] can be applied. However, the stability results in those papers were developed for cascade systems without nonvanishing disturbances. In fact, nonvanishing disturbances may destroy the stability of a cascade system that satisfies all conditions stated in the above papers. Therefore, we will use Lemma 10.1 to investigate the stability of the system (13.78) by verifying all conditions C1–C4.

Verifying Condition C1. Take the Lyapunov function

$$V_3 = \frac{1}{2}d_e^2. \tag{13.79}$$

It is direct to show that C1 holds with

$$c_0 = 0,\, c_1 = c_2 = 0.5,\, c_3 = 1,\, c_4 = k_1. \tag{13.80}$$

Verifying Condition C2. By noting that

$$\left|\cos(\gamma)\tilde{u} + k_1(\cos(\gamma_1) - 1) + \cos(\gamma)(\cos(\gamma_2) - 1))\tilde{d}_e\right| \leq$$
$$|\tilde{u}| + k_1(|\gamma_1| + |\gamma_2|)\left|\tilde{d}_e\right|, \tag{13.81}$$

we have

$$\lambda_1 = 1,\, \lambda_2 = k_1. \tag{13.82}$$

Verifying Condition C3. This condition directly holds from (13.74) and (13.68).

Verifying Condition C4. From (13.80) and (13.82), condition C4 becomes

$$k_1 - \max(\rho_\eta, \rho_u)(k_1 + 0.25\mu_0) > 0. \tag{13.83}$$

From Remark 13.3 and noting that μ_0 is an arbitrarily positive constant, we can see that there always exists k_1 such that (13.83) holds. All conditions of Lemma 10.1 have been verified, and we therefore have

$$|d_e(t)| \leq \alpha_{de}(\cdot)e^{-\sigma_{de}(t-t_0)} + \rho_{de}, \tag{13.84}$$

where α_{de}, σ_{de} and ρ_{de} are calculated as in Lemma 10.1.

(v, w)-dynamics. Expanding (13.55) gives

$$\begin{aligned} q_d = &-\left(J_2^{21}(\boldsymbol{\eta}_2)k_{21}\phi + J_2^{22}(\boldsymbol{\eta}_2)k_{22}\gamma_1 + J_2^{23}(\boldsymbol{\eta}_2)k_{23}\gamma_2\right) - \\ &J_2^{22}(\boldsymbol{\eta}_2)\left(f_1^s\dot{s} + f_1^u u\right) - J_2^{23}(\boldsymbol{\eta}_2)\left(f_2^s\dot{s} + f_2^u u\right) - \\ &\left(J_2^{22}(\boldsymbol{\eta}_2)f_1^v + J_2^{23}(\boldsymbol{\eta}_2)f_2^v\right)v - \left(J_2^{22}(\boldsymbol{\eta}_2)f_1^w + J_2^{23}(\boldsymbol{\eta}_2)f_2^w\right)w, \end{aligned}$$

$$\begin{aligned} r_d = & -\left(J_2^{31}(\boldsymbol{\eta_2})k_{21}\phi + J_2^{32}(\boldsymbol{\eta_2})k_{22}\gamma_1 + J_2^{33}(\boldsymbol{\eta_2})k_{23}\gamma_2\right) - \\ & J_2^{32}(\boldsymbol{\eta_2})\left(f_1^s\dot{s} + f_1^u u\right) - J_2^{33}(\boldsymbol{\eta_2})\left(f_2^s\dot{s} + f_2^u u\right) - \\ & \left(J_2^{32}(\boldsymbol{\eta_2})f_1^v + J_2^{33}(\boldsymbol{\eta_2})f_2^v\right)v - \left(J_2^{33}(\boldsymbol{\eta_2})f_1^w + J_2^{33}(\boldsymbol{\eta_2})f_2^w\right)w, \quad (13.85) \end{aligned}$$

where $J_2^{ij}(\boldsymbol{\eta_2})$ is the element at the ith row and jth of $\boldsymbol{J_2^{-1}}(\boldsymbol{\eta_2})$. To show that the sway and heave velocities are bounded, we take the following quadratic function

$$V_4 = \frac{1}{2}m_{22}^2 v^2 + \frac{1}{2}m_{33}^2 w^2, \quad (13.86)$$

whose derivative along (13.37), (13.54), and (13.85) satisfies

$$\begin{aligned} \dot{V}_4 \leq & -m_{22}d_{v3}v^4 - m_{33}d_{w2}\left|w\right|w^2 - m_{33}d_{w3}w^4 + A_3^{\max}v^2 + A_4^{\max}w^2 + \\ & \frac{1}{4\varepsilon_5}A_1^{\max} + \frac{1}{4\varepsilon_6}A_2^{\max}, \quad (13.87) \end{aligned}$$

where $\varepsilon_i > 0, i = 1,2$, $A_j^{\max}, 1 \leq j \leq 4$ is the maximum value of A_j with

$$\begin{aligned} A_1 = & -m_{11}m_{22}u\left(\tilde{r} - J_2^{31}(\boldsymbol{\eta_2})k_{21}\phi - J_2^{32}(\boldsymbol{\eta_2})k_{22}\gamma_1 - J_2^{33}(\boldsymbol{\eta_2})k_{23}\gamma_2 - \right. \\ & \left. J_2^{32}(\boldsymbol{\eta_2})\left(f_1^s\dot{s} + f_1^u u\right) - J_2^{33}(\boldsymbol{\eta_2})\left(f_2^s\dot{s} + f_2^u u\right)\right) + m_{22}\tau_{wv(t)}, \\ A_2 = & \, m_{11}m_{33}u\left(\tilde{q} - J_2^{21}(\boldsymbol{\eta_2})k_{21}\phi - J_2^{22}(\boldsymbol{\eta_2})k_{22}\gamma_1 - J_2^{23}(\boldsymbol{\eta_2})k_{23}\gamma_2 - \right. \\ & \left. J_2^{22}(\boldsymbol{\eta_2})\left(f_1^s\dot{s} + f_1^u u\right) - J_2^{23}(\boldsymbol{\eta_2})\left(f_2^s\dot{s} + f_2^u u\right)\right) + m_{33}\tau_{ww(t)}, \\ A_3 = & \left(-m_{22}d_{22} - m_{22}d_{v2}\left|v\right| + \varepsilon_5\left|A_1\right| + \frac{m_{11}m_{22}\left|uJ_2^{33}(\boldsymbol{\eta_2})\left(f_1^w + f_2^w\right)\right|}{2} + \right. \\ & \left. \frac{m_{11}m_{33}\left|uJ_2^{22}(\boldsymbol{\eta_2})\left(f_1^v + f_2^v\right)\right|}{2}\right), \\ A_4 = & \left(-m_{33}d_{33} - m_{33}d_{w2}\left|w\right| + \varepsilon_6\left|A_2\right| + \frac{m_{11}m_{22}\left|uJ_2^{33}(\boldsymbol{\eta_2})\left(f_1^w + f_2^w\right)\right|}{2} + \right. \\ & \left. \frac{m_{11}m_{33}\left|uJ_2^{22}(\boldsymbol{\eta_2})\left(f_1^v + f_2^v\right)\right|}{2}\right). \quad (13.88) \end{aligned}$$

It can be seen from (13.88) that $A_i^{\max}$ exist and are finite since their arguments are bounded as shown above. Hence (13.87) and (13.86) guarantee a finite upper bound of the sway and heave velocities.

Initial Conditions for $|\gamma(t)| < \frac{\pi}{2}, |\gamma_i(t)| < \frac{\pi}{2}, i = 1,2, \forall t \geq t_0 \geq 0$**.** Since $|\gamma_i(t)| \leq \left\|\boldsymbol{\eta}_{2\gamma}(t)\right\|$ $i = 1,2$, $\forall t \geq t_0 \geq 0$, from (13.68), it is direct to show that the condition $|\gamma_i(t)| < 0.5\pi$, $i = 1,2$, $\forall t \geq t_0 \geq 0$ holds if the initial conditions are such that

$$\sqrt{2V_1(t_0)} + \sqrt{2\rho_1/\sigma_1} < 0.5\pi, \quad (13.89)$$

which is further equivalent to

$$\sqrt{\left\|\eta_{2\gamma}(t_0)\right\|^2+\lambda_{\max}(M_2)\left\|\tilde{v}_2(t_0)\right\|^2+\sum_{i=1}^{4}\Gamma_i^{-1}\left\|\tilde{\theta}_i(t_0)\right\|^2}+\sqrt{2\frac{\rho_1}{\sigma_1}}<\frac{\pi}{2}, \quad (13.90)$$

for all $t \geq t_0 \geq 0$. It is noted that the terms $\sum_{i=1}^{4} \boldsymbol{\Gamma}_i^{-1}\left\|\tilde{\boldsymbol{\theta}}_i(t_0)\right\|^2$ and ρ_1 can be made arbitrarily small, see Remark 13.3.

From (13.33) and (13.16), the condition $|\gamma(t)| < 0.5\pi$, $\forall t \geq t_0 \geq 0$ holds if the initial conditions are such that

$$\cos(\gamma_1(t)) + \frac{a_e(t)}{d_e(t)}\cos(\theta(t))(\cos(\gamma_2(t)) - 1) > 0. \quad (13.91)$$

Under the assumption that $|\theta(t)| < 0.5\pi$, $\forall t \geq t_0 \geq 0$, the above condition is equivalent to

$$\cos(\gamma_1(t)) + \cos(\gamma_2(t)) > 1. \quad (13.92)$$

From (13.68), the condition (13.92) holds if the initial conditions are such that

$$\sqrt{\left\|\boldsymbol{\eta}_{2\gamma}(t_0)\right\|^2+\lambda_{\max}(\boldsymbol{M}_2)\left\|\tilde{\boldsymbol{v}}_2(t_0)\right\|^2+\sum_{i=1}^{4}\boldsymbol{\Gamma}_i^{-1}\left\|\tilde{\boldsymbol{\theta}}_i(t_0)\right\|^2}+$$
$$\sqrt{2\frac{\rho_1}{\sigma_1}} < \arccos(0.5), \ \forall t \geq t_0 \geq 0. \quad (13.93)$$

Since $\arccos(0.5) < 0.5\pi$, the condition (13.93) covers the condition (13.90).

Initial Conditions for $a_e(t) \geq a_e^* > 0$, $\forall t \geq t_0 \geq 0$**.** Since $a_e^2 = d_e^2 - z_e^2$, we have

$$\begin{aligned}\dot{\overbrace{a_e^2}} &= 2\left(d_e\left(-k_1\cos(\gamma)d_e + k_1\cos(\gamma)\delta_e - \cos(\gamma)\tilde{u}\right) - \right.\\ &\left. z_e\left(\frac{\partial z_d}{\partial s}\dot{s} + \sin(\theta)u - \cos(\theta)\sin(\phi)v - \cos(\theta)\cos(\phi)w\right)\right)\\ &= -2k_1\cos(\gamma)a_e^2 + 2k_1\cos(\gamma)z_e^2 + 2k_1\cos(\gamma)d_e\delta_e - 2\cos(\gamma)d_e\tilde{u} - \\ &2z_e\left(\frac{\partial z_d}{\partial s}\dot{s} + \sin(\theta)u - \cos(\theta)\sin(\phi)v - \cos(\theta)\cos(\phi)w\right). \quad (13.94)\end{aligned}$$

From (13.94), it is not hard to see that under Assumption 13.1 and $d_e(t) \geq d_e^* > 0$, if there exists a strictly positive constant a_e^0 such that

$$a_e(t_0) \geq a_e^0, \quad (13.95)$$

then there exists a strictly positive constant a_e^* such that the condition $a_e(t) \geq a_e^* > 0$, $\forall t \geq t_0 \geq 0$ holds.

In summary, the feasible initial conditions are such that the conditions (13.77), (13.93) and (13.95) hold. Roughly speaking, with the above initial conditions, due to the underactuated configuration in the sway and heave, the sway and heave velocities are not able to push the vehicle to the point $a_e = 0$ and $d_e = 0$.

13.5 Discussion of the Initial Condition

We now discuss how to obtain the initial conditions such that (13.77), (13.93), and (13.95) hold. A close look at these conditions shows that they are always satisfied by selecting the initial value, $s(t_0)$, if the vessel heads toward the conical space containing the initial path to be followed, see Figure 13.2. If the vessel does not, the surge control should be turned off and the yaw and pitch controls should make the vessel turn until (13.77), (13.93), and (13.95) hold before applying the proposed path-following controller. The angle δ_0 (see Figure 13.2) should be increased if the initial velocities $\boldsymbol{v}_1(t_0)$ and $\boldsymbol{v}_2(t_0)$ are large. Otherwise the vessel might cross the edge-line of the subspace in question, which might result in $\gamma_i = \pm 0.5\pi$ and/or $\gamma = \pm 0.5\pi$.

13.6 Parking and Point-to-point Navigation

13.6.1 Parking

Parking Objective. Design the controls $\boldsymbol{\tau}_1$ and $\boldsymbol{\tau}_2$ to park the underactuated underwater vehicle (13.1) from the initial position and orientation $(x(t_0), y(t_0), z(t_0), \phi(t_0), \theta(t_0), \psi(t_0))$ to the desired parking position and orientation of $(x_p, y_p, z_p, \phi_p, \theta_p, \psi_p)$ under the following conditions:

1. There exists a large enough positive constant ϖ_p such that

$$\sqrt{\left(x(t_0)-x_p\right)^2+\left(y(t_0)-y_p\right)^2+\left(z(t_0)-z_p\right)^2} \geq \varpi_p.$$

2. The vessel heads toward the feasible cone containing the desired parking orientation, see Section 13.5.
3. At the desired parking position and orientation, the environmental disturbances are negligible.

The above conditions normally hold for parking practice. However, if the first two conditions do not hold, one can apply the strategy in Section 13.5 to move the vessel until they do hold. Having formulated the parking problem as above, one might claim that the path-following controller proposed in Section 13.3 can be applied by setting u_0 equal zero. However, this will result in an orientation that may be very different from the desired parking one, at the desired parking position, since

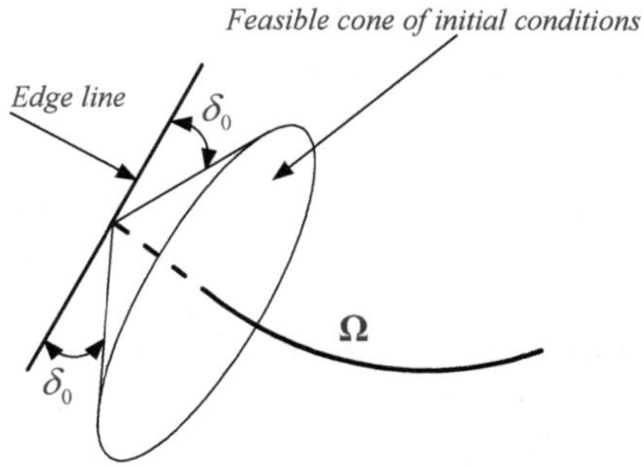

Figure 13.2 Feasible initial conditions

our proposed path-following controller is designed to drive d_e to a small ball, not to zero for reasons of robustness. To resolve this problem, we first generate a regular curve, $\boldsymbol{\Omega}_{\boldsymbol{p}}(x_d, y_d, z_d)$, which goes via the parking position and its orientation at the parking position is equal to the desired parking condition. For simplicity of calculation, the curve can be taken as a straight line in almost all cases of the vessel's initial conditions. Then the proposed path-following controller can be used to make the vessel follow $\boldsymbol{\Omega}_{\boldsymbol{p}}(x_d, y_d, z_d)$. In this case, the velocity u_0 should be chosen such that it goes to zero when the virtual vessel tends to the desired parking position, i.e., $\lim_{d_{ep}\to 0} u_0 = 0$ with $\mathrm{d}_{ep} = \sqrt{(x_d - x_p)^2 + (y_d - y_p)^2 + (z_d - z_p)^2}$. A simple choice can be taken as

$$u_0 = u_0^*(1 - e^{-\chi_1 d_{ep}})e^{-\chi_2 d_e}, \tag{13.96}$$

where $\chi_i > 0, i = 1, 2$. Special care should be taken to choose the initial values of $(x_d(t_0), y_d(t_0), z_d(t_0))$, see Section 13.5, and the sign of u_0 such that it results in a short parking time.

Remark 13.4. Once at the desired parking position and orientation, if there are large environmental disturbances, there will be an oscillatory behavior in the yaw and pitch dynamics, and the vessel might diverge from its desired position. This phenomenon is well known in ship dynamic positioning.

13.6.2 Point-to-point Navigation

As seen in Section 13.1, the requirement that the reference path be a regular curve might be too cumbersome in practice, since this curve has to go via desired points generated by the helmsman and its derivatives are needed in the path-following controller. These restrictions motivate us to consider the point-to-point navigation problem as follows.

Point-to-point Navigation Objective. Design the surge force $\boldsymbol{\tau_u}$ and the yaw moment $\boldsymbol{\tau_r}$ to force the underactuated underwater vehicle (13.1) from the initial position and orientation, $(x(t_0), y(t_0), z(t_0), \phi(t_0), \theta(t_0), \psi(t_0))$, to go via the desired points generated by a path planner.

To achieve this control objective, we first assume that the path planner generates desired points, which are feasible for the vehicle to be navigated through. We then apply the path-following controller proposed in Section 13.3 to each regular curve segments connecting desired points in sequence. The regular curve segments can be straight line, arc, or known regular curve ones. It is, however, noted that a fundamental difference between point-to-point navigation and the proposed smooth path-following is that there are a finite number of "peaks", equal to the number of points, in the orientation errors, γ_1 and γ_2. This phenomenon is because the path is non-smooth in the orientation at the points.

13.7 Numerical Simulations

This section validates the control laws (13.48) and (13.59) by simulating them on a 5.56 m long underwater vehicle whose parameters are given in Section 12.5. The values of the vehicle parameters are assumed to be of the real vessel and are estimated on-line by the adaptation laws (13.49) and (13.60). We assume that these parameters fluctuate around the above values $\pm 15\%$. This fluctuation is chosen here for the purpose of calculating the maximum and minimum values used in the choice of the design constants. Indeed, a different fluctuation of the vehicle parameters results in different maximum and minimum values used in the choice of the design constants. In the simulation, we assume that the environmental disturbances are

$$\tau_{wu} = 0.2m_{11}d(t),\ \tau_{wv} = 0.2m_{22}d(t),\ \tau_{ww} = 0.2m_{33}d(t),\ \tau_{wp} = 0.2m_{44}d(t),$$
$$\tau_{wq} = 0.2m_{55}d(t),\ \tau_{wr} = 0.2m_{66}d(t),$$

where $d(t) = 1 + 0.1\sin(0.2t)$. This choice results in nonzero-mean disturbances. In practice, the environmental disturbances may be different. We take the above disturbances for an illustration of the robustness properties of our proposed controller. It should be noted that only upper bounds of the environmental disturbances are needed in our proposed controller.

In the simulation, based on Section 13.3 the control parameters and initial conditions are taken as

$$\begin{aligned} &k_1 = 0.5, c_1 = 2,\ \boldsymbol{K_2} = \text{diag}(0.05),\ \boldsymbol{K_3} = \text{diag}(2), \boldsymbol{\Gamma_i} = \text{diag}(10),\ \delta_e = 0.2, \\ &\left[\boldsymbol{\eta_1}^T(t_0), \boldsymbol{\eta_2}^T(t_0), \boldsymbol{v_1}^T(t_0), \boldsymbol{v_2}^T(t_0), s(t_0)\right]^T = \\ &\qquad [-145, -15, -5, 0, 0.2, 0.5, 0, 0, 0, 0, 0, 0, 0]^T, \end{aligned}$$

and all initial values of parameter estimates are taken to be 70% of their assumed true ones. The virtual vessel velocity on the path is taken as $u_0(t, d_e) = 5(1 - 0.8e^{-2t})e^{-0.5d_e}$. The reference path is given by $(x_d = -90\cos(s),\ y_d =$

$90\sin(s), z_d = 3s$), i.e., a helix with constant curvature and torsion. Figure 13.3 plots the trajectory of the vessel and the path (dotted line) to be followed in three dimensions. The trajectory of the vessel in the horizontal plane and the path following error are plotted in Figure 13.4. Figure 13.5 plots the control inputs, τ_u, τ_r, τ_p, and τ_q. With nonvanishing environmental disturbances, our proposed controller is able to force the vehicle to follow a predefined path as expected in the control design. As can be seen in Figure 13.5, d_e converges to a nonzero small value, i.e., the sway and heave velocities cannot push the vessel to the point where $d_e = 0$. From Figure 13.5, it can be seen that the control inputs are below their limits. Therefore, we can still further shorten the transient time by increasing the control gains.

13.8 Conclusions

The control scheme developed for path-following of underactuated surface ships in Chapter 11 was extended to design a path-following system for six degrees of freedom underactuated underwater vehicles. The key to the development of the proposed path-following system is the proper selection of the coordinate transformations in Section 13.2. The work presented in this chapter is based on [140, 141].

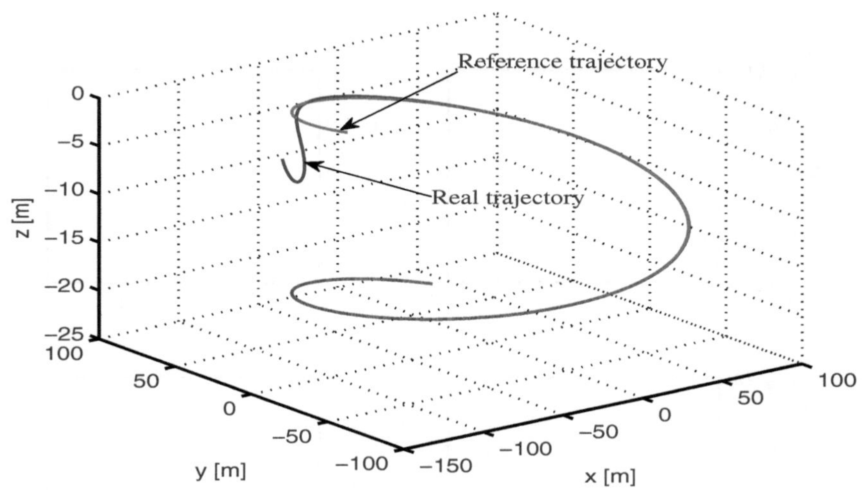

Figure 13.3 Simulation results: Path-following real and reference trajectories in three-dimensional space

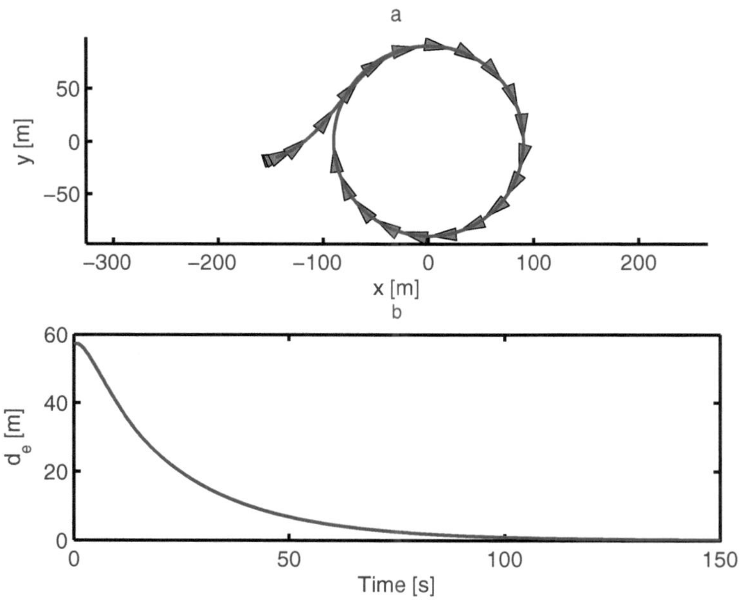

Figure 13.4 Simulation results: **a**. Path-following trajectory in the horizontal plane; **b**. Path-following error d_e

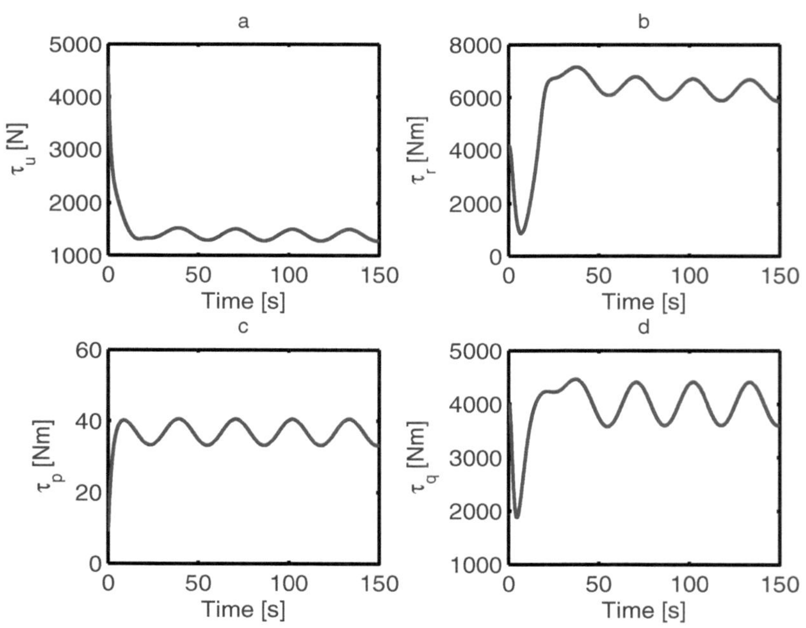

Figure 13.5 Simulation results (control inputs): **a**. Surge force τ_u [N]; **b**. Yaw moment τ_r [Nm]; **c**. Pitch moment τ_p [Nm]; **d**. Roll moment τ_q [Nm]

Part V
Control of Other Underactuated Mechanical Systems

Chapter 14
Control of Other Underactuated Mechanical Systems

This chapter presents several applications of the observer and control design techniques developed in the previous chapters to the control of other underactuated mechanical systems including mobile robots and VTOL aircraft. For mobile robots, a global exponential observer is first designed based on the observer design for underactuated ships in Chapter 7. Output feedback simultaneous stabilization and trajectory-tracking, and path-following controllers are then developed using the control design techniques proposed for underactuated ships in Chapters 6 and 11. For VTOL aircraft, the observer and control design strategies used for underactuated ships in Chapters 5 and 6 are utilized to design a global output feedback trajectory-tracking controller.

14.1 Mobile Robots

14.1.1 Basic Motion Tasks

In order to derive the most suitable feedback controllers for each case, it is convenient to classify the possible motion tasks as follows:

- Point-to-point motion: The robot must reach a desired goal configuration starting from a given initial configuration, see Figure 14.1a.
- Path-following: The robot must reach and follow a geometric path in the Cartesian space starting from a given initial configuration (on or off the path), see Figure 14.1b.
- Trajectory-tracking: The robot must reach and follow a trajectory in the Cartesian space (i.e., a geometric path with an associated timing law) starting from a given initial configuration (on or off the trajectory), see Figure 14.1c.

The three tasks are sketched in Figure 14.1with reference to a three wheel car-like robot. Execution of these tasks can be achieved using either feedforward commands, or feedback control, or a combination of the two. Indeed, feedback solutions exhibit

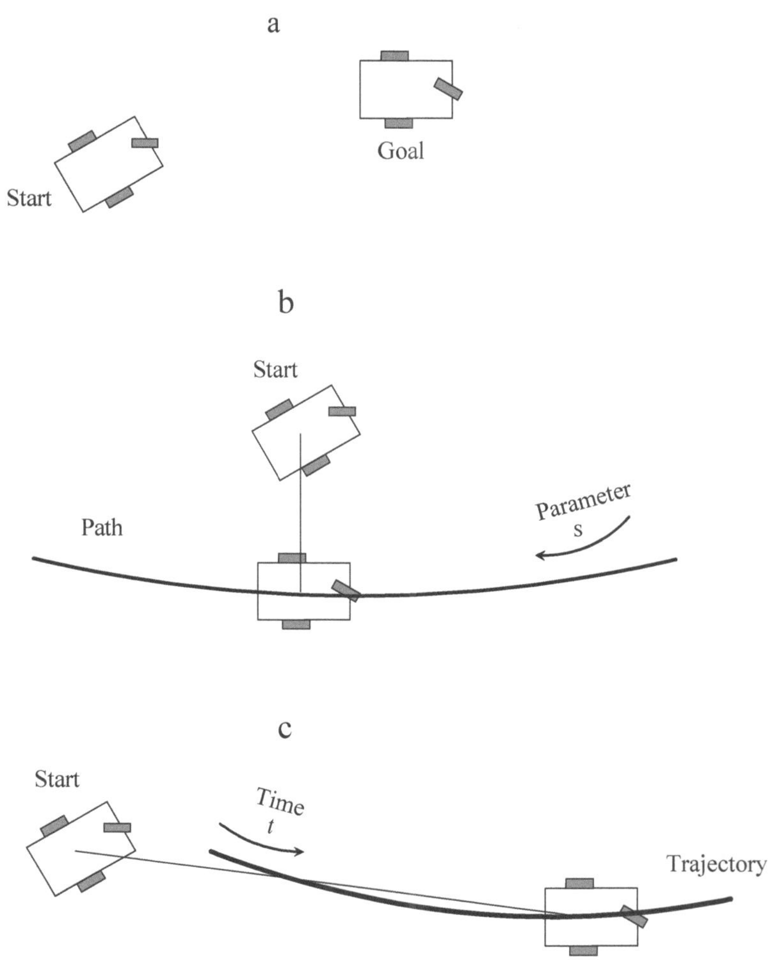

Figure 14.1 Basic motion tasks for a mobile robot

an intrinsic degree of robustness. Explanations of the above motion tasks are similar to those of ocean vehicles, see Section 3.2.

14.1.2 Modeling and Control Properties

14.1.2.1 Modeling

We consider a unicycle-type mobile robot, which under the assumption of no wheel slips has the following dynamics [142]:

$$\dot{\eta} = J(\eta)\omega, \qquad (14.1)$$
$$M\dot{\omega} = -C(\dot{\eta})\omega - D\omega + \tau,$$

where $\eta = [x\ y\ \phi]^T$ denotes the position (x, y), the coordinates of the middle point, P_0, between the left and right driving wheels, and heading angle ϕ of the robot coordinated in the earth-fixed frame OXY, see Figure 14.2, $\omega = [\omega_1\ \omega_2]^T$, with ω_1 and ω_2 being the angular velocities of the wheels of the robot, $\tau = [\tau_1\ \tau_2]^T$, with τ_1 and τ_2 being the control torques applied to the wheels of the robot. The rotation matrix $J(\eta)$, mass matrix M, Coriolis matrix $C(\dot{\eta})$, and damping matrix D in (14.1) are given by

$$J(\eta) = \frac{r}{2}\begin{bmatrix} \cos(\phi) & \cos(\phi) \\ \sin(\phi) & \sin(\phi) \\ b^{-1} & -b^{-1} \end{bmatrix}, \; M = \begin{bmatrix} m_{11} & m_{12} \\ m_{12} & m_{11} \end{bmatrix},$$
$$C(\dot{\eta}) = \begin{bmatrix} 0 & c\dot{\phi} \\ -c\dot{\phi} & 0 \end{bmatrix}, \; D = \begin{bmatrix} d_{11} & 0 \\ 0 & d_{22} \end{bmatrix}, \qquad (14.2)$$

with

$$c = \frac{1}{2b}r^2 m_c a, \; m_{11} = \frac{1}{4b^2}r^2(mb^2 + I) + I_w, \; m_{12} = \frac{1}{4b^2}r^2(mb^2 - I),$$
$$m = m_c + 2m_w, \; I = m_c a^2 + 2m_w b^2 + I_c + 2I_m, \qquad (14.3)$$

where m_c and m_w are the masses of the body and wheel with a motor; I_c, I_w, and I_m are the moments of inertia of the body about the vertical axis through P_c (center of mass), the wheel with the rotor of a motor about the wheel axis, and the wheel with the rotor of a motor about the wheel diameter, respectively; r, a, and b are defined in Figure 14.2; the nonnegative constants d_{11} and d_{22} are the damping coefficients. If these damping coefficients are zero, we have an undamped case. On the other hand, if the damping coefficients are positive, we have a damped case. We take the physical parameters from [142]: $b = 0.75$ m, $a = 0.3$ m, $r = 0.15$ m, $m_c = 30$ kg, $m_w = 1$ kg, $I_c = 15.625$ kgm^2, $I_w = 0.005$ kgm^2, $I_m = 0.0025$ kgm^2, $d_{11} = d_{22} = 5$ kgs^{-1} for numerical simulations. For convenience, we convert the wheel angular velocities (ω_1, ω_2) of the robot to its linear, v, and angular, w, velocities by:

$$\varpi = B^{-1}\omega, \; B = \frac{1}{r}\begin{bmatrix} 1 & b \\ 1 & -b \end{bmatrix}, \qquad (14.4)$$

where $\varpi = [v\ w]^T$ and B is invertible since $\det(B) = -2b/r$. With (14.4), we can write the robot dynamics (14.1) as follows:

$$\dot{x} = v\cos(\phi),$$
$$\dot{y} = v\sin(\phi),$$
$$\dot{\phi} = w, \qquad (14.5)$$
$$\overline{M}\dot{\varpi} = -\overline{C}(w)\varpi - \overline{D}\varpi + \overline{B}\tau,$$

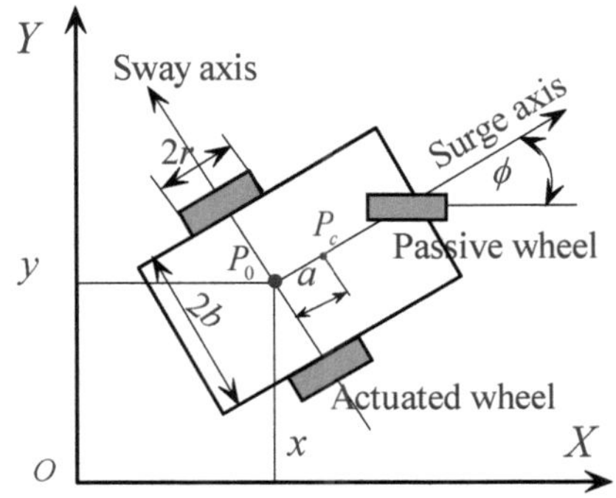

Figure 14.2 Robot parameters

where

$$\begin{aligned}
&\overline{\boldsymbol{M}} = \boldsymbol{B}^{-1}\boldsymbol{M}\boldsymbol{B} = \begin{bmatrix} \bar{m}_{11} & 0 \\ 0 & \bar{m}_{22} \end{bmatrix}, \\
&\overline{\boldsymbol{C}}(w) = \boldsymbol{B}^{-1}\boldsymbol{C}(\dot{\eta})\boldsymbol{B} = \begin{bmatrix} 0 & -bcw \\ \dfrac{c}{b}w & 0 \end{bmatrix}, \\
&\overline{\boldsymbol{D}} = \boldsymbol{B}^{-1}\boldsymbol{D}\boldsymbol{B} = \begin{bmatrix} \bar{d}_{11} & \bar{d}_{12} \\ \bar{d}_{21} & \bar{d}_{22} \end{bmatrix}, \ \overline{\boldsymbol{B}} = \boldsymbol{B}^{-1}, \\
&\bar{m}_{11} = m_{11} + m_{12}, \ \bar{m}_{22} = m_{11} - m_{12}, \\
&\bar{d}_{11} = \frac{1}{2}(d_{11} + d_{22}), \ \bar{d}_{12} = \frac{b}{2}(d_{11} - d_{22}), \\
&\bar{d}_{21} = \frac{1}{2b}(d_{11} - d_{22}), \ \bar{d}_{22} = \frac{1}{2}(d_{11} + d_{22}).
\end{aligned} \tag{14.6}$$

14.1.2.2 Control Properties

Since the last equation of (14.5) is a square system if we consider the robot velocities v and w as outputs and the torques τ_1 and τ_2 as inputs, we only need to investigate control properties of the first three equations of (14.5), i.e., we investigate the control properties of the robot kinematic model:

$$\begin{aligned}
\dot{x} &= v\cos(\phi), \\
\dot{y} &= v\sin(\phi), \\
\dot{\phi} &= w.
\end{aligned} \tag{14.7}$$

From the first two equations of (14.7), the nonholonomic constraint is

$$\dot{x}\sin(\phi) - \dot{y}\cos(\phi) = 0. \tag{14.8}$$

Controllability at a Point. The tangent linearization of (14.7) at any point η_e is the linear system

$$\dot{\tilde{\eta}} = \begin{bmatrix} \cos(\phi_e) \\ \sin(\phi_e) \\ 0 \end{bmatrix} v + \begin{bmatrix} 0 \\ 0 \\ 1 \end{bmatrix} w, \; \tilde{\eta} = \eta - \eta_e , \tag{14.9}$$

which is clearly not controllable. This implies that a linear controller will never achieve posture stabilization, not even in a local sense. In order to study the controllability of the unicycle, we need to use tools from nonlinear control theory [4]. Let us define

$$\boldsymbol{g_1} = \begin{bmatrix} \cos(\phi) \\ \sin(\phi) \\ 0 \end{bmatrix}, \; \boldsymbol{g_2} = \begin{bmatrix} 0 \\ 0 \\ 1 \end{bmatrix}. \tag{14.10}$$

It is easy to check that the accessibility rank condition is satisfied globally (at any η_e), since

$$\text{rank}\,[\boldsymbol{g_1} \quad \boldsymbol{g_2} \quad [\boldsymbol{g_1}, \boldsymbol{g_2}]] = 3, \tag{14.11}$$

where the Lie bracket $[\boldsymbol{g_1}, \boldsymbol{g_2}]$ of the two input vector fields $\boldsymbol{g_1}$ and $\boldsymbol{g_2}$ is

$$[\boldsymbol{g_1}, \boldsymbol{g_2}] = \frac{\partial \boldsymbol{g_2}}{\partial \eta}\boldsymbol{g_1} - \frac{\partial \boldsymbol{g_1}}{\partial \eta}\boldsymbol{g_2} = \begin{bmatrix} \sin(\phi) \\ -\cos(\phi) \\ 0 \end{bmatrix}. \tag{14.12}$$

Since the system is driftless, condition (14.11) implies its controllability. Controllability can also be shown constructively, i.e., by providing an explicit sequence of maneuvers bringing the robot from any start configuration (x_s, y_s, ϕ_s) to any desired goal configuration (x_g, y_g, ϕ_g). Since the unicycle can rotate on itself, this task is simply achieved by an initial rotation on (x_s, y_s) until the unicycle is oriented toward (x_g, y_g), followed by a translation to the goal position, and by a final rotation on (x_g, y_g) so as to align ϕ with ϕ_g. As for the stabilizability of system (14.9) to a point, the failure of the previous linear analysis indicates that exponential stability cannot be achieved by smooth feedback [5]. Things turn out to be even worse: If smooth (in fact, even continuous) time-invariant feedback laws are used, Lyapunov stability is not applicable. This negative result is established on the basis of a necessary condition due to Brockett [21]: Smooth stabilizability of a driftless regular system (i.e., such that the input vector fields are well defined and linearly independent at η_e) requires a number of inputs equal to the number of states. The above difficulty has a deep impact on the control design. In fact, to obtain a posture stabilizing controller it is either necessary to give up the continuity requirement and/or to resort to time-varying control laws.

Controllability About a Trajectory. Given a desired Cartesian motion for the unicycle mobile robot, it may be convenient to generate a corresponding state trajectory $\boldsymbol{\eta_d}(t) = (x_d(t), y_d(t), \phi_d(t))$. In order to be feasible, the latter must satisfy the nonholonomic constraint on the vehicle motion or, equivalently, be consistent with (14.8). The generation of $\boldsymbol{\eta_d}(t)$ and of the corresponding reference velocity inputs $v_d(t)$ and $w_d(t)$ will be addressed properly.

Defining the state tracking error as $\tilde{\boldsymbol{\eta}} = \boldsymbol{\eta} - \boldsymbol{\eta_d}$ and the input variations as $\tilde{v} = v - v_d$ and $\tilde{w} = w - w_d$, the tangent linearization of system (14.7) about the reference trajectory is

$$\dot{\tilde{\boldsymbol{\eta}}} = \begin{bmatrix} 0 & 0 & -v_d\sin(\phi_d) \\ 0 & 0 & v_d\cos(\phi_d) \\ 0 & 0 & 0 \end{bmatrix} \tilde{\boldsymbol{\eta}} + \begin{bmatrix} \cos(\phi_d) & 0 \\ \sin(\phi_d) & 0 \\ 0 & 1 \end{bmatrix} \begin{bmatrix} \tilde{v} \\ \tilde{w} \end{bmatrix} = \boldsymbol{A}(t)\tilde{\boldsymbol{\eta}} + \boldsymbol{B}(t) \begin{bmatrix} \tilde{v} \\ \tilde{w} \end{bmatrix}. \tag{14.13}$$

Since the linearized system is time-varying, a necessary and sufficient controllability condition is that the controllability Grammian is nonsingular. However, a simpler analysis can be conducted by defining the state tracking error as

$$\tilde{\boldsymbol{\eta}}_{\boldsymbol{R}} = \begin{bmatrix} \cos(\phi_d) & \sin(\phi_d) & 0 \\ -\sin(\phi_d) & \cos(\phi_d) & 0 \\ 0 & 0 & 1 \end{bmatrix} \tilde{\boldsymbol{\eta}}. \tag{14.14}$$

Using (14.7), we obtain

$$\dot{\tilde{\boldsymbol{\eta}}}_{\boldsymbol{R}} = \begin{bmatrix} 0 & w_d & 0 \\ -w_d & 0 & v_d \\ 0 & 0 & 0 \end{bmatrix} \tilde{\boldsymbol{\eta}}_{\boldsymbol{R}} + \begin{bmatrix} 1 & 0 \\ 0 & 0 \\ 0 & 1 \end{bmatrix} \begin{bmatrix} \tilde{v} \\ \tilde{w} \end{bmatrix}. \tag{14.15}$$

When v_d and w_d are constant, the above linear system becomes time-invariant and controllable, since matrix

$$\boldsymbol{C} = [\boldsymbol{B} \quad \boldsymbol{AB} \quad \boldsymbol{A}^2\boldsymbol{B}] = \begin{bmatrix} 1 & 0 & 0 & 0 & -w_d^2 & v_d w_d \\ 0 & 0 & -w_d & v_d & 0 & 0 \\ 0 & 1 & 0 & 0 & 0 & 0 \end{bmatrix} \tag{14.16}$$

has rank 3 provided that either v_d or w_d are nonzero. Therefore, we conclude that the kinematic system (14.7) can be locally stabilized by linear feedback about trajectories that consist of linear or circular paths, executed with constant velocity.

14.1.2.3 Feedback Linearizability

Based on the previous discussion, it is easy to see that the driftless nonholonomic system (14.7) cannot be transformed into a linear controllable one using static state feedback. In particular, the controllability condition (14.11) implies that the distribution generated by vector fields $\boldsymbol{g_1}$ and $\boldsymbol{g_2}$ is not involutive, thus violating the necessary condition for full-state feedback linearizability [4]. However, when matrix

$$\boldsymbol{G}(\eta) = \begin{bmatrix} \cos(\phi) & 0 \\ \sin(\phi) & 0 \\ 0 & 1 \end{bmatrix} \tag{14.17}$$

has full column rank, two equations can always be transformed via feedback into simple integrators (input–output linearization and decoupling). The choice of the linearizing outputs is not unique and can be accommodated for special purposes as illustrated in the following example. Define the two outputs as

$$\begin{aligned} y_1 &= x + d\cos(\phi), \\ y_2 &= y + d\sin(\phi), \end{aligned} \tag{14.18}$$

with $d \neq 0$, i.e., the Cartesian coordinates of a point, B, displaced at a distance d along the main axis of the unicycle.

Using the globally defined state feedback

$$\begin{bmatrix} v \\ w \end{bmatrix} = \begin{bmatrix} \cos(\phi) & \sin(\phi) \\ -\sin(\phi)/d & \cos(\phi)/d \end{bmatrix} \begin{bmatrix} u_1 \\ u_2 \end{bmatrix}, \tag{14.19}$$

the unicycle kinematic is equivalent to

$$\begin{aligned} \dot{y}_1 &= u_1, \\ \dot{y}_1 &= u_1, \\ \dot{\phi} &= \frac{u_2\cos(\phi) - u_1\sin(\phi)}{d}. \end{aligned} \tag{14.20}$$

As a consequence, a linear feedback controller for $u = (u_1, u_2)$ will make the point B track any reference trajectory, even with discontinuous tangent to the path (e.g., a square without stopping at corners). Moreover, it is easy to show that the internal state evolution $\phi(t)$ is bounded. This approach, however, will not be pursued in this chapter because of its limited interest, since the robot orientation $\phi(t)$ is not controlled.

14.1.2.4 Chained Forms

The existence of canonical forms for kinematic models of nonholonomic robots allows a general and systematic development of both open loop and closed loop control strategies. The most useful canonical structure is the chained form, which in the case of two-input systems is

$$\begin{aligned}\dot{z}_1 &= u_1,\\ \dot{z}_2 &= u_2,\\ \dot{z}_3 &= z_2 u_1,\\ &\vdots\\ \dot{z}_n &= z_{n-1} u_1.\end{aligned} \tag{14.21}$$

It has been shown that a two-input driftless nonholonomic system with up to $n = 4$ generalized coordinates can always be transformed into a chained form by static feedback transformation [143]. As a matter of fact, most (but not all) wheeled mobile robots can be transformed in a chained form. For the kinematic model (14.7) of the unicycle, we introduce the following globally defined coordinate transformation

$$\begin{aligned}z_1 &= \phi,\\ z_2 &= x\cos(\phi) + y\sin(\phi),\\ z_3 &= x\sin(\phi) - y\cos(\phi),\end{aligned} \tag{14.22}$$

and a static state feedback

$$\begin{aligned}v &= u_2 + z_3 u_1,\\ w &= u_1\end{aligned} \tag{14.23}$$

to give

$$\begin{aligned}\dot{z}_1 &= u_1,\\ \dot{z}_2 &= u_2,\\ \dot{z}_3 &= z_2 u_1.\end{aligned} \tag{14.24}$$

Note that (z_2, z_3) is the position of the unicycle in a rotating left-hand frame having the z_2-axis aligned with the vehicle orientation. Equation (14.24) is another example of static input output linearization, with z_1 and z_2 as linearizing outputs. We note also that the transformation in a chained form is not unique. Mathematical model and control properties of other types of mobile robots are given in [15, 130]. The reader is referred to [22, 33–40, 46, 51, 52, 144–146] for various control design methods ranging from discontinuous to time-varying ones on stabilization and trajectory-tracking control of systems that cover the chained form (14.21).

14.1.3 *Output Feedback Simultaneous Stabilization and Trajectory-tracking*

14.1.3.1 Control Objective

For the reader's convenience, we rewrite the equations of motion here, see (14.1):

$$\begin{aligned}\dot{\eta} &= \boldsymbol{J}(\eta)\boldsymbol{\omega},\\ \boldsymbol{M}\dot{\boldsymbol{\omega}} + \boldsymbol{C}(\dot{\eta})\boldsymbol{\omega} + \boldsymbol{D}\boldsymbol{\omega} &= \boldsymbol{\tau}.\end{aligned} \tag{14.25}$$

We assume that the reference trajectory is generated by the following virtual robot:

$$\begin{aligned}\dot{x}_d &= \cos(\phi_d)u_{1d},\\ \dot{y}_d &= \sin(\phi_d)u_{1d},\\ \dot{\phi}_d &= u_{2d},\end{aligned} \tag{14.26}$$

where (x_d, y_d, ϕ_d) are the position and orientation of the virtual robot, and u_{1d} and u_{2d} are the linear and angular velocities of the virtual robot, respectively.

Control Objective. Under Assumption 14.1, design the control input vector $\boldsymbol{\tau}$ to force the position and orientation, (x, y, ϕ) of the real robot (14.25) to globally asymptotically track (x_d, y_d, ϕ_d) generated by (14.26) with only (x, y, ϕ) available for feedback.

Assumption 14.1.

1. *The reference signals u_{1d}, $\dot{u}_{1d}$, $\ddot{u}_{1d}$, u_{2d}, and $\dot{u}_{2d}$ are bounded. In addition, one of the following conditions holds:*

$$\begin{aligned}&\text{C1.}\ \int_0^\infty (|u_{1d}(t)| + |u_{2d}(t)|)\,\mathrm{d}t \le \mu_{11},\\ &\text{C2.}\ \int_0^\infty |u_{1d}(t)|\,\mathrm{d}t \le \mu_{21}\ and\ |u_{2d}(t)| \ge \mu_{22},\\ &\text{C3.}\ \int_{t_0}^t u_{1d}^2(\tau)\mathrm{d}\tau \ge \mu_{31}(t-t_0) - \mu_{32},\ \forall t \ge t_0 \ge 0,\end{aligned} \tag{14.27}$$

where μ_{11}, μ_{21}, and μ_{32} are nonnegative constants, and μ_{22} and μ_{31} are strictly positive constants.

2. *The robot wheel velocities $\boldsymbol{\omega} = [\omega_1\ \omega_2]^T$ are not available for feedback.*

Remark 14.1. The problem of set-point regulation/stabilization, tracking a path approaching a set-point is included in Condition C1. Tracking linear and circular paths belongs to Condition C3. Condition C2 implies that the case, where the robot linear velocity is zero or approaches zero and its angular velocity is of sinusoidal type, is excluded. The reason is that our control approach introduces a sinusoid signal in the robot angular velocity virtual control to handle set-point stabilization/regulation. Therefore, this case is excluded to avoid two signals canceling each other. If the reference velocity u_{2d} is known completely in advance, the above case can be included. Moreover, item (2) of Assumption 14.1 implies that we need to design an output feedback controller.

Remark 14.2. The problem of simultaneous stabilization and tracking is not only of theoretical interest but also possesses some advantages over the use of separate stabilization and tracking controllers such as only one controller and transient improvement because of the lack of switching. Moreover, if the switching time is unknown, a separate stabilization and tracking control approach cannot be used.

14.1.3.2 Observer Design

We first remove the quadratic velocity terms in the mobile robot dynamics by introducing the following coordinate change:

$$\boldsymbol{X} = \boldsymbol{Q}(\eta)\boldsymbol{\omega}, \tag{14.28}$$

where $\boldsymbol{Q}(\eta)$ is a globally invertible matrix with bounded elements to be determined. Using (14.28), we write the second equation of (14.25) as follows:

$$\dot{\boldsymbol{X}} = [\dot{\boldsymbol{Q}}(\eta)\boldsymbol{\omega} - \boldsymbol{Q}(\eta)\boldsymbol{M}^{-1}\boldsymbol{C}(\dot{\eta})\boldsymbol{\omega}] + \boldsymbol{Q}(\eta)\boldsymbol{M}^{-1}(-\boldsymbol{D}\boldsymbol{\omega} + \boldsymbol{\tau}). \tag{14.29}$$

In [101], $\boldsymbol{Q}(\eta)$ is required with the above properties such that

$$\dot{\boldsymbol{Q}}(\eta) = \boldsymbol{Q}(\eta)\boldsymbol{M}^{-1}\boldsymbol{C}(\dot{\eta}), \ \forall \eta \in \mathbb{R}^3,$$

which does not exist, as a simple calculation shows.

Our method is to cancel the square bracket on the right-hand side of (14.29) for all $(\eta, \boldsymbol{\omega}) \in \mathbb{R}^5$. We assume that $q_{ij}(\eta)$, $i = 1,2$, $j = 1,2$, are the elements of $\boldsymbol{Q}(\eta)$. Using the first equation of (14.25), it can readily be shown that the above square bracket is zero for all $(\eta, \boldsymbol{\omega}) \in \mathbb{R}^5$ if

$$\begin{aligned}
&\frac{\partial q_{i1}}{\partial x}\cos(\phi) + \frac{\partial q_{i1}}{\partial y}\sin(\phi) + \frac{\partial q_{i1}}{\partial \phi}\frac{1}{b} + \frac{n_{12}c}{b}q_{i1} + \frac{n_{11}c}{b}q_{i2} = 0,\\
&\frac{\partial q_{i2}}{\partial x}\cos(\phi) + \frac{\partial q_{i2}}{\partial y}\sin(\phi) - \frac{\partial q_{i2}}{\partial \phi}\frac{1}{b} + \frac{n_{11}c}{b}q_{i1} + \frac{n_{12}c}{b}q_{i2} = 0,\\
&\left(\frac{\partial q_{i1}}{\partial x} + \frac{\partial q_{i2}}{\partial x}\right)\cos(\phi) + \left(\frac{\partial q_{i1}}{\partial y} + \frac{\partial q_{i2}}{\partial y}\right)\sin(\phi) + \\
&\qquad \left(\frac{\partial q_{i2}}{\partial \phi} - \frac{\partial q_{i1}}{\partial \phi}\right)\frac{1}{b} - (n_{11} + n_{12})\frac{c}{b}(q_{i1} + q_{i2}) = 0.
\end{aligned} \tag{14.30}$$

Using the characteristic method to solve the above partial differential equations gives a family of solutions with $i = 1,2$

$$\begin{aligned} q_{i1} &= C_{i1}\sin(c\Delta\phi) + C_{i2}\cos(c\Delta\phi), \\ q_{i2} &= n_{11}^{-1}((C_{i2}\Delta - C_{i1}n_{12})\sin(c\Delta\phi) - (C_{i1}\Delta + C_{i2}n_{12})\cos(c\Delta\phi)), \end{aligned} \tag{14.31}$$

where $n_{11} = m_{11}(m_{11}^2 - m_{12}^2)^{-1}$, $n_{12} = -m_{12}(m_{11}^2 - m_{12}^2)^{-1}$, $\Delta = \sqrt{n_{11}^2 - n_{12}^2}$, and C_{i1} and C_{i2} are arbitrary constants. A choice of $C_{11} = C_{22} = 0$, $C_{12} = C_{21} = n_{11}$ results in

$$\boldsymbol{Q}(\eta) = \begin{bmatrix} n_{11}\cos(a\Delta\phi) & \Delta\sin(a\Delta\phi) - n_{12}\cos(a\Delta\phi) \\ n_{11}\sin(a\Delta\phi) & -n_{12}\sin(a\Delta\phi) - \Delta\cos(a\Delta\phi) \end{bmatrix}. \tag{14.32}$$

This matrix is globally invertible and its elements are bounded. Now we write (14.25) in the $(\eta, \boldsymbol{X})$ coordinates as

$$\begin{aligned} \dot{\eta} &= \boldsymbol{J}(\eta)\boldsymbol{Q}^{-1}(\eta)\boldsymbol{X}, \\ \dot{\boldsymbol{X}} &= -\boldsymbol{D}_\eta(\eta)\boldsymbol{X} + \boldsymbol{Q}(\eta)\boldsymbol{M}^{-1}\boldsymbol{\tau}, \end{aligned} \tag{14.33}$$

where $\boldsymbol{D}_\eta(\eta) = \boldsymbol{Q}(\eta)\boldsymbol{M}^{-1}\boldsymbol{D}\boldsymbol{Q}^{-1}(\eta)$. It can be seen that (14.33) is linear in the unmeasured states. We here use the following passive observer:

$$\begin{aligned} \dot{\hat{\eta}} &= \boldsymbol{J}(\eta)\boldsymbol{Q}^{-1}(\eta)\hat{\boldsymbol{X}} + \boldsymbol{K}_{01}(\eta - \hat{\eta}), \\ \dot{\hat{\boldsymbol{X}}} &= -\boldsymbol{D}_\eta(\eta)\hat{\boldsymbol{X}} + \boldsymbol{Q}(\eta)\boldsymbol{M}^{-1}\boldsymbol{\tau} + \boldsymbol{K}_{02}(\eta - \hat{\eta}), \end{aligned} \tag{14.34}$$

where $\hat{\eta}$ and $\hat{\boldsymbol{X}}$ are the estimates of η and $\boldsymbol{X}$, respectively. The observer gain matrices $\boldsymbol{K}_{01}$ and $\boldsymbol{K}_{02}$ are chosen such that

$$\begin{aligned} \boldsymbol{Q}_{01} &= \boldsymbol{K}_{01}^T\boldsymbol{P}_{01} + \boldsymbol{P}_{01}\boldsymbol{K}_{01}, \\ \boldsymbol{Q}_{02} &= \boldsymbol{D}_\eta^T(\eta)\boldsymbol{P}_{02} + \boldsymbol{P}_{02}\boldsymbol{D}_\eta(\eta) \end{aligned}$$

are positive definite and

$$(\boldsymbol{J}(\eta)\boldsymbol{Q}^{-1}(\eta))^T\boldsymbol{P}_{01} - \boldsymbol{P}_{02}\boldsymbol{K}_{02} = \boldsymbol{0}, \tag{14.35}$$

with $\boldsymbol{P}_{01}$ and $\boldsymbol{P}_{02}$ being positive definite matrices. Since $\boldsymbol{D}_\eta(\eta)$ is positive definite, $\boldsymbol{K}_{01}$ and $\boldsymbol{K}_{02}$ always exist. From (14.34) and (14.33), we have

$$\begin{aligned} \dot{\tilde{\eta}} &= \boldsymbol{J}(\eta)\boldsymbol{Q}^{-1}(\eta)\tilde{\boldsymbol{X}} - \boldsymbol{K}_{01}\tilde{\eta}, \\ \dot{\tilde{\boldsymbol{X}}} &= -\boldsymbol{D}_\eta(\eta)\tilde{\boldsymbol{X}} - \boldsymbol{K}_{02}\tilde{\eta}, \end{aligned} \tag{14.36}$$

where $\tilde{\eta} = \eta - \hat{\eta}$ and $\tilde{\boldsymbol{X}} = \boldsymbol{X} - \hat{\boldsymbol{X}}$. It can now be seen that (14.36) is GES at the origin by taking the Lyapunov function

$$V_0 = \tilde{\eta}^T \boldsymbol{P}_{01}\tilde{\eta} + \tilde{\boldsymbol{X}}^T \boldsymbol{P}_{02}\tilde{\boldsymbol{X}},$$

whose derivative along the solutions of (14.36) and using (14.35) satisfies

$$\dot{V}_0 = -\tilde{\eta}^T \boldsymbol{Q}_{01}\tilde{\eta} - \tilde{\boldsymbol{X}}^T \boldsymbol{Q}_{02}\tilde{\boldsymbol{X}},$$

which in turn implies that there exists a strictly positive constant σ_0 such that

$$\left\|(\tilde{\eta}(t), \tilde{\boldsymbol{X}}(t))\right\| \le \left\|(\tilde{\eta}(t_0), \tilde{\boldsymbol{X}}(t_0))\right\| e^{-\sigma_0(t-t_0)},\ \forall t \ge t_0 \ge 0. \tag{14.37}$$

Define $\hat{\boldsymbol{\omega}} = [\hat{\omega}_1\ \hat{\omega}_2]^T$ as an estimator of the velocity vector $\boldsymbol{\omega}$ as

$$\hat{\boldsymbol{\omega}} = \boldsymbol{Q}^{-1}(\eta)\hat{\boldsymbol{X}}. \tag{14.38}$$

The velocity estimate error vector, $\tilde{\boldsymbol{\omega}} = \boldsymbol{\omega} - \hat{\boldsymbol{\omega}}$, satisfies

$$\tilde{\boldsymbol{\omega}} = \boldsymbol{Q}^{-1}(\eta)\tilde{\boldsymbol{X}}. \tag{14.39}$$

To prepare for the control design in the next section, we convert the wheel velocities ω_1 and ω_2 to the linear, v, and angular, w, velocities of the robot by the relationship:

$$\begin{bmatrix} v \\ w \end{bmatrix} = \boldsymbol{B}^{-1}\begin{bmatrix} \omega_1 \\ \omega_2 \end{bmatrix} \text{ with } \boldsymbol{B} = \frac{1}{r}\begin{bmatrix} 1 & b \\ 1 & -b \end{bmatrix}. \tag{14.40}$$

By defining $\tilde{v} = v - \hat{v}$, $\tilde{w} = w - \hat{w}$, with $\hat{v}$ and $\hat{w}$ being estimates of v and w, we can see from (14.39) and (14.40) that

$$\|(\tilde{v}(t), \tilde{w}(t))\| \le \gamma_0 \left\|(\tilde{\eta}(t_0), \tilde{\boldsymbol{X}}(t_0))\right\| e^{-\sigma_0(t-t_0)},\ \forall t \ge t_0 \ge 0, \tag{14.41}$$

where γ_0 is a positive constant. We now write (14.25) in conjunction with (14.38) and (14.40) as

$$\begin{aligned} \dot{x} &= \cos(\phi)\hat{v} + \cos(\phi)\tilde{v}, \\ \dot{y} &= \sin(\phi)\hat{v} + \sin(\phi)\tilde{v}, \\ \dot{\phi} &= \hat{w} + \tilde{w}, \\ \dot{\hat{v}} &= \tau_{vc} + \Omega_v, \\ \dot{\hat{w}} &= \tau_{wc} + \Omega_w, \end{aligned} \tag{14.42}$$

where Ω_v and Ω_w are the first and second rows of $\boldsymbol{\Omega}$

$$\begin{aligned} \boldsymbol{\Omega} &= \boldsymbol{B}^{-1}\boldsymbol{N}_c\boldsymbol{B}\begin{bmatrix} \hat{v} \\ \hat{w} \end{bmatrix}\tilde{w} + \boldsymbol{B}^{-1}\boldsymbol{Q}^{-1}(\eta)\boldsymbol{K}_{02}\tilde{\eta} \\ \boldsymbol{N}_c &= c\begin{bmatrix} n_{12} & -n_{11} \\ n_{11} & -n_{12} \end{bmatrix}, \end{aligned} \tag{14.43}$$

and we have chosen the control torque

$$\boldsymbol{\tau} = \boldsymbol{M}\boldsymbol{B}\left(\boldsymbol{B}^{-1}\boldsymbol{N}_c\boldsymbol{B}\begin{bmatrix}\hat{v}\\ \hat{w}\end{bmatrix}\hat{w} - \boldsymbol{B}^{-1}\boldsymbol{M}^{-1}\boldsymbol{D}\boldsymbol{B}\begin{bmatrix}\hat{v}\\ \hat{w}\end{bmatrix} + \begin{bmatrix}\tau_{vc}\\ \tau_{wc}\end{bmatrix}\right), \tag{14.44}$$

with τ_{vc} and τ_{wc} being the new control inputs to be designed in the next section.

14.1.3.3 Control Design

We first interpret the tracking errors as

$$\begin{bmatrix}x_e\\ y_e\\ \phi_e\end{bmatrix} = \begin{bmatrix}\cos(\phi) & \sin(\phi) & 0\\ -\sin(\phi) & \cos(\phi) & 0\\ 0 & 0 & 1\end{bmatrix}\begin{bmatrix}x - x_d\\ y - y_d\\ \phi - \phi_d\end{bmatrix}. \tag{14.45}$$

Using (14.45), (14.26), and the kinematic part of (14.42) results in

$$\begin{aligned}\dot{x}_e &= \hat{v} - u_{1d}\cos(\phi_e) + y_e(\hat{w} + \tilde{w}) + \tilde{v},\\ \dot{y}_e &= u_{1d}\sin(\phi_e) - x_e(\hat{w} + \tilde{w}),\\ \dot{\phi}_e &= \hat{w} - u_{2d} + \tilde{w}.\end{aligned} \tag{14.46}$$

Since (14.46) and the last two equations of (14.42) are of a lower triangular structure, we use the backstepping technique [3] to design τ_{vc} and τ_{wc} in two steps.

Step 1

In this step, we consider $\hat{v}$ and $\hat{w}$ as the controls. From (14.46), it can be seen that $\hat{v}$ and $\hat{w}$ can be directly used to stabilize x_e and ϕ_e-dynamics. To stabilize y_e-dynamics, ϕ_e can be used when u_{1d} is persistently exciting. When u_{1d} is not persistently exciting (stabilization/regulation case), we need some persistently exciting signal in $\hat{w}$ to stabilize y_e-dynamics via x_e. With these observations in mind, we define

$$\begin{aligned}\bar{v} &= \hat{v} - \alpha_v,\\ \bar{w} &= \hat{w} - \alpha_w,\\ \bar{\phi}_e &= \phi_e - \alpha_{\phi e},\end{aligned} \tag{14.47}$$

where α_v, α_w, and $\alpha_{\phi e}$ are the virtual controls of $\hat{v}$, $\hat{w}$, and ϕ_e, respectively. From the above discussion, we first choose the virtual controls α_v and $\alpha_{\phi e}$ as

$$\begin{aligned}\alpha_v &= -c_1\Omega_1^{-1}x_e + u_{1d}\cos(\phi_e),\\ \alpha_{\phi e} &= -\arcsin\left(\frac{k(t)}{\Omega_1}y_e\right),\\ k(t) &= \lambda_1 u_{1d} + \lambda_2\cos(\lambda_3 t),\end{aligned} \tag{14.48}$$

where $\Omega_1 = \sqrt{1+x_e^2+y_e^2}$, c_1 is a positive constant, and λ_i, $i = 1,2,3$ are positive constants such that $|k(t)| \leq k_* < 1$, $\forall t$. They will be specified later. For simplicity, the virtual control α_v does not cancel a known term $y_e\hat{w}$ in the x_e-dynamics. It is of interest to note that the choice of (14.48) will result in the global result and bounded virtual velocity controls.

To design α_w, differentiating $\bar{\phi}_e = \phi_e - \alpha_{\phi e}$ along the solutions of (14.46) together with (14.48) yields

$$\begin{aligned}\dot{\bar{\phi}}_e = &\left(1-\frac{k}{\Omega_2}x_e\right)(\alpha_w+\bar{w}+\tilde{w})-u_{2d}-\frac{k}{\Omega_2\Omega_1^2}x_e y_e(\bar{v}+\tilde{v})+\\ &\frac{1}{\Omega_2}\left(\dot{k}y_e+\frac{kc_1}{\Omega_1^3}x_e^2 y_e+\frac{ku_{1d}}{\Omega_1^2}(1+x_e^2)\sin(\phi_e)\right),\end{aligned} \tag{14.49}$$

which suggests that we choose

$$\begin{aligned}\alpha_w = &\frac{1}{1-k\Omega_2^{-1}x_e}\left(-\frac{c_2\bar{\phi}_e}{\sqrt{1+\bar{\phi}_e^2}}+u_{2d}-\frac{1}{\Omega_2}(\dot{k}y_e+\frac{kc_1}{\Omega_1^3}x_e^2 y_e+\right.\\ &\left.\frac{ku_{1d}}{\Omega_1^2}(1+x_e^2)\sin(\phi_e)\right),\end{aligned} \tag{14.50}$$

where $\Omega_2 = \sqrt{1+x_e^2+(1-k^2)y_e^2}$ and c_2 is a positive constant.

Remark 14.3. From (14.48) and (14.50), the virtual controls α_v and α_w, as a simple calculation shows, are bounded by some constants depending on the upper bound of u_{1d}, $\dot{u}_{1d}$, and u_{2d}.

Substituting (14.48) and (14.50) into (14.46) and (14.49) results in

$$\begin{aligned}\dot{x}_e &= -\frac{c_1}{\Omega_1}x_e + y_e(\hat{w}+\tilde{w})+\bar{v}+\tilde{v},\\ \dot{y}_e &= -\frac{ku_{1d}}{\Omega_1}y_e - x_e(\hat{w}+\tilde{w})+\frac{u_{1d}}{\Omega_1}\left(\sin(\bar{\phi}_e)\Omega_2-(\cos(\bar{\phi}_e)-1)ky_e\right),\\ \dot{\bar{\phi}}_e &= -\frac{c_2\bar{\phi}_e}{\sqrt{1+\bar{\phi}_e^2}}+\left(1-\frac{k}{\Omega_2}x_e\right)(\bar{w}+\tilde{w})-\frac{k}{\Omega_2\Omega_1^2}x_e y_e(\bar{v}+\tilde{v}).\end{aligned} \tag{14.51}$$

Step 2

At this step, the control inputs τ_{vc} and τ_{wc} are designed. We note that α_v is a smooth function of x_e, y_e, ϕ_e, and u_{1d}, and that α_w is a smooth function of x_e, y_e, ϕ_e, u_{1d}, $\dot{u}_{1d}$, u_{2d}, and t. By differentiating $\bar{v} = \hat{v} - \alpha_v$ and $\bar{w} = \hat{w} - \alpha_w$ along the solutions of (14.46) and the last two equations of (14.42), and noting the last equation of (14.51), we choose τ_{vc} and τ_{wc} as

$$\begin{aligned}
\tau_{vc} = & -c_3\bar{v} + \frac{\partial\alpha_v}{\partial x_e}(\hat{v} - u_{1d}\cos(\phi_e) + y_e\hat{w}) + \frac{\partial\alpha_v}{\partial\phi_e}(\hat{w} - u_{2d}) + \\
& \frac{\partial\alpha_v}{\partial y_e}(u_{1d}\sin(\phi_e) - x_e\hat{w}) + \frac{\partial\alpha_v}{\partial u_{1d}}\dot{u}_{1d} - \delta_v\left(\hat{v}^2 + \hat{w}^2\right)\bar{v} + \frac{k}{\Omega_2\Omega_1^2}x_e y_e\bar{\phi}_e, \\
\tau_{wc} = & -c_4\bar{w} + \frac{\partial\alpha_w}{\partial x_e}(\hat{v} - u_{1d}\cos(\phi_e) + y_e\hat{w}) + \frac{\partial\alpha_w}{\partial\phi_e}(\hat{w} - u_{2d}) + \\
& \frac{\partial\alpha_w}{\partial y_e}(u_{1d}\sin(\phi_e) - x_e\hat{w}) + \frac{\partial\alpha_w}{\partial u_{1d}}\dot{u}_{1d} + \frac{\partial\alpha_w}{\partial\dot{u}_{1d}}\ddot{u}_{1d} + \frac{\partial\alpha_w}{\partial u_{2d}}\dot{u}_{2d} - \\
& \left(1 - \frac{k}{\Omega_2}x_e\right)\bar{\phi}_e - \delta_w\left(\hat{v}^2 + \hat{w}^2\right)\bar{w}, \qquad (14.52)
\end{aligned}$$

where c_3, c_4, δ_v, and δ_w are positive constants. The terms multiplied by δ_v and δ_w are the nonlinear damping terms to overcome the effect of observer errors, see (14.43). The choice of (14.52) results in

$$\begin{aligned}
\dot{\bar{v}} = & -c_3\bar{v} - \frac{\partial\alpha_v}{\partial x_e}(y_e\tilde{w} + \tilde{v}) - \frac{\partial\alpha_v}{\partial\phi_e}\tilde{w} + \frac{\partial\alpha_v}{\partial y_e}x_e\tilde{w} + \\
& \Omega_v - \delta_v\left(\hat{v}^2 + \hat{w}^2\right)\bar{v} + \frac{k}{\Omega_2\Omega_1^2}x_e y_e\bar{\phi}, \\
\dot{\bar{w}} = & -c_4\bar{w} - \frac{\partial\alpha_w}{\partial x_e}(y_e\tilde{w} + \tilde{v}) - \frac{\partial\alpha_w}{\partial\phi_e}\tilde{w} + \frac{\partial\alpha_w}{\partial y_e}x_e\tilde{w} + \qquad (14.53) \\
& \Omega_w - \left(1 - \frac{k}{\Omega_2}x_e\right)\bar{\phi}_e - \delta_w\left(\hat{v}^2 + \hat{w}^2\right)\bar{w}.
\end{aligned}$$

14.1.3.4 Stability Analysis

To analyze the closed loop system consisting of (14.51) and (14.53), we first consider the $(\bar{\phi}_e, \bar{v}, \bar{w})$-subsystem, then move to the (x_e, y_e)-subsystem.

$(\bar{\phi}_e, \bar{v}, \bar{w})$ **-subsystem.** For this subsystem, consider the Lyapunov function

$$V_1 = \frac{1}{2}(\bar{\phi}_e^2 + \bar{v}^2 + \bar{w}^2), \qquad (14.54)$$

whose derivative along the solutions of the last equation of (14.51) and (14.53) satisfies

$$\begin{aligned}
\dot{V}_1 & \leq -\frac{c_2}{\sqrt{1+\bar{\phi}_e^2}}\bar{\phi}_e^2 - c_3\bar{v}^2 - c_4\bar{w}^2 + (\chi_{11} + \chi_{12}V_1)e^{-\sigma_0(t-t_0)} \\
& \leq (\chi_{11}V_1 + \chi_{12})e^{-\sigma_0(t-t_0)}, \qquad (14.55)
\end{aligned}$$

where χ_{11} and χ_{12} are class-K functions of $\left\|(\tilde{\boldsymbol{\eta}}(t_0), \tilde{\boldsymbol{X}}(t_0))\right\|$. The second line of (14.55) implies that $V_1(t) \leq \chi_{13}$, with χ_{13} being a class-K function of $\left\|(\tilde{\boldsymbol{\eta}}(t_0),\right.$

$\tilde{\boldsymbol{X}}(t_0),\ \bar{\boldsymbol{X}}(t_0))\|$ with $\bar{\boldsymbol{X}}(t) = \left[\bar{\phi}_e(t)\ \bar{v}(t)\ \bar{w}(t)\right]^T$. Substituting this bound into the first line of (14.55) yields

$$\dot{V}_1 \leq -2\min\left(\frac{c_2}{\sqrt{1+2\chi_{13}}}, c_3, c_4\right) V_1 + (\chi_{11} + \chi_{12}\chi_{13}) e^{-\sigma_0(t-t_0)}, \tag{14.56}$$

which implies that there exist $\sigma_1 > 0$ and a class-K function χ_1 depending on $\left\|(\tilde{\eta}(t_0), \tilde{\boldsymbol{X}}(t_0), \bar{\boldsymbol{X}}(t_0))\right\|$ such that $\left\|\bar{\boldsymbol{X}}(t)\right\| \leq \chi_1 e^{-\sigma_1(t-t_0)}$, that is, the $(\bar{\phi}_e, \bar{v}, \bar{w})$-subsystem is GAS.

(x_e, y_e)**-subsystem.** We first prove that the trajectories (x_e, y_e) are bounded by taking the Lyapunov function

$$V_2 = \sqrt{1 + x_e^2 + y_e^2} - 1, \tag{14.57}$$

whose derivative along the solutions of the first two equations of (14.51) satisfies

$$\begin{aligned} \dot{V}_2 &\leq -\frac{c_1}{\Omega_1^2} x_e^2 - \frac{k u_{1d}}{\Omega_1 2} y_e^2 + \chi_{21} e^{-\sigma_{21}(t-t_0)} \\ &\leq \frac{\lambda_2 u_{1d} \cos(\lambda_3 t)}{\Omega_1^2} y_e^2 + \chi_{21} e^{-\sigma_{21}(t-t_0)}, \end{aligned} \tag{14.58}$$

where $\sigma_{21} = \min(\sigma_0, \sigma_1)$ and χ_{21} is a class-K function of $\|(\tilde{\eta}(t_0), \tilde{\boldsymbol{X}}(t_0), \bar{\boldsymbol{X}}(t_0))\|$. Integrating both sides of the second line of (14.58) yields

$$\begin{aligned} V_2(t) &\leq V_2(t_0) + 2\lambda_2 u_{1d}^{\max} + \frac{\chi_{21}}{\sigma_{21}} \\ &\leq \chi_{22}, \end{aligned} \tag{14.59}$$

where $u_{1d}^{\max}$ is the upper bound of $|u_{1d}(t)|$. Therefore, the trajectories (x_e, y_e) are bounded on $[0, \infty)$. To prove convergence of (x_e, y_e) to zero, we consider each case of Assumption 14.1.

Cases C1 and C2. From the first line of (14.58) and noting (14.48), we have

$$\dot{V}_2 \leq -\frac{c_1}{\Omega_1^2} x_e^2 + |\lambda_2 u_{1d}| + \chi_{21}(\cdot) e^{-\sigma_{21}(t-t_0)}. \tag{14.60}$$

Integrating both sides of (14.60) and using Barbalat's lemma, it is straightforward to show that $\lim_{t\to\infty} x_e(t) = 0$. To prove that $\lim_{t\to\infty} y_e(t) = 0$, applying Lemma 2.6 to the first equation of (14.51) yields:

$$\lim_{t\to\infty} (y_e(\alpha_w + \bar{w} + \tilde{w}) + \tilde{v} + \bar{v}) = 0, \tag{14.61}$$

which is equivalent to:

$$\lim_{t\to\infty} \Xi(t) = 0, \tag{14.62}$$

where

$$\Xi(t) = y_e(t)\left(\frac{\dot{k}(t)y_e(t)}{\sqrt{1+(1-k^2(t))y_e^2(t)}} - u_{2d}(t)\right). \tag{14.63}$$

On the other hand, from (14.60), we have

$$\frac{\mathrm{d}}{\mathrm{d}t}\left(V_2 - \int_0^t |\lambda_2 u_{1d}(\tau)|\,\mathrm{d}\tau + \frac{1}{\sigma_{21}}\chi_{21}(\cdot)e^{-\sigma_{21}(t-t_0)}\right) \le 0, \tag{14.64}$$

which means that $V_2 - \int_0^t |\lambda_2 u_{1d}(\tau)|\,\mathrm{d}\tau + \sigma_{21}^{-1}\chi_{21}(\cdot)e^{-\sigma_{21}(t-t_0)}$ is nonincreasing. Since V_2 is bounded from below by zero, V_2 tends to a finite nonnegative constant depending on $\|\bar{X}_e(t_0)\|$ with $\bar{X}_e = (x_e, y_e, \bar{X}, \tilde{\eta}(t_0), \tilde{X}(t_0))$. This implies that the limit of $|y_e(t)|$ exists and is finite, say l_{y_e}. If l_{y_e} were not zero, there would exist a sequence of increasing time instants $\{t_i\}_{i=1}^{\infty}$ with $t_i \to \infty$, such that both of the limits of $\dot{k}(t_i)$ and $\Xi(t_i)$ would not be zero. With this in mind, if we choose $\lambda_i \ne 0$ and such that

$$\frac{\lambda_2\lambda_3}{\sqrt{1-k_*^2}} < \mu_{22}, \tag{14.65}$$

then under conditions (14.27), $\dot{k}(t_i)$ and $\Xi(t_i)$ cannot be nonzero simultaneously for any t_i. Hence l_{y_e} must be zero, which allows us to conclude from (14.62) that $\lim_{t\to\infty} y_e(t) = 0$, i.e., the (x_e, y_e)-subsystem is asymptotically stable.

Case C3. In this case, from (14.58), we have

$$\dot{V}_2 \le -\frac{1}{\Omega_1^2}\left(c_1 x_e^2 + (\lambda_1 u_{1d}^2 - \lambda_2 |u_{1d}|)y_e^2\right) + \chi_{21}e^{-\sigma_{21}(t-t_0)}, \tag{14.66}$$

which means that there exist $\sigma_2 > 0$ and a class-K function χ_2 depending on $\|\bar{X}(t_0)\|$ such that

$$\|(x_e(t), y_e(t))\| \le \chi_2\sigma^{-\sigma_2(t-t_0)}, \tag{14.67}$$

as long as

$$\lambda_1\mu_{31} - \lambda_2 u_{1d}^{\max} \ge \mu_{31}^*, \tag{14.68}$$

where μ_{31}^* is a positive constant. In addition, it can be shown that in this case the closed loop of (14.51) and (14.53) is also locally exponentially stable. Under Assumption 14.1 there always exist λ_i such that (14.65) and (14.68) hold. We have thus proven the following result.

Theorem 14.1. *Under Assumption 14.1 the output feedback control laws consisting of (14.44) and (14.52) force the mobile robot (14.25) to globally asymptotically track the virtual vehicle (14.26) if the constants λ_i, $i = 1,2,3$ are chosen such that $\lambda_i \ne 0$, (14.65) and (14.68) hold.*

14.1.3.5 Simulations

The physical parameters of the robot are given in Section 14.1.2. We perform two simulations. For the first simulation, the reference velocities are chosen as: $u_{1d} = 0.5(\tanh(t_s - t) + 1)$, and $u_{2d} = 0$, where t_s is a positive constant. A switching combination of a tracking controller and a stabilization one in the literature cannot be used to fulfill this task if t_s is unknown in advance. A calculation shows that for $t \leq t_s$ (tracking a curve), condition C3 holds with $\mu_{31} = 0.25$ and for $t > t_s$ (parking) Case C1 holds. Hence, our proposed controller can be applied. We also assume that due to some sudden impact at the time $t_m > t_s$, the robot position is perturbed to $y = y_m \neq 0$ to illustrate the regulation ability of our proposed controller. For the second simulation, the reference velocities are $u_{1d} = 0, u_{2d} = 0.2$, i.e., Case C2 holds with $\mu_{22} = 0.2$. The initial conditions are: $(\boldsymbol{\eta}^T, \boldsymbol{\omega}^T) = ((-2, 2, -0.5), (0, 0))$, $(\hat{\boldsymbol{\eta}}^T, \hat{\boldsymbol{X}}^T) = ((0, 0, 0), (0, 0))$, $(x_d, y_d, \phi_d) = (0, 0, 0)$, and we take $t_s = 20$, $t_m = 30$, $y_m = 1.5$. The control and observer gains are chosen as $c_i = 2$, $1 \leq i \leq 4$, $\delta_v = \delta_w = 0.1$, $\boldsymbol{P_{01}} = \boldsymbol{P_{02}} = \text{diag}(1, 1)$, $\lambda_1 = \lambda_3 = 0.5$, $\lambda_2 = 0.1$, $\boldsymbol{K_{01}} = \text{diag}(1, 1)$, $\boldsymbol{K_{02}} = \left(\boldsymbol{J}(\eta)\boldsymbol{Q}^{-1}(\eta)\right)^T$. The above choice satisfies the requirements in Theorem 14.1. The robot trajectories in the (x, y)-plane are plotted in Figures 14.3 and 14.4. The tracking errors in the form of $\sqrt{x_e^2 + y_e^2 + \phi_e^2}$ are plotted in Figure 14.5. This figure indicates that convergence of the tracking errors for the case of regulation to zero is much slower than those for the other cases, which is a quite well-known effect when using the smooth time-varying controllers. Convergence of tracking errors in Case C2 is slower than that in Case C3, since Case C3 yields local exponential stability but only asymptotic stability for Case C2 (see the proof of Theorem 14.1).

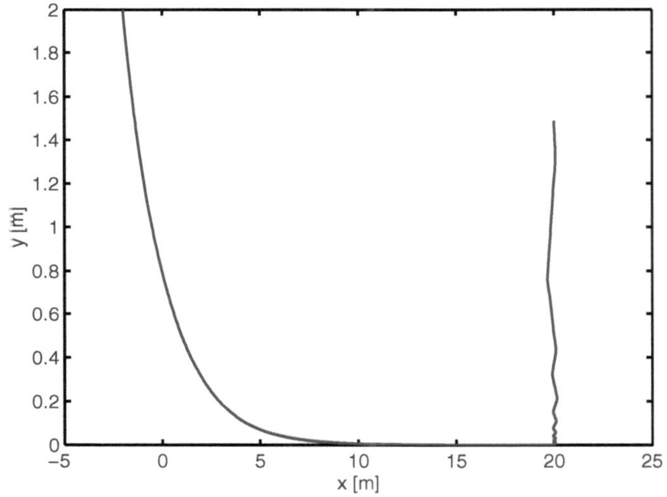

Figure 14.3 First simulation: Robot position in the (x, y)-plane

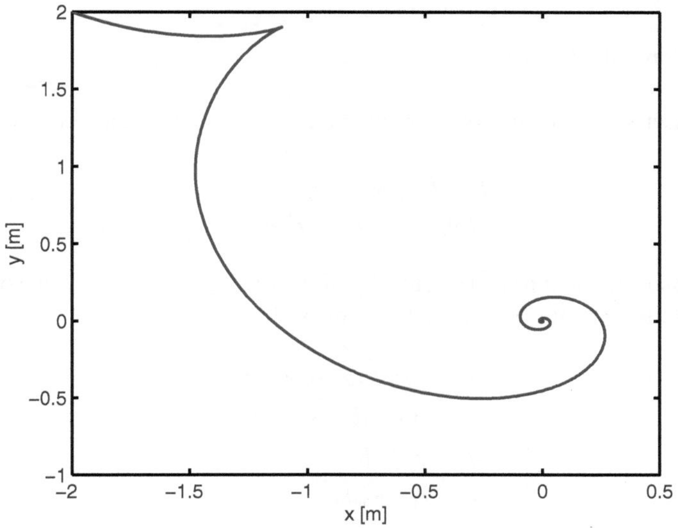

Figure 14.4 Second simulation: Robot position in the (x, y)-plane

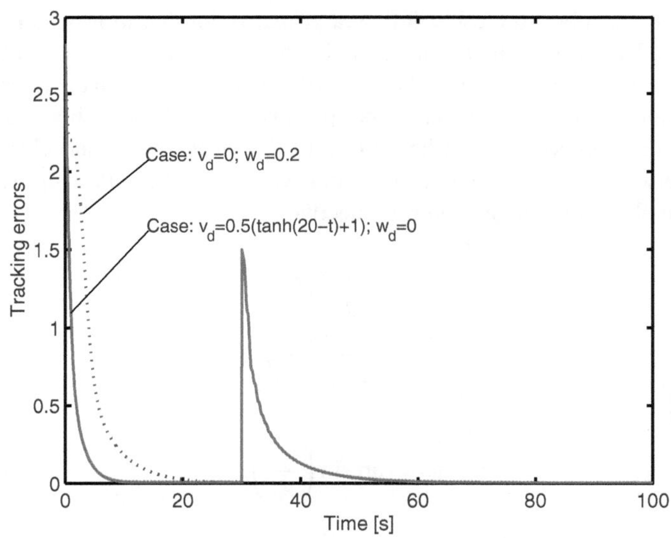

Figure 14.5 Tracking errors with respect to the first and second simulations

14.1.4 Output Feedback Path-following

14.1.4.1 Control Objective

For the reader's convenience, we rewrite the equations of motion here, see (14.1):

$$\begin{aligned} \dot{\boldsymbol{\eta}} &= \boldsymbol{J}(\boldsymbol{\eta})\boldsymbol{\omega}, \\ \boldsymbol{M}\dot{\boldsymbol{\omega}} + \boldsymbol{C}(\dot{\boldsymbol{\eta}})\boldsymbol{\omega} + \boldsymbol{D}\boldsymbol{\omega} &= \boldsymbol{\tau}. \end{aligned} \tag{14.69}$$

After an observer and a primary control design as in Section 14.1.3.2, we only need to consider the robot model (i.e., the dynamics (14.42)):

$$\begin{aligned} \dot{x} &= \cos(\phi)\hat{v} + \cos(\phi)\tilde{v}, \\ \dot{y} &= \sin(\phi)\hat{v} + \sin(\phi)\tilde{v}, \\ \dot{\phi} &= \hat{w} + \tilde{w}, \\ \dot{\hat{v}} &= \tau_{vc} + \Omega_v, \\ \dot{\hat{w}} &= \tau_{wc} + \Omega_w. \end{aligned} \tag{14.70}$$

In this section, we consider a control objective of designing the control vector $\boldsymbol{\tau}$ to force the mobile robot to follow a specified path Γ, see Figure 14.6. If we are able to drive the robot to closely follow a virtual robot that moves along the path with a desired speed v_0, which is tangential to the path, then the control objective is fulfilled, i.e., the robot is in a tube of nonzero diameter centered on the reference path and moves along the specified path at the speed v_0. Roughly speaking, the approach is to steer the robot such that it heads toward the virtual robot and eliminates the distance between itself and the virtual robot. We define the following variables to mathematically formulate the control objective:

$$\begin{aligned} x_e &= x_d - x, \\ y_e &= y_d - x, \\ \phi_e &= \phi - \phi_d, \\ z_e &= \sqrt{x_e^2 + y_e^2}, \end{aligned} \tag{14.71}$$

where

$$\phi_d = \arcsin\left(\frac{y_e}{z_e}\right). \tag{14.72}$$

Control Objective. Under Assumption 14.2, design the controls τ_1 and τ_2 to force the mobile robot (14.70) to follow the path Γ given by

$$\begin{aligned} x_d &= x_d(s), \\ y_d &= y_d(s), \end{aligned} \tag{14.73}$$

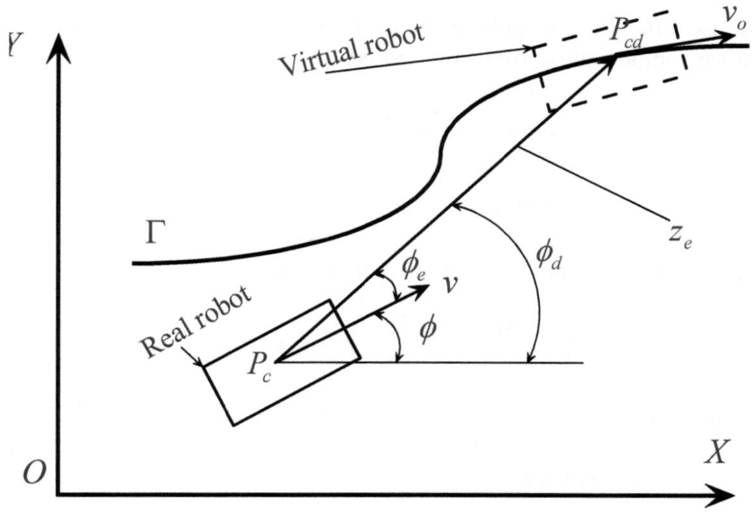

Figure 14.6 General framework of mobile robot path-following

where s is the path parameter variable, such that

$$\lim_{t\to\infty} z_e(t) \leq \bar{z}_e,$$
$$\lim_{t\to\infty} |\phi_e(t)| = 0, \tag{14.74}$$

with $\bar{z}_e$ being an arbitrarily small positive constant.

Assumption 14.2.

1. *The reference path is regular, i.e., there exist positive constants $R_{\min}$ and $R_{\max}$ such that*
$$0 < R_{\min} \leq \left(\frac{\partial x_d}{\partial s}\right)^2 + \left(\frac{\partial y_d}{\partial s}\right)^2 \leq R_{\max} < \infty.$$
2. *The minimum radius of the osculating circle of the path is larger than or equal to the minimum possible turning radius of the robot.*

Remark 14.4.

1. Assumption 14.2 ensures that the path is feasible for the robot to follow.
2. If the reference path is not regular, then we can often split it into regular sections and consider each of them separately.
3. The path parameter, s, is not the arc length of the path in general. For example, a circle with radii of R centered at the origin can be described as $x_d = R\cos(s)$ and $y_d = R\sin(s)$, see [136] for more details.

If one differentiates both sides of $\phi_e = \phi - \phi_d$ to obtain the $\dot{\phi}_e$-dynamics, there will be discontinuity in the $\dot{\phi}_e$-dynamics when x_e changes its sign. This discontinuity will cause difficulties in applying the backstepping technique. To get around this problem, we compute the $\dot{\phi}_e$-dynamics based on

$$\sin(\phi_e) = \frac{x_e \sin(\phi) - y_e \cos(\phi)}{z_e},$$
$$\cos(\phi_e) = \frac{x_e \cos(\phi) + y_e \sin(\phi)}{z_e}. \tag{14.75}$$

We now use (14.71) and (14.75) to transform (14.71) to

$$\begin{aligned}
\dot{z}_e &= -\cos(\phi_e)\hat{v} + \left(\frac{x_e}{z_e}\frac{\partial x_d}{\partial s} + \frac{y_e}{z_e}\frac{\partial y_d}{\partial s}\right)\dot{s} - \cos(\phi_e)\tilde{v},\\
\dot{\phi}_e &= \hat{w} + \left[\left(\frac{\sin(\phi)}{z_e} - \frac{x_e \sin(\phi_e)}{z_e^2}\right)\frac{\partial x_d}{\partial s} - \left(\frac{\cos(\phi)}{z_e} + \frac{y_e \sin(\phi_e)}{z_e^2}\right)\frac{\partial y_d}{\partial s}\right] \times \\
&\quad \frac{\dot{s}}{\cos(\phi_e)} + \frac{\sin(\phi_e)}{z_e}(\hat{v} + \tilde{v}) + \tilde{w},\\
\dot{\hat{v}} &= \tau_{vc} + \Omega_v,\\
\dot{\hat{w}} &= \tau_{wc} + \Omega_w.
\end{aligned} \tag{14.76}$$

It is noted that (14.75) is not defined at $z_e = 0$. However, our controller will guarantee that $z_e(t) \geq z_e^* > 0$, for all $0 \leq t < \infty$ for feasible initial conditions. The second equation of (14.76) is not defined at $\phi_e(t) = \pm 0.5\pi$ but we will design $\dot{s}$ to overcome this problem. Therefore, we will design the controls τ_{vc} and τ_{wc} for (14.76) to yield the control objective. In the next section, a procedure to design a stabilizer for the path-following error system (14.76) is presented in detail. The triangular structure of (14.76) suggests that we design the controls τ_{vc} and τ_{wc} in two stages. First, we design the virtual velocity controls for $\hat{v}$ and $\hat{w}$ and choose $\dot{s}$ to ultimately stabilize z_e and ϕ_e at the origin. Based on the backstepping technique, the controls τ_{vc} and τ_{wc} will be then designed.

14.1.4.2 Control Design

Step 1

The z_e- and ϕ_e-dynamics have three inputs that can be chosen to stabilize z_e and ϕ_e, namely $\dot{s}$, $\hat{v}$, and $\hat{w}$. The input $\hat{w}$ should be designed to stabilize the ϕ_e-dynamics at the origin. Therefore, two inputs, $\dot{s}$ and $\hat{v}$, can be used to ultimately stabilize z_e at the origin. We can either choose the input $\hat{v}$ and $\dot{s}$ and then design the remaining input. If we fix $\dot{s}$, then the virtual robot is allowed to move at a desired speed. The real robot will follow the virtual one on the path by the controller, and vice versa. We here choose to fix $\dot{s}$. This allows us to adjust the initial conditions in most cases without moving the robot. Since the transformed system (14.76) is not defined at

$z_e = 0$, we first assume in the following that $z_e(t) \geq z_e^* > 0, \forall\, 0 \leq t < \infty$. We will then show that there exist initial conditions such that this hypothesis holds.

Define

$$\begin{aligned} v_e &= \hat{v} - \hat{v}_r, \\ w_e &= \hat{w} - \hat{w}_r, \end{aligned} \tag{14.77}$$

where $\hat{v}_r$ and $\hat{w}_r$ are the virtual controls of $\hat{v}$ and $\hat{w}$, respectively. As discussed above, we choose the virtual controls and $\dot{s}$ as follows:

$$\begin{aligned} \hat{v}_r &= k_1(z_e - \delta_e) + \left(\frac{x_e}{z_e}\frac{\partial x_d}{\partial s} + \frac{y_e}{z_e}\frac{\partial y_d}{\partial s}\right)\frac{v_0(t,z_e)}{\sqrt{\left(\frac{\partial x_d}{\partial s}\right)^2 + \left(\frac{\partial y_d}{\partial s}\right)^2}}, \\ \hat{w}_r &= -k_2\phi_e - \left[\left(\frac{\sin(\phi)}{z_e} - \frac{x_e\sin(\phi_e)}{z_e^2}\right)\frac{\partial x_d}{\partial s} - \left(\frac{\cos(\phi)}{z_e} + \frac{y_e\sin(\phi_e)}{z_e^2}\right)\frac{\partial y_d}{\partial s}\right] \\ &\quad \times \frac{v_0(t,z_e)}{\sqrt{\left(\frac{\partial x_d}{\partial s}\right)^2 + \left(\frac{\partial y_d}{\partial s}\right)^2}} - \frac{\sin(\phi_e)}{z_e}\hat{v}_r - \delta_1\left(\frac{\sin(\phi_e)}{z_e}\right)^2\phi_e, \\ \dot{s} &= \frac{\cos(\phi_e)v_0(t,z_e)}{\sqrt{\left(\frac{\partial x_d}{\partial s}\right)^2 + \left(\frac{\partial y_d}{\partial s}\right)^2}}, \end{aligned} \tag{14.78}$$

where $k_1 > 0$, $k_2 > 0$, and $\delta_1 > 0$. The term multiplied by δ_1 is a nonlinear damping term to overcome the observer error effect. $v_0(t,z_e) \neq 0$ for all $t \geq t_0 \geq 0$ and $z_e \in \mathbb{R}$, is the speed of the virtual robot on the path. Indeed, one can choose this speed to be a constant. However, the time-varying speed and position path-following dependence of the virtual robot on the path is more desirable, especially when the robot starts to follow the path. For example, one might choose

$$v_0(t,z_e) = v_0^*(1 - \chi_1 e^{-\chi_2(t-t_0)})e^{-\chi_3 z_e}, \tag{14.79}$$

where $v_0^* \neq 0$, $\chi_i > 0$, $i = 1,2,3$, and $\chi_1 < 1$. The choice of $v_0(t,z_e)$ in (14.79) has the following desired feature: When the path-following error, z_e, is large, the virtual robot will wait for the real one; when z_e is small, the virtual robot will move along the path at the speed closed to v_0^* and the real one follows it within the specified look ahead distance. This feature is suitable in practice because it avoids using a high gain control for a large signal z_e.

Substituting (14.78) into the first two equations of (14.76) results in

$$\begin{aligned} \dot{z}_e &= -k_1\cos(\phi_e)(z_e - \delta_e) - \cos(\phi_e)(\tilde{v} + v_e), \\ \dot{\phi}_e &= -k_2\phi_e - \delta_1\left(\frac{\sin(\phi_e)}{z_e}\right)^2\phi_e + \frac{\sin(\phi_e)}{z_e}(\tilde{v} + v_e) + \tilde{w} + w_e. \end{aligned} \tag{14.80}$$

Step 2

By noting that the virtual control $\hat{v}_r$ is a function of $t,\ x_e,\ y_e$, and s, and the virtual control $\hat{w}_r$ is a function of $t,\ x_e,\ y_e,\ s$, and ϕ, differentiating both sides of (14.77) along the solutions of the last two equations of (14.76) results in

$$\begin{bmatrix}\dot{v}_e\\ \dot{w}_e\end{bmatrix}=\begin{bmatrix}\tau_{vc}+\Omega_v\\ \tau_{wc}+\Omega_w\end{bmatrix}+\begin{bmatrix}\left(\dfrac{\partial\hat{v}_r}{\partial x_e}\cos(\phi)+\dfrac{\partial\hat{v}_r}{\partial y_e}\sin(\phi)\right)\tilde{v}\\ \left(\dfrac{\partial\hat{w}_r}{\partial x_e}\cos(\phi)+\dfrac{\partial\hat{w}_r}{\partial y_e}\sin(\phi)\right)\tilde{v}-\dfrac{\partial\hat{w}_r}{\partial\phi}\tilde{w}-\dfrac{\partial\hat{w}_r}{\partial t}\end{bmatrix}-$$
$$\begin{bmatrix}\dfrac{\partial\hat{v}_r}{\partial t}+\dfrac{\partial\hat{v}_r}{\partial x_e}\left(\dfrac{\partial x_d}{\partial s}\dot{s}-\hat{v}\cos(\phi)\right)+\dfrac{\partial\hat{v}_r}{\partial y_e}\left(\dfrac{\partial y_d}{\partial s}\dot{s}-\hat{v}\sin(\phi)\right)+\dfrac{\partial\hat{v}_r}{\partial s}\dot{s}\\ \dfrac{\partial\hat{w}_r}{\partial x_e}\left(\dfrac{\partial x_d}{\partial s}\dot{s}-\hat{v}\cos(\phi)\right)+\dfrac{\partial\hat{w}_r}{\partial y_e}\left(\dfrac{\partial y_d}{\partial s}\dot{s}-\hat{v}\sin(\phi)\right)+\dfrac{\partial\hat{w}_r}{\partial s}\dot{s}+\dfrac{\partial\hat{w}_r}{\partial\phi}\hat{w}\end{bmatrix}. \tag{14.81}$$

From (14.80) and (14.81), we choose the control inputs with nonlinear damping terms to overcome the observer error effect as follows:

$$\begin{bmatrix}\tau_{vc}\\ \tau_{wc}\end{bmatrix}=-\begin{bmatrix}c_{21}v_e\\ c_{22}w_e\end{bmatrix}-\delta_2\begin{bmatrix}\left(\dfrac{\partial\hat{v}_r}{\partial x_e}\cos(\phi)+\dfrac{\partial\hat{v}_r}{\partial y_e}\sin(\phi)\right)^2 v_e\\ \left(\dfrac{\partial\hat{w}_r}{\partial x_e}\cos(\phi)+\dfrac{\partial\hat{w}_r}{\partial y_e}\sin(\phi)\right)^2 w_e+\left(\dfrac{\partial\hat{w}_r}{\partial\phi}\right)^2\end{bmatrix}+$$
$$\begin{bmatrix}\dfrac{\partial\hat{v}_r}{\partial t}+\dfrac{\partial\hat{v}_r}{\partial x_e}\left(\dfrac{\partial x_d}{\partial s}\dot{s}-\hat{v}\cos(\phi)\right)+\dfrac{\partial\hat{v}_r}{\partial y_e}\left(\dfrac{\partial y_d}{\partial s}\dot{s}-\hat{v}\sin(\phi)\right)+\dfrac{\partial\hat{v}_r}{\partial s}\dot{s}\\ \dfrac{\partial\hat{w}_r}{\partial x_e}\left(\dfrac{\partial x_d}{\partial s}\dot{s}-\hat{v}\cos(\phi)\right)+\dfrac{\partial\hat{w}_r}{\partial y_e}\left(\dfrac{\partial y_d}{\partial s}\dot{s}-\hat{v}\sin(\phi)\right)+\dfrac{\partial\hat{w}_r}{\partial s}\dot{s}+\dfrac{\partial\hat{w}_r}{\partial\phi}\hat{w}\end{bmatrix}-$$
$$\delta_3\begin{bmatrix}(\hat{v}^2+\hat{w}^2)v_e\\ (\hat{v}^2+\hat{w}^2)w_e\end{bmatrix}-\begin{bmatrix}\dfrac{\sin(\phi_e)}{z_e}\phi_e\\ \phi_e-\dfrac{\partial\hat{w}_r}{\partial t}\end{bmatrix}, \tag{14.82}$$

where c_{21}, c_{22}, δ_2, and δ_3 are positive constants. Substituting (14.82) into (14.81) results in

$$\begin{bmatrix}\dot{v}_e\\ \dot{w}_e\end{bmatrix}=-\begin{bmatrix}c_{21}v_e\\ c_{22}w_e\end{bmatrix}+\begin{bmatrix}\Omega_v\\ \Omega_w\end{bmatrix}-\delta_3\begin{bmatrix}(\hat{v}^2+\hat{w}^2)v_e\\ (\hat{v}^2+\hat{w}^2)w_e\end{bmatrix}-\begin{bmatrix}\sin(\phi_e)\phi_e/z_e\\ \phi_e\end{bmatrix}+$$
$$\begin{bmatrix}\left(\dfrac{\partial\hat{v}_r}{\partial x_e}\cos(\phi)+\dfrac{\partial\hat{v}_r}{\partial y_e}\sin(\phi)\right)\tilde{v}\\ \left(\dfrac{\partial\hat{w}_r}{\partial x_e}\cos(\phi)+\dfrac{\partial\hat{w}_r}{\partial y_e}\sin(\phi)\right)\tilde{v}-\dfrac{\partial\hat{w}_r}{\partial\phi}\tilde{w}\end{bmatrix}-$$
$$\delta_2\begin{bmatrix}\left(\dfrac{\partial\hat{v}_r}{\partial x_e}\cos(\phi)+\dfrac{\partial\hat{v}_r}{\partial y_e}\sin(\phi)\right)^2 v_e\\ \left(\dfrac{\partial\hat{w}_r}{\partial x_e}\cos(\phi)+\dfrac{\partial\hat{w}_r}{\partial y_e}\sin(\phi)\right)^2 w_e+\left(\dfrac{\partial\hat{w}_r}{\partial\phi}\right)^2\end{bmatrix}. \tag{14.83}$$

14.1.4.3 Stability Analysis

To analyze the closed loop system consisting of (14.80) and (14.83), we first consider the (ϕ_e, v_e, w_e)-subsystem, then move to the z_e-dynamics.

(ϕ_e, v_e, w_e)-subsystem. For this subsystem, we take the following Lyapunov function:

$$V_1 = \frac{1}{2}\left(\phi_e^2 + v_e^2 + w_e^2\right), \tag{14.84}$$

whose derivative along the solutions of the last equation of (14.80) and (14.83), after some manipulation, satisfies

$$\dot{V}_1 \leq -\rho_1 V_1 + \chi_1(\cdot) e^{-\sigma_0(t-t_0)}, \tag{14.85}$$

where ρ_1 is a positive constant and can be made arbitrarily large by increasing the design constants k_1, k_2, c_{21} and c_{22}, and $\chi_1(\cdot)$ is a class-K function of $\left\|\left(z_e(t_0), \phi_e(t_0), \tilde{\eta}(t_0), \tilde{X}(t_0)\right)\right\|$. From (14.85), it is not hard to show that

$$\|(\phi_e(t), v_e(t), w_e(t))\| \leq \alpha_1(\cdot) e^{-\sigma_1(t-t_0)}, \tag{14.86}$$

where $\alpha_1(\cdot)$ is a class-K function of $\left\|\left(z_e(t_0), \phi_e(t_0), \tilde{\eta}(t_0), \tilde{X}(t_0)\right)\right\|$, and σ_1 is a positive constant. Hence (14.86) implies that the (ϕ_e, v_e, w_e)-subsystem is K-exponentially stable at the origin.

z_e-subsystem: Lower Bound of z_e. Defining $\tilde{z}_e = z_e - \delta_e$, the first equation of (14.80) can be written as

$$\begin{aligned}
\dot{\tilde{z}}_e &= -k_1 \cos(\phi_e)\tilde{z}_e - \cos(\phi_e)(\tilde{v} + v_e) \\
&\geq -k_1 \tilde{z}_e - |\tilde{v} + v_e| \\
&\geq -k_1 \tilde{z}_e - \alpha_2(\cdot) e^{\sigma_2(t-t_0)},
\end{aligned} \tag{14.87}$$

where $\alpha_2(\cdot)$ is a class-K function of $\|(z_e(t_0), \phi_e(t_0), \tilde{\eta}(t_0), \tilde{X}(t_0))\|$, and σ_2 is a positive constant. From (14.87) and the comparison principle, we have

$$\tilde{z}_e(t) \geq \tilde{z}_e(t_0) e^{-k_1(t-t_0)} + \frac{\alpha_2(\cdot)}{\sigma_2 - k_1}\left(e^{-\sigma_2(t-t_0)} - e^{-k_1(t-t_0)}\right). \tag{14.88}$$

Therefore the condition $z_e(t) \geq z_e^*$ holds if

$$\sigma_2 > k_1, \; z_e(t_0) \geq -2\delta_e + \frac{\alpha_2(\cdot)}{\sigma_2 - k_1} + z_e^*. \tag{14.89}$$

z_e-subsystem: Upper Bound of z_e. To estimate the upper bound of z_e, we write the first equation of (14.80) as

$$\dot{z}_e = -k_1 z_e - k_1(\cos(\phi_e) - 1)z_e + k_1 \cos(\phi_e)\delta_e - \cos(\phi_e)(\tilde{v} + v_e). \tag{14.90}$$

By taking the Lyapunov function $V_2 = 0.5z_e^2$ and noting that ϕ_e, v_e, and $\tilde{v}$ exponentially converge to zero, it is not hard to show that

$$|z_e(t)| \leq \alpha_3(\cdot)e^{-\sigma_3(t-t_0)} + \rho_3, \tag{14.91}$$

where $\alpha_3(\cdot)$ is a class-K function of $\|(z_e(t_0), \phi_e(t_0), \tilde{\eta}(t_0), \tilde{\boldsymbol{X}}(t_0))\|$, and σ and ρ_3 are positive constants. The constant ρ_3 can be made arbitrarily small by reducing δ_e.

14.1.4.4 Simulations

To illustrate the effectiveness of the proposed output feedback path-following controller, we perform a numerical simulation. The physical parameters are given in Section 14.1.2. The design constants are:

$$k_1 = 0.5;\ k_2 = 5,\ c_{21} = c_{22} = 2,\ \delta_1 = \delta_2 = \delta_3 = 0.05,\ \delta_e = 0.2.$$

The initial conditions are:

$$(\boldsymbol{\eta}^T, \boldsymbol{\omega}^T) = ((-5, 0, 0.5), (0, 1)),\ (\hat{\boldsymbol{\eta}}^T, \hat{\boldsymbol{X}}^T) = ((0, 0, 0), (0, 0)),\ s(0) = 0.$$

The reference speed of the virtual mobile robot is $v_0 = 2.5$ m / s. The path Γ is chosen to be a sinusoidal path specified by $x_d = s$ and $y_d = 10\sin(0.15s)$. This path is regular and satisfies all requirements in Assumption 14.2. Simulation results are plotted in Figure 14.7. From this figure, it can be clearly seen that the proposed output feedback controller is able to force the mobile robot in question to follow the reference path accurately.

14.1.5 Notes and References

This section has shown that the observer design and control design techniques developed for underactuated ships in Chapters 5, 6, 7, and 11 can be successfully applied to solve the challenging problems of output feedback simultaneous stabilization and trajectory-tracking, and path-following for mobile robots. Indeed, the main difficulty in solving stabilization and trajectory-tracking control of mobile robots is due to the fact that the motion of the systems to be controlled has more degrees of freedom to be controlled than the number of control inputs under nonholonomic constraints. In addition, the cross-terms of the robot velocities due to the Coriolis matrix prevents the success of the traditional observer designs. Stabilization and trajectory-tracking control problems of mobile robots are usually addressed separately in the literature. Discontinuous and time-varying approaches are often used

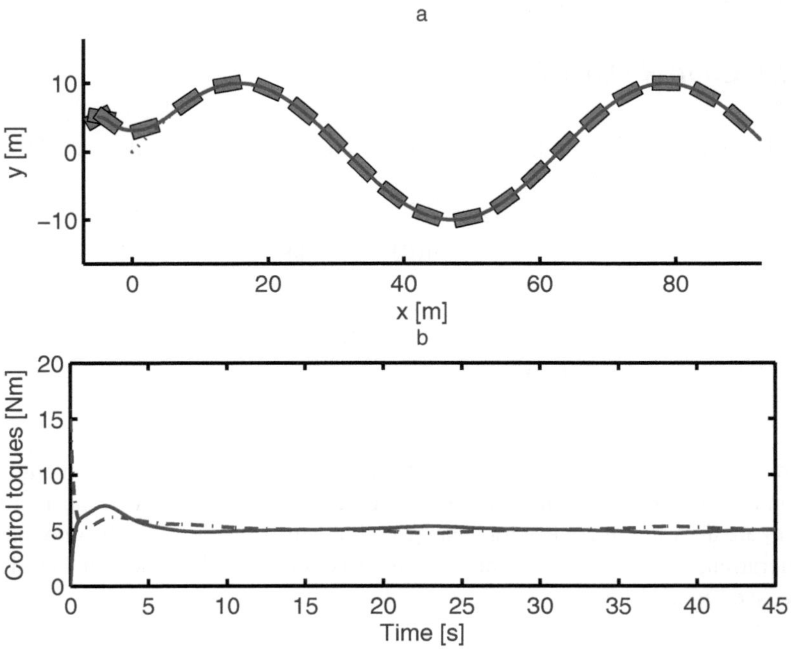

Figure 14.7 **a**. Robot position and orientation in the (x, y)-plane, **b**. Control torques

to solve the stabilization problem where the robot dynamics is transformed to a chained system. Various nonlinear controllers have been developed based on Lyapunov's direct method and the backstepping technique to solve the tracking control problem, see [18, 41–44, 47, 131, 142, 147–150]. The simultaneous stabilization and trajectory-tracking control problem of mobile robots has been solved using the time-varying approach, see [133, 151, 152] and the high-gain (using an oscillator) approach, see [148, 150]. It is also mentioned that all proofs of the main results in this section are based on Lyapunov's direct method for the reader's convenience. Indeed, we can use the stability result for cascade systems given in Section 2.1.3 to analyze stability of the closed loop systems. Moreover, if nonlinear damping terms are included in the robot dynamics, Lemma 2.2 can be used with the proposed coordinate transformation to design an exponential/asymptotic observer.

The work in this section is based on [131, 151].

14.2 Vertical Take-off and Landing Aircraft

14.2.1 Control Objective

A scaled mathematical model of a VTOL aircraft can be described as, see [153]:

$$\begin{aligned}
\dot{x}_1 &= x_2, \\
\dot{x}_2 &= -u_1 \sin(\theta) + \varepsilon u_2 \cos(\theta), \\
\dot{y}_1 &= y_2, \\
\dot{y}_2 &= u_1 \cos(\theta) + \varepsilon u_2 \sin(\theta) - g, \\
\dot{\theta} &= \omega, \\
\dot{\omega} &= u_2,
\end{aligned} \tag{14.92}$$

where x_1, y_1, and θ denote position of the aircraft center of mass and roll angle, x_2, y_2, and ω denote linear and roll angular velocities of the aircraft, respectively, u_1 and u_2 are the vertical control force and rotational moment, $g > 0$ is the gravitational acceleration, and ε is the constant coupling between the roll moment and the lateral force, see Figure 14.8.

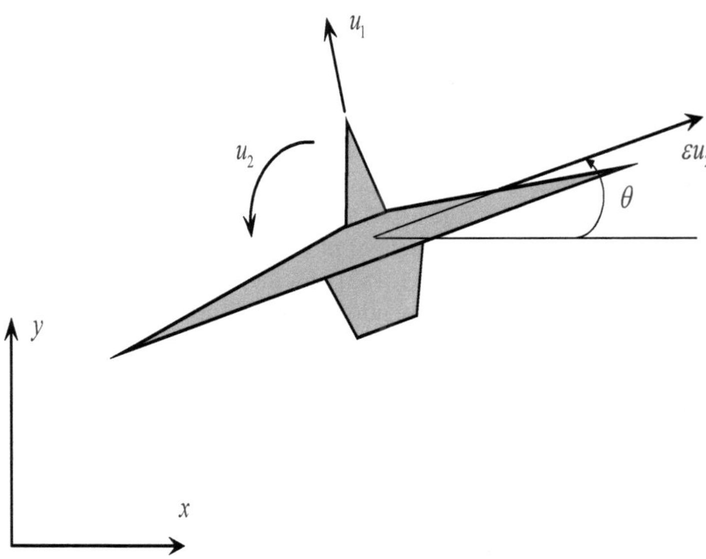

Figure 14.8 The VTOL aircraft

The aircraft model (14.92) is underactuated if we want to use two available control inputs (u_1 and u_2) to control three outputs (x_1, y_1, and θ). Furthermore, the zero-dynamics of the aircraft model (14.92) is nonminimum phase for $\varepsilon \neq 0$ at the steady state when considering (x_1, y_1) as the output and θ as an internal state. This phenomenon can be seen from (14.92) by setting $x_1 = y_1 = x_2 = y_2 = 0$.

We assume that the reference trajectory to be tracked is generated by

$$\begin{aligned}
\dot{x}_{1r} &= x_{2r}, \\
\dot{x}_{2r} &= -u_{1r}\sin(\theta_r) + \varepsilon u_{2r}\cos(\theta_r), \\
\dot{y}_{1r} &= y_{2r}, \\
\dot{y}_{2r} &= u_{1r}\cos(\theta_r) + \varepsilon u_{2r}\sin(\theta_r) - g, \\
\dot{\theta}_r &= \omega_r, \\
\dot{\omega}_r &= u_{2r},
\end{aligned} \tag{14.93}$$

where all of the variables in (14.93) have the same meaning as in (14.92). In this section, we are interested in designing the control inputs u_1 and u_2 to force the aircraft model (14.92) to globally asymptotically track the reference model (14.93) without measurements of the velocities x_2, y_2, and ω under the following assumption:

Assumption 14.3.

1. The reference signals ω_r, u_{1r}, u_{2r}, $\dot{u}_{1r}$, $\ddot{u}_{1r}$ and $\dot{u}_{2r}$ are bounded.
2. There exists a strictly positive constant u_{1r}^* such that

$$u_{1r} - \varepsilon\omega_r^2 \geq u_{1r}^*. \tag{14.94}$$

Remark 14.5. The condition (14.94) covers the stabilization/regulation of the VTOL aircraft and implies that the aircraft is not allowed to land faster than it freely falls under the gravitational force.

14.2.2 Observer Design

From (14.92), a reduced-order observer can indeed be designed. However, such an observer is often noise-sensitive. Here we use the following full-order observer:

$$\begin{aligned}
\dot{\hat{x}}_1 &= \hat{x}_2 + k_{11}(x_1 - \hat{x}_1), \\
\dot{\hat{x}}_2 &= -u_1\sin(\theta) + \varepsilon u_2\cos(\theta) + k_{12}(x_1 - \hat{x}_1), \\
\dot{\hat{y}}_1 &= \hat{y}_2 + k_{21}(y_1 - \hat{y}_1), \\
\dot{\hat{y}}_2 &= u_1\cos(\theta) + \varepsilon u_2\sin(\theta) - g + k_{22}(y_1 - \hat{y}_1), \\
\dot{\hat{\theta}} &= \hat{\omega} + k_{31}(\theta - \hat{\theta}), \\
\dot{\hat{\omega}} &= u_2 + k_{32}(\theta - \hat{\theta}),
\end{aligned} \tag{14.95}$$

where k_{i_j}, $1 \le i \le 3$ and $1 \le j \le 2$ are positive constant observer gains. By defining the observer error as

$$\tilde{X} = \begin{bmatrix} \tilde{x}_1 \\ \tilde{x}_2 \\ \tilde{y}_1 \\ \tilde{y}_2 \\ \tilde{\theta} \\ \tilde{\omega} \end{bmatrix} = \begin{bmatrix} x_1 - \hat{x}_1 \\ x_2 - \hat{x}_2 \\ y_1 - \hat{y}_1 \\ y_2 - \hat{y}_2 \\ \theta - \hat{\theta} \\ \omega - \hat{\omega} \end{bmatrix} \tag{14.96}$$

and subtracting (14.95) from (14.92), we have

$$\dot{\tilde{X}} = A\tilde{X}\ ,\ \ A = \operatorname{diag}(A_i),\ A_i = \begin{bmatrix} -k_{i1} & 1 \\ -k_{i2} & 0 \end{bmatrix},\ i = 1,2,3. \tag{14.97}$$

It is easy to show that

$$\left\|\tilde{X}(t)\right\| \le \varphi_0 \left\|\tilde{X}(t_0)\right\| e^{-\sigma_0(t-t_0)},\ \forall\, t \ge t_0 \ge 0, \tag{14.98}$$

for some positive constants φ_0 and σ_0, which implies that (14.95) is a global exponential observer of (14.92). Therefore, in the following we will design the desired tracking controllers u_1 and u_2 based on the following transformed system:

$$\begin{aligned}
\dot{x}_1 &= \hat{x}_2 + \tilde{x}_2, \\
\dot{\hat{x}}_2 &= -u_1 \sin(\theta) + \varepsilon u_2 \cos(\theta) + k_{12}\tilde{x}_1, \\
\dot{y}_1 &= \hat{y}_2 + \tilde{y}_2, \\
\dot{\hat{y}}_2 &= u_1 \cos(\theta) + \varepsilon u_2 \sin(\theta) - g + k_{22}\tilde{y}_1, \\
\dot{\theta} &= \hat{\omega} + \tilde{\omega}, \\
\dot{\hat{\omega}} &= u_2 + k_{32}\tilde{\theta}.
\end{aligned} \tag{14.99}$$

14.2.3 Coordinate Transformations

Solving u_2 from the last two equations of (14.92) and substituting it into the second and fourth equations of (14.92), we have

$$\begin{aligned}
&\frac{\mathrm{d}}{\mathrm{d}t}(x_1 - \varepsilon \sin(\theta)) = x_2 - \varepsilon \cos(\theta)\omega, \\
&\frac{\mathrm{d}}{\mathrm{d}t}(x_2 - \varepsilon \cos(\theta)\omega) = -\sin(\theta)(u_1 - \varepsilon\omega^2), \\
&\frac{\mathrm{d}}{\mathrm{d}t}(y_1 + \varepsilon \cos(\theta)) = y_2 - \varepsilon \sin(\theta)\omega, \\
&\frac{\mathrm{d}}{\mathrm{d}t}(y_2 - \varepsilon \sin(\theta)\omega) = \cos(\theta)(u_1 - \varepsilon\omega^2) - g.
\end{aligned} \tag{14.100}$$

If we consider $x_1 - \varepsilon\sin(\theta)$ and $y_1 + \varepsilon\cos(\theta)$ as outputs, it can be seen that (14.100) is of a triangular form and does not depend on u_2. It is of interest to note that the above outputs coincide with the aircraft center of oscillation and they are the flat outputs [154]. Motivated by the above discussion, we define the following coordinate changes:

$$\begin{aligned} z_1 &= x_1 - \varepsilon\sin(\theta), \\ \hat{z}_2 &= \hat{x}_2 - \varepsilon\cos(\theta)\hat{\omega}, \\ w_1 &= y_1 + \varepsilon\cos(\theta), \\ \hat{w}_2 &= \hat{y}_2 - \varepsilon\sin(\theta)\hat{\omega}. \end{aligned} \tag{14.101}$$

Applying the above coordinate changes to (14.99) results in

$$\begin{aligned} \dot{z}_1 &= \hat{z}_2 + \tilde{x}_2 - \varepsilon\cos(\theta)\tilde{\omega}, \\ \dot{w}_1 &= \hat{w}_2 + \tilde{y}_2 - \varepsilon\sin(\theta)\tilde{\omega}, \\ \dot{\hat{z}}_2 &= -\bar{u}_1\sin(\theta) + \varepsilon\sin(\theta)\hat{\omega}\tilde{\omega} + k_{12}\tilde{x}_1 - \varepsilon k_{32}\cos(\theta)\tilde{\theta}, \\ \dot{\hat{w}}_2 &= \bar{u}_1\cos(\theta) - g - \varepsilon\cos(\theta)\hat{\omega}\tilde{\omega} + k_{22}\tilde{y}_1 - \varepsilon k_{32}\sin(\theta)\tilde{\theta}, \\ \dot{\theta} &= \hat{\omega} + \tilde{\omega}, \\ \dot{\hat{\omega}} &= u_2 + k_{32}\tilde{\theta}, \end{aligned} \tag{14.102}$$

where $\bar{u}_1 = u_1 - \varepsilon\hat{\omega}^2$. Similarly, applying the coordinate changes

$$\begin{aligned} z_{1r} &= x_{1r} - \varepsilon\sin(\theta_r), \\ z_{2r} &= x_{2r} - \varepsilon\cos(\theta_r)\omega_r \\ w_{1r} &= y_{1r} + \varepsilon\cos(\theta_r), \\ w_{2r} &= y_{2r} - \varepsilon\sin(\theta_r)\omega_r \end{aligned} \tag{14.103}$$

to (14.93) yields

$$\begin{aligned} \dot{z}_{1r} &= z_{2r}, \\ \dot{w}_{1r} &= w_{2r}, \\ \dot{z}_{2r} &= -\bar{u}_{1r}\sin(\theta_r), \\ \dot{w}_{2r} &= \bar{u}_{1r}\cos(\theta_r) - g, \\ \dot{\theta}_r &= \omega_r, \\ \dot{\omega}_r &= u_{2r}, \end{aligned} \tag{14.104}$$

where $\bar{u}_{1r} = u_{1r} - \varepsilon\omega_r^2$.

If we directly design the control inputs $\bar{u}_1$ and u_2 to force (14.102) to track (14.104), it will be very complicated since the angle θ enters in the third and fourth equations of (14.102). Motivated by controlling underactuated ships, we interpret the tracking errors in a frame attached to the virtual aircraft as

$$\begin{bmatrix} z_{1e} \\ w_{1e} \end{bmatrix} = J(\theta_r) \begin{bmatrix} z_1 - z_{1r} \\ w_1 - w_{1r} \end{bmatrix},$$
$$\begin{bmatrix} z_{2e} \\ w_{2e} \end{bmatrix} = J(\theta_r) \begin{bmatrix} \hat{z}_2 - z_{2r} \\ \hat{w}_2 - w_{2r} \end{bmatrix}, \tag{14.105}$$
$$\begin{bmatrix} \theta_e \\ \omega_e \end{bmatrix} = \begin{bmatrix} \theta - \theta_r \\ \hat{\omega} - \omega_r \end{bmatrix},$$

where

$$J(\theta_r) = \begin{bmatrix} -\sin(\theta_r) & \cos(\theta_r) \\ \cos(\theta_r) & \sin(\theta_r) \end{bmatrix}. \tag{14.106}$$

It is seen that the above coordinate changes are globally invertible and that convergence of $(z_{1e}, w_{1e}, z_{2e}, w_{2e})$ to the origin implies that of $(z_1 - z_{1r}, w_1 - w_{1r}, \hat{z}_2 - z_{2r}, \hat{w}_2 - w_{2r})$.

Remark 14.6. If $J(\theta)$ instead of $J(\theta_r)$ is used in the coordinate transformation (14.105), it will be extremely difficult to design the control inputs for the resulting system.

Differentiating both sides of (14.105) along the solutions of (14.102) and (14.104) yields

$$\begin{aligned}
\dot{z}_{1e} &= z_{2e} - w_{1e}\omega_r + O_{z1}, \\
\dot{w}_{1e} &= w_{2e} + z_{1e}\omega_r + O_{w1}, \\
\dot{z}_{2e} &= \bar{u}_1 \cos(\theta_e) - \bar{u}_{1r} - w_{2e}\omega_r - \varepsilon \cos(\theta_e)\omega_e\tilde{\omega} + O_{z2}, \\
\dot{w}_{2e} &= -\bar{u}_1 \sin(\theta_e) + z_{2e}\omega_r + \varepsilon \sin(\theta_e)\omega_e\tilde{\omega} + O_{w2}, \\
\dot{\vartheta}_e &= \omega_e + \tilde{\omega}, \\
\dot{\omega}_e &= u_2 - u_{2r} + k_{32}\tilde{\theta},
\end{aligned} \tag{14.107}$$

where for simplicity of presentation, we have defined the following terms which exponentially converge to zero when $\tilde{X}$ does:

$$\begin{aligned}
O_{z1} &= -\sin(\theta_r)(\tilde{x}_2 - \varepsilon \cos(\theta)\tilde{\omega}) + \cos(\theta_r)(\tilde{y}_2 - \varepsilon \sin(\theta)\tilde{\omega}), \\
O_{w1} &= \cos(\theta_r)(\tilde{x}_2 - \varepsilon \cos(\theta)\tilde{\omega}) + \sin(\theta_r)(\tilde{y}_2 - \varepsilon \sin(\theta)\tilde{\omega}), \\
O_{z2} &= -\sin(\theta_r)(k_{12}\tilde{x}_1 - \varepsilon k_{32} \cos(\theta)\tilde{\theta}) + \\
&\quad \cos(\theta_r)(k_{22}\tilde{y}_1 - \varepsilon k_{32} \sin(\theta)\tilde{\theta}) - \varepsilon \cos(\theta_e)\omega_r\tilde{\omega}, \\
O_{w2} &= \cos(\theta_r)(k_{12}\tilde{x}_1 - \varepsilon k_{32} \cos(\theta)\tilde{\theta}) + \\
&\quad \sin(\theta_r)(k_{22}\tilde{y}_1 - \varepsilon k_{32} \sin(\theta)\tilde{\theta}) + \varepsilon \sin(\theta_e)\omega_r\tilde{\omega}.
\end{aligned} \tag{14.108}$$

We have therefore converted the tracking control problem to a problem of stabilizing (14.107) at the origin.

14.2.4 Control Design

It is observed that (14.107) is of a triangular form. Although there are connection terms between the (z_{1e}, z_{2e})- and (w_{1e}, w_{2e})-subsystems, they do not prevent us from applying the popular backstepping technique [10] to (14.107), see below. To make our control design clear, we divide the control design into two steps. The first step designs $\bar{u}_1$. The control u_2 is designed in the second step.

Step 1

At this step, we design the control input $\bar{u}_1$ and as an application of our approach used in underactuated ship tracking, the angle error θ_e is also used as a "control" to stabilize the $(z_{1e}, z_{2e}, w_{1e}, w_{2e})$-subsystem. We proceed in two substeps.

Substep 1

Define

$$\begin{aligned} \bar{z}_{2e} &= z_{2e} - \alpha_z, \\ \bar{w}_{2e} &= w_{2e} - \alpha_w, \end{aligned} \tag{14.109}$$

where α_z and α_w are virtual controls of z_{2e} and w_{2e}, respectively. Consider the Lyapunov function

$$V_{11} = \sqrt{1 + z_{1e}^2 + w_{1e}^2} - 1, \tag{14.110}$$

whose time derivative along the solutions of (14.109) and the first two equations of (14.107) is

$$\dot{V}_{11} = \frac{z_{1e}}{\Delta_1}(\bar{z}_{2e} + \alpha_z + O_{z1}) + \frac{w_{1e}}{\Delta_1}(\bar{w}_{2e} + \alpha_w + O_{w1}), \tag{14.111}$$

where $\Delta_1 = \sqrt{1 + z_{1e}^2 + w_{1e}^2}$. From (14.111), we choose

$$\alpha_z = -\frac{k_1 z_{1e}}{\Delta_2}, \quad \alpha_w = -\frac{k_2 w_{1e}}{\Delta_2}, \tag{14.112}$$

where $\Delta_2 = \sqrt{1 + z_{1e}^2 + w_{1e}^2 + z_{2e}^2 + w_{2e}^2}$, and k_1 and k_2 are positive constants to be specified later.

Remark 14.7. The choice of the virtual controls α_z and α_w in (14.112) is different from a standard application of the backstepping technique in the sense that α_z and α_w depend on z_{2e} and w_{2e} .This choice together with the Lyapunov function (14.110) instead of a quadratic function will result in a bounded control input $\bar{u}_1$ (see Substep 1.2). This bounded control is crucial to obtain a global result, see Step

2. In addition, the choice of (14.110) and (14.112) can also be considered as an interesting application of the bounded backstepping approaches recently proposed in [155] and [156] to the VTOL aircraft system.

Substituting (14.112) into (14.111) gives

$$\dot{V}_{11} = -\frac{k_1 z_{1e}^2 + k_2 w_{1e}^2}{\Delta_1 \Delta_2} + \frac{z_{1e}\bar{z}_{2e}}{\Delta_1} + \frac{w_{1e}\bar{w}_{2e}}{\Delta_1} + O_{11}, \tag{14.113}$$

where the term O_{11} containing the observer errors as a factor is defined as

$$O_{11} = \frac{z_{1e} O_{z1}}{\Delta_1} + \frac{w_{1e} O_{w1}}{\Delta_1}. \tag{14.114}$$

Substep 2

Define

$$\bar{\theta}_e = \theta_e - \alpha_\theta, \tag{14.115}$$

where α_θ is a virtual control of θ_e. To design $\bar{u}_1$ and α_θ, we take the following Lyapunov function:

$$V_{12} = V_{11} + \frac{1}{2} k_3 \left(\bar{z}_{2e}^2 + \bar{w}_{2e}^2\right), \tag{14.116}$$

where k_3 is a positive constant, which is introduced to enhance the flexibility of choosing the design constants. Differentiating both sides of (14.116) along the solutions of (14.112) and (14.113), the third and fourth equations of (14.107) give

$$\begin{aligned}\dot{V}_{12} = &-\frac{k_1 z_{1e}^2 + k_2 w_{1e}^2}{\Delta_1 \Delta_2} + k_3 \bar{z}_{2e}[b_{11}(\bar{u}_1 \cos(\alpha_\theta) - \bar{u}_{1r}) + b_{12}\bar{u}_1 \sin(\alpha_\theta) + \\ & f_z] + k_3 \bar{w}_{2e}[b_{21}(\bar{u}_1 \cos(\alpha_\theta) - \bar{u}_{1r}) + b_{22}\bar{u}_1 \sin(\alpha_\theta) + f_w] + g_\theta + O_{12},\end{aligned} \tag{14.117}$$

where for simplicity of presentation, we have defined the following terms:
The terms b_{ij}, $i = 1,2$, $j = 1,2$, are

$$\begin{aligned} b_{11} &= 1 - \frac{k_1 z_{1e} z_{2e}}{\Delta_2^3}, \\ b_{12} &= \frac{k_1 z_{1e} w_{2e}}{\Delta_2^3}, \\ b_{21} &= -\frac{k_2 w_{1e} z_{2e}}{\Delta_2^3}, \\ b_{22} &= -1 + \frac{k_2 w_{1e} w_{2e}}{\Delta_2^3}. \end{aligned} \tag{14.118}$$

The terms f_z and f_w, which will be canceled in this step, are

$$
\begin{aligned}
f_z &= \frac{z_{1e}}{k_3\Delta_1} - \alpha_w\omega_r + \frac{k_1}{\Delta_2}(z_{2e} - w_{1e}\omega_r) - \frac{k_1 z_{1e}}{\Delta_2^3}(z_{1e}z_{2e} + w_{1e}w_{2e}),\\
f_w &= \frac{w_{1e}}{k_3\Delta_1} + \alpha_z\omega_r + \frac{k_2}{\Delta_2}(w_{2e} + z_{1e}\omega_r) - \frac{k_2 w_{1e}}{\Delta_2^3}(z_{1e}z_{2e} + w_{1e}w_{2e}).
\end{aligned}
\tag{14.119}
$$

The term g_θ, which will be canceled in the next step, is

$$
\begin{aligned}
g_\theta = k_3\bar{u}_1 \big[&(b_{11}\bar{z}_{2e} + b_{21}\bar{w}_{2e})((\cos(\bar{\theta}_e) - 1)\cos(\alpha_\theta) - \sin(\bar{\theta}_e)\sin(\alpha_\theta)) +\\
&(b_{12}\bar{z}_{2e} + b_{22}\bar{w}_{2e})((\cos(\bar{\theta}_e) - 1)\sin(\alpha_\theta) + \sin(\bar{\theta}_e)\cos(\alpha_\theta))\big].
\end{aligned}
\tag{14.120}
$$

The term O_{12}, which contains the observer error as a factor, is defined as

$$
\begin{aligned}
O_{12} = {} & k_3\bar{z}_{2e}\left[-\varepsilon\cos(\theta_e)\omega_e\tilde{\omega} + O_{z2} + \frac{k_1 O_{z1}}{\Delta_2}\right] +\\
& k_3\bar{w}_{2e}\left[-\varepsilon\sin(\theta_e)\omega_e\tilde{\omega} + O_{w2} + \frac{k_2 O_{w1}}{\Delta_2}\right] -\\
& \frac{k_3(k_1 z_{1e}\bar{z}_{2e} + k_2 w_{1e}\bar{w}_{2e})}{\Delta_2^3}[z_{1e}O_{z1} + w_{1e}O_{w1} +\\
& z_{2e}(-\varepsilon\cos(\theta_e)\omega_e\tilde{\omega} + O_{z2}) + w_{2e}(\varepsilon\sin(\theta_e)\omega_e\tilde{\omega} + O_{w2})] + O_{11}.
\end{aligned}
\tag{14.121}
$$

Since O_{12} is Lipschitz in $\bar{z}_{2e}\omega_e$ and $\bar{w}_{2e}\omega_e$, from (14.117) we choose the control input $\bar{u}_1$ and α_θ such that

$$
\begin{aligned}
b_{11}(\bar{u}_1\cos(\alpha_\theta) - \bar{u}_{1r}) + b_{12}\bar{u}_1\sin(\alpha_\theta) + f_z &= -k_4\bar{z}_{2e}/\Delta_3,\\
b_{21}(\bar{u}_1\cos(\alpha_\theta) - \bar{u}_{1r}) + b_{22}\bar{u}_1\sin(\alpha_\theta) + f_w &= -k_5\bar{w}_{2e}/\Delta_3,
\end{aligned}
\tag{14.122}
$$

where $\Delta_3 = \sqrt{1 + \bar{z}_{2e}^2 + \bar{w}_{2e}^2}$, k_4 and k_5 are positive constants to be selected later. Solving (14.122) results in

$$
\begin{aligned}
\alpha_\theta &= \arctan(\bar{u}_{12}/(\bar{u}_{1r} + \bar{u}_{11})),\\
\bar{u}_1 &= (\bar{u}_{1r} + \bar{u}_{11})\cos(\alpha_\theta) + \bar{u}_{12}\sin(\alpha_\theta),
\end{aligned}
\tag{14.123}
$$

where

$$
\begin{aligned}
\bar{u}_{11} &= \frac{1}{b_{11}b_{22} - b_{12}b_{21}}\left(-\left(\tfrac{k_4\bar{z}_{2e}}{\Delta_3} + f_z\right)b_{22} + \left(\tfrac{k_5\bar{w}_{2e}}{\Delta_3} + f_w\right)b_{12}\right),\\
\bar{u}_{12} &= \frac{1}{b_{11}b_{22} - b_{12}b_{21}}\left(\left(\tfrac{k_4\bar{z}_{2e}}{\Delta_3} + f_z\right)b_{21} - \left(\tfrac{k_5\bar{w}_{2e}}{\Delta_3} + f_w\right)b_{11}\right).
\end{aligned}
\tag{14.124}
$$

From expressions of b_{ij}, $i = 1,2$, $j = 1,2$, given in (14.118), we have

$$b_{11}b_{22}-b_{12}b_{21}=-1+\frac{k_1z_{1e}z_{2e}}{\Delta_2^3}+\frac{k_2w_{1e}w_{2e}}{\Delta_2^3}-\frac{2k_1k_2z_{1e}z_{2e}w_{1e}w_{2e}}{\Delta_2^6}. \quad (14.125)$$

Therefore, there is no singularity in (14.124), if k_1 and k_2 are chosen such that

$$k_1+k_2+2k_1k_2<1. \quad (14.126)$$

Moreover, the virtual control α_θ in (14.123) is well defined if

$$\bar{u}_{1r}+\bar{u}_{11}>0. \quad (14.127)$$

This condition, as a simple calculation shows, is equivalent to

$$u_{1r}^*-\left(\frac{\omega_r^{\max}(k_1+k_2)(1+k_1+k_2)+(1+k_2)(3k_1+1/k_3+k_4)}{1-(k_1+k_2+2k_1k_2)}+\frac{k_1(k_5+1/k_3+3k_2)}{1-(k_1+k_2+2k_1k_2)}\right)>0. \quad (14.128)$$

Indeed, there always exist positive constants k_i, $1\le i\le 5$ such that (14.128) holds. Also it can be shown from (14.123) and (14.124) that $\bar{u}_1$ is bounded by some positive constant, which can be easily calculated from (14.123) and (14.124). Note that under the conditions (14.126) and (14.128), α_θ is a smooth function of z_{1e}, z_{2e}, w_{1e}, w_{2e}, ω_r, and $\bar{u}_{1r}$. Substituting (14.123) into (14.117) yields

$$\dot{V}_{12}=-\frac{k_1z_{1e}^2+k_2w_{1e}^2}{\Delta_1\Delta_2}-\frac{k_3(k_4\bar{z}_{2e}^2+k_5\bar{w}_{2e}^2)}{\Delta_3}+g_\theta+O_{12}. \quad (14.129)$$

Step 2

In this step, we design u_2 to stabilize the (θ_e,ω_e)-system. Define

$$\bar{\omega}_e=\omega_e-\alpha_\omega, \quad (14.130)$$

where α_ω is a virtual control of ω_e. To design α_ω, we take the Lyapunov function

$$V_{21}=V_{12}+0.5\bar{\theta}_e^2. \quad (14.131)$$

By differentiating both sides of (14.131) along the solutions of (14.129) and the fifth equation of (14.107), and using (14.130), the virtual control α_ω is chosen as

$$\alpha_\omega = -k_6\bar{\theta}_e-\frac{g_\theta}{\bar{\theta}_e}+\frac{\partial\alpha_\theta}{\partial z_{1e}}(z_{2e}-w_{1e}\omega_r)+\frac{\partial\alpha_\theta}{\partial z_{2e}}(\bar{u}_1\cos(\theta_e)-\bar{u}_{1r}-w_{2e}\omega_r)+\frac{\partial\alpha_\theta}{\partial w_{1e}}(w_{2e}+z_{1e}\omega_r)+\frac{\partial\alpha_\theta}{\partial w_{2e}}(-\bar{u}_1\sin(\theta_e)+z_{2e}\omega_r)+\frac{\partial\alpha_\theta}{\partial\omega_r}\dot{\omega}_r+\frac{\partial\alpha_\theta}{\partial\bar{u}_{1r}}\dot{\bar{u}}_{1r}, \quad (14.132)$$

where k_6 is a positive constant. Note that $g_\theta/\bar{\theta}_e$ with g_θ being given in (14.120) is well defined because $\frac{\sin(\bar{\theta}_e)}{\bar{\theta}_e} = \int_0^1 \cos(\bar{\theta}_e\lambda)\mathrm{d}\lambda$ and $\frac{(\cos(\bar{\theta}_e)-1)}{\bar{\theta}_e} = -\int_0^1 \sin(\bar{\theta}_e\lambda)\mathrm{d}\lambda$ are smooth functions of $\bar{\theta}_e$. Hence α_ω is a smooth function of z_{1e}, z_{2e}, w_{1e}, w_{2e}, θ_e, ω_r, $\dot{\omega}_r$, $\bar{u}_{1r}$ and $\dot{\bar{u}}_{1r}$. Note that at this stage, we do not need a nonlinear damping term for the observer effects since O_{21}, see (14.134), is Lipschitz in $\bar{\theta}_e\omega_e$ and we wish to design a weak nonlinear control law of α_ω.

Differentiating both sides of (14.131) along the solutions of (14.129) and the fifth equation of (14.107), and using (14.130), (14.132) yields

$$\dot{V}_{21} = -\frac{k_1 z_{1e}^2 + k_2 w_{1e}^2}{\Delta_1\Delta_2} - \frac{k_3(k_4\bar{z}_{2e}^2 + k_5\bar{w}_{2e}^2)}{\Delta_3} - k_6\bar{\theta}_e^2 + \bar{\theta}_e\bar{\omega}_e + O_{21}, \quad (14.133)$$

where the term O_{21}, which contains the observer error as a factor, is defined as

$$\begin{aligned} O_{21} = \; & \bar{\theta}_e\bigg[\tilde{\omega} + \frac{\partial\alpha_\theta}{\partial z_{1e}}O_{z1} + \frac{\partial\alpha_\theta}{\partial w_{1e}}O_{w1} + \frac{\partial\alpha_\theta}{\partial z_{2e}}(-\varepsilon\cos(\theta_e)\omega_e\tilde{\omega} + O_{z2}) + \\ & \frac{\partial\alpha_\theta}{\partial w_{2e}}(\varepsilon\sin(\theta_e)\omega_e\tilde{\omega} + O_{w2})\bigg] + O_{12}. \end{aligned} \quad (14.134)$$

We are now ready to design the control input u_2 by considering the Lyapunov function

$$V_{22} = V_{21} + \frac{1}{2}\bar{\omega}_e^2. \quad (14.135)$$

By differentiating both sides of (14.135) along the solutions of (14.133) and the sixth equation of (14.107), and using (14.130) and (14.132), the control input u_2 is chosen as

$$\begin{aligned} u_2 = \; & -k_7\bar{\omega}_e - \bar{\theta}_e + u_{2r} + \frac{\partial\alpha_\omega}{\partial\omega_r}\dot{\omega}_r + \frac{\partial\alpha_\omega}{\partial\dot{\omega}_r}\ddot{\omega}_r + \frac{\partial\alpha_\omega}{\partial\bar{u}_{1r}}\dot{\bar{u}}_{1r} + \\ & \frac{\partial\alpha_\omega}{\partial\dot{\bar{u}}_{1r}}\ddot{\bar{u}}_{1r} + \frac{\partial\alpha_\omega}{\partial z_{1e}}(z_{2e} - w_{1e}\omega_r) + \frac{\partial\alpha_\omega}{\partial z_{2e}}(\bar{u}_1\cos(\theta_e) - \bar{u}_{1r} - w_{2e}\omega_r) + \\ & \frac{\partial\alpha_\omega}{\partial w_{1e}}(w_{2e} + z_{1e}\omega_r) + \frac{\partial\alpha_\omega}{\partial w_{2e}}(-\bar{u}_1\sin(\theta_e) + z_{2e}\omega_r) + \frac{\partial\alpha_\omega}{\partial\theta_e}\omega_e + u_{2\mathrm{dam}}, \end{aligned} \quad (14.136)$$

where k_7 is a positive constant and the damping term $u_{2\mathrm{dam}}$ to take care of observer error effects is given by

$$\begin{aligned} u_{2dam} = \; & -\delta\bigg[\left(\frac{\partial\alpha_\omega}{\partial z_{1e}}\right)^2 + \left(\frac{\partial\alpha_\omega}{\partial w_{1e}}\right)^2 + \left(\frac{\partial\alpha_\omega}{\partial z_{2e}}\right)^2(\omega_e^2 + 1) + \\ & \left(\frac{\partial\alpha_\omega}{\partial w_{2e}}\right)^2(\omega_e^2 + 1) + \left(\frac{\partial\alpha_\omega}{\partial\theta_e}\right)^2\bigg]\bar{\omega}_e, \end{aligned} \quad (14.137)$$

where δ is a positive constant. Differentiating both sides of (14.135) along the solutions of (14.133) and the sixth equation of (14.107), and using (14.130), (14.132), and (14.136) yields

$$\begin{aligned}\dot{V}_{22} = & -\frac{k_1 z_{1e}^2 + k_2 w_{1e}^2}{\Delta_1 \Delta_2} - \frac{k_3(k_4 \bar{z}_{2e}^2 + k_5 \bar{w}_{2e}^2)}{\Delta_3} - \\ & k_6 \bar{\theta}_e^2 - k_7 \bar{\omega}_e^2 + u_{2\text{dam}} \bar{\omega}_e + O_{22},\end{aligned} \tag{14.138}$$

where the term O_{22} containing the observer error as a factor, is defined as

$$\begin{aligned}O_{22} = \; & \bar{\omega}_e \left[k_{32} \tilde{\theta} + \frac{\partial \alpha_\omega}{\partial z_{1e}} O_{z1} + \frac{\partial \alpha_\omega}{\partial w_{1e}} O_{w1} + \frac{\partial \alpha_\omega}{\partial z_{2e}} (-\varepsilon \cos(\theta_e) \omega_e \tilde{\omega} + O_{z2}) + \right. \\ & \left. \frac{\partial \alpha_\omega}{\partial w_{2e}} (\varepsilon \sin(\theta_e) \omega_e \tilde{\omega} + O_{w2}) + \frac{\partial \alpha_\omega}{\partial \theta_e} \tilde{\omega} \right] + O_{21}.\end{aligned} \tag{14.139}$$

By replacing ω_e by $\bar{\omega}_e + \alpha_\omega$ in all of the terms O_{12}, O_{21}, and O_{22}, and noting that $\bar{u}_1$ is bounded by some positive constant, after a lengthy but simple calculation by completing squares, it is readily shown that

$$\begin{aligned}\dot{V}_{22} \leq \; & -\frac{k_1 z_{1e}^2 + k_2 w_{1e}^2}{\Delta_1 \Delta_2} - \frac{k_3(k_4 \bar{z}_{2e}^2 + k_5 \bar{w}_{2e}^2)}{\Delta_3} - k_6 \bar{\theta}_e^2 - k_7 \bar{\omega}_e^2 + \\ & (\chi_1(\cdot) V_{22} + \chi_2(\cdot)) e^{-\sigma_0 (t - t_0)} \\ & \leq (\chi_1(\cdot) V_{22} + \chi_2(\cdot)) e^{-\sigma_0 (t - t_0)},\end{aligned} \tag{14.140}$$

where $\chi_1(\cdot)$ and $\chi_2(\cdot)$ are some class-K functions of $\left\| \tilde{X}(t_0) \right\|$. To prove convergence of $(z_{1e}, w_{1e}, \bar{z}_{2e}, \bar{w}_{2e}, \bar{\theta}_e, \bar{\omega}_e)$ to zero based on (14.140), we need the following lemma.

Lemma 14.1. *Consider the following first-order scalar differential equation*

$$\dot{x} = (ax + b) e^{-c(t - t_0)}, \; \forall t \geq t_0 \geq 0, \tag{14.141}$$

where $a \geq 0, b \geq 0$ and $c > 0$ are constants. The solution of (14.141) is bounded and satisfies

$$|x(t)| \leq |x(t_0)| e^{a/c} + b a^{-1} \left(e^{a/c} - 1 \right) := \pi \left(|x(t_0)| \right), \; \forall t \geq t_0 \geq 0. \tag{14.142}$$

Proof. Solving (14.141) for x then taking the norm of x gives (14.142). □

Applying the above lemma to the second inequality of (14.140), we have

$$V_{22}(t) \leq \pi_{22} \left(\left\| (\tilde{X}(t_0), z_{1e}(t_0), w_{1e}(t_0), \bar{z}_{2e}(t_0), \bar{w}_{2e}(t_0), \bar{\theta}_e(t_0), \bar{\omega}_e(t_0)) \right\| \right),$$

with $\pi_{22}(\cdot)$ being calculated from the above lemma. This implies that z_{2e}, w_{2e}, θ_e, ω_e, $\bar{u}_1$, and u_2 are also bounded for all $t \in [0, \infty)$. With this observation in mind, the first line of (14.140) yields

$$\dot{V}_{22} \leq -\frac{k_1 z_{1e}^2 + k_2 w_{1e}^2}{\Delta_{1m}\Delta_{2m}} - \frac{k_3(k_4 \bar{z}_{2e}^2 + k_5 \bar{w}_{2e}^2)}{\Delta_{3m}} - k_6 \bar{\theta}_e^2 - k_7 \bar{\omega}_e^2 + (\chi_1(\cdot)\pi_{22}(\cdot) + \chi_2(\cdot))e^{-\sigma_0(t-t_0)}, \quad (14.143)$$

where Δ_{im}, $i = 1, 2, 3$ are Δ_i with the arguments being replaced by their upper bounds. Applying Babarlat's lemma to (14.143), it is readily shown that

$$\lim_{t\to\infty} \left(z_{1e}(t), w_{1e}(t), \bar{z}_{2e}(t), \bar{w}_{2e}(t), \bar{\theta}_e(t), \bar{\omega}_e(t)\right) = 0. \quad (14.144)$$

By construction, see (14.109), (14.112), (14.115) and (14.123), the limit (14.144) implies that

$$\lim_{t\to\infty} (z_{1e}(t), w_{1e}(t), z_{2e}(t), w_{2e}(t), \theta_e(t), \omega_e(t)) = 0.$$

We now summarize the main result of this section in the following theorem.

Theorem 14.2. *Under Assumption 14.3, the dynamic output feedback control law consisting of (14.99), (14.123), and (14.136) forces the aircraft (14.92) to globally asymptotically track the reference model (14.93) if the design constants k_i, $1 \leq i \leq 5$, are chosen such that (14.126) and (14.128) hold.*

14.2.5 Simulations

In this section, we perform a numerical simulation to illustrate the effectiveness of the proposed controller with $\varepsilon = 0.8$. The observer and control gains are chosen as: $k_{11} = k_{21} = k_{31} = 1$, $k_{12} = k_{22} = k_{32} = 2$, $k_1 = k_2 = 0.15$, $k_3 = 5$, $k_4 = k_5 = 0.4$, $k_6 = k_7 = 5$, $\delta = 0.1$, and initial conditions are $x_1(0) = y_1(0) = 5$, $\theta(0) = 0.2, x_2(0) = y_2(0) = \omega(0) = 0.2$, $\hat{x}_1(0) = \hat{y}_1(0) = 3$, $\hat{\theta}(0) = 0.1$, $x_{1r}(0) = y_{1r}(0) = \theta_r(0) = x_{2r}(0) = y_{2r}(0) = \omega_r(0) = 0$, $\hat{x}_2(0) = \hat{y}_2(0) = \hat{\omega}(0) = 0$. The goal is to force the aircraft to track a sinusoid signal of $5(\sin(0.1t) + 1.2)$ in the vertical plane generated by (14.93). It is shown that there exists u_{1r} for this case with $u_{1r}^* > 2$ (see Assumption 14.3) and that conditions (14.126) and (14.128) hold. The simulation results are plotted in Figures 14.9, 14.10, and 14.11. It is seen that the proposed controller forces the aircraft to track the reference trajectory very well. The oscillation in the control input u_2 is due to the initial conditions and the above selection of the control and observer gains. By further tuning these gains, we can improve the transient response. In general, higher control and observer gains result in a shorter transient response time but more overshoot in the velocities and control inputs of the aircraft.

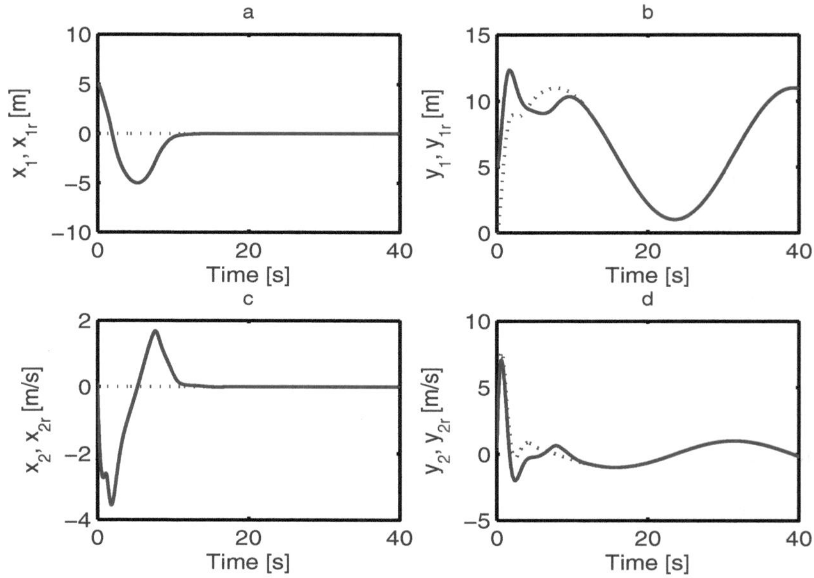

Figure 14.9 Reference (*dotted line*) and real (*solid line*) position trajectories: **a**. Position in x direction; **b**. Position in y direction; **c**. Velocity in x direction; **d**. Velocity in y direction

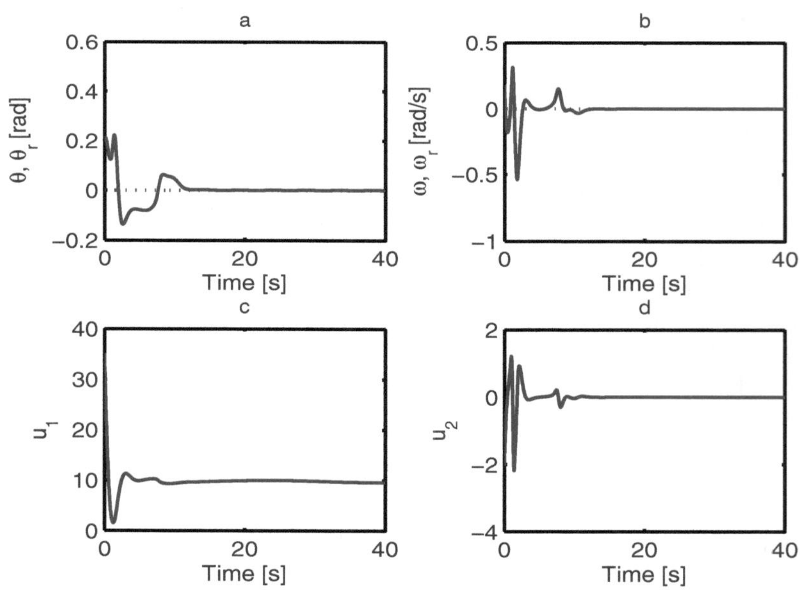

Figure 14.10 Reference (*dotted line*) and real (*solid line*) angular trajectories, and control inputs: **a**. Angles; **b**. Angular velocities; **c**. Control input u_1; **d**. Control input u_2

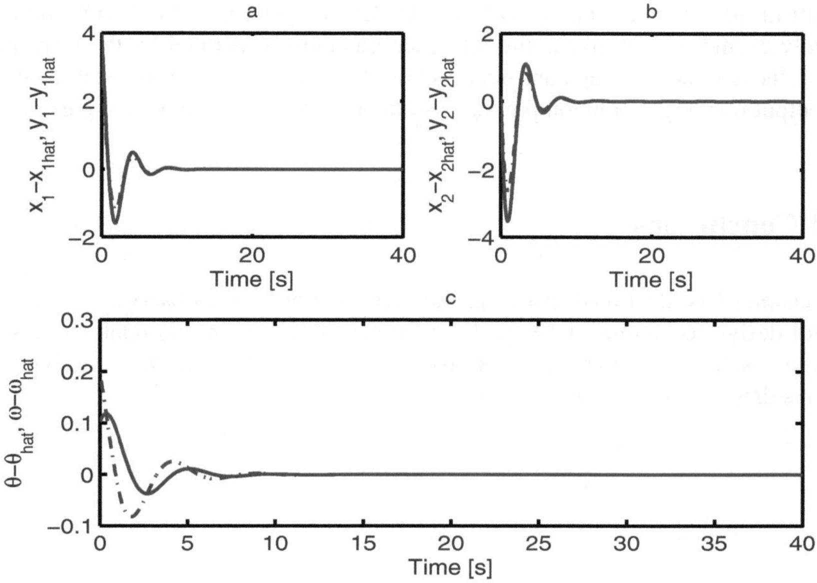

Figure 14.11 Observer errors: **a.** Errors $x_1 - \hat{x}_1$ (*solid line*) and $y_1 - \hat{y}_1$ (*dash-dotted line*); **b.** Errors $x_2 - \hat{x}_2$ (*solid line*) and $y_2 - \hat{y}_2$ (*dash-dotted line*); **c.** Errors $\theta - \hat{\theta}$ (*solid line*) and $\omega - \hat{\omega}$ (*dash-dotted line*)

14.2.6 Notes and References

The main difficulty in controlling VTOL aircraft is that they are underactuated and nonminimum phase. An approximate input–output linearization approach was used in [1, 153, 157–159] to develop a controller for stabilization and output tracking/regulation of a VTOL aircraft. In these papers, the controller was initially designed by ignoring the coupling between roll moment and thrust. The controller parameters were then selected to take into account the effects of the coupling. In [154], by noting that the output at a fixed point with respect to the aircraft body (Huygens center of oscillation) can be used, an interesting approach was introduced to design an output tracking controller. However, the proposed controller was not defined in the whole space. A simple approach was developed in [160] to provide a global controller for the stabilization of a VTOL aircraft. An optimal controller was provided in [6] for robust hovering control of a VTOL aircraft. In [161], dynamic inversion and robust control techniques were used to deal with the nonminimum phase dynamics. However, this approach imposed restrictions on the desired reference trajectories. Recently, a dynamic high-gain approach was used in [162] to design a controller to force the VTOL aircraft to globally practically track a reference trajectory generated by a reference model. In all of the aforementioned papers, all of the VTOL aircraft states are required for feedback. The output feedback tracking controller for a VTOL aircraft in this section is based on [163]. Under this controller, the VTOL aircraft globally asymptotically tracks the reference trajectory generated by the reference model. Indeed, the tracking controller proposed in this section covered stabilization and output tracking/regulation problems studied in the above-mentioned papers.

14.3 Conclusions

This chapter has illustrated that a careful investigation of the observer design and control design techniques developed for the ocean vessels in this book can result in various solutions of many control problems for other underactuated mechanical systems that are common in practice.

Chapter 15
Conclusions and Perspectives

15.1 Summary of the Book

This monograph has concentrated on the control of underactuated ocean vessels including ships and underwater vehicles. These vessels have more degrees of freedom to be controlled than the number of independent control inputs. Ships do not have an independent sway actuator while for underwater vehicles there are no independent sway and heave actuators. The book started with a review of the necessary background on ocean vessel dynamics and nonlinear control theory. The authors then demonstrated a systematic approach based on various nontrivial coordinate transformations together with advanced nonlinear control design methodologies founded on the basis of Lyapunov's direct method, backstepping, and parameter projection techniques for the development and analysis of a number of ocean vessel control systems to achieve advanced motion control tasks. These tasks include stabilization, trajectory-tracking, path-tracking, and path-following. The book has offered new knowledge regarding the nonlinear control of underactuated ocean vessels, efficient controllers for practical control of underactuated ocean vessels, and general methods and strategies to solve nonlinear control problems of other underactuated systems, including underactuated land and aerial vehicles. Numerical simulations and real-time implementations of the designed control systems on a scaled model ship have been included to illustrate the effectiveness and practical guidance of the control systems developed in the book. As an illustration of the constructive feature of the book, the observer design, control design, and stability analysis techniques developed for underactuated ocean vessels have been applied to control of other underactuated mechanical systems including mobile robots and VTOL aircraft.

The book consists of 15 Chapters. Chapter 1 presented a brief review of the development in nonlinear control theory and its applications, and difficulties in the control of underactuated ocean vessels. The main contributions of the book are organized into five parts.

In Part I (Chapter 2), various mathematical tools such as Barbalat-like lemmas, Lyapunov stability theory, and the backstepping technique were presented for control design and stability analysis of controlled systems designed in the book.

Part II consists of Chapters 3 and 4. This part addressed modeling, motion control tasks, and control properties of ocean vessels. The existing literature on the control of underactuated ocean vessels was also reviewed in this part. This review motivated contributions of the book on new solutions for the motion control of underactuated ships and underwater vehicles.

Part III consists of Chapters 5, 6, 7, 8, 9, 10, and 11. This part focused on underactuated surface ships by proposing a number of solutions for various control problems including stabilization, trajectory-tracking, path-tracking and path-following. Chapter 5 addressed the problem of trajectory-tracking control of underactuated ships. These ships do not have independent actuators in the sway axis. The reference trajectory was generated by a suitable virtual ship. The control development was based on an elegant coordinate transformation, Lyapunov's direct method, and the backstepping technique, and utilized passivity properties of ship dynamics and their interconnected structure. Chapter 6 examined the problem of designing a single controller that achieved stabilization and trajectory-tracking simultaneously for underactuated ships. In comparison with the preceding chapter, a path approaching the origin and a set-point were included in the reference trajectory, that is, stabilization/regulation was also considered. The control development was based on several special coordinate transformations plus the techniques in the preceding chapter. Chapter 7 presented global partial-state feedback and output feedback control schemes for trajectory-tracking control of underactuated ships. For the case of partial-state feedback, measurements of the ship sway and surge velocities were not needed, while for the case of output feedback, no ship's velocities were required for feedback. Global nonlinear coordinate changes were introduced to transform the ship dynamics to a system affine in the ship velocities. This affine form allowed us to design observers that globally exponentially estimate unmeasured velocities. These observers plus the techniques in Chapters 4 and 5 facilitated the development of the controllers. Chapter 8 dealt with the problem of path-tracking control of underactuated ships. In comparison with Chapters 5, 6, and 7, the requirement of the reference trajectory generated by a suitable virtual ship was relaxed. Both full state feedback and output feedback cases were considered. The control development was based on a series of nontrivial coordinate transformations plus the techniques in the previous chapters. Chapter 9 addressed the problem of way-point tracking control of underactuated ships. Both full state feedback and output feedback controllers were designed. The controllers in this chapter could be regarded as an advanced version of the conventional course-keeping controllers in the sense that in addition to maintaining the desired heading, both the sway displacement (lateral distance) and sway velocity were controlled. Chapter 10 developed full state feedback and output feedback controllers that forced underactuated ships to follow a predefined path. The control development was motivated by an observation that it is practical to steer a vessel such that the vessel is on the reference path and its total velocity is tangent to the reference path, and that the vessel forward speed is controlled separately by the main thruster control system. The techniques in the previous chapters plus the use of the Serret–Frenet frame facilitated the results. Chapter 11 presented a different approach from Chapter 10 to solve a path-following problem for underactuated ships. Unlike Chapter 10, the control development was based on the method of generating reference paths by the helmsman. The path-following errors were first interpreted in polar coordinates, then the techniques developed in the previous chapters were used to design path-following controllers. Interestingly, some situations of practical importance such as parking and point-to-point navigation were covered in this chapter as a by-product of the developed path-following system.

Part IV consists of Chapters 12 and 13. This part focused on underactuated underwater vehicles. Chapter 12 addressed the problem of trajectory-tracking control of underactuated underwater vehicles. These vehicles do not have independent actuators in the sway and heave axes. The controller development was built on the techniques developed for underactuated ships in Chapters 5, 6, and 7. Due to complex dynamics of the underwater vehicles in comparison with that of the ships, the control design and stability analysis required more complicated and coordinate transformations and control design than those already developed for underactuated ships in Chapters 5, 6, and 7. Chapter 13 extended the results of Chapter 11 to the design of a path-following control system for underactuated underwater vehicles. A series of path-following strategies for the vehicles was first discussed. A practical approach was then chosen to facilitate the control development. We also addressed the parking and point-to-point navigation problems for the vehicles in this chapter.

Part V (Chapter 14) presented several applications of the observer and control design techniques developed in the previous chapters to other underactuated mechanical systems including mobile robots, and VTOL aircraft.

Finally this chapter concludes the book by briefly summarizing the main results presented in the previous chapters and presenting related open problems for further investigation.

15.2 Perspectives and Open Problems

Control of underactuated ocean vessels in particular and of underactuated mechanical systems in general is far from being well completed. There are in fact numerous important problems, which are still open to further investigation. In the previous chapters, we provided a number of control objectives and their solutions. However, this book has not covered all aspects of dynamics and control of the ocean vessels. Here, we highlight some open problems with detailed discussion. The problems are mainly chosen based on the following two criteria. First, to our knowledge these problems are still open and have not been solved in the literature. Second, we think that these problems potentially have important impacts on the development of control theory for underactuated systems and significant applications in practice. These criteria are highly subjective and mainly reflect our personal experience and points of view. However, the discussions are usually made in an elementary and intuitive manner.

15.2.1 *Further Issues on Control of Single Underactuated Ocean Vessels*

15.2.1.1 Theoretical issues

Robust and Adaptive Issues. Although the trajectory-tracking controllers in this book possess a certain robustness, more research is needed to design such a controller that explicitly takes the environmental disturbances into account. Disturbances entering the actuated dynamics (e.g., surge and yaw for ships) can be easily taken care of. However, disturbances that enter the unactuated dynamics (e.g., sway for ships) are more difficult to handle. A possible way to reject the constant part of these disturbances on the unactuated dynamics is to introduce a small angle to the heading error to compensate for the disturbances.

Since the controllers in the book require exact knowledge of the system parameters, it is of interest to examine the problem of entirely unknown vessel parameters. As opposed to the approach based on time-varying linear system stability, the proposed control designs in this book generate explicit Lyapunov functions. This feature can be further exploited in conjunction with adaptive control algorithms in the literature to address the adaptive tracking problem.

Output Feedback Issues. The global coordinate transformations in the book bring the ship and mobile robot systems to an affine form for which a global exponential nonlinear observer can be easily designed. It is worth investigating the possibility of extending this type of transformation to a certain class of Lagrange systems. Indeed, the most difficult task is to obtain a solution of a set of partial differential equations, which are in general not easy to solve. Furthermore, an adaptation rule should be included in the observers to estimate the constant components (bias) of the environmental disturbances.

15.2.1.2 Real Application Issues

Although several controllers presented in this book have been successfully implemented on a scaled model ship, the following problems of technology transfer to real application studies should be considered:

1. assessment of benefits obtainable from implementing nonlinear controllers on real ocean-going vessels;
2. the issues of controller implementation in practice;
3. water-tank trials of controllers with scaled models;

4. sea-going trials: a full set of application trials;
5. commercial implementation.

15.2.2 Coordination Control of Multiple Underactuated Ocean Vessels

Coordination control involves controlling positions of a group of ocean vehicles such that they perform desired tasks such as optimizing objective functions from measurements taken by each vessel, and stabilization/tracking desired locations relative to reference point(s). Therefore from a reference trajectory/path point of view, we can divide coordination control of a network of vessels into two main classes: (1) Known reference trajectories in advance and (2) unknown reference trajectories in advance, for the agents in the network. Three popular approaches to the known reference trajectory in advance class are leader-following (e.g. [164, 165]), behavioral (e.g., [166, 167]), and use of virtual structures (e.g., [168, 169]). Most research work investigating formation control utilize one or more of these approaches in either a centralized or a decentralized manner. Centralized schemes ([165, 170, 171]) use a single controller that generates collision free trajectories in the workspace. Although these guarantee a complete solution, centralized schemes require high computational power and are not robust. Decentralized schemes (e.g., [57, 172]) require less computational effort, and are relatively more scalable to the team size. However, it is very difficult to predict and control the critical points. For the unknown reference trajectory in advance class, coordination control often involves optimizing objective functions to give reference trajectories for the agents in the network to track. An application of formation control based on the potential field method [165] and Lyapunov's direct method [173] to gradient climbing was addressed in [174]. Geometric formation based on Voronoi's partition optimization is given in [175]. Some other work belonging to this class includes [176] on geometric formation, [177, 178] on pattern formation, [57] on flocking, [179] on swarm aggregation, and [180] on deployment and task allocation. The main problem with this class is that the final arrangement of the agents cannot be foretold due to the fact that the aforementioned results use local optimization methods to optimize nonconvex objective functions. This means that only local results can be obtained. For both classes, the aforementioned works assume that the vehicles have very simple dynamics such as single or double integrators. Vehicle dynamics are usually complex in the sense of being nonlinear and are subject to disturbances. Therefore, motion coordination control systems and deployment algorithms for networked vehicles should consider these complex features of the vehicle dynamics. It is suggested to combine the control developments for the single underactuated ocean vessels in this book with the available coordination control algorithms to achieve the aforementioned coordination objectives. Indeed, this combination is not a trivial task. It requires a careful examination of the algorithms developed for single vessels and coordination objectives. The reader is referred to [181–185] for several coordination control results based on the control of single mobile robots in Chapter 14 of this book and the formation control results in [186].

References

1. P. Kokotovic and M. Arcak, "Constructive nonlinear control: a history perspective," *Automatica*, vol. 37, no. 5, pp. 637–662, 2001.
2. H. Nijmeijer and A.J. van der Schaft, *Nonlinear Dynamical Control Systems.* New York: Springer, 1990.
3. M. Krstic, I. Kanellakopoulos, and P. Kokotovic, *Nonlinear and Adaptive Control Design.* New York: Wiley, 1995.
4. A. Isidori, *Nonlinear Control Systems.* London: Springer, 3rd ed., 1995.
5. M.A. Kaashoek, J.H. van Schuppen, and A.C.M. Ran, *Robust Control of Linear Systems and Nonlinear Control.* Boston MA: Birkhauser, 1990.
6. H. Khalil, *Nonlinear Systems.* Englewood Cliffs, NJ: Prentice Hall, 2002.
7. R. Sepulchre, M. Jankovic, and P. Kokotovic, *Constructive Nonlinear Control.* New York: Springer, 1997.
8. R. Marino and P. Tomei, *Nonlinear Control Design: Geometric, Adaptive and Robust.* New York: Prentice Hall, 1995.
9. G. Tao and P. Kokotovic, *Adaptive Control of Systems with Actuator and Sensor Nonlinearities.* New York: Wiley, 1995.
10. G. Tao, *Adaptive Control Design and Analysis.* John Wiley & Sons, 2003.
11. T.I. Fossen, *Marine Control Systems.* Trondheim, Norway: Marine Cybernetics, 2002.
12. T.I. Fossen, *Guidance and Control of Ocean Vehicles.* New York: Wiley, 1994.
13. N. Xiros, *Robust Control of Diesel Ship Propulsion.* New York: Springer, 2002.
14. H. Nijmeijer and T.I. Fossen, *New Directions in Nonlinear Observer Design.* London: Springer, 1999.
15. C.d.W. Canudas, B. Siciliano, and G. Bastin, *Theory of Robot Control.* London: Springer, 1996.
16. R. Ortega and W. C. Rheinboldt, *Solutions of Nonlinear Equations in Several Variables.* New York: Academic Press, 1970.
17. E. Panteley and A. Loria, "On global uniform asymptotic stability of nonlinear time-varying non-autonomous systems in cascade," *Systems and Control Letters*, vol. 33, no. 2, pp. 131–138, 1998.
18. Z.P. Jiang and H. Nijmeijer, "Tracking control of mobile robots: a case study in backstepping," *Automatica*, vol. 33, no. 7, pp. 1393–1399, 1997.
19. Z.P. Jiang, "Global tracking control of underactuated ships by Lyapunov's direct method," *Automatica*, vol. 38, no. 2, pp. 301–309, 2002.
20. W.S. Levine, *The Control Handbook.* Boca Raton, FL: CRC Press, 1996.
21. R.W. Brockett, "Asymptotic stability and feedback stabilization," in *Differential Geometric Control Theory* (R.W. Brockett, R.S. Millman, and H.J. Sussmann, eds.), pp. 181–191, Boston: Birkhauser, 1983.

22. O.J. Sordalen and O. Egeland, "Exponential stabilization of nonholonomic chained systems," *IEEE Transactions on Automatic Control*, vol. 40, no. 1, pp. 35–49, 1995.
23. J.P. Lasalle and S. Lefschetz, *Stability by Liapunov's Direct Method.* New York: Academic Press, 1961.
24. A.M. Lyapunov, *Stability of Motion.* New York: Academic Press, 1966.
25. E.M. Lewandowski, *The Dynamics of Marine Craft.* Singapore: World Scientific, 2004.
26. SNAME, The Society of Naval Architects and Marine Engineers. Nomenclature for Treating the Motion of a Submerged Body Through a Fluid. In: *Technical and Research Bulletin* No. 1-5. 1950.
27. O.M. Faltinsen, *Sea Loads on Ships and Offshore Structures.* Cambridge University Press, 1990.
28. W. Blendermann, "Parameter identification of wind loads on ships," *Journal of Wind Engineering and Industrial Aerodynamics*, vol. 51, no. 3, pp. 339–351, 1994.
29. K.Y. Wichlund, O.J. Sordalen, and O. Egeland, "Control properties of underactuated vehicles," *Proceedings of the IEEE International Conference on Robotics and Automation*, pp. 2009–2014, 1995.
30. J.T.-Y. Wen, *Control Handbook*, ch. Control of nonholonomic systems, pp. 1359–1368. Boca Raton, FL: CRC Press, 1996.
31. I. Kolmanovsky and N.H. McClamroch, "Developments in nonholonomic control problems," *IEEE Control Systems Magazine*, no. 6, pp. 20–36, 1995.
32. R. Murray and S. Sastry, "Nonholonomic motion planning: Steering using sinusoids," *IEEE Transactions on Automatic Control*, vol. 38, no. 5, pp. 700–716, 1993.
33. A. Bloch, M. Reyhanoglu, and N. McClamroch, "Control and stabilization of nonholonomic dynamic systems," *IEEE Transactions on Automatic Control*, vol. 37, no. 11, pp. 1746–1757, 1992.
34. C. Canudas and O. Sodalen, "Exponential stabilization of mobile robots with noholonomic constraints," *IEEE Transactions on Automatic Control*, vol. 37, no. 11, pp. 1791–1797, 1992.
35. A. Astolfi, "Discontinuous control of the brockett integrator," *European Journal of Control*, vol. 4, no. 1, pp. 49–63, 1998.
36. A. Astolfi and W. Schaufelberger, "State and output feedback stabilization of multiple chained systems with discontinuous control," *Systems and Control Letters*, vol. 32, no. 1, pp. 49–56, 1997.
37. A.M. Bloch and S. Drakunov, "Stabilization and tracking in the nonholonomic integrator via sliding mode," *Systems and Control Letters*, vol. 29, no. 2, pp. 91–99, 1996.
38. Z. Sun, S.S. Ge, W. Huo, and T.H. Lee, "Stabilization of nonholonomic chained systems via nonregular feedback linearization," *Systems and Control Letters*, vol. 44, no. 4, pp. 279–289, 2001.
39. J.P. Hespanha, D. Liberzon, and A. Morse, "Logic based switching control of a nonholonomic system with parametric modeling uncertainty," *Systems and Control Letters*, vol. 38, no. 3, pp. 167–177, 1999.
40. W. Dong and W. Huo, "Adaptive stabilization of uncertain dynamic nonholonomic systems," *International Journal of Control*, vol. 72, no. 18, pp. 1689–1700, 2000.
41. B. dAndrea-Novel, G. Bastin, and G. Campion, "Control of nonholonomic wheeled mobile robots by state feedback linearization," *International Journal of Robotics Research*, vol. 14, no. 6, pp. 543–559, 1995.
42. A.D. Luca and M.D.D. Benedetto, "Control of nonholonomic systems via dynamic compensation," *Kybernetica*, vol. 29, no. 6, pp. 593–608, 1993.
43. C. Samson and K. Ait-Abderrahimn, "Feedback control of a nonholonomic wheeled cart in cartesian space," *Proceedings of 1991 IEEE International Conference on Robotics and Automatation*, Sacramento, CA, pp. 1136–1141, 1991.
44. C. Samson, "Velocity and torque feedback control of a nonholonomic cart," *In Advanced Robot Control*, C. Canudas de Wit (ed.), *Lecture Notes in Control and Information Sciences*, Berlin: Springer, pp. 125–151, 1991.

45. R. McCloskey and R. Murray, "Exponential stabilization of driftless nonlinear control systems using homogeneous feedback," *IEEE Transactions on Automatic Control*, vol. 42, no. 5, pp. 614–628, 1997.
46. R. Colbaugh and K. Glass, "Learning control for nonholonomic mechanical systems," *NOLCOS'98*, pp. 771–776, 1998.
47. T.C. Lee, K.T. Song, C.H. Lee, and C.C. Teng, "Tracking control of unicycle-modelled mobile robots using a saturation feedback controller," *IEEE Transactions on Control Systems Technology*, vol. 9, no. 2, pp. 305–318, 2001.
48. Z.P. Jiang, "Iterative design of time-varying stabilizers for multi-input systems in chained form," *Systems Control Letters*, vol. 28, no. 5, pp. 255–262, 1996.
49. V.I. Arnold and S. Novikov, *Dynamical Systems VII*. Berlin: Springer, 1994.
50. Z. Qu, *Robust Control of Nonlinear Systems*. New York: Wiley, 1998.
51. Z.P. Jiang, "Robust exponential regulation of nonholonomic systems with uncertainties," *Automatica*, vol. 36, no. 2, pp. 189–209, 2000.
52. K.D. Do and J. Pan, "Adaptive global stabilization of nonholonomic systems with strong nonlinear drifts," *Systems and Control Letters*, vol. 46, no. 3, pp. 195–205, 2002.
53. K.Y. Pettersen and O. Egeland, "Exponential stabilization of an underactuated autonomous surface vessel," *Proceedings of 35th IEEE Conference on Decision and Control*, pp. 967–971, 1996.
54. K.Y. Pettersen, *Exponential Stabilization of Underactuated Vehicles, PhD thesis*. Trondheim, Norway: Norwegian University of Science Technology, 1996.
55. M. Reyhanoglu, "Exponential stabilization of an underactuated autonomous surface vessel," *Automatica*, vol. 33, no. 12, pp. 2249–2254, 1997.
56. K.Y. Pettersen and T.I. Fossen, "Underactuated dynamic positioning of a ship-experimental results," *IEEE Transactions on Control Systems Technology*, vol. 8, no. 5, pp. 856–863, 2000.
57. R. Olfati-Saber, "Flocking for multi-agent dynamic systems: algorithms and theory," *IEEE Transactions on Automatic Control*, vol. 51, no. 3, pp. 401–420, 2006.
58. A.P. Aguiar and A.M. Pascoal, "Regulation of a nonholonomic autonomous underwater vehicle with parametric modeling uncertainty using Lyapunov functions," *Proceedings of 40th IEEE Conference on Decision and Control*, vol. 40, pp. 4178–4183, 2001.
59. K.Y. Pettersen and H. Nijmeijer, "Global practical stabilization and tracking for an underactuated ship-a combined averaging and backstepping approach," *Proceedings of IFAC Conference on Systems Structure Control*, pp. 59–64, 1998.
60. M. Aicardi, G. Casalino, A. Bicchi, and A. Balestrino, "Closed loop steering of unicycle-like vehicles via Lyapunov technique," *IEEE Robotics and Automation Magazine*, vol. 2, no. 1, pp. 27–35, 1995.
61. F. Mazenc, K.Y. Pettersen, and H. Nijmeijer, "Global uniform asymptotic stabilization of an underactuated surface vessel," *IEEE Transactions on Automatic Control*, vol. 47, no. 10, pp. 1759–1762, 2002.
62. N.E. Leonard, "Control synthesis and adaptation for an underactuated autonomous underwater vehicle," *IEEE Journal of Oceanic Engineering*, vol. 20, no. 2, pp. 211–220, 1995.
63. N.E. Leonard, "Periodic forcing, dynamics and control of underactuated spacecraft and underwater vehicles," *Proceeding of 34th IEEE Conference on Decision and Control*, pp. 3980–3985, 1995.
64. O. Egeland, M. Dalsmo, and O. Sordalen, "Feedback control of a nonholonomic underwater vehicle with a constant desired configuration," *The International Journal of Robotics Research*, vol. 15, no. 1, pp. 24–35, 1996.
65. A. Astolfi, D. Chhabra, and R. Ortega, "Asymptotic stabilization of some equilibria of an underactuated underwater vehicle," *Systems and Control Letters*, vol. 45, no. 3, pp. 193–206, 2002.
66. R. Ortega, M.W. Spong, F. Gomez-Estern, and G. Blankenstein, "Stabilization of a class of underactuated mechanical systems via interconnection and damping assignment," *Systems and Control Letters*, vol. 47, pp. 1218–1233, 2002.
67. F. Bullo, "Stabilization of relative equilibria for underactuated systems on Riemannian manifolds," *Automatica*, vol. 36, no. 12, pp. 1819–1834, 2000.

68. E. Lefeber, K.Y. Pettersen, and H. Nijmeijer, "Tracking control of an underactuated ship," *IEEE Transactions on Control Systems Technology*, vol. 11, no. 1, pp. 52–61, 2003.
69. E. Panteley and A. Loria, "Growth rate conditions for uniform asymptotic stability of cascaded time varying systems," *Automatica*, vol. 37, no. 3, pp. 453–460, 2001.
70. Z.P. Jiang and H. Nijmeijer, "A recursive technique for tracking control of nonholonomic systems in chained form," *IEEE Transactions on Automatic Control*, vol. 44, no. 2, pp. 265–279, 1999.
71. K.Y. Pettersen and H. Nijmeijer, "Underactuated ship tracking control: theory and experiments," *International Journal of Control*, vol. 74, no. 14, pp. 1435–1446, 2001.
72. V. Jurdjevic and J. Quinn, "Controllability and stability," *Journal of Differential Equations*, vol. 28, no. 3, pp. 381–389, 1979.
73. S. Gopalswamy and J. K. Hedrick, "Tracking nonlinear nonminimum phase systems using sliding mode control," *International Journal of Control*, vol. 57, no. 5, pp. 1141–1158, 1993.
74. R. Zhang, Y. Chen, Z. Sun, F. Sun, and H. Xu, "Path control of a surface ship in restricted waters using sliding mode," *IEEE Transactions on Control Systems Technology*, vol. 8, no. 4, pp. 722–732, 2000.
75. J.M. Godhavn, T.I. Fossen, and S. Berge, "Nonlinear and adaptive backstepping designs for tracking control of ships," *International Journal of Adaptive Control and Signal Processing*, vol. 12, no. 8, pp. 649–670, 1998.
76. R. Olfati-Saber, *Nonlinear Control of Underactuated Mechanical Systems with Application to Robotics and Aerospace Vehicles*. PhD. thesis, Massachusetts Institute of Technology, 2001.
77. A. Behal, D.M. Dawson, W.E. Dixon, and F. Yang, "Tracking and regulation control of an underactuated surface vessel with nonintegrable dynamics," *IEEE Transactions on Automatic Control*, vol. 47, no. 3, pp. 495–500, 2002.
78. W.E. Dixon, D.M. Dawson, and E. Zergeroglu, *Nonlinear Control of Wheeled Mobile Robots*. London: Springer, 2001.
79. G.J. Toussaint, T. Basar, and F. Bullo, "H_∞-optimal tracking control techniques for nonlinear underactuated systems," *Proceedings of the 39th IEEE Conference on Decision and Control*, pp. 2078–2083, 2000.
80. G.J. Toussaint, T. Basar, and F. Bullo, "Tracking for nonlinear underactuated surface vessels with generalized forces," *Proceedings of the 2000 IEEE International Conference on Control Applications*, pp. 355–360, 2000.
81. H. Sira-ramirez, "On the control of the underactuated ship: A trajectory planning approach," *Proceedings of the 38th Conference on Decision and Control*, pp. 2192–2197, 1999.
82. G.J. Toussaint, T. Basar, and F. Bullo, "Motion planning for nonlinear underactuated vehicles using techniques," *Proceedings of the American Control Conference*, pp. 4097–4102, 2001.
83. I. Kaminer, A. Pascoal, E. Hallberg, and C. Silvestre, "Trajectory tracking for autonomous vehicles: An integrated approach to guidance and control," *Journal of Guidance, Control, and Dynamic Systems*, vol. 21, no. 1, pp. 29–38, 1998.
84. A. Wahl and E. Gilles, "Model predictive versus linear quadratic control for the tracking problem of automatic river navigation," *Proceedings of ECC'99, European Control Conference*, vol. 37, 1999.
85. P. Encarnacao, A. Pacoal, and M. Arcak, "Path following for autonomous marine craft," *Proceedings of the 5th IFAC Conference on Manoeuvring and Control of Marine Craft*, pp. 117–122, 2000.
86. M. Aicardi, G. Casalino, G. Indiveri, A. Aguiar, P. Encarnacao, and A. Pascoal, "A plannar path following controller for underactuated marine vehicles," *Preprint*, 2001.
87. R. Skjetne and T.I. Fossen, "Nonlinear maneuvering and control of ships," *Proceedings of OCEANS 2001 MTS/IEEE Conference and Exhibition*, pp. 1808–1815, 2001.
88. M. Egerstedt, X. Hu, and A. Stotsky, "Control of mobile platforms using a virtual vehicle approach," *IEEE Transaction on Automatic Control*, vol. 46, no. 11, pp. 1777 –1782, 2001.
89. M. Egerstedt and X. Hu, "A hybrid control approach to action coordination for mobile robots," *Automatica*, vol. 38, no. 1, pp. 125–130, 2002.

90. C. Silvestre, A. Pascoal, and I. Kaminer, "On the design of gain-scheduling trajectory tracking controllers," *International Journal of Robust and Nonlinear Control*, vol. 12, no. 9, pp. 797–839, 2002.
91. T.I. Fossen and M. Blanke, "Nonlinear output feedback control of underwater vehicle propellers using feedback from estimated axial flow velocity," *IEEE Journal of Oceanic Engineering*, vol. 25, no. 2, pp. 241–255, 2000.
92. O.M. Aamo, M. Arcak, T.I. Fossen, and P. Kokotovic, "Global output tracking control of a class of Euler–Lagrange systems with monotonic nonlinearities in the velocities," *International Journal of Control*, vol. 74, no. 7, pp. 649–658, 2001.
93. A. Loria and K. Melhem, "Position feedback global tracking control of EL systems: A state transformation approach," *IEEE Transactions on Automatic Control*, vol. 47, no. 5, pp. 841–847, 2002.
94. K.D. Do, Z.P. Jiang, and J. Pan, "Output feedback tracking control of surface ships," *Proceedings of IFAC World Congress, Barcelona, Spain*, 2002.
95. H. Berghuis and H. Nijmeijer, "A passivity approach to controller-observer design for robots," *IEEE Transactions on Robotics and Automation*, vol. 9, no. 6, pp. 740–754, 1993.
96. S. Nicosia and P. Tomei, "Robot control by using only joint position measurement," *IEEE Transactions on Automatic Control*, vol. 35, no. 9, pp. 1058–1061, 1995.
97. F. Zhang, D.M. Dawson, M.S. Queiroz, and W. Dixon, "Global adaptive output feedback tracking control of robot manipulators," *IEEE Transactions on Automatic Control*, vol. 45, pp. 1203–1208, 2000.
98. F. Mazenc and A. Astolfi, "Robust output feedback stabilization of the angular velocity of a rigid body," *Systems and Control Letters*, vol. 39, no. 3, pp. 203–210, 2000.
99. W. Lohmiller and J.J.E. Slotine, "Control system design for mechanical systems using contraction theory," *IEEE Transactions on Automatic Control*, vol. 45, no. 5, pp. 984–989, 2000.
100. W. Lohmiller and J.J.E. Slotine, "On contraction analysis for nonlinear systems," *Automatica*, vol. 34, no. 6, pp. 683–696, 1998.
101. G. Besancon, "Global output feedback tracking control for a class of Lagrangian systems," *Automatica*, vol. 36, no. 12, pp. 1915–1921, 2000.
102. G. Besancon, S. Battilotti, and L. Lanari, "State transformation and global output feedback disturbance attenuation for a class of mechanical systems," *The 6th Mediterranean Control Conference, Italy*, vol. 36, pp. 516–566, 1998.
103. A. Loria, "Global tracking control of one-degree-of-freedom Euler Lagrange systems without velocity measurement," *European Journal of Control*, vol. 2, no. 2, pp. 144–151, 1996.
104. Z.P. Jiang and I. Kanellakopoulos, "Global output feedback tracking for a benchmark nonlinear system," *IEEE Transactions on Automatic Control*, vol. 45, no. 5, pp. 1023–1027, 2000.
105. K.D. Do, Z.P. Jiang, J. Pan, and H. Nijmeijer, "A global output-feedback controller for stabilization and tracking of underactuated odin:a spherical underwater vehicle," *Automatica*, vol. 40, no. 1, pp. 117–124, 2004.
106. E. Lefeber, *Tracking Control of Nonlinear Mechanical Systems, PhD thesis.* The Netherlands: University of Twente, 2000.
107. K.Y. Pettersen and E. Lefeber, "Way-point tracking control of ships," *Proceedings of the 40th IEEE Conference on Decision and Control*, pp. 940–945, 2001.
108. D.E. Kodistchek, "Adaptive techniques for mechanical systems," *Proceedings of the 5th Yale Workshop on Adaptive Systems*, pp. 259–265, 1987.
109. K.D. Do, Z.P. Jiang, and J. Pan, "Underactuated ship global tracking under relaxed conditions," *IEEE Transactions on Automatic Control*, vol. 47, no. 9, pp. 1529–1536, 2002.
110. K.D. Do, Z.P. Jiang, and J. Pan, "Global exponential tracking control of underactuated ships in the body frame," *Proceedings of American Control Conference*, pp. 4702–4707, 2002.
111. K.D. Do, Z.P. Jiang, and J. Pan, "Global tracking control of underactuated ships in the body frame," *Proceedings of IFAC World Congress*, pp. 367–372, 2002.
112. Z.P. Jiang and I. Mareels, "A small-gain control method for nonlinear cascaded systems with dynamic uncertainties," *IEEE Transactions on Automatic Control*, vol. 42, no. 3, pp. 296–308, 1997.

113. K. D. Do, Z. P. Jiang, and J. Pan, "Universal controllers for stabilization and tracking of underactuated ships," *Systems and Control Letters*, vol. 47, no. 4, pp. 299–317, 2002.
114. K.D. Do, Z.P. Jiang, J. Pan, and H. Nijmeijer, "Global output feedback universal controller for stabilization and tracking of underactuated ODIN-an underwater vehicle," *Proceedings of the 41st IEEE Conference on Decision and Control*, pp. 504–509, 2002.
115. K.D. Do, Z.P. Jiang, and J. Pan, "Universal saturation controller design for mobile robots," *Proceedings of the 41st IEEE Conference on Decision and Control*, pp. 2044–2049, 2002.
116. K.D. Do, Z.P. Jiang, and J. Pan, "Robust global output feedback stabilization of underactuated ships on a linear course," *Proceedings of the 41st IEEE Conference on Decision and Control*, pp. 1687–1692, 2002.
117. K.D. Do, Z.P. Jiang, and J. Pan, "Global partial-state feedback and output-feedback tracking controllers for underactuated ships," *Systems and Control Letters*, vol. 54, no. 10, pp. 1015–1036, 2005.
118. J.B. Pomet and L. Praly, "Adaptive nonlinear regulation: Estimation from the Lyapunov equation," *IEEE Transactions on Automatic Control*, vol. 37, no. 6, pp. 729–740, 1992.
119. M.T. Rashed, "Numerical solutions of functional integral equations," *Applied Mathematics and Computation*, vol. 156, no. 2, pp. 507–512, 2004.
120. A.R. Teel, "Global stabilization and restricted tracking for multiple integrators with bounded control," *Systems and Control Letters*, vol. 18, no. 3, pp. 165–171, 1992.
121. K.D. Do and J. Pan, "Global trajectory tracking control of underactuated ships with off-diagonal terms," *Proceedings of the 42nd IEEE Conference on Decision and Control*, pp. 1250–1255, 2003.
122. K.D. Do and J. Pan, "Global tracking of underactuated ships with nonzero off-diagonal terms," *Automatica*, vol. 41, no. 1, pp. 87–95, 2005.
123. K.D. Do and J. Pan, "Underactuated ships follow smooth paths with integral actions and without velocity measurements for feedback: Theory and experiments," *IEEE Transactions on Control Systems Technology*, vol. 14, no. 2, pp. 308–322, 2006.
124. K.D. Do and J. Pan, "Robust path-following of underactuated ships: Theory and experiments on a model ship," *Ocean Engineering*, vol. 33, no. 10, pp. 1354–1372, 2006.
125. J.Pan and K.D. Do, "Active vibration and motion control of surface ships," *6th International Symposium on Active Noise and Vibration Control*, pp. 5354–5359, 2006.
126. K.D. Do, Z.P. Jiang, and J. Pan, "Robust global stabilization of underactuated ships on a linear course: State and output feedback," *International Journal of Control*, vol. 76, no. 1, pp. 1–17, 2003.
127. K.D. Do, Z.P. Jiang, and J. Pan, "Robust adaptive control of underactuated ships on a linear course with comfort," *Ocean Engineering*, vol. 30, no. 17, pp. 2201–2225, 2003.
128. P. Encarnacao, A. Pacoal, and M. Arcak, "Path following for autonomous marine craft," *Proceedings of 5th IFAC Conference on Manoeuvring and Control of Marine Craft*, pp. 117–122, 2000.
129. R. Skjetne and T.I. Fossen, "Nonlinear maneuvering and control of ships," *Proceedings of OCEANS 2001 MTS/IEEE Conference and Exhibition*, pp. 1808–1815, 2001.
130. Y.F. Zheng, *Recent Trends in Mobile Robots*. World Scientific Publisher, 1993.
131. K. D. Do, Z.P. Jiang, and J. Pan, "A global output-feedback controller for simultneous tracking and stabilization of unicycle-type mobile robots," *IEEE Transactions on Robotics and Automation*, vol. 20, no. 3, pp. 589–594, 2004.
132. K.D. Do and J. Pan, "Global output-feedback path tracking of unicycle-type mobile robots," *Robotics and Computer-Integrated Manufacturing*, vol. 22, no. 2, pp. 166–179, 2006.
133. K.D. Do, Z.P. Jiang, and J. Pan, "Simultaneous stabilization and tracking control of mobile robots: An adaptive approach," *IEEE Transactions on Automatic Control*, vol. 49, no. 7, pp. 1147–1151, 2004.
134. T. Perez, *Ship Motion Control, Course Keeping and Roll Stabilisation using Rudders and Fins*. Berlin: Springer, 2005.
135. K.D. Do, Z.P. Jiang, and J. Pan, "State and output feedback controllers for path following of underactuated ships," *Ocean Engineering*, vol. 31, no. 5-6, pp. 587–613, 2004.

136. J. Oprea, *Differential Geometry and its Applications*. Englewood Cliffs, NJ: Prentice Hall, 1997.
137. K.D. Do, Z.P. Jiang, and J. Pan, "Robust adaptive path following of underactuated ships," *Proceedings of the 41st IEEE Conference on Decision and Control*, pp. 3243–3248, 2002.
138. K.D. Do, Z.P. Jiang, and J. Pan, "Robust and adaptive path following for underactuated ships," *Automatica*, vol. 40, no. 6, pp. 929–944, 2004.
139. R. McEwen and K. Streitlien, "Modeling and control of a variable-length AUV," in *12th International Symposium on Unmanned Untethered Submersible Technology*, (University of New Hampshire, Durham, NH), 2001.
140. K.D. Do and J. Pan, "Robust and adaptive path following for underactuated autonomous underwater vehicles," *Proceedings of American Control Conference*, pp. 1994–1999, 2003.
141. K.D. Do, Z.P. Jiang, and J. Pan, "Robust and adaptive path following for underactuated autonomous underwater vehicle," *Ocean Engineering*, vol. 31, no. 16, pp. 1967–1997, 2004.
142. T. Fukao, H. Nakagawa, and N. Adachi, "Adaptive tracking control of a nonholonomic mobile robot," *IEEE Transactions on Robotics and Automation*, vol. 16, no. 5, pp. 609–615, 2000.
143. R.M. Murray, "Control of nonholonomic systems using chained forms," *Fields Institute Communications*, vol. 1, p. 219245, 1993.
144. Z.P. Jiang, "Iterative design of time-varying stabilizers for multi-input systems in chained form," *Systems and Control Letters*, vol. 28, no. 5, pp. 255–262, 1996.
145. Z.P. Jiang, "Lyapunov design of global state and output feedback trackers for nonholonomic control systems," *International Journal of Control*, vol. 73, no. 9, pp. 744–761, 2000.
146. S.S. Ge, Wang, T.H. Lee, and G. Y. Zhou, "Adaptive robust stabilization of dynamic nonholonomic chained systems," *Journal of Robotic Systems*, vol. 18, no. 3, pp. 119–133, 2001.
147. R. Fierro and F. Lewis, "Control of a nonholonomic mobile robot: backstepping kinematics into dynamics," in *Proceedings of the 34th IEEE Conference on Decision and Control*, vol. 4, (New Orleans, LA), pp. 3805 – 3810, 1995.
148. W.E. Dixon, D.M. Dawson, and E. Zergeroglu, *Nonlinear Control of Wheeled Mobile Robots*. London: Springer, 2001.
149. W.E. Dixon, D.M. Dawson, F. Zhang, and E. Zergeroglu, "Global exponential tracking control of mobile robot system via a pe condition," *IEEE Transactions on Systems Man and Cybernetics, B*, vol. 30, no. 1, pp. 129–142, 2000.
150. W. . Dixon, M.S. de Queiroz, D.M. Dawson, , and T.J. Flynn, "Adaptive tracking and regulation of a wheeled mobile robot with controller/update law modularity," *IEEE Transactions on Control Systems Technology*, vol. 12, no. 1, pp. 138–147, 2004.
151. K.D. Do, *Nonlinear Control of Underactuated Ocean Vehicles, PhD thesis*. School of Mechanical Engineering, The University of Western Australia, 2003.
152. K.D. Do, Z.P. Jiang, and J. Pan, "Simultaneous tracking and stabilization of mobile robots without velocity measurement," *Proceedings of the 42nd IEEE Conference on Decision and Control*, pp. 3852–3857, 2003.
153. J. Hauser, S. Sastry, and G. Meyer, "Nonlinear control design for slightly non-minimum phase systems: application to v/stol aircraft," *Automatica*, vol. 28, no. 4, pp. 665–679, 1992.
154. P. Martin, S. Devasia, and B. Paden, "A different look at output tracking control of a VTOL aircraft," *Automatica*, vol. 32, no. 1, pp. 101–107, 1996.
155. F. Mazenc and A. Iggidr, "Backstepping with bounded feedbacks for systems not in feedback form," *5th IFAC SYMPOSIUM "Nonlinear Control Systems"*, 2001.
156. R. Freeman and L. Praly, "Integrator backstepping for bounded controls and rates," *IEEE Transactions on Automatic Control*, vol. 43, no. 2, pp. 258–262, 1998.
157. C.S. Huang and K. Yuan, "Output tracking of a nonlinear non-minimum phase PVTOL aircraft based on nonlinear state feedback," *International Journal of Control*, vol. 75, no. 6, pp. 466–473, 2002.
158. F. Lin, W. Zhang, and R. Brandt, "Robust hovering control of a PVTOL aircraft," *IEEE Transactions on Control Systems Technology*, vol. 7, no. 3, pp. 343–351, 1999.
159. C. J. Tomlin and S.S. Sastry, "Bounded tracking for non-minimum phase nonlinear systems with fast zero dynamics," *International Journal of Control*, vol. 68, no. 4, pp. 819–847, 1997.

160. R. Olfati-Saber, "Global configuration stabilization for the VTOL aircraft with strong input coupling," *IEEE Transactions on Automatic Control*, vol. 47, no. 11, pp. 1949–1952, 2002.
161. S. Al-Hiddabi and N. McClamroch, "Output tracking for non-minimum phase VTOL aircraft," *Proceedings of the 37th IEEE Conference on Decision and Control.*
162. P. Setlur, D.M. Dawson, Y. Fang, and B. Costic, "Nonlinear tracking control of the VTOL aircraft," *Proceedings of the 40th IEEE Conference on Decision and Control, Florida USA*, pp. 4592–4597, 2001.
163. K.D. Do, Z.P. Jiang, and J. Pan, "Global tracking control of a VTOL aircraft without velocity measurements," *IEEE Transactions on Automatic Control*, vol. 48, no. 12, pp. 2212–2217, 2003.
164. A. Das, R. Fierro, V. Kumar, J. Ostrowski, J. Spletzer, and C. Taylor, "A vision based formation control framework," *IEEE Transactions on Robotics and Automation*, vol. 18, no. 5, pp. 813–825, 2002.
165. N.E. Leonard and E. Fiorelli, "Virtual leaders, artificial potentials and coordinated control of groups," *Proceedings of IEEE Conference on Decision and Control, Orlando, FL*, pp. 2968–2973, 2001.
166. R.T. Jonathan, R.W. Beard, and B. Young, "A decentralized approach to formation maneuvers," *IEEE Transactions on Robotics and Automation*, vol. 19, no. 6, pp. 933–941, 2003.
167. T. Balch and R.C. Arkin, "Behavior-based formation control for multirobot teams," *IEEE Transactions on Robotics and Automation*, vol. 14, no. 6, pp. 926–939, 1998.
168. M.A. Lewis and K.H. Tan, "High precision formation control of mobile robots using virtual structures," *Autonomous Robots*, vol. 4, no. 4, pp. 387–403, 1997.
169. R. Skjetne, S. Moi, and T.I. Fossen, "Nonlinear formation control of marine craft," *Proceedings of IEEE Conference on Decision and Control, Las Vegas, NV*, pp. 1699–1704, 2002.
170. E. Rimon and D. E. Koditschek, "Robot navigation functions on manifolds with boundary," *Advances in Applied Mathematics*, vol. 11, no. 4, pp. 412–442, 1990.
171. H. G. Tanner and A. Kumar, "Towards decentralization of multi-robot navigation functions," *Proceedings of IEEE International Conference on Robotics and Automation*, pp. 4143–4148, 2005.
172. D. M. Stipanovica, G. Inalhana, R. Teo, and C.J. Tomlina, "Decentralized overlapping control of a formation of unmanned aerial vehicles," *Automatica*, vol. 40, no. 8, pp. 1285–1296, 2004.
173. P. Ogren, M. Egerstedt, and X. Hu, "A control Lyapunov function approach to multi-agent coordination," *IEEE Transactions on Robotics and Automatation*, vol. 18, no. 5, pp. 847–851, 2002.
174. P. Ogren, E. Fiorelli, and N.E. Leonard, "Cooperative control of mobile sensor networks: Adaptive gradient climbing in a distributed environment," *IEEE Transactions on Automatic Control*, vol. 49, no. 8, pp. 1292–1302, 2002.
175. J. Cortes, S. Martinez, T. Karatas, and F. Bullo, "Coverage control for mobile sensing networks," *IEEE Transactions on Robotics and Automation*, vol. 20, no. 2, pp. 243–255, 2004.
176. I. Suzuki and M. Yamashita, "Distributed anonymous mobile robots: Formation of geometric patterns," *SIAM Journal on Computing*, vol. 28, no. 4, pp. 1347–1363, 1999.
177. A. Jadbabaie, J. Lin, and A.S. Morse, "Coordination of groups of mobile autonomous agents using nearest neighbor rules," *IEEE Transactions on Automatic Control*, vol. 48, no. 6, pp. 988–1001, 2003.
178. C. Belta and V. Kumar, "Abstraction and control for groups of robots," *IEEE Transactions on Robotics*, vol. 20, no. 5, pp. 865–875, 2004.
179. V. Gazi and K.M. Passino, "A class of attraction/repulsion functions for stable swarm aggregations," *International Journal of Control*, vol. 77, no. 18, pp. 1567–1579, 2004.
180. J. Cortes, S. Martinez, and F. Bullo, "Spatially-distributed coverage optimization and control with limited-range interactions," *ESAIM. Control, Optimisation and Calculus of Variations*, vol. 11, pp. 691–719, 2005.
181. K.D. Do, *Formation Control of Mobile Robots.* Hanoi: Science and Technics Publishing House, 2007.

182. K.D. Do, “Formation tracking control of unicycle-type mobile robots with limited sensing ranges,” *IEEE Transactions on Control Systems Technology*, vol. 16, no. 3, pp. 527–538, 2008.
183. K.D. Do, “Bounded controllers for decentralized formation control of mobile robots with limited sensing,” *International Journal of Computers, Communications and Control*, vol. II, no. 4, pp. 340–354, 2007.
184. K.D. Do and J. Pan, “Nonlinear formation control of unicycle-type mobile robots,” *Robotics and Autonomous Systems*, vol. 55, no. 3, pp. 191–204, 2007.
185. K.D. Do, “Output-feedback formation tracking control of unicycle-type mobile robots with limited sensing ranges,” *Robotics and Autonomous Systems*, vol. 57, pp. 34–47, 2009.
186. K.D. Do, “Bounded controllers for formation stabilization of mobile agents with limited sensing ranges,” *IEEE Transactions on Automatic Control*, vol. 52, no. 3, pp. 569–576, 2007.

Index

Other titles published in this series (continued):

Soft Sensors for Monitoring and Control of Industrial Processes
Luigi Fortuna, Salvatore Graziani, Alessandro Rizzo and Maria G. Xibilia

Adaptive Voltage Control in Power Systems
Giuseppe Fusco and Mario Russo

Advanced Control of Industrial Processes
Piotr Tatjewski

Process Control Performance Assessment
Andrzej W. Ordys, Damien Uduehi and Michael A. Johnson (Eds.)

Modelling and Analysis of Hybrid Supervisory Systems
Emilia Villani, Paulo E. Miyagi and Robert Valette

Process Control
Jie Bao and Peter L. Lee

Distributed Embedded Control Systems
Matjaž Colnarič, Domen Verber and Wolfgang A. Halang

Precision Motion Control (2nd Ed.)
Tan Kok Kiong, Lee Tong Heng and Huang Sunan

Optimal Control of Wind Energy Systems
Iulian Munteanu, Antoneta Iuliana Bratcu, Nicolaos-Antonio Cutululis and Emil Ceangă

Identification of Continuous-time Models from Sampled Data
Hugues Garnier and Liuping Wang (Eds.)

Model-based Process Supervision
Arun K. Samantaray and Belkacem Bouamama

Diagnosis of Process Nonlinearities and Valve Stiction
M.A.A. Shoukat Choudhury, Sirish L. Shah, and Nina F. Thornhill

Magnetic Control of Tokamak Plasmas
Marco Ariola and Alfredo Pironti

Real-time Iterative Learning Control
Jian-Xin Xu, Sanjib K. Panda and Tong H. Lee

Deadlock Resolution in Automated Manufacturing Systems
ZhiWu Li and MengChu Zhou

Model Predictive Control Design and Implementation Using MATLAB®
Liuping Wang

Predictive Functional Control
Jacques Richalet and Donal O'Donovan

Fault-tolerant Flight Control and Guidance Systems
Guillaume Ducard

Fault-tolerant Control Systems
Hassan Noura, Didier Theilliol, Jean-Christophe Ponsart and Abbas Chamseddine

Detection and Diagnosis of Stiction in Control Loops
Mohieddine Jelali and Biao Huang (Eds.)
Publication due October 2009

Stochastic Distribution Control System Design
Lei Guo and Hong Wang
Publication due November 2009

Advanced Control and Supervision of Mineral Processing Plants
Daniel Sbárbaro and René del Villar (Eds.)
Publication due December 2009

Active Braking Control Design for Road Vehicles
Sergio M. Savaresi and Mara Tanelli
Publication due January 2010

MIX
Papier aus verantwortungsvollen Quellen
Paper from responsible sources
FSC® C105338

If you have any concerns about our products,
you can contact us on
ProductSafety@springernature.com

In case Publisher is established outside the EU,
the EU authorized representative is:
Springer Nature Customer Service Center GmbH
Europaplatz 3, 69115 Heidelberg, Germany

Printed by Libri Plureos GmbH
in Hamburg, Germany